Advances in Food Chemistry

O. P. Chauhan
Editor

Advances in Food Chemistry

Food Components, Processing and Preservation

Springer

Editor
O. P. Chauhan
Defence Food Research Laboratory,
DRDO
Mysore, India

ISBN 978-981-19-4795-7 ISBN 978-981-19-4796-4 (eBook)
https://doi.org/10.1007/978-981-19-4796-4

This Springer imprint is published by the registered company Springer Nature Singapore Pte Ltd.
The registered company address is: 152 Beach Road, #21-01/04 Gateway East, Singapore 189721, Singapore

Dedicated to:
My beloved father

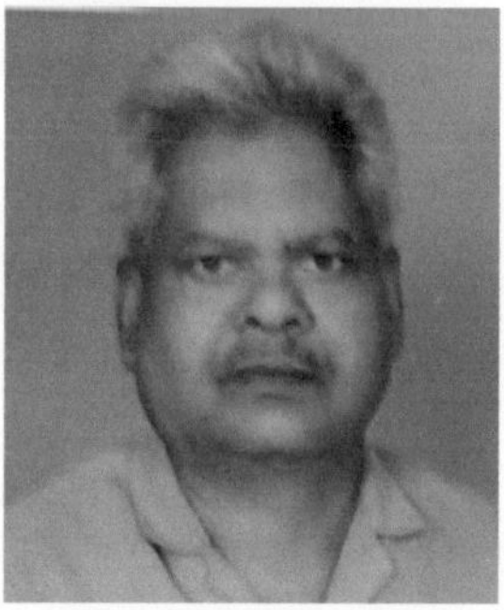

Shri R.D. Chauhan
(01.08.1940–04.05.2020)

Preface

The book pertains to advances in chemistry of foods and changes in food components during food processing, preservation and storage. As such the book deals with different food components such as water, protein, fat, carbohydrates, minerals, vitamins, pigments, flavors, chemistry of plant and animal tissues, and milk. Effect of different food processing operations on the food components has also been discussed in the book. The book is a compilation of various chapters authored by eminent personalities working in the area of Food Science and Technology. The book is a comprehensive compilation of recent advances in food chemistry and will be useful to students, researchers, and faculty as well as to food professionals.

- The book provides latest information with regard to the food chemistry.
- The proposed book provides information about the effect of most of the food processing technologies on food components.
- The book also provides latest information available in research papers with regard to the food chemistry.
- The chapters have been designed in such a way that almost all the aspects of food chemistry have been covered.

Mysore, India O. P. Chauhan

Contents

About the Editor

O. P. Chauhan Dr. OP Chauhan belongs to Defence Research Development Service (DRDS) under Defence Research and Development Organization (DRDO), Ministry of Defence, Government of India. He is currently working as a Scientist and Head of Department of the Fruits and Vegetables Technology Division at Defence Food Research Laboratory, Mysore, India. Dr. Chauhan holds M.Sc. and Ph.D. in Food Technology from G B Pant University of Agriculture and Technology, Pantnagar. His research interest includes postharvest handling of fruits and vegetables including high pressure processing, pulsed electric field processing, microwave dehydration, and modified/controlled atmosphere packaging/storage. His research findings have appeared in more than 100 international and national peer-reviewed journals. He has also published several book chapters in various national and international books. He also has 15 patents to his credit. He has presented more than 100 papers in various national and international conferences. He has served as chief editor of American Journal of Food Technology as well as editorial board member of Journal of Food Science and Technology, Indian Food Industry, International Journal of Food and Fermentation Technology, Pantnagar Journal of Research, Journal of Food and Agriculture Research, etc. Dr. Chauhan has supervised 3 Ph.D. and several M.Sc., M.Tech., and B.Tech. students. He is a recipient of INSA fellowship to work at German Institute of Food Technologies (DIL), Quakenbruck, Germany, on Advanced Food Technologies besides being trained on high pressure processing technology in UK. He has transferred several technologies to more than 40 firms for commercialization. Dr. Chauhan is recipient of DRDO Young Scientist Award, AFSTI Young Scientist Award, DRDO Laboratory Scientist of the Year Award, Laljee Godhoo Smarak Nidhi Award (AFSTI), Prof. G.S. Bains Award (AFSTI), DRDO Technology Group Award, FICCI Best Postharvest Technology Innovation Award, DRDO Technology Spin-off Award, DRDO Technology Absorption Award, Fellow of Bioved Research Society (India), Fellow of Society of Tropical Agriculture (India), Dr. JS Pruthi Award (AFSTI & AIFPA), as well as several Best Paper and Poster Awards from different associations. He is a Life Member of the Association of Food Scientists and Technologists (India), the Nutrition Society of India, and the Indian Science Congress.

Chemical Composition of Foods 1

Shruti Sethi, Alka Joshi, Bindvi Arora, and O. P. Chauhan

1.1 Introduction

Food is an essential component required by the body to fulfill the requirements of growth, energy, and regulatory functions. Apart from imparting physiological functions, food also provides a sense of security and is a symbol of love and affection. Food comprises of several nutrient groups including carbohydrates, fats, proteins, vitamins, minerals, pigments, and enzymes along with water. Each individual group further comprises of several nutrients. Proteins are also known as "body building foods," carbohydrates and fats are classified as "energy-giving foods," vitamins and minerals are considered as "protective foods," and water and roughage are considered as "regulatory foods." An adequate supply of food, nutrients, and their proper utilization is what is required by the body to remain physically active, mentally healthy, and disease-free. Any imbalance in the intake of right quantity or quality of food leads to a health state termed as malnutrition. A low intake of nutrients due to causes such as poverty and nonaccessibility results in undernutrition manifested by deficiency diseases. Excessive intake of a particular nutrient, is termed as over nutrition, due to urbanization and changing lifestyle is a major concern as it causes obesity and cardiovascular disorders. Therefore, Recommended Dietary Allowances (RDA) have been set up for all essential nutrients that are adjudged to be safe and sufficient levels to meet the nutrient requirement of almost 98% of the particular age or gender group. In India, nutritional requirements of different groups of people (according to age, sex, and work profile) have been computed and compiled by the Indian Council of Medical Research (ICMR). A recent report of

S. Sethi (✉) · A. Joshi · B. Arora
Division of Food Science and Postharvest Technology, ICAR-Indian Agricultural Research Institute, New Delhi, India

O. P. Chauhan
Defence Food Research Laboratory, DRDO, Mysore, India

O. P. Chauhan (ed.), *Advances in Food Chemistry*,
https://doi.org/10.1007/978-981-19-4796-4_1

the Expert Group has been drafted by ICMR-National Institute of Nutrition on Recommended Dietary Allowances and Estimated Average Nutrient Requirements for Indians in 2020 that tabulates the nutrient requirements and dietary allowances of 17 important nutrients according to age and category of work along with the tolerable upper limit of these individual nutrients.

The foods that we consume daily are complex substances composed of numerous chemical components termed as nutrients. The major food constituents include carbohydrates, proteins, fats, minerals, vitamins, enzymes, pigments, fiber, and water. Each nutrient performs specific function for sustenance of life.

1.2 Water

Water (moisture) is the most common constituent of all agricultural products, foods, and other biological objects affecting properties of foods such as taste, texture, appearance, shape, weight, and shelf life. Water acts as the medium for chemical reactions, enzyme activity, and microbial growth besides having a physical role in maintaining texture of foods. Moisture content of foods vary from >90% as in some fruits to <0.5% as in edible oils. Moisture content in foods also has legal implications and is highly regulated in the food trade. Table 1.1 lists the moisture content of some common foods.

1.2.1 Water Activity

Water activity (a_w) is a measure of the availability of free water for biological reactions. Water activity of products ranges from 0 to 1. In 1952, Scott concluded that the storage quality of food does not depend on the water content but on water activity. It is the ratio of the vapor pressure of water in a solution (P_s) to the vapor pressure of pure water at a given temperature (P_w):

Table 1.1 Moisture content of common foods

Food	Average moisture content (%)
Apple	84
Broccoli	91
Cucumber	96
Raw chicken	69
Bread	36
Jams	30
Wheat flour	11
Biscuits/cookies	6
Chips	3
Ghee	0.5

$$a_w - P_s/P_w$$

In terms of equilibrium relative humidity (ERH), water activity may also be represented as:

$$a_w = \%\text{ERH} \times 100$$

Water activity determines the microbiological stability of foods. For example, a water activity of 0.80 means the vapor pressure of food is 80% of that of pure water at the same temperature and pressure.

1.2.2 Relationship Between Moisture Content and Water Activity

Information about moisture content is not enough to determine the microbiological stability of the food products or to predict their shelf life. Water activity is a better indicator as it provides information about the free available water for the biological reactions to occur. A simple relationship between moisture content and water activity does not exist and is dependent upon the relative humidity of the food and environment. Although it is easy to assume that foods with higher water content will have a higher water activity than dry foods, this is not always correct. It is also possible to have products with the same water content but very different water activities. For example, salami and cooked beef have similar water content of approximately 60%. However, the water activity of salami is 0.82 and cooked beef is approximately 0.98, and, hence, a salami will have better microbiological stability and a longer shelf life than cooked beef.

1.2.3 Water Activity and Food Deterioration

As given in Table 1.2, different groups of microorganisms have varied requirements of minimum a_w to grow and cause spoilage. Thus, foods containing high moisture ($a_w > 0.8$) such as fruits and vegetables are prone to spoilage by bacteria and molds. Dried foods on the other hand are prone to xerophilic yeasts. No microbial proliferation is observed below a water activity of 0.60 as there is not sufficient water for the microbe to maintain cell integrity.

As mentioned earlier, water may be bound to food constituents or may be available as free water in the tissues. This water-binding behavior can be studied by the sorption isotherms which are a graphical representation of the partial pressure of water in the food and its water content at a constant temperature. When dry food is placed in atmospheres of increasing relative humidity, an increase in the sample weight is observed owing to the absorption of water. When plotted on a graph it yields a sorption isotherm as shown in Fig. 1.1. The moisture sorption isotherm can be divided into three zones (A, B, C) depending on the state of water present in foods. The initial rise in moisture content with the corresponding increase in relative

Table 1.2 Requirements of a_w by microorganisms

Microorganisms	Minimum a_w
Bacteria	0.91
Clostridium botulinum	0.94
Listeria monocytogenes	0.92
Pseudomonas fluoroscens	0.97
Staphylococcus aureus	0.86
Escherichia coli	0.95
Vibrio parahaemolyticus	0.94
Bacillus subtilis	0.91
Halophilic bacteria	0.75
Yeasts	0.88
Saccharomyces cerevisiae	0.90
Debarymoces hansenii	0.83
Osmophilic yeasts	0.61
Molds	0.80
Aspergillus niger	0.77
Aspergillus flavus	0.78
Rhizopus nigricans	0.93
Xerophilic molds	0.65

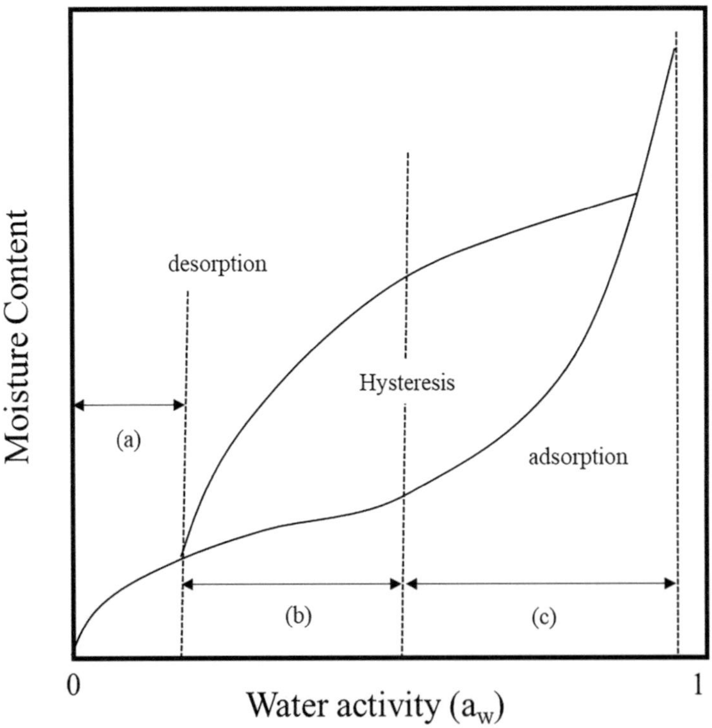

Fig. 1.1 Sorption isotherm for a typical food product. (*Source*: Andrade et al. 2011)

humidity (zone A) indicates a monolayer layer of water adsorbed on food. The second region (zone B) depicting very gradual increase in moisture content is indicative of the adsorption of additional layers of water. Zone C shows the water present in capillaries and pores and is indicative of the free water present in the food.

Alternatively, desorption isotherms are obtained when the moist samples are placed in similar relative humidity conditions with a gradual decline in weight observed indicating loss of moisture. For any food sample, the desorption and adsorption curves do not overlap showing differences in moisture content of the samples. The difference between adsorption and desorption is termed as hysteresis.

This knowledge is useful for developing protocols of concentration and dehydration, standardizing packaging and storage requirements for different foods and predicting the shelf life of products.

Lowering the water activity of foods is one of the strategies for food preservation. Methods employed to reduce the water activity include:

1. Dehydration or evaporation of water
2. Addition of hydrocolloids such as agar, pectin, and humectants such as sugar and salt
3. Freezing of foods

These three methods reduce the water activity either by removal of the water or making it unavailable to the spoilage microorganisms so as to inactivate them.

The effect of water activity on the reaction kinetics of food deterioration was presented in 1980 by Labuza. The plots of water activity with the predominance of reaction contributing to food deterioration were widely accepted and are known as Labuza plots. Figure 1.2 depicts the relationship of food deterioration as a function of water activity (Labuza plots).

Figure 1.2 explains that low moisture foods ($a_w < 0.25$) are more prone to deterioration by lipid oxidation reactions, whereas high moisture foods ($a_w > 0.75$) are rapidly spoiled by microbial growth (yeast, molds, and bacteria). Water activity lower than 0.75 is sufficient enough to prevent microbial spoilage due to lack of free water, but foods can still be deteriorated by enzymes such as polyphenol oxidase, peroxidase and catalase. Non-enzymatic browning reactions are also a cause of food spoilage during processing. Most microbial forms are inactivated at $a_w < 0.75$ owing to the lack of free water for microbial metabolism to continue. Fats present in foods can be oxidized at low a_w (<0.25) due to dense availability of substrates or at high a_w (>0.65) owing to higher mobility of substrate for the reaction to occur. Most enzymes are inactivated at $a_w < 0.65$, and, hence, browning by polyphenol oxidases and texture deterioration by peroxidase and catalase is inhibited. Non-enzymatic browning reactions such as caramelization and Maillard reaction require higher temperatures and a_w over a wide range between 0.25 and 0.75 to occur. Due to this, non-enzymatic browning reactions especially Maillard reactions are not possible to control once initiated.

Thus, moisture content and water activity both significantly impact the food quality as well as safety and information about these is of utmost importance to

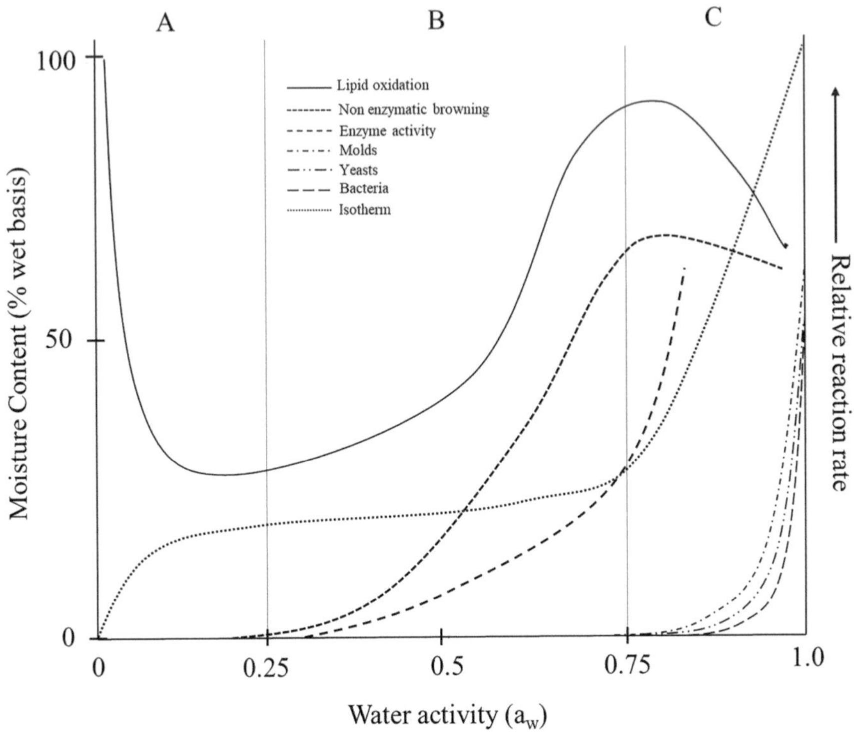

Fig. 1.2 Relationship of food deterioration as a function of water activity. (Adapted from Rahman and Labuza 2007)

decide the processing and preservation techniques to be applied to foods. It is also very interesting to note that mere knowledge of moisture and water activity of a food does not ensure the absence of microbial growth. For example, in case of chocolate, although there is low moisture and water activity, but frequent spoilage occurs due to microbial attack due to dense availability of nutrients (Pandey and Singh 2011).

1.2.4 Classification of Food on Basis of Moisture Content and Water Activity

Food products based on their moisture content can be classified into three categories, viz., low, intermediate, and high moisture foods. Moisture content of foods varies by a great extent and can be as low as <1% as in fats and oils, that can be stored at ambient conditions for fairly long periods to as high as >90% as in fresh fruits and vegetables, that have extremely low shelf stability. Table 1.3 lists the examples of foods with low, intermediate, and high moisture content.

Although the distinction between low and high moisture foods is clear, intermediate moisture foods have a large range of moisture content and water activity.

Table 1.3 Classification of food on the basis of moisture content

Type of food	Moisture content (%)	Examples
Low moisture content	5–15	Dried foods, grains, flour
Intermediate moisture content	15–40	Cakes, breads, waffles
High moisture content	>40	Fresh fruits, vegetables, milk

Prabhakar (2014) defined intermediate moisture foods (IMF) as foods that are moist enough to be consumed without rehydration and can be preserved easily at ambient temperatures by restricting the water mobility such as breads, cakes, and waffles. Preservation of intermediate moisture foods require addition of ingredients such as humectants (salt and sugar) that prevent moisture migration and reduce the water activity. Food preserved by these added natural ingredients has an extended shelf life even at ambient atmospheric conditions without providing any other processing or chemical preservatives (Li et al. 2014).

1.3 Carbohydrates

Carbohydrates are hydrates of carbon composed of carbon, hydrogen, and oxygen usually present in the ratio 1:2:1 and with a general formula of $C_n(H_2O)_n$. They are an important source of energy and provide 17 kJ energy per gram and are utilized before fats and protein in the body to meet the energy requirements. They are abundantly found in nature in plants, animals, and microorganisms in the form of simple sugars such as glucose, fructose, and sucrose and as complex sugars such as cellulose, pectin, and hemicellulose.

1.3.1 Classification

Carbohydrates are broadly classified into monosaccharides, oligosaccharides, and polysaccharides. The term saccharide means "*sugar*." *Monosaccharides* are simple sugars composed of a short carbon chain with an aldehyde group (called aldoses) or a ketose group (called ketoses). The number of carbon atoms may vary from three (trioses), four (tetroses), five (pentoses), six (hexoses), and so on (Table 1.4). Simplest three-carbon sugars are glyceraldehyde (aldose) and dihydroxy acetone (ketone). The most abundant monosaccharide found in nature is glucose also referred to as dextrose. It is the major constituent of starch and cellulose, the polysaccharides most abundant in nature. Glucose present in the bloodstream of animals is the major source of their energy. Monosaccharides are easily digestible, impart sweetness to the products, and are hygroscopic in nature.

Oligosaccharides are formed by the condensation of 2–9 monosaccharide units through glycosidic bonds. Depending on the number of saccharide units joined together, oligosaccharides are termed as disaccharide (2), trisaccharide (3),

Table 1.4 Classification of monosaccharides

No. of carbon atoms	Aldose	Ketose
3C Triose	Glyceraldehyde	Dihydroxyacetone
4C Tetrose	Erythrose	Erythrulose
5C Pentose	Ribose, Xylose	Ribulose, Xylullose
6C Hexose	Glucose, Galactose, Mannose	Fructose
7C Heptose	Mannoheptulose (in avocado)	Sedoheptulose (intermediary in lipid biosynthesis)

Table 1.5 Examples of oligosaccharides found in nature

Oligosaccharide	Constituent monosaccharides	Foods
Disaccharide		
Sucrose	Glucose–Fructose	Sugarcane, sugar beet, maple syrup
Maltose	Glucose–Glucose	Formed upon depolymerization of starch
Lactose	Glucose–Galactose	Milk
Trisaccharide		
Raffinose (Melitose)	Galactose–Glucose–Fructose	Beans, asparagus, cabbage, broccoli, Brussels sprouts, sweet potatoes, lupin, and whole grains
Tetrasaccharide		
Stachyose	Galactose–Galactose–Glucose–Fructose	Beans, peas
Pentasaccharide		
Verbascose	Galactose–Galactose–Galactose–Glucose–Fructose	Leguminous plants

tetrasaccharide (4), and so on. Examples of most common oligosaccharides found in nature are given in Table 1.5.

Sugars may be classified as reducing or non-reducing based on their ability or inability to reduce other substrates in the reaction mixture having higher redox potential than respective sugars. Accordingly, they are termed as reducing or non-reducing sugar. Due to this ability, reducing sugars can convert cupric ion to cuprous ion which is the basis of their qualitative as well as quantitative estimation. This property is based on the availability of the free carbonyl group in the monosaccharides that can react with cupric ion to form a brick red cuprous oxide. All monosaccharides including glucose, fructose, galactose, etc., and a few disaccharides such as lactose and maltose are reducing sugars. Sucrose and other polysaccharides are non-reducing in nature. Based on their physiological properties, these carbohydrates are grouped as digestible or non-digestible. Sucrose, maltose, and lactose are examples of digestible oligosaccharides. Raffinose, stachyose, and verbascose are examples of non-digestible fermentable oligosaccharides which, when consumed in excess, can trigger abdominal bloating, excessive gas formation, and diarrhea related problems. These non-digestible oligosaccharides serve as dietary fiber and prebiotics and promote the proliferation of gut microbiota. The oligosaccharides are about 0.3–0.6 times as sweet as sucrose. This attribute is exploited in food industry to develop formulations with the replacement of sucrose. For example, lactose, milk sugar is about one-sixth times sweeter than normal sugar (sucrose). So many sugar replacer formulations are commercially available now-a-days, but hardly any formulation is having sugars in original form either whole or in fraction, despite their multifold sweetness expression than common sugar.

Polymers consisting of more than nine monosaccharide units are termed as *polysaccharides*. They are high molecular weight compounds and are insoluble in water. They do not impart sweet taste to the products and are bland. Some polysaccharides have linear carbon chain such as cellulose, while some polysaccharides such as starch (in plants) and glycogen (in animals) are more complex and mixtures of branched and linear moieties. Polysaccharides may also be classified as homopolysaccharides and heteropolysaccharides.

- Homopolysaccharides are made of repeating units of a single type of monosaccharide (e.g., cellulose and starch).
- Heteropolysaccharides comprise of more than two types of monosaccharides (e.g., pectin, hemicelluloses, hydrocolloids, chitin).

Polysaccharides may further be classified according to the function performed. Storage polysaccharides are those that are used for storage. For instance, plants store simple sugar (glucose) in the form of starch while animals store in the form of glycogen. Structural polysaccharides are carbohydrates that have a structural role, for example, cellulose as a constituent of plants and chitin as a constituent of exoskeleton in animals.

Plant polysaccharides such as cellulose, hemicelluloses, gums, and lignin are resistant to hydrolysis by the digestive enzymes in human body and are referred to as dietary fibers. These are also termed as non-starch polysaccharides. Although, they can be digested by animals, when ingested by humans, they cannot be digested as the body lacks the enzymes to break it down. Fiber has the ability to absorb water, develop softer stools, ensure efficient elimination of waste and improve bowel health. An average daily intake of 25–30 g (max 40 g/2000 kcal) of fiber is recommended to ensure a good bowel system. Thus, it is advisable to consume fruits along with peel, salad, and whole grains to get an adequate supply of fiber. Interest in fiber-rich food products has paved way for the addition of fiber from various sources into the food products. Different wastes from food processing industry are being utilized as a source of fiber enrichment, especially in the bakery industry. Sethi and Gupta (2016), Singh (2016), and Hernández-Ortega et al. (2013) successfully developed fiber-rich tortilla chips, cakes, and cookies that could cater to the fiber requirement in the diet. Addition of defatted soy flour also enhances fiber content in compound chocolate with a simultaneous increase in protein content has

Table 1.6 Examples of gums used for food applications

Gum	Source	Levels at which used	Application
Guar gum	Guar beans	0.5–1.0%	Ice cream, yogurt
Locust bean (carob) gum	Seeds of carob tree	0.05–0.25%	Dairy and frozen desserts
Gum arabic	Acacia tree	1.0%	Essential oil emulsification in soft drinks, prevention of bloom and sugar crystallization in confections, glazing of candies
Xanthan gum	Bacterium: *Xanthomonas campestris*	0.5–1.0%	Thickener and stabilizer for salad dressings, chocolate syrup, gravies, yogurt, frozen desserts
Gellan gum	Bacterium: *Sphingomonas elodea*	GMP[a]	Bakery mixes, nutrition bars, diet beverages
Curdlan	Bacterium: *Agrobacterium* Biovar1	GMP	Gelling agent, texture modifier in noodles, pasta and fish paste products, water-holding agent in sausages, hams, and hamburgers
Carrageenan	Red seaweed (*Chondrus crispus*)	0.015–0.025%	Chocolate milk, ice cream, evaporated milk, infant formulas, coating for meat
Algins	Brown seaweed (*Phaeophyceae*)	0.05–0.5%	Foaming, emulsifying, and stabilizing in protein dispersions, thickener in salad dressings
Pectin	Plant cell wall	1.0%	Thickener in ketchup, gelling agent in jams, sugar replacer in low-calorie jam

[a]Good Manufacturing Practices

also been standardized by Pandey et al. (2012). Even in dairy products, such as *paneer*, addition of fibers from coconut is possible (Chauhan et al. 2016).

Gums are examples of thickening heteropolysaccharides used as food additives. They can be derived from many natural sources as shown in Table 1.6 and are used in food as thickening, gelling, emulsification, stabilizing, and surface coating agents.

1.3.2 Properties of Carbohydrates

Length of the carbon chain and molecular weight govern the properties and functionality of carbohydrates. All simple sugars (monosaccharides) possess good solubility and are hygroscopic in nature. This property is sometimes used in the food industry where these are used as humectants to retain moisture of the foods. As the molecular weight increases, solubility of carbohydrates in water decreases. Thus, simple sugars form highly uniform solutions while polysaccharides such as starch are insoluble in water. Simple sugars are sweet in taste with fructose having maximum sweetness while polysaccharides are tasteless.

Mutarotation: Carbohydrates are optically active compounds and exhibit mutarotation. When dissolved in water the α and β forms of glucose tend to go towards an equilibrium. The α form has a specific rotation of +112.2° while the β form has a value of +18.7°. Upon dissolution in water, the specific rotation shifts to +52.7°. This gradual change in optical rotation is termed as mutarotation that results from an interconversion between the two forms.

Inversion: Inversion is the process in which sucrose is hydrolyzed into its two constituent sugars, glucose and fructose in 1:1 ratio resulting in the production of invert sugar, a liquid sugar with an equal proportion of the two monosaccharides. Of the two monosaccharides, glucose is dextrorotatory, i.e., it rotates the plane polarized light to right with a value of +52.7° and fructose is highly laevorotatory (rotation of plane polarized light to left) with value of −92.0°. Sucrose has an optical rotation of +66.5° while the invert sugar has value of −19.7°. As a result, the mixture is levorotatory. Since there is a reversal in the direction of rotation, the phenomenon is called "inversion" and the resultant mixture is called invert sugar. Since the intensity of the sweetness of fructose is more than sucrose or glucose, invert sugar is sweeter than sucrose. Additionally, both these sugars cannot form crystals; therefore, inversion of sugars is highly desirable for inhibition of crystallization in processed products such as candies, jam, and jelly.

Browning: Sugars melt upon heating above their melting points. Heating further, results in their dehydration, decomposition, and polymerization to form a brown mass called caramel. This process is called caramelization and the brown pigment formed imparts new flavors to the product, for example, in case of chocolate manufacture. If the browning takes place as a result of reaction between an aldehyde group (reducing sugar) and an amino group (protein), it is termed as Maillard reaction. This type of browning is visible during baking of bread, roasting of peanuts, and frying of foods.

Gelatinization: Starch is a native state that is able to absorb a maximum of 30% water. Upon heating, more water can be imbibed into the matrix by the breakdown of the intermolecular hydrogen bonds. This causes the starch suspension to swell resulting in loss of birefringence. This phenomenon is termed as gelatinization that is manifested by an increase in the viscosity of the suspension. The temperature at which this occurs is called gelatinization temperature and it varies according to starch origin.

Retrogradation: When the thickened mixture of starch is allowed to cool without disturbance, there is a tendency for reassociation of intermolecular hydrogen bonding. This is usually accompanied by increased viscosity and turbidity of pastes, exudation of water, and increased degree of crystallinity. The process is termed as retrogradation. The development of hard and tough crystalline structure of retrograded amylopectin fraction of starch leads to undesirable changes in texture as in case of staling of bread and cakes. Retrogradation is desirable in some food applications due to the slower enzymatic digestion of retrograded starch and moderated release of glucose into the bloodstream such as food for geriatric and obese people.

Modification of starch: The gelatinizing property of starches is used in the food industry to develop thick gels and also to increase the viscosity of flowable foods such as soups. Properties of native starch can be altered to achieve variable paste viscosity/consistency, clarity, and stability. For this purpose different chemical and physical treatments may be employed such as cross-linking, oxidization, depolymerization, and pregelatinization. Modified starches show increased solubility and stability, improved gelatinization and pasting characteristics, better freeze-thaw stability and paste clarity, and reduced gel syneresis.

Modified cellulose: Cellulose, a water-insoluble linear polysaccharide of glucose units can be modified as per the requirement to yield soluble powders. Spray-dried microcrystalline cellulose (MCC) is the soluble derivative obtained after partial hydrolysis of cellulose-rich wood pulp. It is stable to both heat and acid and is used as an anti-caking agent, flavor carrier, and stabilization of foams and emulsions. Alkali cellulose is obtained upon treatment of the wood pulp with 18% sodium hydroxide which on further reaction with sodium salt of chloroacetic acid forms the sodium salt of carboxy methyl cellulose (CMC). It is used as a stabilizer for protein dispersions. Methyl cellulose (MC) and hydroxyl propyl methyl cellulose (HPMC) are other cellulose derivatives obtained after the reaction of alkali cellulose with methyl chloride alone or in conjunction with propylene oxide, respectively. These are cold water-soluble and can be used as binders, fat replacers, and stabilizers. Cellulose and all its modified derivatives contribute to the dietary fiber component of food and improve bowel health.

Resistant starch: Resistant starch is indigestible by body enzymes due to strong binding with other food components and retrogradation. Native starch has a tendency to develop resistant starch by induced physical and chemical modifications. Higher resistant starch content in a food is indicative of its lower digestibility and reduced glycemic index. Azizi et al. (2020) could successfully alter the resistant starch content of potato tubers by mere boiling followed by cooling of the tubers. Resistant starch forms a functional food ingredient for lowering the calorific value of foods and as laxatives. Reduction in the glycemic index of multigrain bread can also be achieved by sourdough process without affecting the loaf characteristics. The use of the pure starter culture in propagation of sourdough has been emphasized in the study conducted by Singh et al. (2019) to yield bread with a low GI and good organoleptic attributes.

1.4 Proteins

Proteins are another important class of organic macromolecules essential in biological systems. They are characterized by the presence of carbon, hydrogen, oxygen, nitrogen, and sometimes also sulfur. Conjugated proteins are complexed with non-protein moieties called as prosthetic group. This class of protein includes glycoproteins, nucleoproteins, lipoproteins, phosphoproteins, and metalloproteins complexed with carbohydrates, nucleic acid, lipids, phosphorus, and metal, respectively. Both plant and animal foods are good sources of proteins. Plant foods such as cereals, pulses, nuts, and oilseeds contribute an appreciable quantity of proteins to the diet. Likewise, foods of animal origin such as milk, egg, fish, poultry, and meat are good sources of high-quality protein. Non-conventional sources of proteins include microorganisms such as *Spirulina* and *Candida* that are termed as Single Cell Protein (SCP). They have limited demand owing to dietary restrictions and high amount of nucleic acid content. The product "Quorn" approved for use as human food by the UK Ministry of Agriculture, Fisheries and Food in 1983 utilizes molasses or glucose for the cultivation of *Fusarium venenatum* and has a final protein content of 45%. The mycoprotein has reduced RNA content (<2% (w/w)) owing to the thermal shock (68 °C for 20 min) given to make the organism non-viable.

Proteins are formed by the condensation of amino acids through peptide bonds. An amino acid molecule comprises of a carboxyl group, an amino group, and a side chain (R) joined to a central carbon atom. The linkage between the carboxyl group of one amino acid and the amino group of the other is called a peptide bond. The R-group may vary in size, shape, charge, and reactivity giving rise to more than 20 different amino acids found in nature. Condensation of two or more different amino acids gives rise to compounds called peptides. Peptides comprising of 3–10 amino acids are termed as oligopeptides. Very long chain of amino acids is called a polypeptide. Polypeptides showing bio functionality are known as proteins. These are synthesized during the translation stage on ribosomes. The difference in the sequence of amino acids in the polypeptide chain, their ratio, and chain length gives rise to numerous types of proteins found in nature.

Amino acids that a human body can synthesize are termed as non-essential while those which have to be taken through diet are referred to as essential amino acids. Of these 20 amino acids; 10 amino acids, i.e., methionine, arginine (essential for young ones), threonine, tryptophan, valine, isoleucine, leucine, phenylalanine, histidine, and lysine are essential. The remaining amino acids are non-essential. Classification of amino acids is done according to the nature of R-group. Amino acids with aliphatic (valine, alanine, methionine, leucine, isoleucine) and aromatic (tryptophan, tyrosine, phenylalanine) side chains are hydrophobic in nature and are not readily soluble in water. Charged (arginine, aspartic acid, glutamic acid, histidine, lysine) and uncharged (serine, threonine, cysteine, asparagines, glutamine) amino acids are hydrophilic and water-soluble. The net charge of the protein depends on the ratio of basic and acidic amino acids constituent of the particular protein.

1.4.1 Functions of Proteins

Proteins perform varying functions in the body. They act as biocatalysts (as enzymes) for chemical reactions. Storage proteins are also found in plant seeds and eggs. Storage proteins such as ferritin store iron in the body. Hormones such as insulin perform the regulatory functions in the body. Myosin and actin are examples of contractile proteins helping in muscle contraction. Proteins also play a structural, transporting, and protective role in the body. Keratin and collagen are examples of proteins giving support and structure. Hemoglobin acts as a transport medium for nutrients and waste in the body while antibodies give the body the ability to fight against diseases.

Proteins markedly affect the sensory quality of foods during processing, storage, and consumption. Plant proteins such as wheat significantly influence the dough forming and handling properties and the organoleptic quality of baked products. Milk proteins influence the gelling characteristics in coagulated dairy products and crumb texture of cakes. In case of meat, contractile proteins influence the succulence and textural attributes. Ripening changes in cheese and aging of rice during storage affect the textural and cooking characteristics, respectively.

1.4.2 Properties of Proteins

Isoelectric pH: Amino acids contain both an amino and a carboxylic group, and, thus, they behave both as base and acid, i.e., they are ampholytes. The R-group governs the overall charge on the amino acid/protein. At certain pH, amino acids possess a net neutral charge and are called as zwitterion. This pH is called isolectric pH or p*I*. Isoelectric point is the pH at which the protein has minimum solubility and maximum precipitation, the basis for the extraction of protein concentrates and isolates from protein-rich food bases. Proteins containing acidic amino acids (aspartic, glutamic) have a low p*I*. Presence of basic amino acids (lysine, arginine) results in a higher p*I* of proteins. This property is used in electrophoresis to separate the individual proteins. Under the influence of an electric field, the zwitterion does not move. At a pH lower than the p*I*, the protein migrates towards the cathode while at pH greater than p*I* the protein moves towards the anode.

Denaturation: Alteration of the structure of proteins under the influence of pH, chemicals, temperature, or enzymes without cleavage of the peptide bonds is termed as denaturation. It causes the unfolding of 3D helical structure of proteins, and thus exposing the hydrophobic groups which result in a reduction in water solubility of proteins. It may negatively affect the functional properties of proteins and reduce their solubility. The positive influence of denaturation is the improved digestibility of protein owing to the increased availability of peptide bonds for reaction and better absorption into the body. Moreover, during processing, the partial denaturation of protein improves the emulsifying and foaming properties of dispersions. Also, the thermal destruction of anti-nutritive proteins such as trypsin inhibitors improves the digestibility of pulse proteins.

Solubility: The net charge on the proteins governs its solubility in different solvents under varying pH and temperature conditions. As discussed earlier, denatured proteins are less soluble as compared to their native counterparts.

Proteins serve as an important ingredient in developing foods for all segments of population; however, the functionality of proteins also assists in texture designing of foods. The functional attributes of proteins like gelation, foaming, emulsification, and thickening also drive the incorporation of isolated proteins in various foods like mayonnaise, baked foods, and beverages. However, the manner of converting the isolated proteins into powders also determines their functional properties (Swanson 1990; Rudra et al. 2016). Understanding the composition of proteins, their processing methods, and their effects on the functional properties provides better insights for promoting the application of the protein isolates in new food formulations.

Enzymes are biocatalysts that are proteinaceous in nature. They are essential to maintain the metabolic activities and overall health of the body. A non-protein moiety is attached to the enzyme for its proper functioning. It might be a metal ion (cofactor) such as Mg^{2+}, Zn^{2+}, Mn^{2+}, and Fe^{2+} or an organic compound (coenzyme) such as vitamins. The protein part along with the non-protein part is termed as holoenzyme. Loss of enzymatic activity results in dissociation of the nonprotein part from the protein part and the enzyme is called as an apoenzyme. Activity of individual enzymes is pH and temperature-dependent. In human body, amylases in the saliva (salivary amylase) in the mouth get activated upon ingestion of food and break down the complex carbohydrates into smaller fragments. As the food travels in the gut, prevailing pH conditions result in enhanced activity of the other enzymes such as proteases, lipases, and maltases. They further break down the food into smaller digestible fragments. The digested material is absorbed through the intestinal wall into the bloodstream and utilized for the maintenance of health.

Enzymes bring about desirable or undesirable changes in foods. Enzymes such as phenolases may bring about undesirable browning (as in case of cut apple, potato), while polygalacturonase and pectin methyl esterases are responsible for softening of fresh produce. Desirable effects of enzymes may also be brought about in foods/processed products such as:

- Ripening: Pectinases play a significant role in the ripening of fruits.
- Bakery industry: Enzymes (amylases) are used to improve dough handling properties.
- Wine industry: Removal of haziness by pectinases by forming a complex and precipitation.
- Coffee industry: Pectinases are useful in digesting the mucilaginous matrix during coffee manufacture.
- Corn syrup industry: High fructose corn syrup of varying sweetness is prepared by the action of glucose isomerase enzyme on the corn syrup that is further used as a sweetener.
- Cheese manufacture: Addition of renin enzyme facilitates curdling of milk during cheese manufacture. Lipases may also be added to develop desired flavor profile of cheeses.

1.5 Lipids

Lipids are a diverse group of compounds that are soluble in organic solvents such as benzene, ether, and chloroform. Lipids include fats, oils, waxes, cholesterol, etc. Lipids are a concentrated form of energy and yield 9 kcal/g upon oxidation. Animal foods, e.g., meat, milk, liver and plant foods such as oilseeds and nuts are rich sources of lipids. They are stored in seeds as a form of energy in the germ portion. Waxes provide a protective role both on plant leaf and fruit surface and on skin. They act as an insulating medium to prevent change in temperatures. In dietary sources, fats are the carriers of fat-soluble vitamins and help in their absorption. In the scenario, when there is excess intake of lipids and demand for energy is low then the fats are stored as the adipose tissue in the body. Lipids form one of the major constituents of the cell structure and influence the permeability of membranes. In foods, lipids influence the calorific value, textural attributes, flavor, mouthfeel, and storage stability.

The basic structural unit of any lipid is the fatty acid that comprises of a long carbon chain with a carboxylic acid. Naturally occurring fatty acids mostly have an even number of carbon atoms and are linear. The absence of any double bond in the linear fatty acid molecule yields a saturated fatty acid. Unsaturated fatty acids may contain one to six double bonds in their structure. The placement of the double bond in the structure gives rise to the different fatty acids. Table 1.7 shows some examples of different fatty acids along with their short-hand description. In this, the two numbers divided by a colon represent the carbon chain length of the fatty acid and the number of double bonds. Usually, the numbering of carbon atoms is done from carboxylic acid end being the highest oxidized form as per IUPAC system. In case of

Table 1.7 Different fatty acids existing in nature

Fatty acid	No. of carbon atoms and short-hand description	Source
Saturated (SFA)		
Butyric acid	4 (4:0)	Butter, hard cheese, milk, sauerkraut, fermented soy products
Caproic acid	6 (6:0)	Butter, cheese, coconut oil
Caprylic	8 (8:0)	Palm oil, coconut oil, human, and bovine milk
Palmitic acid	16 (16:0)	All animal and plant fats
Stearic acid	18 (18:0)	All animal and plant fats
Arachidic acid	20 (20:0)	Groundnut oil
Monounsaturated (MUFA)		
Palmitoleic	16 (16:1)	Blue-green algae, macadamia nuts, sea buckthorn oil
Oleic acid	18 (18:1)	Olive oil and other plant oils
Polyunsaturated (PUFA)		
Linoleic acid	18 (18:2ω6)	Sunflower oil, corn oil
Linolenic acid	18 (18:3ω3)	Soybean, flax seed, hemp seeds
Arachidonic acid	20 (20:4ω6)	Peanut oil, groundnut oil
Eicosapentaenoic acid	20 (20:5ω3)	Corn and linseed oil
Docosahexaenoic acid	22 (22:6ω3)	Seafood, fish oil

omega fatty acids, such as linoleic and linolenic acids, carbon atom numbering is done from the methyl end instead of the carboxylic acid end (unlike in IUPAC system). Both linoleic and linolenic acids cannot be synthesized in the body and hence are designated as essential fatty acids.

Lipids naturally occur as esters of fatty acid with alcohol. Based on their physical state under ambient conditions, lipids are classified as fats (solid) and oils (liquid). These are also referred to as simple lipids. Condensation of three fatty acids with the glycerol moiety results in the formation of triglycerides. Animal tissues are predominantly composed of saturated fats. Plants on the other hand mainly comprise of unsaturated fatty acids (oils). The point of unsaturation is highly prone to oxidation and results in rancidity which causes undesirable flavor changes in the oil-containing products. Esters of fatty acid along with alcohol and another moiety such as phosphate group or carbohydrate yield complex lipids and are named as phospholipids and glycolipids, respectively.

The growing market demand for reduced-fat (≥25 lower than control), low fat (≥3 g per 30 g of serving), or fat-free (≥0.5 g per serving) food products has led to addition of fat replacers in the food product (Potter and Hotchkiss 2006). To prevent drastic change in the mouthfeel, texture, and viscosity caused due to reduction in fat, different fat replacers have been developed including caprenin (ester of glycerol with caprylic, capric, and behenic fatty acids), salatrim (mixture of short-chain and long-chain fatty acids), olestra (sucrose polyester), simplesse (protein-based), and carbohydrate-based replacers (Avicel, carrageenan, inulin), hydrocolloids (C-trim30) (Ognean et al. 2006; Food Safety Network 2014).

1.5.1 Properties of Lipids

Polymorphism: This is the phenomenon exhibited by triglycerides and is their ability to exist in different crystalline forms with varied molecular packing. The three polymorphic forms are denoted as α (hexagonal packing), β′ (orthorhombic packing), and β (triclinic packing), with each crystalline form having different melting point. The different polymorphic forms influence the physical properties of food systems markedly. For example, fine β′ crystals are desirable in spreads and margarine to impart smoothness, spreadability, mouthfeel, and glossiness. The larger β· polymorphic crystals are required in bakery shortening and stabilizing cocoa in chocolate. A unique property of cocoa butter is its polymorphism (phenomenon of multiple melting points when the fat has several possible crystal packing). Six polymorphs or crystal forms exist, but only three forms, namely, form IV (β′-2), form V (β2-3), and form VI (β1-3) are important in the commercial production of chocolate. Form V is the characteristic form of chocolate, produced when it is tempered (controlled crystallization) during production. It may be considered a stable form for practical purposes. Form IV is characteristic of untempered chocolate, and form VI is a transformation of form V associated with bloom (the greyish-white discoloration), sometimes found on the surface of chocolate (Stewart and Timms 2004).

Interesterification: The rearrangement of acyl groups on the glycerol backbone in the triglycerides without change in the fatty acid composition is termed as interesterification. The process is chemically or enzymatically induced by altering the melting point and crystallization behavior of the fats. Enzyme-aided position-specific interesterification can be targeted to get triglycerides that mimic human milk and can be included in infant formulas (Spurgeon et al. 2003). The lipases obtained from microbial source (*Candida rugosa*) are used for the purpose. Sodium methoxide is used as a chemical catalyst for interesterification. Interesterified fats can be mixed with liquid oils in varying proportions to achieve desired functionality of mouthfeel and melt-in-mouth characteristics. Additionally, the blending allows lowering of saturated fatty acid content thus giving a healthy alternative, for example, combining 75% interesterified palm oil with a liquid rapeseed oil can yield the same melt profile as "native" palm oil, while reducing the SFA content by 20% (Berry et al. 2019).

Hydrogenation: In the food industry, liquid oils are converted to solid or semisolid fats by the process of hydrogenation. Hydrogen is added to the double bonds to alter the melting point and crystallization behavior of lipids and make them more stable towards oxidative changes. Nickel is used as a catalyst for the transformation and is added at the rate of 0.01–0.02%. Hydrogenation is carried out at a temperature range of 250–300 °C and takes about 40–60 min to complete. Fat constants (Reichert Meissl value and Polenske value) are used to judge the end point of hydrogenation reaction. They are widely used as shortenings in baked products owing to the capacity to shorten the gluten network. In the process, the natural *cis*-configuration of the fatty acid is changed to the *trans*-configuration resulting in fats called as *trans*-fatty acids. These *trans*-fatty acids have been found to have a direct link with cardiovascular diseases and regulatory limits are being set for their use.

Rancidity: This is the term describing the process of deterioration of lipids. Liberation of fatty acids from triglycerides by action of enzyme (lipases) or in presence of water at high temperature (as in case of post-frying storage of oil) is termed as hydrolytic rancidity. This reduces lipid stability and the smoke point, produces off-flavor, and results in foaming. On the other hand, production of short-chain volatile fatty acids that cause off-flavor as a result of the reaction of lipids with oxygen is termed as oxidative rancidity. Fat oxidation, although undesirable, is beneficial for certain processed foods such as ripened cheeses wherein they contribute to flavor development.

Flavor reversion: This phenomenon is exhibited by linoleate-rich oils, especially soybean oil. It occurs at low levels of oxidation and yields peroxide value of below 10 as against the oxidative rancidity that yields a peroxide value of greater than 10. The reaction results in the production of flavors other than the characteristic of the crude oil, and thus the misnomer as "reversion." The various reversed flavors developed are documented in Table 1.8.

Table 1.8 Chemical compounds responsible for reversed flavor of soybean oil

Reversed flavor	Contributing compound
Green beany	3-*cis*hexenal
Beany	2-pentyl furan
Grassy	2-*trans*hexenal
Cucumber	2-*trans*, 6-*cis*nonadienal
Fishy	2-*trans*,4-*cis*,7-*cis*-decatrienal

1.6 Vitamins and Minerals

Vitamins and minerals are together classified as micronutrients which are essential for human growth, development, and maintenance. Vitamins are majorly organic compounds having direct role in functionality of metabolic enzymes whereas minerals are inorganic nutrients that form the bulk of bodies such as bones and also have metabolic functions as cofactors and prosthetic groups for enzymes in many metabolic reactions in the body.

1.6.1 Vitamins

Vitamins are organic compounds required by humans through diet in small quantities to sustain life. Most vitamins need to come from food as the human body does not have the mechanism to synthesize them. Exceptions exist such as vitamin D that can be synthesized by humans through its precursors. Vitamins form essential nutrients and have a direct role in many metabolism-related reactions. Their deficiency has been directly related to occurrence of metabolic diseases. There are 13 known vitamins which are classified into two categories as water-soluble and fat-soluble. Table 1.9 lists the vitamins essential for humans, dietary sources, and the corresponding metabolic diseases that may result in its deficiency.

Eating a balanced diet is the best way to maintain a desirable vitamin balance in the body. Supplements can be provided to persons with physiological conditions or specific target groups to meet vitamin requirements if necessary. As water-soluble vitamins can be excreted in urine, their regular replenishment through diet is necessary. On the other hand, fat-soluble vitamins can be stored in the body, and, thus, excessive consumption of fat-soluble vitamins can be a health risk to consumers due to vitamin toxicity.

1.6.1.1 Effect of Processing on Vitamins

Depending upon the chemical structure, different vitamins have different tolerance to food processing and storage conditions. Blanching, for example, results in leaching losses of vitamins. Pasteurization and commercial sterilization can lead to destruction of many heat-labile vitamins. Some vitamins are more stable (less labile) than others. Water-soluble vitamins (B-group and C) are more labile than fat-soluble vitamins (K, A, D, and E) during food processing and storage. The most unstable

Table 1.9 List of vitamins, their major sources, and associated deficiency diseases

Vitamin	Solubility	Major sources	Deficiency disease
A (Retinol)	Fat	Green leafy vegetables, citrus fruits, tomatoes, carrots, guava, milk, liver, carrots, broccoli, and watermelon	Poor vision, night blindness, prolonged deficiency can lead to complete blindness
D (Calciferol)	Fat	Fish, beef, cod liver oil, egg yolk, liver, chicken breast, and cereals, fortified milk, butter, and other fat-rich products	Rickets, osteomalacia, osteoporosis
E (Tocopherol)	Fat	Potatoes, pumpkin, guava, mango, milk, nuts, and seeds	Fertility disorders
K (Phytonadione)	Fat	Green leafy vegetables	Bleeding diathesis, increased blood coagulation time
B_1 (Thiamine)	Water	Fresh fruits, whole grains	Beri-beri, muscle weakness
B_2 (Riboflavin)	Water	Meat, eggs, mushrooms	Ariboflavinosis
B_3 (Niacin)	Water	Meat, eggs, fish, milk products, fresh fruits, mushrooms, whole grains	Pellagra
B_5 (Pantothenic acid)	Water	Meat, kidney, egg yolk, fish, chicken, milk, yogurt, legumes, mushrooms	Paresthesia
B_6 (Pyridoxine)	Water	Pork, chicken, fish, whole grain cereals, eggs, vegetables, soya beans	Anemia, peripheral neuropathy
B_7 (Biotin)	Water	Walnuts, peanuts, cereals, milk, egg yolks, salmon, pork, mushroom	Dermatitis
B_9 (Folic acid)	Water	Citrus fruits, green leafy vegetables, whole grains, legumes	Megaloblastic anemia
B_{12} (Cyanocobalamine)	Water	Fish, meat, poultry, eggs, milk	Anemia
C (Ascorbic acid)	Water	Fresh citrus fruits, broccoli, goat milk, black currant, and chestnuts	Scurvy

vitamins include folate, thiamine, and ascorbic acid whereas the most stable vitamins include niacin, phytonadione, calciferol, biotin, and pantothenic acid. The deterioration of fat-soluble vitamins is distinctly different from that of water-soluble. Most fat-soluble vitamins are prone to auto-oxidation processes similar to lipids and thus are labile to light, oxidation, metals especially iron acting as catalysts and free radicals to promote the reaction (Reddy and Love 1999). Alternatively, water-soluble minerals can be lost due to leaching in processing water or due to structural deterioration during cooking.

1.6.2 Minerals

Minerals are the inorganic elements which are essential for human body for growth, development, and maintenance. Minerals that are essential for health include calcium, phosphorus, potassium, sodium, chloride, magnesium, iron, zinc, iodine, chromium, copper, fluoride, molybdenum, manganese, and selenium.

Minerals can be classified into two categories as major and minor minerals, depending upon the amount required to be consumed in the diet. Major minerals include calcium, phosphorus, chlorine, sodium, potassium, magnesium, and sulfur. Minor minerals include the ones that are required by the body in trace amounts such as chromium, copper, fluoride, iodine, iron, manganese, selenium, and zinc. Table 1.10 lists the major food sources of minerals and their function in human body.

Table 1.10 Minerals with major food sources and body function

Mineral	Sources	Function
Macrominerals		
Calcium	Yogurt, cheese, milk, salmon, leafy green vegetables	Bone and teeth development and maintenance, muscle contraction, nerve signal transmission, blood clotting
Phosphorus	Meat, fish, poultry, eggs	For bone and teeth health, acid–base balance, signal transmission
Chlorine	Salt	Fluid balance in the body
Sodium	Salt, soy sauce, vegetables	Fluid balance, muscle contraction, nerve signal transmission
Potassium	Meat, milk, fruits, vegetables, grains, legumes	Fluid balance, muscle contraction, nerve signal transmission
Magnesium	Spinach, broccoli, legumes, seeds, whole-wheat bread	Bone development, protein synthesis, nerve signal transmission, and muscle contraction
Microminerals		
Chromium	Meat, poultry, fish, nuts, cheese	Blood sugar maintenance
Copper	Shellfish, nuts, seeds, whole grain products, beans	Cofactor with many enzymes, iron metabolism
Fluoride	Fish, tea	Cavity protection, needed for healthy teeth and bones
Iron	Red meat, poultry, eggs, fruits, green vegetables, fortified bread	Energy metabolism, component of hemoglobin
Iodine	Iodized salt, seafood	Component of thyroid hormone, important for growth regulation
Manganese	Nuts, legumes, whole grains, tea	Cofactor for many enzymes
Selenium	Organ meat, seafood, walnuts	Antioxidant
Zinc	Meat, shellfish, legumes, whole grains	Protein and DNA synthesis

1.6.2.1 The Bioavailability and Absorption of Minerals

Bioavailability of a mineral defines how easily can it be absorbed and utilized in the body. A diet rich in minerals of low bioavailability may not be sufficient to maintain a healthy mineral balance. Bioavailability is impacted by many factors such as the chemical form of the mineral, the composition of the diet, the individual's gut pH, and the physiological need of the consumer.

The dietary composition significantly affects the bioavailability of certain minerals. For example, heme iron (from animal source) is more bioavailable than non-heme iron (from plant source), and phytates and oxalates present in whole grain cereals can bind with minerals such as calcium and iron resulting in reduced bioavailability. Presence of vitamin C can improve iron absorption. Minerals also compete with each other to be absorbed, for example, excess iron in the diet can prevent absorption of zinc in the gut making it less bioavailable. The human body also has a system of regulating mineral absorption by feedback mechanisms such as, if enough iron is stored, more iron will be absorbed by the gut even though it is present in the diet. Compared to vitamins, minerals are generally stable during food processing and storage, but physical processes such as milling of whole grains may lead to the removal of minerals along with germ and bran layers.

As different minerals have different functional roles in human body, a varied amount of each of them is required. Recommended Dietary Allowance (RDA) values vary depending on the age, gender, state of health, and physiological state (such as pregnant or lactating women). British Nutrition Foundation defines the Reference Nutrient Intake (RNI) as the amount of a nutrient that will satisfy the needs of 97.5% of the population. Lower RNIs have also been established and are defined as nutrient levels that will be sufficient for only 2.5% of a given population, everyone else will require more.

1.7 Pigments

Pigments impart color to food products. Visual appearance and aesthetic appeal of foods is largely governed by the color/pigment intensity. End point of many biochemical reactions including deteriorative reactions, processing treatments can be judged by specific color tinge. For example, color change is a major predictor of end point of baking, roasting, and mold growth.

In this chapter, pigments are described in terms of food colorants. According to a survey, the global food color market size will be of USD 2.97 billion by 2025. Food colorants can be divided into two categories on the basis of origin: Artificial colorants/dyes and natural colorants. Artificial colorants are developed synthetically while natural colorants, also known as biocolorants, are of biological origin. Since synthetic food colorants are associated with hyperactivity in children and also cause various health problems therefore their utilization must be regularized. European Union has characterized food colorants by giving specific codes called the E numbers that are universally accepted. By looking at the particular E number, one can know which colorant has been added to the food product. Maximum limit of use

Table 1.11 Artificial colorants permitted in India

S. No.	Predominant hue	Permitted artificial colourant	E number
1.	Red	Carmoisine	E122
		Ponceau 4 R	E124
		Erythrosine	E127
2.	Yellow	Tartrazine	E102
		Sunset yellow FCF	E110
3.	Blue	Brilliant Blue FCF	E133
		Indigo Carmine	E132
4.	Green	Fast green	E143

of any artificial colorant or a mixture of them should not exceed 100 ppm at the time of consumption. Some of the colorants not requiring any certification when used in food include annatto (E160b), carotenes (E160a), beetroot betanins (E162), caramel (E150), riboflavin, saffron, turmeric, and its oleoresins.

Inherent solubility characteristics divide colorants into two categories (a) water-soluble and (b) fat-soluble. List of permitted colorants whether synthetic or natural along with their respective E numbers have been well documented by Fellow (2005). Table 1.11 gives the list of artificial colorants permitted in India.

Chlorophylls (E140): Green pigment associated with the photosynthesis of plant cells is known as chlorophyll. It is a fat-soluble pigment. Chlorophyll may be of many types but higher plant spices contain type "a" and "b" in the ratio of 3:1. Four pyrrole rings with magnesium ion in center is the characteristic of chlorophyll molecules. Acidic environment changes the chlorophyll molecules to pheophytin which causes change in color to olive-brown. Presence of chlorophyllase and removal of magnesium also lead to brown color generation.

Flavonoids: These are glycosides containing polyphenolic compounds. Flavonoids include a number of compounds such as flavones, flavonols, flavanones (and isoflavones), and flavoxanthin (E161(a)). Yellow-coloured flavonoids such as kaempferol, quercetin, and myricetin are most widely present in all fruits and vegetables. Besides that, tea especially green tea is considered as the most concentrated and convenient source of flavonoids. Citrus fruits are also a rich source of flavonoids. Soybean isoflavones are commercially available in stable isolated form.

Anthocyanins (E163): Water-soluble pigment with various shades of color ranging from red to purple is exceptional characteristic of anthocyanins. The true flavonoids consist of the anthocyanins which are red, blue, and purple pigments; the anthoxanthins which are yellow; and the catechins and leucoanthocyanins which are colorless and readily change to brown color. The color of the anthocyanin results from the structure of the anthocyanidin which is combined with the monosaccharides—glucose, galactose, rhamnose, and occasionally with the pentose molecules. Anthocyanidins usually found in the plant tissue are cyanidin, delphinidin, and pelargonidin of which cyanidin is the most common (Ranganna 2007). Acetylated anthocyanin has remarkable heat stability. However, instability at lower pH range limits its utilization as a colorant, especially in low acidic foods. Foods containing sulfur dioxide and KMS as preservatives also bleach anthocyanins pigments. Ascorbic acid and anthocyanins are also not compatible molecules in terms of stability, when coexist. Degraded product of ascorbic acid causes degradation of anthocyanins. Therefore, delivery of anthocyanins from ascorbate-rich foods is possible when stability and amount of ascorbic acid is assured to be on the higher side such as in Indian gooseberry (*Phyllanthus emblica* L.). Similarly, degradation products of sugars like furfurals also lead to the degradation of anthocyanins.

Tannins: Tannins are complex mixtures of polymeric phenols and are also termed as tannic acid or gallotannic acid. The tannins can be divided into two categories: condensed tannins and hydrolyzable tannins. Tannins mainly contribute to the astringency and enzymatic browning reactions of foods. Tannins as anti-nutrient factor can also bind with iron divalent ions and reduces their bioavailability. Tannic acid has also been reported as mordant to color cellulosic fibers and natural dyes (Hong 2018).

Betalains (E162): Due to thermal and light sensitivity, betalains are relatively underexploited as bio pigments. They are often confused with anthocyanin pigments due to similar water-soluble nature and color tinge. Natural coexistence of both the pigments in same matrix has never been reported. Although, to date, around 75 types of betalains have been identified (Khan and Giridhar 2015), still these pigments have not been scientifically studied much. Interestingly, many plants accumulate betalains, but only two sources namely *Beta vulgaris* L. and *Opuntiaficus-indica* are approved by the European Union to be used in food as natural colorants (Delgado-Varges and Lopez 2002). Bio functionality of betalains is mainly governed by its two pigments: purple betacyanins (Latin Beta, beet; Greek kyanos, blue) and yellow betaxanthins (Latin Beta; Greek xanthos, yellow).

Quinones and Xanthones: Some of them are found in flowering plants while majority of these pigments are available in fungi, bacteria, and algae. Quinones are mainly used in the synthesis of artificial colorants/dyes. A derivative of quinoline yellow (E104) has been permitted as food colorant. However, the best example of xanthones is the pigment present in mango (*Mangifera indica*) known as mangiferin.

Carotenoids: Lipid soluble hydrocarbons are known as carotenoids which upon oxygenation get converted into xanthophylls. Similar to flavonoids, they are also widely present in plant kingdom. Along with plant carotenoids, they are also present in animal foods including milk, egg yolk, and some fish. Carotenoids are important, from the standpoint of human nutrition due to its bio-conversion into vitamin A. Beta carotene, found in milk, potato, carrot, and pumpkin is a precursor of vitamin A yielding two molecules of vitamin A. Lutein [E161(b)] of marigold, lycopene [E160(d)] of tomato and guava, and bixin in annatto color are the examples of carotenoids molecules. *Trans*-configuration of carotenoids is the stable form. A well-accepted pigment from saffron (*Crocus sativus*) known as "*kesar*" is chemically a carotenoid pigment, crocin, imparts a rich golden-yellow hue to dishes and textiles. Carotenoids are oxygen, light, and heat-sensitive therefore hot pulping/break of tomato is always recommended for better lycopene retention over cold pulping/break. Acidic environment at high temperature facilitates the conversion of respective *trans* form of carotenoids to *cis* form.

Curcuminoids/curcumin (E100): Curcuminoids namely curcumin are used for the enhancement of color and additionally for prolonged storage period of food products due to its antimicrobial capacity (Joe et al. 2004). George and Rastogi (2017) used curcuminoids as a color enhancer for pineapple slices from turmeric (*Curcuma longa* L.). Several physiological active roles of curcuminoids have also been reported such as antioxidant, anti-inflammatory, antidiabetic, antibacterial, and anticancer properties.

Myoglobin and Haemoglobin: Animal muscle and blood, respectively, contain these pigments, characteristic of iron in the center of the structure. Both pigments have the capacity to bind with iron. Bright expression of both the pigments are highly oxygen-dependent. Myoglobin gets oxidized to oxymyoglobin which gets converted to metmyoglobin, a brown color pigment, upon prolonged exposure to oxygen. However, the reduction of metmyoglobin also induces the conversion of metmyoglobin to oxymyoglobin. During curing, nitrosomyoglobin is formed which remains as a stable red color pigment even after thermal processing (Manay and Shadaksharaswamy 2008).

Caramel (E150a): At a temperature of 170 °C or above, sugar is converted into caramel, a brown water-soluble pigment which is generally used as a colorant in various soft drinks and confectionery. It is a chief ingredient of sugar boiled confectionery mainly of amorphous candies. Caramel can be produced by a variety of carbohydrate sources such as corn starch with a wide range of DE and electrocharges. For example, electronegative caramel (E150c) is used in soft drink industries. Caramel is a mixture of low and high molecular weight colored sugar derivatives which also contain a variety of volatiles therefore it imparts color as well as flavor to the product.

1.7.1 Antioxidant Potential of Phytopigments

Most of the phytopigments along with imparting color exert antioxidant activity too. Therefore, imparting color in food matrix with biopigments serves dual advantages; improving visual appeal as well as providing an additional physiological benefit. Anthocyanins, carotenoids, betanins, and lycopene are well known for their antioxidant and pharmacological importance. For example, betacyanins are a group of compounds exhibiting antioxidant and radical-scavenging activities (Escribano et al. 1998). Betaxanthins have been used as a food supplement in order to fortify processed food products with essential amino acids, giving rise to an "essential dietary colorant" (Leathers et al. 1992). Anthocyanins are naturally occurring flavonoid compounds, associated with a wide range of biological properties including antioxidant, anti-inflammatory, anti-cancerous, and antidiabetic effects (Norberto et al. 2013). Chlorophyll exerts lower antioxidant potential than other documented pigments and has not been explored much. However, prooxidant activity of chlorophyll has been demonstrated by Lanfer-Marquez et al. (2005) who further emphasized the role of intactness of porphyrin ring for expression of antioxidant effect of chlorophyll. Eleven conjugated double bonds contribute significantly to the antioxidant capacity of lycopene. Heber and Lu (2002) reported lycopene as the most powerful antioxidant with the ranking as lycopene > α-tocopherol > -α-carotene > β-cryptoxanthin > zeaxanthin = β-carotene > lutein. Arora et al. (2018) recently compiled more than 75 studies on lycopene in a comprehensive manner highlighting its bioavailability and health-boosting effects. Several pigments such as anthocyanins and betanins have also been reported for their antidiabetic potential and help in managing type-2 diabetes. Betanins and anthocyanins also exhibit α-glucosidase inhibitory activity. Quantification of various plant pigments such as carotene, lycopene, chlorophyll, and anthocyanins have been explained by Ranganna (2007) while that of the antioxidant activity has been compiled by Sethi et al. (2013).

1.7.2 Technological Interventions

As described earlier, the presence of biocolorants provides dual benefits to the food matrix; firstly, they provide a natural tinge to the product and secondly they improve the antioxidant potential of the product. Technologically, water-soluble pigments such as flavonoids, betanins, and anthocyanins are easy to infuse into the matrix but their respective temperature and pH sensitivity limit their utilization in selected food matrix only. Infusion of betanins in papaya candy and microwaved potato chips has been successfully attempted by various researchers earlier (Joshi et al. 2019a, b). On a similar concept, a unique product has been developed by ICAR-Indian Agricultural Research Institute, India named as "Pusa Nutra Candy" which has a rare combination of ascorbic acid and anthocyanins and it has been transformed into a shelf-stable product. Infusion of curcuminoids in coconut slices, pineapple slices, and apple slices has been documented by CFTRI, Mysore, India for color enhancement and additional health benefits.

Utilization of natural colors as a replacer of their commercial counterparts is a very challenging task due to their high cost and reduced stability under various processing conditions. Upon heating, the colorants may be degraded due to hydrolysis, decarboxylation, and autoxidation process. During processing, the stability of color pigments are chiefly affected by various physical factors (temperature, pH, light, aeration, water activity) and chemical factors (enzyme activity: β-glucosidases, polyphenoloxidases, and peroxidises; antioxidants, preservatives, sugars and their derivatives, salt) (Herbach et al. 2006). The susceptibility of colorants to these factors restricts their use as food colorants (Ravichandran et al. 2013). Therefore, both intrinsic and extrinsic factors that affect the stability of natural colorants/pigments need to be considered to ensure optimum pigment and color retention in foods. Co-pigmentation has been widely explored to give stability to natural pigments. For example, Kumar et al. (2020) have found that through acylated black carrot and black soybean anthocyanins thermal stability of jamun/plum/strawberry and pomegranate juice can be enhanced. Co-pigmentation, as confirmed by hyperchromic and bathochromic shift in comparison to the native pigment and stability, can further be confirmed through higher half-life and lower degradation rate constant under extreme temperature/light/pH conditions. Mu et al. (2017) described the phenomenon of co-pigmentation in detail in sweet potato anthocyanins. In fact, co-pigmentation is a phenomenon in which the pigments and other colorless organic compounds, or metallic ions, form molecular or complex associations, generating a change or an increment in the color intensity. It is interesting to note that, the co-pigmentation interaction may be due to intramolecular, intermolecular, self-association, and interaction with a metal association.

Minor change in color intensity must be quantified objectively for concrete results and conclusion. Color quantification can be done through spectrophotometer, Munsell system, Hunter Colour chart, CIE system, and Lovibond tintometer (DeMan 2008). Image analysis by RGB methodology can extract the color information from food too (Joshi et al. 2016). This technique is also helpful for prediction of food composition as well on the basis of color differences among the ingredients (Pandey et al. 2011).

1.8 Flavoring Compounds

Flavor is a combined perception of taste and odor by the consumers. It is one of the most influencing factors that decide product sustainability in terms of market place and in consumer preference. Compounds responsible for odor and taste do not impart any nutritious value but the success of a product in the market for instance and consumer preferences are largely associated with flavor intensity, type of flavor, cleanliness, and dominance of the flavor compounds, flavor compatibility, and interaction with the predominant taste of the food, presence of any off-flavor and after taste compounds and most importantly with the temperature of the foods. Whole mouth and nasal cavity perceive the fullness of flavor. That means tongue and nose are the two distinguished sensory organs, associated with qualitative evaluation of this important sensory attribute as a joint sensory evaluation tool.

Flavors are volatile, generally heat-labile compounds, very specific to their molecular configuration, required in a very minute amount for the desirable functionality. They may be natural flavors, nature-identical flavors, artificial flavors, etc. As per the description given by DeMan (2008), natural flavors and flavoring substances are preparations or single substances obtained exclusively by physical processes from raw materials in their natural state or processed for human consumption. Nature-identical flavors are produced by chemical synthesis or from aromatic raw material; they are chemically identical to natural products used for human consumption. Artificial flavors on the other hand are flavorings that are synthesized chemically.

Whatsoever the flavor or combination of flavors used as additives, it should not exceed 300 ppm in food preparation. Safety evaluation protocols for flavors are also described earlier by Smith et al. (1996). Most of the flavors are self-limiting compounds, overdose of flavours is spontaneously restricted by organoleptic properties of the food itself. Some of the substances like monosodium glutamate (MSG) (E621), mono potassium glutamate (E622), calcium glutamate (E623), sodium guanylate (E627), and sodium inosinate (E631) are permitted to be used in the foods as flavor enhancers.

Taste Perception: Generally four tastes (1) sweet, (2) salt, (3) sour, and (4) bitter are considered as the basic tastes. Some countries also categorize "*Umami*" taste as the fifth basic taste. Some flavor and taste enhancers which are considered to be safe are often confused with the term "basic," which means their expression should be independent of the presence of other compounds and whose receptors are present on the tongue, as is the case with sweet, salty, sour, or bitter taste. However, flavor enhancers need a basic compound for their noticeable expression such as for delivering "*Umami*" which are glutamates by nature, and require a salty taste for its full expression. A recently discovered "*Kokumi*" taste is being considered as the sixth taste (Ueda et al. 1989). It is chemically glutathione in nature. However, sometimes a few terpene molecules such as γ-glutamyl-valyl-glycine exhibit a strong "*Kokumi*" taste. Similar to "*Umami*" it can also be considered as a taste enhancer (Maruyama et al. 2012).

Compounds required for the taste characteristics of the product are non-volatile in nature. The sensitivity to a taste decreases gradually with age which may be either due to decrease in the number of taste buds or decrease in the nerve signal transmission with age. "*Ageusic*" and "*anosmic*" conditions also hinder the capacity to evaluate aroma and taste properly in the foods. Taste buds are not uniformly distributed in foods; however, it is documented that tip of the tongue is more sensitive to sweet taste, while posterior end of the tongue is more sensitive to bitterness; and lateral ends identify sour and salty tastes. The after taste is also felt by posterior end of the tongue; however, some off flavors such as beany flavor of soybean is felt by whole oral cavity. These ambiguous findings concluded that there cannot be any clear-cut demarcation or boundary line existing for taste bud distribution in the tongue but significant overlapping might exist or may be the same taste buds can perceive more or all the tastes.

Taste compounds: Numerous compounds are responsible for aroma but relatively a narrow range is responsible for taste attributes. Pure salty taste is provided by sodium chloride also known as common salt. It has more than 99% sodium chloride content. Aliphatic hydroxyl compounds along with artificial sweeteners such as saccharin and duclin, peptides (aspartame), and cyclamates give sweetness to the products. Acidic taste is mainly attributed to proton donors. A wide range of compounds like caffeine and alkaloids provide bitterness to the product. Saponins are considered as bitterness-contributing compounds in bitter gourd. Again prediction of molecules for particular taste expression can not be judged accurately, for example, sugars have similar reducing potential while those that exhibit different degree of sweetness; equal proton donating potential exhibit different degrees of sourness. In general, sensitivity for taste exist as:

sweet < salt < sour < bitter. Generally, during orientation and training schedules for sensory evaluation in food laboratories, screening of panelists is done using 1%, 0.2%, 0.03%, and 0.008% solution of sucrose (sweet taste), sodium chloride (salty taste), citric acid (sour taste), and quinine hydrochloride (bitter taste), respectively. The respective taste threshold of these compounds are 0.03 M, 0.001 M, 0.003 M, and 0.00001 M, respectively (Manay and Shadaksharaswamy 2008). Sometimes some sensory errors are also associated with flavor, since high intensity of color is psychologically related with high intensity of flavor or taste, such as dark pink or orange color is related with intense strawberry or orange flavor, respectively.

As already mentioned, a large range of compounds are responsible for foods' flavor; therefore, for the ease of readers, most of the predominant flavor compounds are tabulated in Table 1.12.

1.9 Anti-nutritional Compounds

Apart from the nutritional components, foods also comprise of toxic and anti-nutritional components. These will be dealt in subsequent chapters. However, a brief description on naturally occurring anti-nutrients and toxic compounds are mentioned below.

Anti-nutrients are compounds which do not have any nutritional advantage. Moreover, they reduce the nutritional property (quality) of foods and sometimes reduce the bioavailability of vital nutritive compounds. They may be heat stable or heat labile (sensitive to heat). A brief description of anti-nutrient factors is mentioned in Table 1.13.

Besides these processing remedies for anti-nutrient depletion, varietal modification is also a continuous activity of plant breeders to reduce toxicant level upto a safe limit. For example, canola oil is a type of mustard oil developed from mustard variety having low erucic acid and glucosinolate content: 2% v/s 35.7–51.4% and 30 v/s 120.3 μmol/g than the conventional mustard oil extracted from expellers known as "*Kacchhi Ghani.*" Erucic acid impairs myocardial conduction and thioglucosides affect the iodine uptake.

1.10 Natural and Developed Toxins

Marine toxicants: These are produced by microalgae and get accumulated in fish that consume these toxic microalgae such as shellfish, crustacean, and finfish. Tetrodotoxin is one kind of such marine neurotoxin, found in puffer fish and may cause acute lethality. Similarly, more than 250 species of coral reef fish cause seafood poisoning due to accumulated ciguatoxins, which are lipophilic and highly thermostable in nature. Ciguatoxins have the capacity to cross the blood–brain barrier and manifest as both central and peripheral neurologic symptoms. Under

Table 1.12 Compounds responsible for flavor profile of the foods

S. No.	Class	Chemical nature	Subclass	Example	Remarks
1.	Terpenoids	Linking of the five-carbon unsaturated hydrocarbon isoprene	Limonene	Lemon, Citrus oil, Grapefruit	Citrus juice gets bittered in presence of air due to terpenoids
			Neral		
			Geranial		
			Nootkatone		
2.	Flavonoids	Polyphenolic compounds having a benzo-γ-pyrone structure	Hesperidin (tasteless)	Orange, Lemon, Grapefruit	Exhibit antioxidant and copigment effects
			Naringin (bitter)		
3.	Sulfur Compounds	–	Allium (cysteine sulfoxide), allinase (*S*-2-propenyl(allyl) cysteine sulfoxide), *S*-methyl-cysteine sulfoxide, and thioglucosides	Onion, Garlic, Mustard	Uniquely present in genus *Brassica* and *Allium*
4.	Esters	Acid and ethanol reaction leads to ester formation	Pentyl acetate, butyl acetate, octyl acetate, ethyl butyrate, pentyl valerate, ethyl salicilate	Banana, Raspberries, Strawberries, Oranges, Apples, Grapes	Fruity aroma depends on the type and molecular weight of acid and alcohol
5.	Aldehydes	Oxidation of alcohol leads to formation of aldehydes	Isopentylacetate, Citral, Benzaldehyde, acetaldehyde, geranial, 5-methyl-2-phenyl-2-hexenal, *cis*-4-heptanol, *trans*-2-*cis*-6-nonadienal	Banana, Lemon, Almonds, Butter, Lemon, Chocolate, Cream, Cucumber	Single aromatic compound dominates the flavor
6.	Fattyacids	Acid derivatives of fats	2-methyl butyrate	Apple	Also used in cosmetic industry
7.	Hydroxy-compounds	Alcohol derivatives	*cis*-3-hexenol, octenol-3, geosmin	Tomatoes, Mushrooms, Beetroots	Also responsible for off-flavor and earthy flavor
		Phenols derivatives	Vinylguaiacol, eugenol, thymol, 1-(*p*-hydroxy-phenyl)-3-butanone, thymol	Cheese, Cloves, Raspberries, Tangerines	–
8.	Ketones	Oxidation products of secondary alcohols	2,3-butanedione	Butter, Celery	–

9.	Acids	Oxidation products of alcohol and ketone	Acetic acid, 2-methyl-butyric acid	Vinegar, Cranberries	Dominance of single flavor compounds
10.	Lignin	Carbohydrate derivatives	Vanillin	Vanilla	Most widely used flavor in bakery and confectionery industry

Table 1.13 Anti-nutritional factors in foods

S. No.	Anti-nutritional factor	Action	Nature	Remedy
1.	Trypsin inhibitors	Inhibition of activity of trypsin enzyme	Heat labile	Thermal treatment
2.	Hemaglutenins	Agglutination of red blood cells	Heat labile	Thermal treatment
3.	Goitrogens	Interfere with iodine uptake by thyroid gland	Heat labile	Thermal treatment
4.	Gossypol	Reduces bioavailability of lysine	–	Solvent extraction
5.	Phytate	Reduces availability of divalent ion calcium, zinc, etc.	–	Germination
6.	Flatulence factors	Indigestion	Heat stable	Soaking followed by washing
7.	Cyanogens	Produces HCN upon hydrolysis	–	Repeated washing with fresh water
8.	Saponins	Cause nausea and vomiting	–	Washing followed by soaking
9.	Aflatoxins	Hepatotoxins, Hepatocarcinogens	Heat stable	Eliminate infested grains/ contaminated milk from diet and animal feed
10.	Tannins	Reduces iron absorption	–	Removal of seed coat
11.	Lathyrogens	Causes lathyrism (neuro or osteo)	Heat labile	Soaking followed by parboiling
12.	Alkaloid glycoside (vicin)	Hemolyticanaemia (Favism)	–	Germination and boiling
13.	Glycoalkaloids (chaconine and solanine)	Inhibit acetylcholine esterase	Heat stable	Avoid consumption of sprouted and green potato

low temperature, storage of seafoods and fish, histidine, and amino acid gets converted to histamine, a biogenic amine, from the action of histidine decarboxylase enzyme produced by bacterial action. Histamine is thermostable in nature and causes extreme food allergic reactions. These toxins may produce acute toxicity (reflected within an hour of ingestion) or chronic toxicity (after long-term exposure and ingestion).

Mushroom toxicity: Majority of poisonings and hallucination effects in mushrooms are possessed by *Amanita phalloides* and *Amanita pantherina*, respectively. Mushroom poisoning caused by alpha-amanitin and orellanine results in severe liver damage and kidney failure. Toxin muscarine causes respiratory failure and the toxin gyromitrin has the ability to convert stomach acid to monomethylhydrazine and affects multiple body systems by blocking the neurotransmitter GABA. *Amanita muscaria* is also known as "*Alice in Wonderland*" mushroom, due to its hallucinatory effects attributed to the presence of ibotenic acid. Therefore, collection and production of edible mushrooms should be done by skilled and experienced personnel for eliminating the toxicity from human platter.

Toxins developed during processing: Some of the toxins may be produced during processing of foods. Very lethal example of developed toxins is the nitrosamines found in cured meat. They are formed by the reaction of nitrites (added as a preservative) with secondary amines present in meat. Acrylamide (prop-2-enamide) in baked and fried products, polycyclic aromatic hydrocarbons such as benzopyrenes formed during smoking from wood smoke are other examples of developed toxins. Smoke reduces the nutritional value of protein-rich foods since phenolic and polyphenolic compounds of smoke react with the sulfhydryl group of the proteins, whereas, the carboxyl group from smoke reacts with the amino groups. Grilling of foods at a higher temperature also leads to generation of HCAs (heterocyclic amines) and PAHs (polycyclic aromatic hydrocarbons) that cause DNA damage in the cells. Excessive hydrogenation causes generation of *trans*-fatty acids. Likewise, repeated oil frying facilitates formation of free radicals which are also potential carcinogens.

1.11 Food Allergens

Allergens in food are either polypeptides or glycoproteins. Food allergens are generally water-soluble and resistant to heat, acid, and proteolysis, which enables them to affect the gastrointestinal tract of the host. Allergenicity can be associated with the type of protein structure (primary, secondary, or tertiary). Allergenicity in case of tertiary proteins often disappears on denaturation, whereas it is retained in the case of primary structures. Moreover, protein should be large enough to be recognized by the immune system as a foreign compound to induce an allergic response. Maximum allergenicity is induced by the molecules with a molar mass between 10 and 70 kDa. In general, food allergies are more common in people whose family members have allergies, suggesting involvement of a genetic factor. Some foods contain multiple allergenic protein moieties such as milk that has whey protein as well as casein. Individuals may be allergic to any one of the fractions.

Foods associated with allergic reactions are usually a major dietary component and the immune response differs according to geographic regions, cultural eating patterns, and age.

More than 160 different foods have been reported to cause allergies in various parts of the world depending on patterns of consumption. For example, in Europe, there is a high prevalence of allergies to mustard and celery. Sesame seed is a common food allergen in Israel and allergy to bird's nest is frequent in Singapore, but peanut allergy is rare in these countries. The most common allergenic foods, sometimes known as "the Big 8" or "priority food allergens," are milk, eggs, fish (salmon, mackerel), crustacean (e.g., shrimp), peanuts, tree nuts (almonds, walnuts, etc.), soybeans, and wheat. These eight foods or food groups account for more than 90% of all food allergic reactions.

Cow's milk allergy (CMA) is the most common allergy in infants as it contains numerous proteins (allergens) that can cause allergic reactions. Attempts to modify the protein components of cow milk is an effort to reduce their allergenic potential have included the application of prolonged heat, enzymatic treatment, infant formula, extensively hydrolyzed formula (EHF), amino acid-based formula (AABF), and soy formula. Wheat-dependent exercise-induced anaphylaxis (WDEIA) is a rare anaphylactic response in adults occurring after 1–4 h of ingestion of wheat that is triggered by ω-5 gliadin fraction of the wheat protein. "Bakers' asthma" is a wheat allergy that causes asthma. It results after inhaling small quantities of fine wheat flour. People with this allergy do not have to avoid eating wheat. Celiac disease and wheat allergy are two distinct conditions. Celiac disease or gluten sensitivity is a permanent adverse reaction to gluten protein that causes diarrhea and loss of appetite. It requires a lifelong restriction of gluten by the affected individuals. Antihistamines may help to relieve mild symptoms of allergy, such as itching.

Food allergies may affect the skin (urticaria, hives, eczema), gastrointestinal tract (nausea, cramps, diarrhea, colitis, esophagitis), respiratory (allergic rhinitis, asthma, shortness of breath), and circulatory (blood pressure drop, loss of consciousness systems). In some cases, it may also result in multisystem disorders, termed as anaphylaxis and may lead to multisystem failure and death.

A food allergy is an adverse immune response towards an allergen (protein) in food. This is in contrast to a larger number of non-immune mediated adverse reactions (food intolerance) to food termed as food intolerance. The abnormal physiological responses may be due to inherent properties of the food or physiology of the individual and are often related to the amount of offending chemicals in food. These reactions involve food malabsorption due to intestinal enzyme deficiency, as in lactose intolerance where the activity of β-galactosidase enzyme is diminished. It may also be caused by naturally occurring chemicals in foods (salicylates, histamine, serotonin) or due to the presence of preservatives (sodium benzoate and sulfites), flavorants (MSG, aspartame), or colorants (tartrazine).

1.12 Summary

Food is essential to sustain life. The nutrients present in food help to maintain a healthy, active, and balanced life. Consumption of the right kind of food at each stage of life is essential to keep diseases at bay. Carbohydrates, proteins, fats, vitamins, minerals, and water are the essential nutrients required to carry out vital processes in the body. Meeting the recommended dietary intake of nutrients as per the individual's age, gender, level of activity, and health conditions are important to avoid health complications. The constituents of the food can be modified to yield products with modified properties that cater to specific needs. Further, these nutrients can be altered to achieve desirable change in functionality that can be exploited in food processing to change product characteristics. Processing protocols can be standardized on basis of the elimination of undesirable factors that are inherently present or that may get developed during the process.

References

Andrade RD, Lemus R, Pérez CE (2011) Models of sorption isotherms for food: uses and limitations. Vitae 18:325–334

Arora B, Sethi S, Joshi A et al (2018) Potential role of lycopene as health boosting food ingredient: a review. Ind Food Packer 72:50–58

Azizi AF, Sethi S, Joshi A et al (2020) Biochemical and functional attributes of raw and boiled potato flesh and peel powder for suitability in food applications. J Food Sci Technol 57:3955–3965

Berry SE, Bruce JH, Steenson S et al (2019) Interesterified fats: what are they and why are they used? A briefing report from the Roundtable on Interesterified Fats in Foods. Nutr Bull 44:363–380. https://doi.org/10.1111/nbu.12397

Chauhan S, Chandra R, Shukla S (2016) Process optimisation and shelf life extension of fiber enriched paneer. Pharma Innov J 5:77–79

Delgado-Varges F, Lopez OP (2002) Betacyanins and phenolic compounds from *Beta vulgaris* L. roots. Food Chem 58:255–258

DeMan JM (2008) Principles of food chemistry, 3rd edn. Springer, New York. (978-1-4614-6390-0)

Escribano J, Pedreño MA, García-Carmona F et al (1998) Characterization of the antiradical activity of betalains from *Beta vulgaris* L. roots. Phytochem Anal 9:124–127

Fellow P (2005) Food processing technology: principles and practice. Woodhead Publishing Limited, Cambridge, UK. ISBN 1855735334

Food Safety Network (2014) Providing reliable information to help keep food safe and healthful. University of Guelph, Guelph. https://www.uoguelph.ca/foodsafetynetwork/fat-substitutes

George JM, Rastogi NK (2017) Impact of high pressure on the infusion of curcuminoids in pineapple slices. J Food Sci Nutr. https://doi.org/10.24966/FSN-1076/100027

Heber D, Lu QY (2002) Overview of mechanisms of action of lycopene. Exp Biol Med 10:920

Herbach KM, Stintzing FC, Carle R (2006) Betalain stability and degradation-structural and chromatic aspects. J Food Sci 71:41–50

Hernández-Ortega M, Kissangou G, Necoechea-Mondragón H et al (2013) Microwave dried carrot pomace as a source of fiber and carotenoids. Food Nutr Sci 4:1037–1046

Hong KH (2018) Effects of tannin mordanting on coloring and functionalities of wool fabrics dyed with spent coffee grounds. Fash Text 5:33

Joe B, Vijaykumar M, Lokesh BR (2004) Biological properties of curcumin-cellular and molecular mechanisms of action. Crit Rev Food Sci Nutr 44:97–111

Joshi A, Sethi S, Arora B et al (2016) Colour quantification of food through RGB colour cube concept. ICAR Newsl 22:2–4

Joshi A, Prakash J, Sagar VR et al (2019a) Betanin rich papaya candy (tutti-fruity): an innovative fusion product with enhanced functionality. Int J Trop Agric 37:459–465

Joshi A, Sethi S, Sagar VR et al (2019b) Betanins rich purple potato chips: an innovative fusion product with enhanced functionality. J Agric Eng Food Technol 6:1–3

Khan MI, Giridhar P (2015) Plant betalains: chemistry and biochemistry. Phytochemistry 117:267–295

Kumar M, Dahuja A, Sachdev A et al (2020) Black carrot (*Daucuscarota* ssp.) and black soybean (*Glycine max* (L.) Merr.) anthocyanin extract: a remedy to enhance stability and functionality of fruit juices by copigmentation. Waste Biomass Valorization 11:99–108

Labuza T (1980) The effect of water activity on reaction kinetics of food deterioration. Food Technol 34:36

Lanfer-Marquez UM, Barros RM, Sinnecker P (2005) Antioxidant activity of chlorophylls and their derivatives. Food Res Int 38:885–891

Leathers RR, Davin C, Zryd JP (1992) Betalain producing cell cultures of *Beta vulgaris* L. var. Bikores Monogerm (red beet). In Vitro Cell Dev Biol 28:39–45

Li H, Fu X, Bima Y et al (2014) Effect of the local microenvironment on survival and thermal inactivation of salmonella in low- and intermediate-moisture multi-ingredient foods. J Food Protect 77:67–74

Manay NS, Shadaksharaswamy M (2008) Foods: facts and principles. New Age International (P) Limited, New Delhi

Maruyama Y, Yasuda R, Kuroda M, Eto Y (2012) Kokumi substances, enhancers of basic tastes, induce responses in calcium-sensing receptor expressing taste cells. PLoS One 7:1–8

Mu T, Sun H, Zhang M et al (2017) Sweet potato processing technology. In: Sweet potato anthocyanins. Academic, San Diego, pp 279–355. https://doi.org/10.1016/B978-0-12-812871-8.00006-4

Norberto S, Sara S, Manuela M et al (2013) Blueberry anthocyanins in health promotion: a metabolic overview. J Funct Foods 5:1518–1528

Ognean CF, Darie N, Ognean M (2006) Fat replacers: review. J Agroaliment Process Technol 12: 433–442

Pandey A, Singh G (2011) Development and storage study of reduced sugar soy containing compound chocolate. J Food Sci Technol 48:76–82

Pandey A, Thakur M, Jain SC (2011) Visible image analysis and its applications for foods. Ind Food Ind 30:44–48

Pandey A, Singh G, Jain SC (2012) Development of soy containing compound chocolates. LAP LAMBERT Academic Publishing, Sunnyvale, CA

Potter NN, Hotchkiss JH (2006) Food science, 5th edn. CBS Publishers and Distributers, New Delhi, India

Prabhakar K (2014) Intermediate moisture foods. In: Batt CA, Tortorello ML (eds) Encyclopedia of food microbiology, 2nd edn. Academic, Oxford

Rahman MS, Labuza T (2007) Water activity and food preservation. In: Handbook of food preservation. Taylor and Francis LLC, Milton Park, UK

Ranganna S (2007) Handbook of analysis and quality control for fruit and vegetable products, 2nd edn. Tata McGraw Hill, New Delhi

Ravichandran K, Saw NMMT, Mohdaly A (2013) Impact of processing of red beet on betalain content and antioxidant activity. Food Res Int 50:670–675

Reddy MB, Love M (1999) The impact of food processing on the nutritional quality of vitamins and minerals. In: Jackson LS, Knize MG, Morgan JN (eds) Impact of processing on food safety, Advances in experimental medicine and biology, vol 459. Springer, Boston, MA. https://doi.org/10.1007/978-1-4615-4853-9_7

Rudra SG, Sethi S, Jha SK, Kumar R (2016) Physico-chemical and functional properties of cowpea protein isolate as affected by dehydration technique. Legume Res. https://doi.org/10.18805/lr.v0iOF.9441

Sethi S, Gupta S (2016) High fibre tortilla chips. ICAR Newsl 22:4

Sethi S, Singh S, Rudra SG et al (2013) Protocols for analysis of antioxidants and functional quality in food (Part-I). Indian Agricultural Research Institute, New Delhi. TB-ICN:112/2013

Singh R (2016) Development of fiber enriched bakery products by incorporating fruit pulp waste powder and their acceptability evaluation. Int J Adv Res Biol Sci 3:222–226

Singh S, Sethi S, Gupta S et al (2019) Fermentation of multigrain dough - an approach to reduce glycemic index for healthy bread. Eureka Life Sci 5:19–31

Smith RL, Newberne P, Adams TB et al (1996) GRAS flavouring substances 17. Food Technol 50: 72–81

Spurgeon MJ, Palmer AK, Hepburn PA (2003) An investigation of the general, reproductive and postnatal developmental toxicity of Betapol, a human milk fat equivalent. Food Chem Toxicol 41:1355–1366

Stewart IM, Timms RE (2004) Fats in food technology. In: Fats for chocolate and sugar confectionery. Blackwell, Hoboken, NJ, pp 159–191

Swanson BG (1990) Pea and lentil protein extraction and functionality. J Am Oil Chem Soc 67: 276–280

Ueda Y, Sakaguchi M, Hirayama K et al (1989) Characteristic flavor constituents in water extract of garlic. Food Nutr 54:163–169

Role of Water Activity in Food Preservation

2

E. J. Rifna, Madhuresh Dwivedi, and O. P. Chauhan

2.1 Introduction

Food sustentation comprises the actions taken to preserve the preferred food attributes, within a particular time structure, so that it continues to be secure and enjoyable to consume. A constant food kind could be prepared by using various processing methods and by storing it in a suitable environment. Assessment of food stability on a systematic basis relatively than empiricism is a major confront to food engineers and technologists (Labuza and Rahman 2007). Water is a key component in food matrix. Continually, agribusiness and food processing industries have acknowledged how significant it is to assess free water percentage in fresh and processed food. The measurement of water activity (a_w) forms the foundation for this and offers vital details on the product quality. The significance of measurement of a_w in food matrix could not be exaggerated. All through times past, food industry have applied techniques such as freezing, drying, or incorporation of salt or sugar and freezing to manage water activity values. These techniques avoid microbial deterioration and preserve food properties (Sandulachi 2012). Water activity is defined as the "ratio of the partial vapor pressure of water in equilibrium within a food (P) to the partial saturation vapor pressure of water vapor in air at the same temperature (P_0)," which resembles the relative humidity of ambient air that is in equilibrium with matrix of food. The a_w within food product demonstrates energy rate of water in a particular food product, in addition to its potential to work as a key component and contribute to microbial growth and biochemical reactions. This is a vital attribute which is applied to foresee the stability, robustness, and food product safety with

E. J. Rifna · M. Dwivedi (✉)
Department of Food Process Engineering, National Institute of Technology Rourkela, Rourkela, Odisha, India

O. P. Chauhan
Defence Food Research Laboratory, DRDO, Mysore, India

O. P. Chauhan (ed.), *Advances in Food Chemistry*,
https://doi.org/10.1007/978-981-19-4796-4_2

regards to growth of microbial cells, order of decaying reactions, and physical/ biochemical attributes (Rifna et al. 2019).

The rising identification of the water activity concept is demonstrated through its amalgamation with regulations of United States Department of Agriculture (USDA), Good Manufacturing Practices (GMP), US Food and Drug Administration (FDA) and Hazard Analysis and Critical Control Points (HACCP) necessities, moreover very lately in National Sanitation Foundation/American National Standards Institute, International Standard Draft 75 (Tapia et al. 2020). The objective of these policies is to feature the detailed necessities and monitor whether standard operation procedures are practiced by food industry to guarantee that food samples are manufactured in appropriate sanitary environment and are wholesome, safe, and clean. Novel device methodologies have greatly enhanced the accuracy and speed of assessment of a_w and are certainly a desirable instrument for product quality and safety.

The outcome of a microbiological breakdown, predominantly regards to recall of products, could be significantly expensive. Brand identification and sales could eventually endure as an outcome of customers concerning the call back to various food kinds produced by a peculiar industry (Rahman 2010). In the current scenario of growing constraints with decreasing resources, the requirement to build up assurance programs for food microbial safety is one of the major challenges. Furthermore, there is increasing strain than before on microbiological quality and safety management. Food quality needs to be restricted during the period of manufacturing process from commencement to last part, fairly than allotting on identification of harms in the end product (Rahman et al. 2009).

Hardly, a small number of intrinsic attributes are vital as a_w in forecasting the continued existence of microbial growth in a food product. Scott (1957) demonstrated that microbes possess a restrictive water activity value beneath which they cannot survive. The lowest value of water activity at which an enormous populace of food spoilage organisms can grow is around 0.90. *Cronobacter sakazakii* under an anaerobic environment is reserved to a water activity of 0.92–0.93; whereas, aerobically the level is about 0.86. The a_w limit for mold growth particularly for mycotoxigenic molds is about 0.78 a_w (Sandulachi 2012).

In the twentieth century, among the Ibero-American Programme on Science and Technology, a project was proposed for the production of intermediate moisture foods (IMF) in about ten nations, by gathering data on more than 200 traditional high moisture and intermediate moisture foods (Van den Berg and Bruin 1978). The dual amalgamation of water activity and acidity value was identified to act as a significant hurdle in numerous chosen food samples preventing propagation of harmful microorganisms whereas the remaining (antimicrobials, heat applications, etc.) play a secondary function, chiefly against pathogenic microbes (Amit et al. 2017). This book chapter describes particulars that attempt to explain why the concept of water activity has been effectively used to attain microbial safeguarding of food products, its mechanism of action, and its function in food preservation.

2.2 Concept of Water Activity (a_w)

The notion of water activity has been significantly applicable in preservation of food and on this regard various methods have been productively modified and novel products have been tailored. Water is termed as the universal solvent since it is vital for metabolism, multiplication, and to sustain various chemical reactions happening in food substances (Reid 2020). Free water in food product is the kind of water obtainable for biochemical processes to occur, to gear up microbial growth, and to transport components within the medium. Whereas, the water in bound phase is confronted to take part in the biochemical processes as it gets restricted using various soluble ingredients say salt, sugars, and through the surface impact of the substrate-binding medium (Rahman and Labuza 2020a, 2020b). Figure 2.1 depicts the schematic representation of free, semi-bound, and bound water within a food system:

$$a_w = \frac{P}{P_0} = \frac{n_1}{n_1 + n_2} \tag{2.1}$$

where P and P_0 are the partial vapor pressures of water in equilibrium above the food and partial saturation vapor pressure of water vapor in air at the same temperature; n_1 and n_2 are the number of moles of solute and solvent, respectively (Table 2.1).

The aforementioned water-binding properties decrease the food vapor pressure in accordance with Raoult's Law. Collating this vapor pressure to pure water produces an outcome in ratio defined as water activity (a_w) (Sandulachi and Tatarov 2012; Slade et al. 1991). Water activity value of pure water is 1, and one molar solution of

Moisture content	**Water activity**
•Quantitative amount of free water and bound water in a food system	•Quantitative and qualitative amount of free water in a food system
•Measured using moisture meter	•Measured using water activity meter
•Can be measured either in wet basis or dry basis	•Microbes use free water for their growth

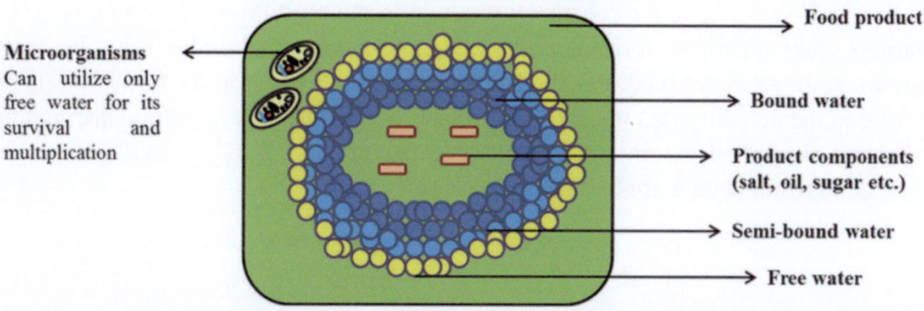

Fig. 2.1 Schematic representation of free, semi-bound, and bound water present within a food system

Table 2.1 Vapor pressure and vapor pressure ratios of pure ice and water at varied super cooling temperature

	Vapor pressure					
	Water		Food product with ice		P_{ice}/P_{water}	Reference
Temperature (°C)	torr	Pa	torr	Pa		
0	4.58	611	4.58	611	1.00	Sandulachi (2012)
−5	3.16	421	3.02	402	0.95	Slade et al. (1991)
−10	2.15	287	1.95	260	0.91	Slade et al. (1991)
−15	1.43	191	1.24	165	0.86	Slade et al. (1991)
−20	0.94	125	0.77	103	0.82	Sandulachi (2012)
−30	0.38	50.9	0.29	38	0.75	Sandulachi (2012)
−40	0.14	18.9	0.098	13	0.69	Sandulachi (2012)
−50	0.05	6.4	0.029	3.9	0.61	Slade et al. (1991)

sodium chloride has an a_w value of 0.9669, saturated solution of NaCl has an a_w value of 0.755, and one molar solution of sugar possesses a_w of 0.98. This same one molar sugar solution in a sealed vessel will build up an equilibrium relative humidity (ERH) within head space of value 98%. A relationship henceforth occurs between a_w and ERH as both are based on vapor pressure (Barbosa-Cánovas et al. 2007):

$$a_w = \frac{\text{ERH}}{100} \tag{2.2}$$

The equilibrium relative humidity of a food matrix is demonstrated as the RH of air neighboring the product upon which the food either loses or acquire its normal value of moisture and is in balance with the surroundings. The description of moisture surroundings wherein pathogenic microorganisms cannot multiply is of great significance to food safeguarding. It had been significantly proven that every microbial cell has a particular a_w beneath which multiplication is back-breaking (Labuza and Altunakar 2020).

Food matrix are significantly stable at their "BET-monolayer water activity" or "BET-monolayer moisture" content and uneven over or beneath BET-monolayer (Labuza and Rahman 2007). Furthermore, experimental verification revealed that optimal moisture value for constancy was within the multilayer adsorption area. In various other situations, it has been revealed that the most favorable water percentage for stability is not precisely the BET-monolayer. The cause for this disparity is owing to the actuality that the BET hypothesis of adsorption was formulated regard to many simplified assumptions which are not applicable when food surface is regarded (Syamaladevi et al. 2009).

2.3 Triple Point of Water

The solitary grouping of temperature and pressure where water could exist in solid, liquid, and ice at stable equilibrium at the order of 273.1575 K and a VP of 0.00603659 atm is known as triple point of water. At triple point, it is probable to alter every matter to water, ice, and vapor by building subjectively minute alterations in temperature and pressure. Though the net system pressure is higher than the water triple point, given that the partial pressure of vapor is 0.00603659 atm, subsequently the system could yet be shifted to the triple point of water. Stringently explaining, the surface straightening out the various states need to be absolutely flat, to counteract the involvement of surface tension (Guildner et al. 1976).

The solid-liquid-gas water triple point describes to smallest pressure at which water in liquid state could subsist. At values of pressures underneath triple point, ice crystals upon being heated at steady pressure is transformed straight to water vapor in a course called sublimation (Murphy and Koop 2005). At region above triple point, ice when heated at stable pressure initially melts and gets transformed into liquid state, which then boils to outline as vapor at elevated temperature.

For majority of food products solid-liquid-gas triple point is the least point of temperature wherein the liquid state could occur. However, with regards to water this statement cannot be accepted as the melting point of normal ice lowers as function of pressure, as described through the diagram. At values of temperatures just beneath the triple point, solidity at stable temperature transforms water vapor initially to solid and from there to liquid state (water in ice has decreased density compared to water in liquid state, so rising pressure results in liquefaction). Figure 2.2 illustrates at what phase the triple point occurs and its reaction when equilibrium is altered.

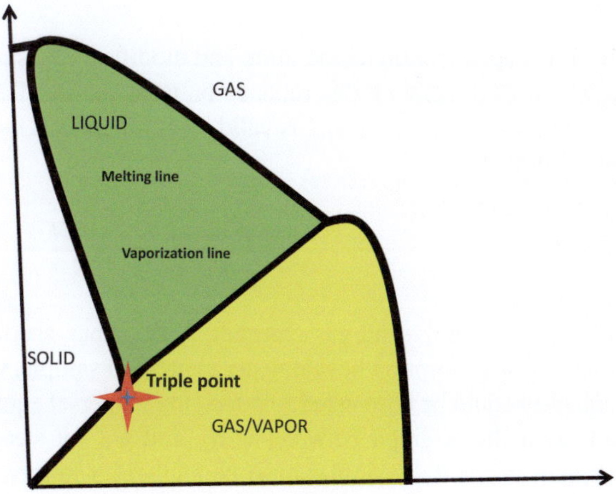

Fig. 2.2 The schematic diagram showing at what phase the triple point occurs and its reaction when equilibrium is altered

2.4 Water Activity and Vapor Pressure

Out of the numerous criterions employed to study food quality, indisputably a_w is the major intended parameter for calculating the water necessity for growth of microbes and enzyme action. The quantity of water removed or gained from the food relies on the concentration and characteristics of water vapor in the environment. The absorbable or removable water existing within the food product could be defied with respect to relative vapor pressure (RVP) or equilibrium relative humidity (ERH).

$$\mathrm{ERH} = \left(P^{\mathrm{eq}}/P^{\mathrm{sat}}\right)_T, P = 760\ \mathrm{mmHg} \tag{2.3}$$

where P^{eq} is vapor pressure of water at equilibrium state and P^{sat} is the saturation vapor pressure of water at 760 mmHg and at temperature T, respectively. Alternatively, it is an assessment of water quantity really available in the atmosphere at equilibrium, divided by the quantity that could be available if the atmosphere was saturated. The term temperature is significant as equilibrium relative humidity is greatly dependent on temperature. Generally, in food processing units, equilibrium relative humidity is generally mentioned as percentage. A food component in wet environment could swap over moisture till equilibrium vapor pressure at a particular temperature equates to the vapor pressure of water in wet air; henceforth, vapor pressure or ERH value is a straight assessment of whether water would be absorbed or removed (Kou and Schmidt 1999). Determination of a_w do not provide any suggestion on the form (solid, liquid, bound, or free) of the water existing. To respond to this query other methodologies have been applied, for instance nuclear magnetic resonance (NMR), differential thermal analysis (DTA), and relaxation dielectric technique.

2.4.1 Relative Vapor Pressure (RVP) and Its Temperature Dependence

RVP is dependent on temperature and modified Clausius-Clapeyron equation offers a way for determining this reliance on temperature. The below described equation, even though is respect to a_w, is valid for RVP and has the subsequent form (Fennema and Berny 1974):

$$\frac{d\ \ln\left(a_w\right)}{d\left(\frac{1}{T}\right)} = \frac{-\Delta H}{R} \tag{2.4}$$

wherein R is universal gas constant, T describes absolute temperature, and ΔH is total isosteric sorption heat at water content of sample. On reorganization, the above equation could be completed to match the universal equation for linear line. It is thus apparent that a graph between ln a_w and $1/T$ (at steady water content) will be a straight line and the similar must be factual for ln p/p_0 against $1/T$. With respect to studics of various investigators, it was observed that for food products rich in proteins and carbohydrates the temperature coefficients varied from value 0.003 to 0.02 $°C^{-1}$ in temperature variation of 5–50 °C as a result, it can be demonstrated that relying on the food product, a rise or fall in 10 °C of temperature could root minor variations (less than 0.05) in value of p/p_0 (Teeter 1991). This property could be vital for packaged food products as it could endure a variation in RVP as the temperature varies, elucidating that temperature reliance of constancy to be higher when compared to that of similar product lose (not packed) in bulk. In addition, it has also been reported that Plots of ln p/p_0 v/s T^{-1} will not at all times be a straight line at extensive temperature variations, and could usually display spiky breaks with beginning of development of ice. Prior to viewing statistics at temperatures in subfreezing levels, it is suitable to judge the meaning of RVP since it is applicable to subfreezing temperatures. It is obligatory since a query rises that the denominator term (p_0) needs to be equated to vapor pressure of water at subfreezing temperature or to that of ice vapor pressure. The VP of cooled water falls to be in proper choice as (a) data value obtained for RVP at super cooled state can be precisely compared with values of RVP super freezing point of temperatures and (b) selection of ice vapor pressure as p_0 could infer, for food products with ice crystals, in a worthless condition where RVP value will be 1 at every chosen subfreezing values of temperatures. The second point arouses as the water partial pressure in cold food is highly comparable to ice vapor pressure at an identical temperature (Fennema and Berny 1974). As the VP of supercooled water could be calculated downward to value of −15 °C, and ice vapor pressure could be determined to much lesser temperature values, it is probable to precisely analyze RVP in case of cold foods, which is evidently clear while one considers the subsequent connection:

$$a_w = \frac{p_f}{p_{0(SCW)}} = \frac{p_i}{p_{0(SCW)}} \tag{2.5}$$

where p_f is defined as partial pressure of water in cold solid foods, $p_{0(SCW)}$ is VP of pure frozen water, and p_i is the VP of ice. Table 2.1 explains the values of RVP determined for cold water and pure ice, and investigators identified that these values were indistinguishable from that of cold foods at that particular point of temperature. Through various constant studies (Storey and Stainsby 1970) on plot of log p/p_0 against $1/T$, researchers observed that (a) the association is always straight lie at subfreezing values of temperature, (b) the association of temperature on RVP is characteristically much superior at temperatures below freezing points than at temperatures greater than freezing points, and finally (c) a spiky shatter happens in the plot at freezing point of food product. Two significant points need to be renowned in relating RVP at below and above freezing temperatures. Primarily, at temperatures above freezing point, RVP depends on the composition of food product taken and temperature, with previous aspect prevailing.

At temperatures falling to subfreezing state, RVP is free of composition of sample and is totally dependent on temperature; that is, in occurrence of ice crystals RVP cannot be prejudiced by values of solutes at hand. As an outcome, every subfreezing occurrence which is inclined through the variety of solutes at hand might not be precisely predicted with respect to RVP value (Fennema and Berny 1974). Therefore, RVP at subfreezing points is insignificant important index of physicochemical procedures than are RVP values at greater than freezing points. It describes that comprehension of RVP at subfreezing value of temperature could not be applied to envisage RVP at temperatures above freezing points. Following, as the temperature is altered adequately to structure or dissolve ice, the denotation of RVP, with respect to food stability, also get modified. For instance, a product held at −15 °C ($p/p_0 = 0.86$) shows a reduced level of microbial growth and chemical reactions. On the other hand, at a temperature of 20 °C (p/p_0 0.86), few chemical changes can take place quickly and few microbial strains will cultivate at reasonable levels. Table 2.1 describes the vapor pressure and vapor pressure ratios of pure ice and water at varied supercooling temperature.

2.4.2 The Significance of Vapor Pressure or Equilibrium Relative Humidity

Numerous biological elements in touch with atmospheric air incessantly regulate its moisture percentage by removing or absorbing water. In food and agriculture sector, equally the hindrance of moisture interchange (in storage and transportation of food products) as well as the restricted elimination of moisture to decrease mass and prolong the shelf life (in drying and dehydration are significant). The moisture percentage in the food product is vital, moderately since it is a weight that needs to be paid during the period when food is acquired or vended. Unquestionably most noteworthy of all, the lesser the moisture value, the processed food and other agricultural commodities would store up devoid of spoilage (Stamp et al. 1984). The growth rate of mites and microbes enhances extremely with moisture value. The speed of these methods is usually enhanced with rising temperature, so that further water content could be supported at reduced temperatures. The quantity of removable water at a specified temperature does not rely on actual value of moisture existing in sample; however, a responsive function of concentration and characteristics of hydrophilic components are present. The quantity and speed of gain or elimination chiefly relies on the value water vapor existing in air. This concentration is diverse above a broad series and is essential to encompass a directory of whether the food product gains or removes water content in particular storage environment. Food scientists from before are recognizable with the detail that the capacity of microbes to proliferate within a food substance is linked greatly to the form of the water existing. Microbiologists on examining the bacterial growth on meat products observed that RH of cold storage compartment had a chief effect with an association among a_w of food product and capability of microbes to proliferate over them. Through such works, the notion of a_w has generated a foundation for an escalating knowledge of the effect of various attributes on the growth of microbes at low values of a_w. The a_w could be applied to alter the metabolism and cell activity within a microbe. Furthermore, it is probable to reduce the development of microbes in food matrix by inhibiting the accessibility of "free" water. The a_w computes the accessibility of moisture in an equilibrium state. On the other hand, in nonequilibrium unit, for instance, in intermediate moisture products, the moisture is efficiently disabled using elevated activation hurdles, such that the moisture is not accessible to microbes in a time period commensurate with its shelf life.

2.5 Mechanism of Food Preservation

Water is necessary for microbial cells to maintain the cellular structure, movement of metabolites, discard the cellular waste, and for preserving the cellular structure. In every food product, water exists in a free as well as in bound state (Troller 2012). Wherein, free water necessitates microbial growth and bound water aids the biological functions. Microorganisms preserve turgor pressure within microbial cells by maintaining considerably decreased water percentage within the living cells compared to ambient condition. This process in cells is known as osmoregulation and is vital for cell growth. The water present in free state in environment if removed either by adding solutes (salts, flours, sugars, etc.) or hydrophilic colloids; the free water within the cells (hypotonic) flows outside (hypertonic environment) till the equilibrium point is attained (Miyawaki 2018). This water loss will aid in plasmolysis of cells and osmotic shock, and inhibits cell growth. A 0.005 diminution in a_w from 0.955 to 0.95 in the surrounding decreases the water content present within a cell by around 50% and decreases the volume of cell to around 45%. This demonstrates the susceptibility of microbial cells to minor a_w variation. Microbial cells may either stay alive or expire in the surroundings with the change in a_w value. Few microbial cells carry effectual mechanisms to conquer the process of plasmolysis and to recover turgor pressure. These microorganisms safeguard their cells to a definite level of drop in a_w (Tapia et al. 2020; Rifna and Dwivedi 2021). Furthermore, microbial cells confirm comparative resistance against osmotic stress (decreased a_w) by shifting solutes from surroundings (e.g., amino acids, polyols). Variations in a_w among the environment and microbial cells trigger osmoregulatory capacities of microbial cells (Barbosa-Cánovas et al. 2020). Resistance to reduce a_w varies among each microbial kind and species owing to variations in osmoregulation levels to preserve a steady state of cellular internal, chemical, and physical conditions. Halo-tolerant microbial cells and xero-tolerant cells accumulate polyols molecules (e.g., erythritol, arabitol, and glycerol) in cells against reduced a_w conditions. Water activity ranges of various food systems and growth of microorganisms are given in Table 2.2.

Table 2.2 Water activity ranges of various food systems and growth of microorganisms

Values of a_w	Microbial cells usually inhibited within this limit	Applied foods within this limit
1.00–0.95	*Escherichia*, *Bacillus*, *Pseudomonas*, Gram-negative bacteria, few species of yeasts, *Clostridium*	Fresh foods, canned foods (with 7% salt or 40% sucrose), meat, fish, milk, fruits, and vegetables
0.95–0.91	*Clostridium*, *Salmonella*, *Pedicoccus*, *Lactobacillus*, few species of molds, yeasts, Cocci, *Serratia*	Ham, mayonnaise, hard cheese, jams, cured meat, Swiss cheese, provolone cheese
0.91–0.87	Micrococcus and most species of yeasts	Hard cheese, fruit juices, condensed milk, sweetened syrups, fermented meat products
0.86	Multiplication of *S. aureus*	Rice flour, wheat flour, all-purpose flour, rice, cakes, grains
0.86–0.80	*Pencillia*, *S. aureus*, yeast, *Debaryomyces*, molds	Condensed milk, fruit juices, fermented sausages, brined meat, cake
0.80–0.74	Molds, yeasts, and halophilic bacteria	Jam, molasses, marmalade, brined fish, and meat
0.74–0.65	*Aspergillus chevalieri*, *Wallemia sebi*, *Saccharomyces*, Xerophilic microbes	Dry raisins, dates, oats, jelly, table sugar, molasses, dry nuts
0.68	Most of fungus	Milk powder, cheese, creams, butter
0.65–0.60	*Saccharomyces rouxii*, *Aspergillus*, molds, osmophilic yeast	Raisins, nuts, dates, milk powder, honey, caramel, honey
Less than 0.60	No microbial cell multiplication	Instant tea powder, powdered infant formula, instant coffee, flour, cereals, egg yolk powder

2.6 Techniques to Lower Water Activity in Food Products

Drying is an indispensable technique applied to food preservation. A variety of drying methods and dryers are applied by the food processing industry to eliminate water present within the food products. Elimination of molecules of water lowers a_w, this produces a negative impact on the survival and development of various microbes in the food matrix, retarding the spoilage of food product. With respect to food safety, few drying techniques comprise the use of thermal energy on food products. Use of dry heat is merely effectual when compared to moist heat with regards to the inactivation of microorganisms (Mathlouthi 2001; Zeuthen and Bøgh-Sørensen 2003). As few microbial cells could survive in decreased a_w environments, eradication of such microbes at the process of drying is a significant process regards to the safety of food matrix. The a_w value of food products can be decreased by applying one of the below-mentioned three primary techniques: (1) dehydration, (2) addition of solutes, or (3) crystallization. The most general technique applied to attain increased solid content in foodstuff is the technique of evaporation or boiling of some amount of water using the technique of heat (Ochoa-Martínez et al. 2012). Other techniques for drying of food products include a method of membrane separation, hot air drying, smoking, freeze-drying, osmotic dehydration (using a suitable solute), mechanical drying, and foam drying.

2.7 Attributes Affecting Preservation of Food by Decreasing Water Activity

2.7.1 Attributes Related to Foods

Values of a_w do not alter in a uniform matrix within a food package. In a varied matrix of food holding products of various a_w (e.g., fast foods say burger with various ingredients in the same lot), a moisture variation is recorded. This might direct to the growth of microbial cells. Food ingredients could also impact the microbial death rate at a similar a_w (Maldonado-Astudillo et al. 2019).

2.7.2 Attributes Related to Water Activity

Water activity of each food kind varies. In every food matrix, water activity will always be lower than the whole quantity of water present. This statement is applicable even in case of pure water. The association between the a_w and moisture content of a food product could be assessed using a sorption isotherm (Peromingo et al. 2016). Researchers have identified that the sorption isotherm develops a loop as the water gets removed (desorption) or through the process of addition of water (adsorption) into a food (Witczak et al. 2020). Condensation of water results in the drenching of moisture to foods at a period of storage with temperature variations. This could modify secure a_w levels to an unsecure condition.

2.7.3 Attributes Related to Microorganisms

Water activity necessities of microbial cells differ considerably. A decline in a_w enhances the initial phase of microbial cells and diminish the growth velocity. Values of a_w for diverse foods and retarding microbes have been studied through various works. Microorganisms cultivate, sporulate, and develop at diverse a_w values. In particular, fungi can rise at reduced a_w levels than bacteria. Gram-negative bacteria need considerably increased a_w values when compared to Gram-positive organisms (Syamaladevi et al. 2016). Most favorable a_w values for the nourishment of major bacteria in food products are 0.95–0.98. At a_w value of 0.97, gram-negative organisms dominate, and at a_w value of 0.96, gram-positive microbes dominate. Significantly dangerous gram-negative organisms, for instance, *Pseudomonas* spp. and majority members of the family cronobacter, could proliferate even at a_w value between 0.26 and 0.30, respectively (Asafo-Adjaye 2019). Gram-positive non-spore-forming organisms are significantly less susceptible to decreased a_w. Whereas, numerous microbes belonging to the family Lactobacillaceae have bare minimum a_w

value of around 0.94, and microbes belonging to Micrococcaceae family shows the potential of multiplying beneath a_w value of 0.90. The least amount of a_w for the multiplication of *Staphylococcus aureus* is around 0.86. Majority of spore-forming organisms do not multiply below a_w value of 0.93. Decreased a_w limit for *Clostridium* species is around 0.94 and toxin generation happens at this a_w value. As a_w value falls below 0.86, xerophilic and osmiophilic mold spores predominate, and the microbial development ceases below 0.60. Halophilic organisms, for example, *Hansenula anomala*, *Candida pseudotropicalis*, and *Debaryomyces hansenii* may breed well on treated pickles with NaCl solutions upto 11% ($a_w = 0.93$) (Smith et al. 2016). Few osmophilic yeasts are capable of nourishing food products with increased sugar percentage (e.g., syrups, jams, etc.). Germination of mold spore could be restricted by lowering a_w to 0.93. The least a_w for growth of molds (*Penicillium digitatum*, *Fusarium oxysporum*, *Aspergillus oryzae*) varies from 0.70 to >0.9 whereas the optimum a_w value is around 0.96. Xerophilic molds generally do not multiply at a_w values higher than 0.97. The smallest a_w necessities for multiplication and release of mycotoxin in few toxigenic molds have also been studied by various researchers (Mannaa and Kim 2017).

2.8 Calculation of Water Activity in Realistic Applications

Water activity (a_w) could be prejudiced majorly in three ways throughout the preparation of high moisture, intermediate, and dried foods (Jayaraman 1995):

(a) Water might be eliminated by the process of concentration, evaporation, or dehydration process.
(b) Supplementary solute particles might be integrated.
The diffusion of solute may be assessed either through dry infusion or else by moist infusion. Moist infusion comprises soiling the targeted products in a mixture of reduced a_w whereas dry infusion comprises direct integration of food particles and solute molecules in necessary ratios. As solid-rich products, such as food products, are subjected to dry or moist infusion, the following three flows occur:
    - An outflow of the food product's own solutes
    - A flow of solute, from the environment to the product
    - An outflow of water, from product to the environment

The above flow process is regarded as "osmotic dehydration" and permits the concoction of not only the solute particles applied to manage a_w but also the preferred percentage of antimicrobial and anti-browning molecules or any solute for enhancing nutritional and sensory properties. By regulating these aforementioned intricate exchanges, it is probable to envisage diverse combinations of solid gain and water elimination from an effortless dewatering method to salting or candying (a process which preserves fruits, nuts, etc., by saturating or coating it with sugar syrup) technique (wherein penetration of solute is favored and water elimination is restricted) (Caballero-Cerón et al. 2015). For a food matrix of porous nature, moist infusion could also be performed in vacuum conditions. The liquid trapped within the open pores is interchanged for an external liquid phase (of restricted composition) owing to pressure modifications.

(c) Combining (a) and (b): The food products are permeated into the chosen solutes and are then moderately dried. The recompense available with this amalgamation with respect to only drying is an augment in the pigment stability accountable for the color, flavor enhancement, and improved texture. No matter what the method applied to lower a_w, it is obligatory to identify the water activity-moisture content association within the food matrix. Significant assistance has been completed in the ground of water activity value prediction over the precedent 50 years and a complete investigation of the measures conventionally in use to determine a_w has been carried out by Sereno et al. (2001). In every situation, the use of a variety of empirical and theoretical equations was evaluated, describing some descriptive instances.

Till date, there has been no model with an effortless mathematical configuration competent of describing the a_w or sorption lowering aspects of foods or their ingredients in the complete choice of water activities, as the fall of a_w in foods is owing to an amalgamation of mechanisms each of that could be prime in a known choice of water activity value. In intermediate and high moisture foods, a_w is chiefly monitored by the character of soluble solute molecules (i.e., NaCl, sugars, amino acids, other salts, and organic molecules) within the aqueous state of food (Labuza and Rahman 2007; McMeekin and Ross 1996). Numerous equations, with respect to the thermodynamic attributes of multicomponents nonelectrolyte and electrolyte solutions, have been researched experimentally and theoretically for evaluating or calculating the water activity value of these foods.

Even though the theoretical presumption is mistaken for mixed food surface relations, for practical applications the proposed assumption was found to be very significant in evaluating the best possible moisture content percentage (i.e., the one which is analogous to the monolayer water) for preserving chemical stability of dried food matrix (Chirife and Fontana 2020; Labuza and Altunakar 2020). The Guggenheim, Anderson, and Boer (GAB) equation (value $0 < a_w < 0.9$), recently regarded as the most adaptable sorption model and suggested by the European COST 90 Project improves the BET model to take into consideration the interaction energies among the initial and outlying sorbet compound molecules at the individual sorption locations. Researchers have described that this would also permit computation of the monolayer water molecule (Tapia et al. 2020). Figure 2.3 explains various empirical and theoretical models recommended for the estimation of a_w in food with high and intermediate moisture contents.

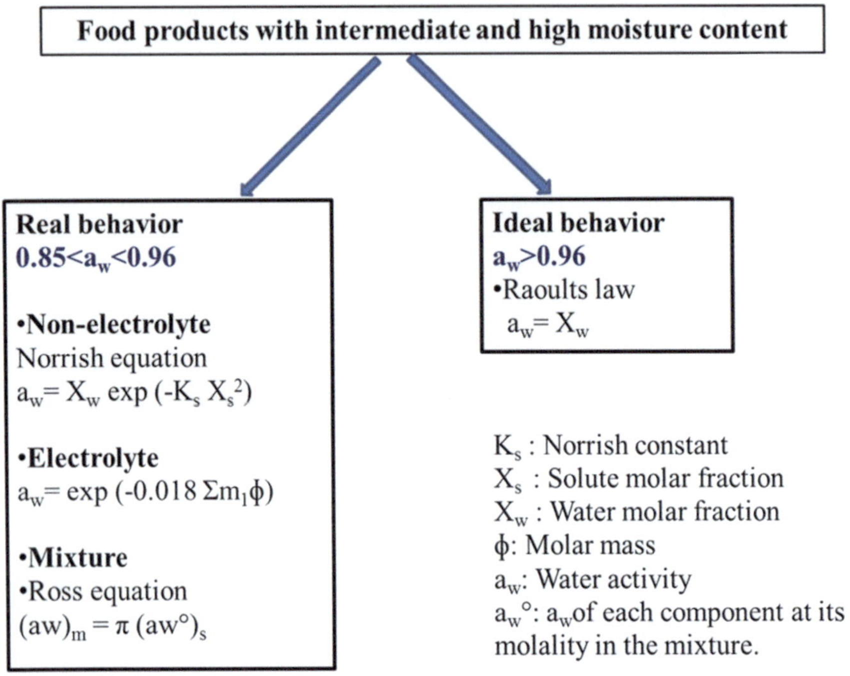

Fig. 2.3 Applicability of empirical and theoretical models recommended for the estimation of water activity in moist and semi-moist food

2.9 Water Sorption Isotherms

The water sorption isotherm elucidates the association among a_w and EMC of food compounds at a particular pressure and temperature. The idea and concept of water sorption isotherms are greatly significant in food processing industries to devise and optimize processing equipments, packaging needs, projection of product quality, and calculation of moisture availability and shelf-life owing to period of storage. Various preservation techniques have been proposed so as to extend the life period of food sample by reducing the water available for the microbial cell to grow and pause few chemical actions (Arslan-Tontul 2020). The exact sketch of isotherm suggests the means through which the water molecule cohere the food particles. As the association between water molecule weakens, the food product turns out to be highly unstable. The a_w relies chiefly on chemical composition, physical properties, and temperature of the substances.

It is from data of an adsorption process or a desorption process the water sorption isotherm is drawn. The differentiation between adsorption curve and desorption curve is elucidated as hysteresis, as depicted in Fig. 2.4. The process in which molecules of water gradually combine together with food particles through

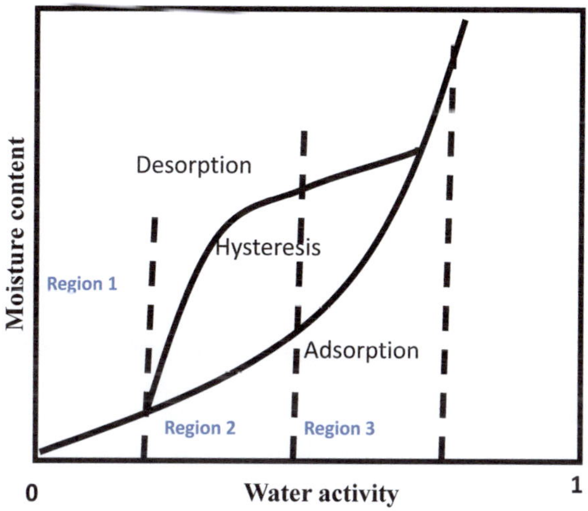

Fig. 2.4 Water sorption isotherm showing hysteresis for food products

condensation or adsorption (physical or chemical) is known as water adsorption (Bastıoğlu et al. 2017). Any isotherm graph can be characteristically classified into three sections; the water in region 1 signifies water molecules that are bound strongly to the polar and hydrophilic particles (proteins, carbohydrates, and so forth) in the food. Furthermore, this robustly bound water in region 1 does not freeze and will not be accessible for any biochemical reactions within the food. However, in region 2, the molecules of water attach less strongly compared to region 1. This category of element water could be identified as uninterrupted conversion from a bound state to free state of water. The characteristics of water molecules in region 3 are identical to those of free molecules of water positioned in voids and crevices; in addition, the water molecules in this portion easily cohere with the food particles (Gichau et al. 2020). Furthermore, hysteresis is connected with nature and the state of food and structural rearrangements that modifies the availability of vigorously active polar units. The existence of capillaries in food matrix outcomes to a significant decline in a_w. The justification for the generation of water sorption hysteresis includes capillary condensation, swelling fatigue theory, and ink bottle theory (Maigalura and Suleiman 2020).

2.9.1 Classification of Sorption Isotherms

Water sorption isotherm is classified into various categories depending on the processes and shapes (Brunauer et al. 1940). Category 1: Langmuir isotherms which depict a distinctive augment in a_w compared to higher water content; the initial derivative of this isotherm augments with water percentage and formed curves are convex to upside. This kind of isotherm is usually appropriate in course of loading a monomolecular water sheet at interior region of the substance. Category 2: sigmoidal isotherms, wherein curves are concave upside; it considers to account for the subsistence of various layers at interior portion of the substance. Category 3: called as Flory-Huggins isotherm, it is applicable for a solvent that is over glass transition temperature. Category 4: explains when utmost hydration point is reached adsorption of expanding hydrophilic compounds initiates. Category 5: Brunauer-Emmett-Teller (BET) multilayer isotherm is studied in water vapor adherence on charcoal and is associated with isotherm category two and three. The most regularly studied isotherm respect to food matrix is isotherm category 2 and 4 (Tian et al. 2020).

2.9.2 Determination of Water Sorption Isotherms

With respect to food products, the water sorption isotherm could be calculated with regard to three diverse measuring methods (Arslan-Tontul 2021): manometric, hygrometric, and gravimetric. In technique of manometric, the vapor pressure is determined at a point it is in equilibrium with the chosen food product at particular moisture content. In the hygrometric technique, the ERH of targeted sample at chosen moisture content is determined. In the gravimetric technique, the food product weight is calculated using a weighing balance. The novel and nondestructive methods to estimate the moisture percentage in a sample is spectroscopic tools, say near infrared spectroscopy, hyperspectral imaging, FTIR, magnetic resonance, and so forth.

2.9.3 Numerical Models for Developing Water Sorption Isotherms in Food Products

Focusing to scientifically articulate the association among the a_w and moisture value of food products, various numerical models had been applied to date say linear, nonlinear, and regressional, comprised within their variable through two to six fractional regression coefficients, that describe all of the three regions that sorption isotherm of humidity falls to. In the majority of situations, model which is appropriate for a definite food matrix will not be appropriate for another kind, that is, each particular model displays an appropriate analytical facility for definite moisture activity levels.

Numerous numerical models had been projected to explain sorption isotherms. Few among them were proposed using a hypothetical foundation to explain adsorption principle (Yan et al. 2008), while the remaining are presently generalization of other complex models. In some values of a_w, water sorption isotherms could be equated to linear equations. However, there exist a few equations comprising of 2–3 fitting attributes to explain water sorption isotherms. The major widespread equations which are applied to demonstrate the sorption process in food compounds are BET model, Langmuir equation, Henderson model, Oswin model, and Halsey model.

2.9.3.1 Langmuir Equation

Langmuir projected the subsequent adsorption model with respect to unimolecular layers with indistinguishable and independent sorption points that are described as displayed through below equation:

$$a_{\mathrm{w}}\left(\frac{1}{M_{\mathrm{e}}}-\frac{1}{M_{\mathrm{i}}}\right)=\frac{1}{DM_{\mathrm{i}}} \tag{2.6}$$

where M_e is EMC (kg water/kg dry matter), M_i is monolayer sorbate percentage (kg water/kg dry matter), and D is a constant.

The value of M_i is of great significance since it resembles the sum of moisture, i.e., robustly adsorbed in particular point, and is regarded to be the worth at which a food is mainly steady. Langmuir's sorption model is a major model among the numerical models that is with respect to forces developed among the water condensed as a monomolecular layer and food surface (Basu et al. 2006). The add-on of Langmuir's model underpinning the notion about multimolecular surface outcomes in GAB and BET sorption models that could explain sigmoidal-shaped sorption models commonly identified in case of biological material and food products. Values of a few regression coefficients (R^2) and coefficients (C) for Langmuir sorption isotherm is given in Table 2.3.

2.9.3.2 Brunauer-Emmett-Teller (BET) Sorption Isotherm Model

The BET equation that is a broadly applied model in food products was initially projected by three researchers namely Brunauer, Emmett, and Teller (Brunauer et al. 1938). It designates an elementary highlight in understanding multilayer sorption curves, predominantly the category 2 and 3. BET is an effectual technique to calculate the quantity of bound moisture in particular polar points of dried food products.

$$M_{\mathrm{e}}=\frac{M_{\mathrm{L}}Da_{\mathrm{w}}}{(1-a_{\mathrm{w}})(1+(D-1)a_{\mathrm{w}})} \tag{2.7}$$

Table 2.3 Predicted values of Langmuir equation for food products

Food sample	Temperature (°F)	M_i	R^2	C	Kind	Reference
Pear	104	1596.92	0.83	0.00038	Desorption	Park et al. (2001)
Mint	86	5.75	0.93	0.055	Desorption	Park et al. (2002)
Garden mint leaves	104	9.25	0.98	0.029	Desorption	Park et al. (2002)
Dried pear	104	1015.78	0.79	0.000593	Adsorption	Park et al. (2001)

Table 2.4 Predicted values of BET equation for food products

Food sample	Temperature (°F)	M_e	R^2	C	Kind	Reference
Blueberry	104	0.1	–	101.4	Desorption	Vega-Gálvez et al. (2009)
Blueberry	104	0.06	–	5.73	Adsorption	Vega-Gálvez et al. (2009)
Tomato	86	0.16	–	14.05	Adsorption	Goula et al. (2008)
Potato	86	0.08	–	7.34	Desorption	Iguedjtal et al. (2008)
Yam	113	0.71	0.91	2.91	Desorption	Montes et al. (2009)

wherein, M_L is monolayer moisture percentage that denotes the water content at which the molecule of water positioned to every ionic bond initiates to perform as a liquid system. D is an energy constant also called as Langmuir constant associated with total sorption heat which is associated with dissimilarity among molecules which absorb energy from primary layer and subsequent layers. In most cases, the variation of linearity of obtained plots represents that, at increased vapor pressure, the quantity adsorbed by means of sorbent is lesser compared to those described using sorption isotherm (Van den Berg 1985).

The BET equation could be regarded as the major sorption model to study the best moisture condition for safe storage, particularly for dried food samples. The variables of BET model for various food substances are mentioned in Table 2.4.

2.9.3.3 Oswin Model

Oswin model comprises a sequence extension for sigmoid fashioned plots and was proposed by Oswin, 1946 (McLaughlin and Magee 1998). Oswin empirical model is elucidated and described through below equation:

$$M_e = C\left(\frac{a_w}{1 - a_w}\right)n \tag{2.8}$$

wherein, n and C are constants.

The Oswin model was applied to compare the moisture percentage of milk powder and freeze-dried hot beverages to a_w of 0.5, in addition to other foods.

2.9.3.4 Halsey Model

Halsey model explains an equation for condensation of moisture at a comparatively great distance from exterior, presuming that energy of water molecule differs as inverse pth power of the distance from the surface. Halsey equation is an excellent demonstration of sorption statistics concerning isotherms category 1, 2, and 3. Furthermore, the equation explains the sorption properties of food substances which holds starch (Samapundo et al. 2007). Halsey model is explained in below equation:

Table 2.5 Predicted values of Halsey equation for food products

Food sample	Temperature (°F)	p	R^2	C	Kind	Reference
Banana	20	0.75	0.98	0.18	Desorption	de Gouveia et al. (2004)
Blueberry	104	0.88	–	0.10	Adsorption	Vega-Gálvez et al. (2009)
Corn flour	22	2.52	0.96	0.002	Desorption	Gálvez et al. (2006)
Blueberry dried	104	1.67	–	0.05	Desorption	Vega-Gálvez et al. (2009)

$$M_e = M_0\left(-\frac{A}{RT\ \ln a_w}\right)1/p \tag{2.9}$$

where, R is the universal gas constant, n and A are constants, T is temperature, and M_0 is moisture percentage.

As the application of RT terms does not eradicate the dependency of temperature term on n and A, the Halsey isotherm equation was tailored by Iglesias and Chirife, 1976 (Andrade et al. 2011) to below mentioned form:

$$M_e = \left(\frac{D}{\ln a_w}\right)1/p \tag{2.10}$$

where n and C are constants. The Halsey sorption model has been applied for numerous food substances, as it is shown in Table 2.5.

2.9.3.5 Henderson Model

Henderson model is the most generally used model and which could be demonstrated as explained through below equation:

$$M_e = \left(-\frac{\ln(1-a_w)}{D}\right)1/p \tag{2.11}$$

where D and p are constants.

As per this model, a graph between ln (−ln (1 − a_w)) and ln M_e produces a linear line. On the other hand, it has also been studied by a few scientists that observed this model lacked in providing accurate data on physical phases of water (Peng et al. 2007). The important parameters of Henderson model for few food products are described in Table 2.6.

In general, moisture sorption isotherms are imperative tools for foretelling the associations among food and water molecule. Of the above-discussed molecule, the most widely applied mathematical model for food substances is BET model. However, the limitation of BET model is that it provides excellent outcomes in $a_w < 0.45$ only, and in addition its use is limited to surface area determination.

Table 2.6 Predicted values of Henderson equation for food products

Food sample	Temperature (°F)	p	R^2	C	Kind	Reference
Cashew apple	86	1.08	0.99	0.05	Adsorption	Alcântara et al. (2009)
Yam	113	0.13	0.93	0.13	Desorption	Montes et al. (2009)
Walnut	77	1.78	0.98	0.04	Desorption	Toğrul and Arslan (2007)
Pineapple peel	77	0.59	0.99	2.16	Desorption	Oliveira et al. (2006)

2.10 Recommended Techniques and Equipments in the Measurement of Water Activity

Various techniques and equipments are accessible for laboratory determination of a_w in food products. Techniques have relied on the characteristics of solutions. Water activity might be projected by assessing the following: osmotic pressure, vapor pressure, depression in freezing point, elevation of boiling point, or by applying the isopiestic technique, digital hygrometers, etc. (Troller 2012). However, to date there has been no instrument that could be placed directly into the food component to measure the water activity. Relatively, a_w is calculated applying an indirect technique as elaborated below. The chosen sample is located within a small box and the water in the surrounding environment is assessed once it equilibrates to the sample. Techniques for the determination of a_w are described in the *AOAC International*. Novel instrument methodologies have immensely enhanced the speed, reliability, and robustness of measurements. Consistent laboratory instrumentation is necessary to promise the security of the food matrix and impose administration policy. Two diverse kinds of a_w equipments are commercially obtainable. One applies chilled mirror dew-point methodology, while the latter one estimates the RH value with sensors that alter electrical capacitance or resistance. Each technique has its own merits and demerits. The techniques differ in repeatability, rapidity of measurement, calibration stability, linearity, and convenience to use (Gee et al. 1992; Gómez et al. 1990).

2.11 Influence of a_w on Enzyme Activity

Enzyme reactions generally take place in an aqueous matrix even though in vivo enzyme activity can initiate in cytoplasm, cell membranes, and lipid depots. The three chief techniques of observing the influence of a_w on enzyme activity are as mentioned below. The first way is to cautiously remove the moisture from food material possessing active enzymes, followed by equilibrating it using numerous a_w and determining the velocity of reaction of enzymes. An instance of abovementioned technique is studied in case of lecithin (Blain 1962). It was observed through the research that below a_w 0.35 (less than 1% of total water) no phospholipase activity was observed on lecithin. However, at water activity above 0.35 a nonlinear uplift in enzyme activity was observed. However, the greatest activity was yet not attained at a_w of 0.9 (around 12% net water). The enzyme β-Amylase possessed zero activity on starch till a_w of 0.8 (approximately 2% total water); however, enzyme activity then augmented 15 times as a_w enhanced to 0.95 (approximately 12% total water). From these work, it could be inferred that net water should be between 1% and 2% to avert enzyme activity.

The second technique of deciding the amount of water required for initiating enzyme activity is to substitute some quantity of water using organic solvents. Substitution of water with water-soluble glycerol decreases the reaction of lipoxygenase and peroxidase as the water level is concentrated beneath 75%. At water concentration of 20% and 10% lipoxygenase and peroxidase possessed zero activity (Troller 2012).

The viscosity effects of glycerol possess a significant role on the above findings. In the last technique, majority of water could be substituted by immiscible solvents in lipase-catalyzed trans-esterification of tributyrin with a range of alcohols. The "desiccated" lipase molecules (approximately 0.48% total water), poised in dry butanol at water concentration 1.1, 0.9, 0.6, and 0.3% (w/w) total, provided original speeds of 4, 5, 3.5, and 0.8 μmol transesterification/h 100 mg lipase. Henceforth, pancreatic lipase possessed the greatest transesterification velocity at water concentration of 0.9%. Organic liquids could show two chief influences on enzyme-assisted activities: influence on stability and influence on path of reversible reactions. These impacts are diverse in hydrophilic and hydrophobic solvents. In hydrophobic solvents, a transition has been observed from hydrolysis to synthesis (Chamouleau et al. 2001). Speed of lipase-assisted transesterification activities are enlarged greater than six times whereas about 16-fold decline in the level of hydrolysis as "dry" (approximately 1% total water) enzyme molecules are poised in the hydrophobic solvent. Alkylation of enzymes to formulate them to be further immiscible has comparable impacts in changing from hydrolysis to synthesis as do the application of hydrophobic liquids. The speed of sucrose esterification using oleic acid was amplified six times through alkylation of a few amino groups. An alteration in the stereospecificity of compounds developed in organic solvents was also observed (Rahman and Labuza 2020a, b).

Enzymes could be highly steady in organic solutions compared to aqueous solutions. Lysozyme and ribonuclease turn out to be highly robust as moisture level is reduced, though this is performed by dehydrating or by mixing it with hydrophobic solvents. When water content is around 6%, ribonuclease shows up a transition temperature of 124 °C with a half-life of 2.0 h at a temperature above 150 °C. The constancy lowers as water concentration amplifies; an attenuated mixture of ribonuclease shows a transition temperature of 61 °C with a half-life that is too less to measure. Lysozyme, another important enzyme is basically as steady in hydrophobic organic solvents in dry condition, with reduced half-life period. Enzymes generally do not alter its structure when located in hydrophobic organic solvents. The catalytic action and stability of enzymes in hydrophilic solvent units are diverse from that in hydrophobic systems. Leung (2017) observed that protease-assisted reaction of casein with water, in 5% ethanol, 5% acetonitrile, and 95% aqueous buffer resulted in augmentation of the transition temperature, decline in velocity, and stability when associated with controls in aqueous solutions. It is a renowned finding that any solvent that holds a labile H^+, for example, amines, alcohols, and so forth contend with a water molecule in hydrolytic enzyme reactions.

2.12 Application of the Water Activity Conception

Through the above section, it is understood that how significantly water activity value is applicable in forecasting food quality and safety with regards to microbial growth, physical attributes, and chemical reaction rates. Through assessing and regulating the water activity of food components, it is probable to envisage which microbes will be prospective sources of deterioration and infection; preserve the chemical steadiness of foods; reduce nonenzymatic browning and impulsive auto-catalytic oxidation reactions; extend the action of enzymes in food; and control the physical attributes of foods, say shelf life and texture (Alzamora et al. 2003; Leistner 1992). A worldwide stability map of food products describes these attributes as a purpose of water activity.

The increasing acknowledgment of the water activity theory is elucidated by its integration into USDA regulations, GMP, FDA, and HACCP laws, and most lately in NSF International Standard Draft 75. Water activity is a chief critical control point for hazard scrutiny as distinct by the HACCP idea (Sandulachi 2012). These policies and necessities are regard to the present FDA Food Code description of potentially risky foods. Potentially dangerous foods are those which necessitate temperature regulators as they hold the swift and gradual development of pathogenic microbes. Potentially harmful food does not comprise items with a_w values of 0.85 or below, food within a pH level of 4.6 or below, food in hermetically closed vessels, which preserve commercial barrenness under non-refrigerated cargo space and allotment, or those in which quick and progressive development of harmful microbes might not occur.

2.13 Quality and Food Security

Hypothetical and experimental analysis of water activity notion permits us to assess the selection of techniques for safeguarding food. However, research wants to be emphasized on the stability in every micro and macro area in order to investigate further basic regulations for constancy. Safety and food quality relies on water activity and pH in food surrounding (Fontana 1988). Foods containing high water activity value are easily fragile. Water activity, temperature, and other attributes have a through effect on the growth of microbial cells, consequently water activity and pH are the two vital characteristics. Free water which is obtainable to yeasts, bacteria, and molds is accountable for cellular growth and associated production of toxin. This could initiate participation in chemical/biochemical reactions that may damage flavor, texture, color, nutritional profile of a food matrix, and its constancy → shelf-life point (Leistner and Gould 2012; Novasima 2005; Ramirez-Jiménez et al. 2003; Sandulachi 2012; Jerome et al. 2019). Water activity has its major practical relevance in forecasting the development of bacteria, molds, and yeasts. For a food product to encompass a functional shelf life devoid of depending on refrigerated storage space, it is essential to manage either its acidity level or the value of (a_w) or an appropriate arrangement of the above two properties. This might successfully augment the product's constancy and compose its potential to envisage the shelf life under recognized ambient storage space surroundings (Pittia and Antonello 2016).

2.14 Conclusion

Anticipation will prolong to be fundamental to food security with respect to both the government and industrial processor perspective. Numerous food industries are following HACCP regulations willingly, in accumulation to the obligatory poultry, meat, and horticultural produce related programs. Incorporated in HACCP and GMP programs are processes necessitating warning method in the early hours, risk evaluation, enhanced recognition/control techniques, and enhanced examination and acquiescence. Evaluating water activity is a critical control point for a lot of producers and is supposed to be integrated into numerous other food safety initiations. The number one precedence is shielding the consumer. Recalls can charge trillions of dollars in operational delays and product losses, all along with fatalities to customer assurance and an industry's reputation. Combining water activity monitoring techniques and further science-based evaluations into a food safety program aids to assure the premier quality and secured food supply.

References

Alcântara SR, Almeida F d A, da Silva FL et al (2009) Adsorption isotherms of the dry cashew apple. Rev Bras de Eng Agricola e Ambient 13(1):81–87

Alzamora S, Tapia M, López-Malo A et al (2003) The control of water activity. In: Food preservation techniques. Woodhead, Cambridge, UK, pp 126–153

Amit SK, Uddin MM, Rahman R et al (2017) A review on mechanisms and commercial aspects of food preservation and processing. Agric Food Secur 6(1):1–22

Andrade PRD, Lemus MR, Perez CCE et al (2011) Models of sorption isotherms for food: uses and limitations. Vitae 18(3):325–334

Arslan-Tontul S (2020) Moisture sorption isotherm, isosteric heat and adsorption surface area of whole chia seeds. LWT 119:108859

Arslan-Tontul S (2021) Moisture sorption isotherm and thermodynamic analysis of quinoa grains. Int J Heat Mass Transf 57(3):543–550

Asafo-Adjaye G (2019) Water activity profile and microbial safety of hot pepper sauce "Shito". Kwame Nkrumah University of Science and Technology. p 56

Barbosa-Cánovas GV, Fontana A, Schmidt SJ et al (2007) Water activity in foods. Fundamentals and applications. Wiley, Hoboken, NJ, p 405
Barbosa-Cánovas GV, Fontana AJ, Schmidt SJ et al (2020) Water activity in foods: fundamentals and applications. Wiley, Hoboken, NJ, p 439
Bastıoğlu AZ, Koç M, Ertekin FK (2017) Moisture sorption isotherm of microencapsulated extra virgin olive oil by spray drying. J Food Meas Charact 11(3):1295–1305
Basu S, Shivhare U, Mujumdar A (2006) Models for sorption isotherms for foods: a review. Dry Technol 24(8):917–930
Blain J (1962) Moisture levels and enzyme activity. Recent Adv Food Sci 2:41–45
Brunauer S, Emmett PH, Teller E (1938) Adsorption of gases in multimolecular layers. J Am Chem Soc 60(2):309–319
Brunauer S, Deming LS, Deming WE et al (1940) On a theory of the van der Waals adsorption of gases. J Am Chem Soc 62(7):1723–1732
Caballero-Cerón C, Guerrero-Beltrán J, Mújica-Paz H et al (2015) Moisture sorption isotherms of foods: experimental methodology, mathematical analysis, and practical applications. In: Water stress in biological, chemical, pharmaceutical and food systems. Springer, New York, pp 187–214
Chamouleau F, Coulon D, Girardin M et al (2001) Influence of water activity and water content on sugar esters lipase-catalyzed synthesis in organic media. J Mol Catal B Enzym 11(4–6):949–954
Chirife J, Fontana AJ (2020) Introduction: historical highlights of water activity research. In: Water activity in foods: fundamentals and applications. Wiley, Hoboken, NJ, pp 1–11
de Gouveia JP, Nascimento J, Almeida FA et al (2004) Mathematical models for adjustment of desorption isotherms of banana variety silver. Engenharia Agrícola 24(3):799–806
Fennema O, Berny L (1974) Equilibrium vapour pressure and water activity of food at subfreezing temperature. Paper presented at the Proceedings of IV International Congress of Food Science and Technology. p 12
Fontana AJ (1988) Water activity: why it is important for food safety. In: International conference on food safety, pp 177–185
Gálvez AV, Aravena EL, Mondaca RL (2006) Isotermas de adsorción en harina de maíz (Zea mays L.). Food Sci Technol 26(4):821–827
Gee G, Campbell M, Campbell G et al (1992) Rapid measurement of low soil water potentials using a water activity meter. Soil Sci Soc Am J 56(4):1068–1070
Gichau AW, Okoth JK, Makokha A (2020) Moisture sorption isotherm and shelf life prediction of complementary food based on amaranth–sorghum grains. J Food Sci Technol 57(3):962–970
Gómez R, Cabezas L, Alcalá M et al (1990) Estimation of water and water activity in different samples of honey. Alimentaria 27(210):33–36
Goula AM, Karapantsios TD, Achilias DS et al (2008) Water sorption isotherms and glass transition temperature of spray dried tomato pulp. J Food Eng 85(1):73–83
Guildner L, Johnson D, Jones F (1976) Vapor pressure of water at its triple point. J Res Natl Inst Stand Technol 80(3):505
Iguedjtal T, Louka N, Allaf K (2008) Sorption isotherms of potato slices dried and texturized by controlled sudden decompression. J Food Eng 85(2):180–190
Jayaraman K (1995) Critical review on intermediate moisture fruits and vegetables. In: Food preservation by moisture control (fundamentals and applications), pp 411–442
Jerome RE, Singh SK, Dwivedi M (2019) Process analytical technology for bakery industry: a review. J Food Process Eng 42(5):e13143
Kou Y, Schmidt SJ (1999) Vapor pressure and water activity measurements of saturated salt solutions made with D2O at 20° C. Food Chem 66(2):253–255
Labuza TP, Altunakar B (2020) Water activity prediction and moisture sorption isotherms. In: Water activity in foods: fundamentals and applications. Wiley, Hoboken, NJ, pp 161–205
Labuza T, Rahman M (2007) Water activity and food preservation. In: Handbook of food preservation. CRC Press, Boca Raton, FL, pp 447–476
Leistner L (1992) Food preservation by combined methods. Food Res Int 25(2):151–158

Leistner L, Gould GW (2012) Hurdle technologies: combination treatments for food stability, safety and quality. Springer Science & Business Media, Cham

Leung HK (2017) Influence of water activity on chemical reactivity. In: Water activity: theory and applications to food. Routledge, New York, pp 27–54

Maigalura K, Suleiman J (2020) Moisture sorption isotherm characteristics of awara food. FUDMA J Sci 4(2):620–626

Maldonado-Astudillo YI, Jiménez-Hernández J, Arámbula-Villa G et al (2019) Effect of water activity on extractable polyphenols and some physical properties of Hibiscus sabdariffa L. calyces. J Food Meas Charact 13(1):687–696

Mannaa M, Kim KD (2017) Influence of temperature and water activity on deleterious fungi and mycotoxin production during grain storage. Mycobiology 45(4):240–254

Mathlouthi M (2001) Water content, water activity, water structure and the stability of foodstuffs. Food Control 12(7):409–417

McLaughlin C, Magee T (1998) The determination of sorption isotherm and the isosteric heats of sorption for potatoes. J Food Eng 35(3):267–280

McMeekin TA, Ross T (1996) Shelf life prediction: status and future possibilities. Int J Food Microbiol 33(1):65–83

Miyawaki O (2018) Water and freezing in food. Food Sci Technol Res 24(1):1–21

Montes E, Torres R, Andrade R et al (2009) Models of desorption isotherms of yam (Dioscorea rotundata). Dyna Rev Fac Nac Minas 76(157):145–152

Murphy D, Koop T (2005) Review of the vapour pressures of ice and supercooled water for atmospheric applications. Q J Royal Meteorol Soc 131:1539

Novasima A (2005) The meaning of water activity. Novasima, Lachen

Ochoa-Martínez C, Quintero P, Ayala A (2012) Drying characteristics of mango slices using the Refractance Window™ technique. J Food Eng 109(1):69–75

Oliveira MM, Campos ARN, Dantas JP (2006) Desorption isotherms of passion fruit peel (Passiflora edulis Sims): experimental determination and mathematical model evaluation. Ciência Rural 36(5):1624–1629

Park KJ, Bin A, Brod FPR (2001) Sorption isotherms data and mathematical models for pear bartlett (Pyrus sp.) with and without osmotic dehydration. Food Sci Technol 21(1):73–77

Park KJ, Vohnikova Z, Brod FPR (2002) Evaluation of drying parameters and desorption isotherms of garden mint leaves (Mentha crispa L.). J Food Eng 51(3):193–199

Peng G, Chen X, Wu W et al (2007) Modeling of water sorption isotherm for corn starch. J Food Eng 80(2):562–567

Peromingo B, Rodríguez A, Bernáldez V et al (2016) Effect of temperature and water activity on growth and aflatoxin production by Aspergillus flavus and Aspergillus parasiticus on cured meat model systems. Meat Sci 122:76–83

Pittia P, Antonello P (2016) Safety by control of water activity: drying, smoking, and salt or sugar addition. In: Regulating safety of traditional and ethnic foods. Elsevier, Amsterdam, pp 7–28

Rahman MS (2010) Food stability determination by macro–micro region concept in the state diagram and by defining a critical temperature. J Food Eng 99(4):402–416

Rahman MS, Labuza TP (2020a) Fundamentals of water activity concept. In: Handbook of food preservation. CRC Press, Boca Raton, FL, pp 473–486

Rahman MS, Labuza TP (2020b) Water activity and food preservation. In: Handbook of food preservation. CRC Press, Boca Raton, FL, pp 487–506

Rahman MS, Al-Belushi RM, Guizani N et al (2009) Fat oxidation in freeze-dried grouper during storage at different temperatures and moisture contents. Food Chem 114(4):1257–1264

Ramírez-Jiménez A, Guerra-Hernández E, García-Villanova B (2003) Evolution of non-enzymatic browning during storage of infant rice cereal. Food Chem 83(2):219–225

Reid DS (2020) Water activity: fundamentals and relationships. In: Water activity in foods: fundamentals and applications. Wiley, Hoboken, NJ, pp 13–26

Rifna EJ, Dwivedi M (2021) The microbiological safety of food powders. In: Food powders properties and characterization. Springer, New York, pp 169–193

Rifna EJ, Singh SK, Chakraborty S et al (2019) Effect of thermal and non-thermal techniques for microbial safety in food powder: recent advances. Food Res Int 126:108654
Samapundo S, Devlieghere F, De Meulenaer B (2007) Sorption isotherms and isosteric heats of sorption of whole yellow dent corn. J Food Eng 79(1):168–175
Sandulachi E (2012) Water activity concept and its role in food preservation
Sandulachi E, Tatarov P (2012) Water activity concept and its role in strawberries food. Chem J Mold 7(2):103–115
Scott W (1957) Water relations of food spoilage microorganisms. In: Advances in food research. Elsevier, Amsterdam, pp 83–127
Sereno AM, Hubinger M, Comesana J et al (2001) Prediction of water activity of osmotic solutions. J Food Eng 49(2–3):103–114
Slade L, Levine H, Reid DS (1991) Beyond water activity: recent advances based on an alternative approach to the assessment of food quality and safety. Crit Rev Food Sci Nutr 30(2–3):115–360
Smith DF, Hildebrandt IM, Casulli KE et al (2016) Modeling the effect of temperature and water activity on the thermal resistance of Salmonella enteritidis PT 30 in wheat flour. J Food Prot 79(12):2058–2065
Stamp J, Linscott S, Lomauro C et al (1984) Measurement of water activity of salt solutions and foods by several electronic methods as compared to direct vapor pressure measurement. J Food Sci 49(4):1139–1142
Storey RM, Stainsby G (1970) The equilibrium water vapour pressure of frozen cod. Int J Food Sci 5(2):157–163
Syamaladevi RM, Sablani SS, Tang J et al (2009) State diagram and water adsorption isotherm of raspberry (Rubus idaeus). J Food Eng 91(3):460–467
Syamaladevi RM, Tang J, Villa-Rojas R et al (2016) Influence of water activity on thermal resistance of microorganisms in low-moisture foods: a review. Compr Rev Food Sci 15(2): 353–370
Tapia MS, Alzamora SM, Chirife J (2020) Effects of water activity (aw) on microbial stability as a hurdle in food preservation. In: Water activity in foods: fundamentals and applications. Wiley, Hoboken, NJ, pp 323–355
Teeter MM (1991) Water-protein interactions: theory and experiment. Annu Rev Biophys 20(1): 577–600
Tian Y, Chen Q, Yan C (2020) Classification of adsorption isotherm curves for shale based on pore structure. Petrophysics 44:417
Toğrul H, Arslan N (2007) Moisture sorption isotherms and thermodynamic properties of walnut kernels. J Stored Prod Res 43(3):252–264
Troller J (2012) Water activity and food. Elsevier, Amsterdam
Van den Berg C (1985) Development of BET-like models for sorption of water on foods, theory and relevance. In: Properties of water in foods. Springer, New York, pp 119–131
Van den Berg C, Bruin S (1978) Water activity and its estimation in food systems. In: Proceedings int. symp. properties of water in relation to food quality and stability, Osaka
Vega-Gálvez A, López J, Miranda M et al (2009) Mathematical modelling of moisture sorption isotherms and determination of isosteric heat of blueberry variety O'Neil. Int J Food Sci 44(10): 2033–2041
Witczak T, Stępień A, Gumul D (2020) The influence of the extrusion process on the nutritional composition, physical properties and storage stability of black chokeberry pomaces. Food Chem 334(2):127548
Yan Z, Sousa-Gallagher MJ, Oliveira FA (2008) Sorption isotherms and moisture sorption hysteresis of intermediate moisture content banana. J Food Eng 86(3):342–348
Zeuthen P, Bøgh-Sørensen L (2003) Monitoring the effect of food preservation. In: Food preservation techniques. Elsevier, Amsterdam, pp 552–565

Starch Gelatinization and Modification 3

Swati Sethi, Poonam Choudhary, Prerna Nath, and O. P. Chauhan

3.1 Introduction

Starch occurs in the form of granules which varied in shape and size but are highly ordered structures. Starch is abundantly present in plant materials, especially cereal grains. Quantitatively, the starch content varied between 56% and 80% among different cereals. The grain with high starch content, i.e., maize is used for the extraction of starch at the commercial level. Apart from this, the tuber and root crops namely potato, tapioca, and arrowroot are also used for the extraction of starch. The international starch market is estimated to reach 156.3 million metric tons by 2025. The market is expected to grow by 47.3 million metric tons at a compounded growth of 5.3%. Starch is used indispensably in many industrial applications including both food and non-food applications. The food processing sector stands as the major recognized domain for starch utilization, supported by augmented demand for convenience packaged foods. In the food industry, starch is used as a versatile, natural "Clean label" additive for improving functional properties such as texture, viscosity, and stability in processed food products. Starch also finds applications as a pharmaceutical excipient due to its non-toxic nature. Another important area of application is the packaging industry. Starch-based polymers are increasingly gaining interest owing to their recyclability, easy solubility, increased stability and shelf life, excellent heat resistance, and cost benefits. Therefore, it is important to understand the behavior of starch during various processing operations, especially heat. In the presence of heat and moisture, starch undergoes various transformations known as glass transition, gelatinization, and melting. Gelatinization is an

S. Sethi (✉) · P. Choudhary · P. Nath
Central Institute of Post-Harvest Engineering and Technology, ICAR, Ludhiana, India

O. P. Chauhan
Defence Food Research Laboratory, DRDO, Mysore, India

O. P. Chauhan (ed.), *Advances in Food Chemistry*,
https://doi.org/10.1007/978-981-19-4796-4_3

endothermic reaction involving starch, moisture, and heat. The gelatinization properties of starches such as gelatinization temperature, change in enthalpy, and melting of amylose–lipid complex vary substantially depending on botanical sources and genomic credentials. In its native form, starch does not support various processing operations due to its poor shear and heat stability (Jayakody and Hoover 2008). Moreover, they have a tendency to retrograde and undergo syneresis. Therefore, various modification techniques are required to apply to escalate their commercial applications. Starches are commonly altered by physical, chemical, and enzymatic techniques to improve their particular functional properties. To convert the natural starches into appropriate form for commercial applications, these starches are altered structurally to obtain the anticipated properties such as thermal stability, enhanced solubility, texture, and adhesion. Substantial developments in starch science such as the development of innovative starch extraction and modification techniques, emerging new uses of starch and starch derivatives such as starch-based biodegradable plastics/starch-composite plastics, and starch-based binders for metal injection molding are making starch an affordable alternative. This chapter gives insight into structural aspects, transformations such as gelatinization, retrogradation, various methods for modification of starches, and methods to monitor and estimate the degree of alteration.

3.2 Starch Structure

Starch is present in the form of granules in plant material which varies in shape and size depending on the botanical source. The commonly occurring shapes of starch granules are spheres, ellipsoids, polygon, platelets, and irregular tubules. The size (diameter) of granules ranged between 0.1 and 200 μm (Tester et al. 2004). The starch granules are composed of two major components; linear chain polysaccharide amylose α-(1–4) and branched-chain complex polysaccharide amylopectin α-(1–4) and α-(1–6). Amylose creates tough layers and strong gels and also has the affinity to retrograde. In contrast, amylopectin creates fragile films, and weak gels when dispersed in water, but is more stable. The structural differences in amylose and amylopectin fractions of starch are depicted in Table 3.1. Starches vary widely with respect to their gelatinization temperatures. Chemical structures, amylose to amylopectin ratio, molecular sizes, branched-chain lengths, and distributions also affect the gelatinization properties of starches. Other factors such as the water: starch ratio, pH, and compositional constituents including salt, sugar, fat, and protein in food formulation also affect the gelatinization temperature. Gelatinization temperature is defined as the temperature at which starch granules transit from an ordered shape to a disordered shape. Under polarized light, starch granules display unique optical properties known as birefringence or maltese cross which indicates a greater degree of ordered structure. In anisotropic material, upon passing through a crystalline material a light ray decomposes into two rays which are known as birefringence or double refraction. In a starch granule, amylose and amylopectin polysaccharides are arranged in a radial anisotropic, semi-crystalline structure responsible for maltese

Table 3.1 Physicochemical characteristics of amylose and amylopectin

Characteristic	Amylose	Amylopectin
Molecular structure/branches	Mainly linear/primarily α-1–4	Highly branched/α-1–4; α-1–6
Molecular weight	105–106 Da	107–109 Da
Iodine bonds/stain	20%/blue–black	<1%/red–purple
β-amylolysis	100%	Approx. 60%
Dilatation in aqueous solutions	Unstable	Stable
Solubility	Low/barely soluble	High
Gelatinization temperature	Low	High
Melting temperature	Low	High
Amylose–lipid complex	Very high amount	No
Gel formation	Firm, irreversible	Soft, reversible
Films	Coherent	Not readily formed
Viscosity	Low	High
Thickener	Poor	Good
Shear stability	Relative stable	Unstable
Adhesive forces	Weak	Strong
Freezing–thawing stability	Unstable	Stable
Retrogradation rate	High	Low

Adapted from Schirmer et al. (2015)

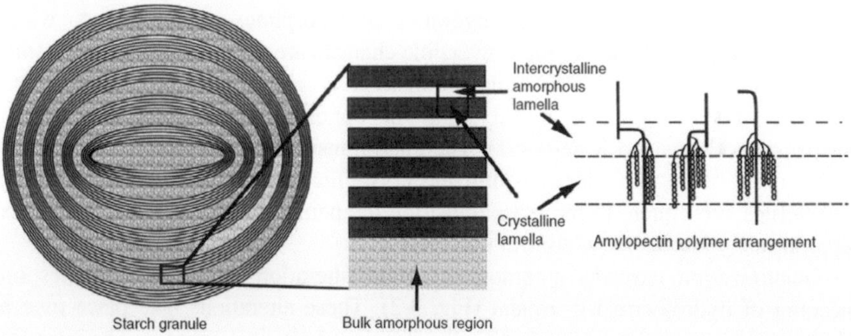

Fig. 3.1 Schematic diagram of starch granule structure. (Adopted from Donald et al. 1997)

cross (Greenwood 1979). Starch granule is composed of two domains, crystalline region and amorphous region (Fig. 3.1). Amylopectin polysaccharide is responsible for the crystallinity in the starch structure. Majority of outer branches of amylopectin polymer and a small number of amylose polymer units organize themselves into crystallites forming a crystalline structure within the granule. Whereas, the amorphous region is constituted by amylose units and some noncrystalline regions of amylopectin. These crystalline and noncrystalline structures and their interactions within starch granules control the properties of starch during processing. Granules enclose alternate concentric rings of the amorphous and crystalline region forming a semi-crystalline atmosphere inside (Tester et al. 2004). Based on polymorphism, the crystalline arrangements within a starch granule are categorized into three discrete groups namely, A, B, and C. The main differentiation among these polymorphs is how the double helices are organized with each other, and the number of units of water bound inside the crystalline structure. The A-type polymorph structure crystallizes in an orthogonal unit cell with somewhat partial hexagonal packing and 8 water molecules per unit cell, while the B-type polymorph crystallizes in a hexagonal unit cell with 36 water molecules per unit cell. The C polymorphic structure is a mixture of A and B polymorphs and consequently is known to be intermediate between A and B types in packing density and structure (Sarko and Wu 1978).

3.3 Gelatinization of Starch

Gelatinization is defined as the irreversible structural transformation of a starch granule during thermal treatment in presence of water. When starch is heated in presence of water, interactions between amylose and amylopectin alter the structure of starch. This alteration leads to various changes known as glass transition, gelatinization, and melting (Schirmer et al. 2015).

The glass transition is defined as a temperature at which a glassy or amorphous state is converted into a rubbery state. This conversion of state during glass transition is a reversible phenomenon. The hydration of amorphous regions due to water absorption is possible before the irreversible changes are completed during gelatinization (Benczedi et al. 1998). Glass transition is affected by permeability and porosity of starch granule, density, and the cohesive energy density of starch suspensions (Roos and Karel 1991). The glass transition enables the hydration and detachment of double helices within the crystallites, whereas detachment of the crystallites subsequent to the glass transition temperature of amorphous regions is appropriately known as gelatinization.

Gelatinization explains common structural alterations in starch granules on account of hydrothermal treatment (Fig. 3.2). These alterations take place over a sharp temperature range representing the different botanical sources. The onset of gelatinization is initiated by the absorption of water by the amorphous region of the starch granule and concurrent destruction of hydrogen bonds (Belitz et al. 2009). After a certain extent of hydration, the destabilizing and disruptive pressures persuaded by the swelling of the amorphous region are conveyed to the lamellar crystallites. This progression is facilitated by the long chains of amylopectin that interconnect the side chain groups inside the semi-crystalline structure along with the alternating semi-crystalline and amorphous regions. Subsequently, additional heating promotes irreversible alteration such as melting of starch crystallites, starch solubilization, and leaching out of starched granules which results in loss of birefringence and increased viscosity (Jenkins and Donald 1998). Gelatinization also creates amylose–lipid complexes and some are even present naturally (Morrison

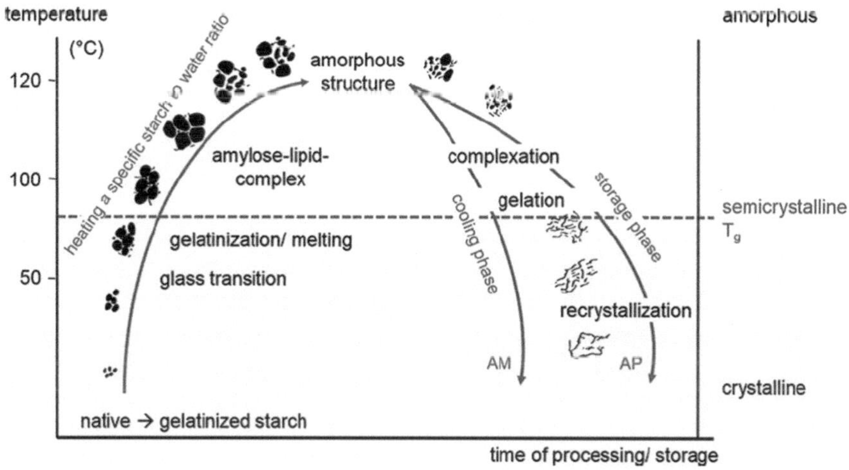

Fig. 3.2 Phase transition diagram of starch when applied with a temperature profile where Tg is glass transition temperature, AM is amylose, and AP is amylopectin. (Adopted from Schirmer et al. 2015)

et al. 1993). These complexes are categorized into two types based on how orderly the structure is. Type I is a less ordered structure while type II denotes a semi-crystalline structure. Type II complexes are formed at 95 °C while type I complexes dissociate in a temperature range of 90–105 °C (Cooke and Gidley 1992).

3.4 Factors Affecting Gelatinization of Starches

Several factors influence the gelatinization of starches. These factors include water: starch ratio, presence of solutes such as sugar, salt, lipids, nonionic constituents, electrolytes, and processing methods. Individually, these factors can exert a substantial effect on starch gelatinization characteristics subjected to the type of the starch, for example, amylose to amylopectin ratio. Likewise, various pretreatments affect the starch gelatinization characteristics to a wider extent.

3.4.1 Water:Starch Ratio

Water:Starch ratios play an important role in governing the gelatinization characteristics. As depicted by differential scanning calorimetry, if the water:starch ratio is below 1.5:1, the gelatinization endotherm (ΔH_1) shows a trailing shoulder with a noticeable decline in the enthalpy. This shouldering effect might be attributed to the destabilizing effect of water and heat treatments on the amorphous and crystalline regions of starch granules. With lower water:starch ratio a shift towards the higher side was observed with the onset temperature and gelatinization temperature range. This is because the shoulder peak forms its own endotherm which consequently moves to a higher temperature.

3.4.2 Presence of Lipids

Polar substances containing large hydrophobic groups profoundly alter the physical characteristics of starch pastes. In general, it was observed that short-chain fatty adjuncts (4–5 carbon atoms), e.g., butyl and amyl alcohols promote the hydration of corn starch granules which in turn results in lowering down the gelatinization temperature and increased solubility. However, fatty adjuncts with longer chains (more than 12 carbon atoms) showed a tendency to form complexes with the amylose at a temperature below 85 °C which inhibits swelling, and, hence, decreases solubilization. Furthermore, this complex breaks down at higher temperatures and may promote hydration afterward. Fatty adjuncts with 18 carbon atoms inhibit hydration and solubilization over the complete gelatinization temperature range as they form stable complexes that can sustain high temperatures up to 95 °C (Gray and Schoch 1962). The temperature of cooking, time of addition, and time for diffusion of fatty adjuncts into the starch granule are important as they can affect the characteristics of the cooked starch or grains.

3.4.3 Ionic and Non-ionic Solutes

Ionic and non-ionic constituents significantly affect the gelatinization properties of starches. Largely, non-ionic constituents cause a delay in gelatinization and increased temperature of gelatinization. This influence is elucidated mainly on the fact that solutes affect the water activity of the starch-water mixture. Sucrose exerts a limiting effect on the gelatinization process as depicted by ΔH_G. Sugar delays gelatinization which is mainly explained by two mechanisms; (1) by reducing water activity of the solution and (2) stabilization of the amorphous region by sucrose-starch interactions (Spies and Hoseney 1982). Sugar competes with starch for water absorption; therefore, it binds with water and lowers the water activity, which drops the chemical potential of water, and, hence, more energy is required to accomplish the reactions. This increased energy requirement leads to higher gelatinization temperature of starches. Also, the sugar molecule interacts with the starch in the amorphous region of the granule. Longer sugar moieties can link more gaps between the chains, and, thus, create more linkages than shorter molecules (Lund and Lorenz 1984). Salt exerts a more complex effect on starch gelatinization as compared to sugars. Different salt concentrations affected the temperature of onset, peak, and final gelatinization in the gelatinization endotherm.

Starch phase transition characteristics were studied by Ratnayake et al. (2009) in maize, wheat, and potato starches. Pretreated starch samples in excess water, dilute sodium chloride, and dilute sucrose solutions were analyzed for their effect on gelatinization properties using differential scanning calorimetry (DSC). The changes

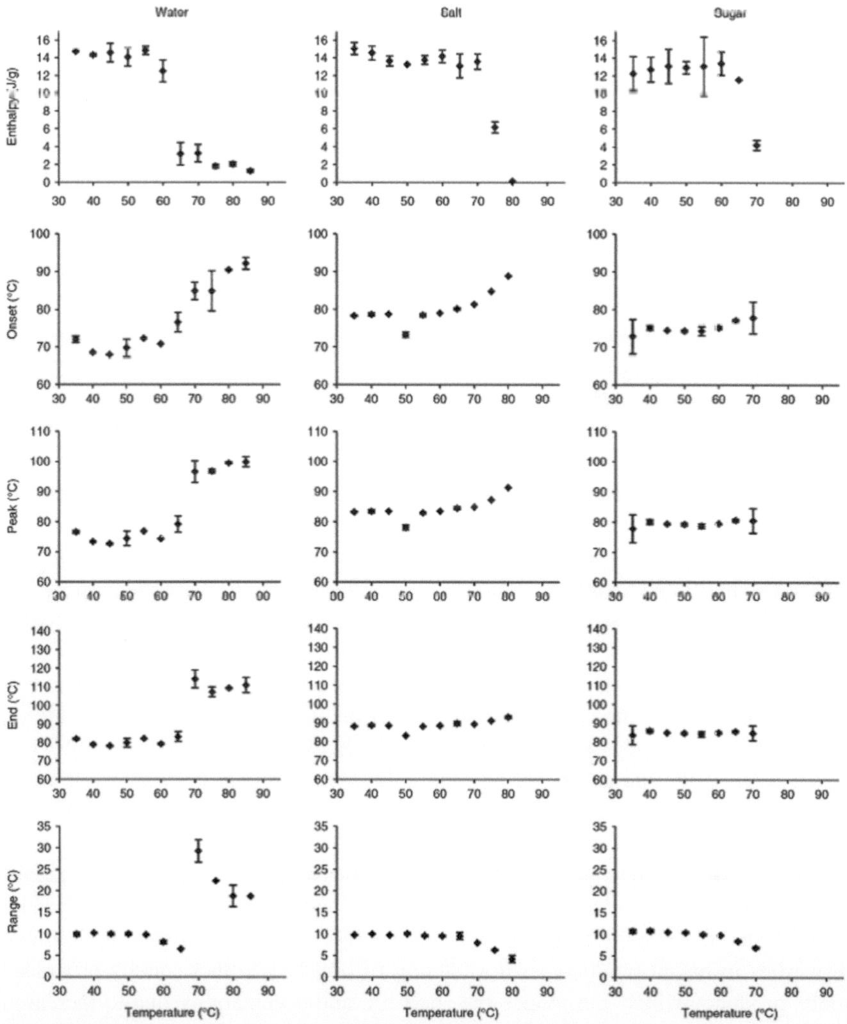

Fig. 3.3 DSC parameters of maize starch. (Adopted from Ratnayake et al. 2009)

were dependent on the nature of starch and the type of the solvent, phase transition-related enthalpic changes initiated at low temperatures and continued to high temperatures resulting in a sequence of structural alterations and wide transformations in endothermic shape from low to high temperatures. DSC parameters of maize starch are depicted in Fig. 3.3.

3.5 Starch Retrogradation

Gelatinized starch upon subsequent cooling promotes the reassociation of disrupted chains of amylose and amylopectin as they are not in an equilibrium state. This reassociation results in the formation of a differently ordered structure in a process called retrogradation. In the course of the retrogradation process, the polysaccharides amylose and amylopectin function in different ways. Amylose retrogradation was found to be a fast occurring process while amylopectin takes a longer time to retrograde (Van Soest et al. 1994). The difference in retrogradation behavior shown by the two polymers was attributed to the highly branched structure, close proximity, and attachment of the formed double helices to the same polymer backbone in amylopectin which resulted in the formation of aggregates or crystals. Due to this, researchers have explained retrogradation as a two-stage process wherein the first stage reassociation of linear chains of amylose occurs while the second stage accompanies the reassociation of branched-chain amylopectin (Goodfellow and Wilson 1990). The reassociation of the outmost amylopectin branches (DP about 15) can give rise to the B-type polymorphs (Hoover 2001); whereas, detached amylose units form double-helical structures (40–70 glucose units) via hydrogen bonding (Leloup et al. 1992). Starch retrogradation is not favored by chains in a size range of about 14–24 DP (Shi and Seib 1992). Retrogradation of starch goes along with a sequence of changes in functional properties of starches which implicate a greater extent of crystallinity due to the formation of B-type polymorphs, increased viscosity, gel formation, increased pastes turbidity, and oozing out of the water (Hoover et al. 2010). After retrogradation, non-waxy starch pastes alter into a firm gel forming a three-dimensional network while waxy starch pastes convert into soft gel-forming aggregates (Tang and Copeland 2007).

The changes that occur during retrogradation are the main contributing factor in deciding its industrial applications. These properties regulate the quality, acceptability, nutritional characteristics, and shelf life of the food product (Wang and Copeland 2013). Starch retrogradation is frequently deliberated to impart undesirable properties in baked products such as staling of bread and other starchy processed foods products, which can reduce the shelf life and acceptability due to increased crumb firmness, change in flavor and aroma, and loss of crispness (Collar and Rosell 2013). These changes occur most promptly at lower temperatures (0–4 °C). Nevertheless, starch retrogradation is beneficial in some cases where such modifications in the structural, nutritional, and sensory attributes are required (Karim et al. 2000). These applications include breakfast cereals, parboiled rice, dehydrated mashed potatoes, etc. The degree of the retrogradation of the starches is dependent on the parameters such as the temperature of storage and the composition of the food matrix (moisture content, sugars, fats, salts, and anti-staling enzymes). The process of retrogradation can be interpreted by light microscopy and X-ray diffraction. Retrogradation also has nutritional significance; it helps in slower enzymatic digestion and moderated release of glucose into the blood (Wang and Copeland 2013).

3.6 Retardation of Retrogradation

Several additives are used to impede the process of retrogradation in starch-rich processed food products. Depending on nature, the additives are grouped into carbohydrates, salts, amino acids/proteins/peptides, lipids, and other food components such as polyols, emulsifiers, citric acid, and amylase.

3.6.1 Carbohydrates

Carbohydrates are the most commonly used additives to retard starch retrogradation. It includes monosaccharides (glucose, ribose, and fructose), oligosaccharides (sucrose, maltose, and lactose), polysaccharides (pectin, β-glucan, glucomannan, carboxymethyl cellulose), gums (gum arabic, iota-carrageenan, xanthan gum, and guar gum), and β-cyclodextrin and hydroxypropyl-β-cyclodextrin. Carbohydrates compete with starch for water, and, hence, retard the process of retrogradation. The inclusion of carbohydrates has been shown to inhibit retrogradation in starch gels (Chang and Liu 1991). The degree of inhibition is principally dependent on the type and concentration of carbohydrates. Different carbohydrates expressed a varied degree of inhibition depending on the type of food matrix. Hydroxypropyl β-cyclodextrin worked better with amylose-rich rice starch (Tian et al. 2010). Hydrocolloids appear to stimulate immediate retrogradation while they delay long-standing retrogradation depending on the concentration. At first, they affect amylose–amylose interactions followed by amylopectin–amylopectin associations (BeMiller 2011).

3.6.2 Lipids

Lipids work by delaying water diffusion into granules and granule swelling which, in turn, reduces amylose leaching during heating. This inhibits the mobility of amylose molecules, and, hence, slower amylose retrogradation (Becker et al. 2001). Furthermore, the amylose–lipid complexes formed during heating and/or storage could prevent the cross-linking and formation of double-helical structures between amylose units. The amylose–lipid complexes may perhaps delay the crystallization of amylopectin. Among lipids, the shorter chain fatty acids are more effective retarding additives (Germani et al. 1983).

3.6.3 Polyols

Polyols with more hydroxyl groups displayed greater retardation in the recrystallization of potato starch (Smits et al. 2003). Emulsifiers such as sodium dodecyl sulfate, cetyltrimethyl ammonium bromide, and monoglycerides function as anti-staling agents. They form complexes with amylopectin fraction, and, hence, retard retrogradation (Gudmundsson and Eliasson 1990). Emulsifiers/surfactants function as anti-staling agents in bread or cakes (Fadda et al. 2014). Thermostable amylases form low molecular weight dextrins which interfere in the reassociation of starch chains (Fu et al. 2015). These are comprehensively used by commercial bread manufacturers.

Salt like sodium chloride significantly retards the degree of starch retrogradation. Protein also retard starch retrogradation to some extent. Soy and pea protein hydrolysates delayed retrogradation in maize starch (Ribota et al. 2012; Lian et al. 2013).

3.7 Starch Modification

Starch forms complexes with other food constituents such as lipids and proteins and can influence the physicochemical properties of starch depending upon the source. Functional properties such as viscosity, texture, water absorption, adhesion, binding, and gel formation are affected by the type of starch molecule. Starch is mainly used as filler, thickener, stabilizer, coating agent in bakery products, snack foods, meat products, and dairy confectionary. In its native form, starch has restricted applications; however, this can be improved by various modification techniques (Kaur et al. 2012). Food companies prefer modified starches over native starches for their better functional characteristics. Various approaches are used for the modification of native starches. These approaches include physical methods, chemical methods, enzymatic, and genetic methods (Table 3.2). Modification partially degrades the starch molecule which enhances specific functional properties such as thickening, water-holding, heat resistance, and reduced syneresis. The selection of modified starch for a particular food application depends on the technical and consumer requirement. Consumers consider organoleptic properties, esthetics, and stability, whereas, technical requirements include viscosity, resistance to temperature and shear, pH, etc. The processing parameters or reaction conditions of different types of modified starches and the associated alterations are depicted in Table 3.3. Modified starches are used as food additives under defined limits to improve product properties. Limits of starch modification, their use, and labeling requirements are markedly defined in the US Code of Federal Regulation. Based on functional applications the modified starches are grouped in different categories; (1) fat replacers/fat mimetics, (2) texture improvers, (3) for high nutritional claims, (4) for stability to high shear and temperature, and (5) for encapsulation.

3.8 Physical Modification

Physical modification of starch is a common approach and can be accomplished through commonly used methods such as pre-gelatinization, heat treatment, and annealing (Zia-ud-Din and Fei 2017). However, innovative techniques are being progressed for the physical modification of starch which can help in maintaining the

Table 3.2 Methods for modification of starch

Type of method	Advantages	Nature of treatment	Method
Physical modification	The techniques preserve structural integrity during the process. The modified starches are free from chemical residues.	**Thermal treatment**	Oven heating Extrusion Fluidized bed heating Spray drying Superheating
		Hydrothermal treatment	Heat moisture treatment Annealing
		Radiation treatment	Microwave heating Gamma irradiation Ultra-violet irradiation
		Nonthermal treatment methods	
		Pressure treatment	High hydrostatic pressure treatment Osmotic-pressure treatment Instantaneous controlled pressure and drop (DIC) process
		Mechanical treatment	Milling/grinding Activation by stirring Micronization in vacuum ball mill
		Other methods	Ultrasonication Pulsed electric field treatment Cold plasma Ozone treatment Corona electrical discharges
		Low-temperature treatment	Deep freezing Multiple freezing and thawing
Chemical modification	The morphology and granular size distribution remain unchanged during the chemical modifications. Different methods can be adopted to get desired functional properties in modified starches.	Oxidation	Oxidation
		Esterification	Acetylation Fatty acylation Phosphorylation Succinylation

(continued)

Table 3.2 (continued)

Type of method	Advantages	Nature of treatment	Method
		Etherification	Methylation Hydroxymethylation Carboxymethylation Hydroxypropylation
		Hydrolysis	Acid hydrolysis Alkaline hydrolysis
		Cross-linking	Cross-linking with dicarboxylic acids Alkaline cross-linking
		Cationization	Dry cationization Semi-dry cationization Wet cationization
Enzymatic modification	The specificity of enzymatic reactions offer higher yields and the resultant altered starches are safe as no residues are left in the process.	Hydrolysis	Enzyme catalyzed hydrolysis
Dual modification	The process can overcome the limitations of single modified starches and may provide desirable functional characteristics.		Homo-dual modification Hetero-dual modification
Genetic modification	Genetic engineering can alleviate the effects of environmentally hazardous modification techniques.		Transgene technology

integrity of native starch. These methods include multiple deep freezing and thawing, osmotic-pressure treatment, superheated starch, corona electrical discharges, pulsed electric fields treatment, high-pressure processing, ultrasonication, ozone treatment, cold plasma treatment, micronization in vacuum ball mill and drop (DIC) process, etc. (Raghunathan et al. 2021).

3.8.1 Pre-gelatinized Starches

Pre-gelatinized starches are often referred to as instant starches as they are precooked and dried. Extrusion cooking, drum drying, and spray drying are commonly used methods for the pre-gelatinization of starches (Hodge and Osman 1976). The technique used strongly influences the functional properties of modified starches. The temperature of drying and residence time plays an important role in governing the properties of starch (Bonazzi et al. 1996). These types of starch create moderately stable emulsion upon dispersion in cold water and are broadly used as thickeners in different food products. At the commercial level, drum drying is widely used to create a variety of textures with different porosity which influences the functional

Table 3.3 Functional properties and applications of modified starches

Type of starch	Process parameters	Source	Functional properties	References
Physical modification				
Pre-gelatinized starches	Suspension 37% (w/v) Drum drying temperature: 150 °C, pressure 5 bar	Wheat starch	Increased solubility, swelling capacity, and viscosity	Li et al. (2014)
Heat-moisture treatment	Moisture levels: 22–27% Temperature: 100–120 °C Time: 16 h	Smooth pea, wrinkled pea, navy bean, and lentil	Reduced amylose leaching, granular swelling and peak viscosity, and increased thermal stability, gelatinization temperatures and susceptibility towards α-amylase and acid hydrolysis	Chung et al. (2009), Hoover and Vasanthan (1994b)
Annealing	Moisture: <65% w/w or 40–50% w/w Temperatures: below the onset temperature of gelatinization	Lentil, smooth pea, and wrinkled pea starch	Decrease granular swelling and amylose leaching Increased gelatinization temperatures, thermal stability, and susceptibility towards α-amylase	Chung et al. (2009), Hoover and Vasanthan (1994a)
Microwave treatment	Oxidant: hydrogen peroxide Catalyst: sodium tungstate Microwaves power: 90, 190 W	Potato-starch paste	An increment in starch gelation temperature and decrease of the process enthalpy was observed in microwave-treated starches	Łukasiewicz et al. (2011)
Osmotic-pressure treatment	Starch solution suspended in sodium sulfate	Potato starch	Potato starch changed from a B to an A-type, increased gelatinization temperatures	Pukkahuta et al. (2007)
High-pressure processing	Air dry potato starch: 84.9% (d.s.) Pressure: 50–2000 MPa Treatment time: 1 h	Air dry potato starch	Increased bulk density of potato starch, change in the contours of the granules from oval to polyhedral, increased surface roughness, reduced enthalpy of starch gelatinization, enhanced hydrolytic susceptibility to the amylolytic enzyme	Słomińska et al. (2015)

(continued)

Table 3.3 (continued)

Type of starch	Process parameters	Source	Functional properties	References
	Pressure: 600 MPa Treatment time: 30 min Temperature: 30 °C	Waxy and non-waxy rice starch	Slow digestibility	Tian et al. (2014)
Ultrasonication	Starch-water dispersions: 10% Ultrasonication: 15–30 min Frequency: 30 kHz Treatment: pulsed mode (0.8 Hz or 80% duty cycle)	Wheat starch	A-type crystalline remained unaffected, increase in relative crystallinity	Karwasra et al. (2020)
Pulsed electric field processing	Potato starch-water suspensions: 8.0%, w/w PEF: 30–50 kV/cm	Potato starch	Altered intragranular molecular structure	Han et al. (2009a, b)
Ozone processing	Ozone gas concentration in water: 4.2 mg ozone/L Treatment time: 1 h Treatment temperature: 5 °C	Corn and potato starch	DSC gelatinization temperatures (i.e., onset, peak and conclusion) of starch samples increased, decrease in gelatinization enthalpies, decrease in the RVA viscosities	Çatal and Ibanoğlu (2012)
Cold plasma	Starch:10 g, db, 10% moisture Plasma under atmosphere: operated at 20 kV and frequency of 1 kHz Treatment time: 30 s with 2 mm discharge distance	Tatary buckwheat, quinoa, and sorghum starches	Higher solubility, lower swelling capacity, increased relative crystallinity, more accessible to amylolytic enzymes, increased gelatinization temperature, significant reduction in viscosity, enhanced starch digestibility	Gao et al. (2019)
	Cold plasma: air generated 15 kHz, 250 W Treatment time: 5, 10, and 20 min	Rice	α-amylase activity and water uptake increased, and hardness decreased	Lee et al. (2016)

(continued)

Table 3.3 (continued)

Type of starch	Process parameters	Source	Functional properties	References
Chemical methods				
Acetylation	Hydrothermal pretreatment: Suspension at 65% (w/v) Temperature: 48–62 °C Reaction time: 3 h Chemical reactive: Octenyl succinic anhydride (OSA) Catalyst: Sodium hydroxide	Commercial corn starch	Pretreatment increased the swelling capacity of granules and the peak of viscosity. Decrease in paste temperature	Chen et al. (2014)
Cross-Linking	Cross-linking agent: $POCl_3$ Reaction time: 30 min	Maize starch	Decreased peak viscosity with increasing amounts of cross-linking agent, reduced swell, higher viscosity, more rigid external surface area, with hard crust	Hirsch and Kokini (2002)
Oxidation	Chemical reactive: sodium hypochlorite	Oat Starch	Apparent increase in linear molecules due to amylopectin depolymerization. Reduction in viscosity at high temperature and faster development of viscosity upon cooling	Berski et al. (2011)
Acid Hydrolysis	Acid solutions: chlorhydric acid (0.36 N) and sulfuric acid (0.72 N) Starch suspension: 40% Temperature: 40 °C Treatment time: 4 and 24 h	Wheat (30% starch) Potato (25% starch) Peas starch	Reduced paste viscosity. The amylopectin depolymerization increased linear chains which favored gel formation and enhanced strength	Ulbrich et al. (2014)
Cationization	Chemical reactive: 2,3-epoxypropil trimethyl ammonium chloride pH: 9.0	Waxy maize starch	Reduction in starch granule size and fragmentation	Liu et al. (2015)

(continued)

Table 3.3 (continued)

Type of starch	Process parameters	Source	Functional properties	References
Enzymatic hydrolysis	B-amylase and transglucosidase	Maize starch	Increased amount of α-1,6 linkage imparted slow digestibility of starches (33.5%)	Miao et al. (2014)

properties. Upon reconstitution, pre-gelatinized starches prepared by the drum dryer form a weak paste.

Spray drying is another important technique for the preparation of pre-gelatinized starches. The advantage of using spray drying is that the method maintains the granular integrity of the starch molecule. Spray drying is generally used to alter the starches physically by atomizing starch solution. The spray-dried modified starch granules exhibit a hollow structure inside the granule and present a uniform shape (Yan and Zhengbiao 2010). This method improves the fluidity of starch granules.

Extrusion processing uses high-temperature short-time treatment for the preparation of pre-gelatinized starches. The high temperature, pressure, and shear bring structural modifications to the starch molecule. Under high pressure and shear, the semi-crystalline starch molecules are compressed into a viscous plastic material (Brummer et al. 2002).

3.8.2 Hydrothermal Treatment

Heat moisture treatment and annealing treatment involves the use of moisture and heat, and are commonly used physical modification approaches. In these methods, the starch is heated below the gelatinization temperature of starch granules which helps in maintaining the granular integrity. In contrast to extrusion treatment, the starch remains in a rubbery state after hydrothermal treatment. Moisture content and temperature are critical parameters in these processes as they strongly affect the properties. These treatments bring structural and functional modifications to the starch molecule (Jayakody and Hoover 2008).

The process of annealing is used for starches having low to intermediate moisture content. This process improves crystallinity and weakens structural relaxation. The starches modified by this method exhibit improved mechanical properties and thermal stability (Lv et al. 2015). This process reorganizes amylose units which support the long-range interaction of amylopectin molecules. These enhanced interactions eventually caused uniformity in crystal arrangement and stability of modified starches (Hoover and Vasanthan 1994a). The method guards the integrity of granules by augmenting the gelatinization temperature and constricting the thermal transition temperature range. In annealing, the glass transition temperature is approached for better mobility of starch granule and at the same time gelatinization is prevented by controlling the temperature and moisture levels. During this process, the amorphous region absorbs moisture which enhances its mobility. As the process progresses, the weaker crystallites progressively diminish while the other crystallites become more perfect due to recrystallization and fusion (Jayakody and Hoover 2008). Annealed starches display higher gelatinization temperatures and can be used as an alternative to chemical processes (Tester and Debon 2000).

High moisture treatment treats the starch granules with low moisture content at high temperatures ranging from 80 to 140 °C and low gelatinization temperatures (Hoover et al. 2010; Hoover and Vasanthan 1994b). This method maintains the integrity of the granular structure and modifies the properties such as reduced starch solubility, swelling power, amylose leaching, and peak viscosity but increased the pasting temperature (Pinto et al. 2012). This method alters both the amorphous and crystalline regions of starch molecules; however, the alteration in the crystalline region considerably affects the properties of starch (Hoover and Manuel 1996a, b). The parameters which affect the starch properties include the temperature of treatment, moisture content of starch, exposure time, source of heating, cooling process, and botanical source (Vermeylen et al. 2006). Commonly, these starches find applications in the preparation of infant foods. Also, these starches are known to improve the quality of baked food and freeze-thaw stability (Collado and Corke 1999).

3.8.3 Microwave-Assisted Modification of Starch

Microwaves work in the electromagnetic radiation range of 3000 MHz–30 GHz. It is considered as one of the important physical modification methods to impart desired alterations in the structural and functional characteristics of starches. The notable advantages of microwaves over conventional physical modification methods include rapid transmission of the energy inside the material which allows selective and volumetric heating and shorter processing duration (Muzimbaranda and Tomasik 1994). Regardless of the associated benefits, the process also has some limitations which include an additional cost of specialized equipment, optimization of process parameters, and inadequate information on the dielectric properties of heated materials. Microwave heating may result in nonuniform heating in cases where the dielectric constant increases with the temperature and can lead to the thermal degradation of the material. Therefore, the process requires continuous monitoring for attaining uniform temperature. The effectiveness of microwave-assisted modification of starches depends on the botanical origin of starch, moisture content of starch, dielectric properties, penetration depth, frequency, power, and treatment duration (Brasoveanu and Nemtanu 2014). The dielectric properties of starches are governed by the nature of starch, moisture content, frequency of microwaves, and resultant temperature of the treatment. The dielectric constant describes the penetration of the microwave into the starch molecule, while the dielectric loss factor depicts the capability of starch to accumulate energy. As a result of microwave application, the relationship of temperature rise over time is nonlinear as the increase in temperature depends on the moisture content of starch and the power of applied microwaves. The moisture content of the starch is a critical parameter in microwave processing; therefore, starches with varied moisture content display different alterations concerning microwave-assisted modification. Starches with low moisture content (1–5%) exhibited a prompt increase in temperature, while higher moisture starches (7–15%) displayed a less prominent rise in temperature, and starches with 20% moisture content, showed a plateau period, which is directly proportional to the increase in moisture content. However, these associations are unpredictable with the fact that microwave heating works effectively with high moisture content. In such cases, the moderate temperature increase in the samples with higher moisture content is attributed to the high specific heat capacity of water, which marks it a great cooling-heating channel (Brasoveanu and Nemtanu 2014).

Microwave processing can present substantial alterations in the process of gelatinization and rheological properties of the starch and also alters the starch structure and crystalline arrangement of the granule. An increment in starch gelation temperature and decrease of the process enthalpy was observed in microwave-treated starches. Also, a lower viscosity was observed in microwave-heated starches. A lower viscosity was observed in wheat and corn starch (30% moisture content) treated with a microwave power of 0.5 W/g. Amylopectin-rich waxy corn starch did not demonstrate any deviations in the swelling ability under the action of microwave processing (Lewandowicz et al. 1998). The extent of alteration in solubility and reduction ability of air dry potato and corn starches upon microwave processing was at par with the conventional heating (130–200 °C) (Lewandowicz et al. 1998; Pałasínski et al. 2000). The microwave treatment of potato starch contributed to significant structural alterations (Xie et al. 2013). Native potato starch demonstrates a regular elliptical shape and smooth surface while the effect of microwave treatment on structural alterations was directly proportional to the treatment duration. Five seconds (40 °C) of exposure created visible defects and small cracks on the surface of some of the starch granules, at 10 s (55 °C) flaws or fractures appeared, at 15 s (80 °C) deformation of the starch granules initiated, whereas at 20 s exposure (95 °C) cracks, deformation, and destruction of most of the starch granules was observed (Fig. 3.4). These modifications were owed to the destruction triggered by the rise in temperature or due to the reorganization of crystalline structure caused by loss of water.

Microwave-assisted starch oxidation is an important area of starch modification. Łukasiewicz et al. (2011) demonstrated strong interactions of starch with microwaves. Microwave-assisted oxidation of starch paste instigated noteworthy deviations in viscosity and also no yield stress was observed. Microwave processing creates starches with a high degree of oxidation and an extensive range of possible applications. Microwave-assisted starch esterification is another popular method of starch modification. Starches are being esterified with a catalyst in presence of microwave radiation to get the desired modifications. Tapioca starch was esterified with succinic anhydride, the resultant succinates demonstrated increased viscosity and water absorption capacity in comparison to native starch. A high degree of substitution was achieved in microwave-assisted starch esterification with acetic anhydride and iodine (0.16–2.5 mol%) as a catalyst, in a reaction duration of

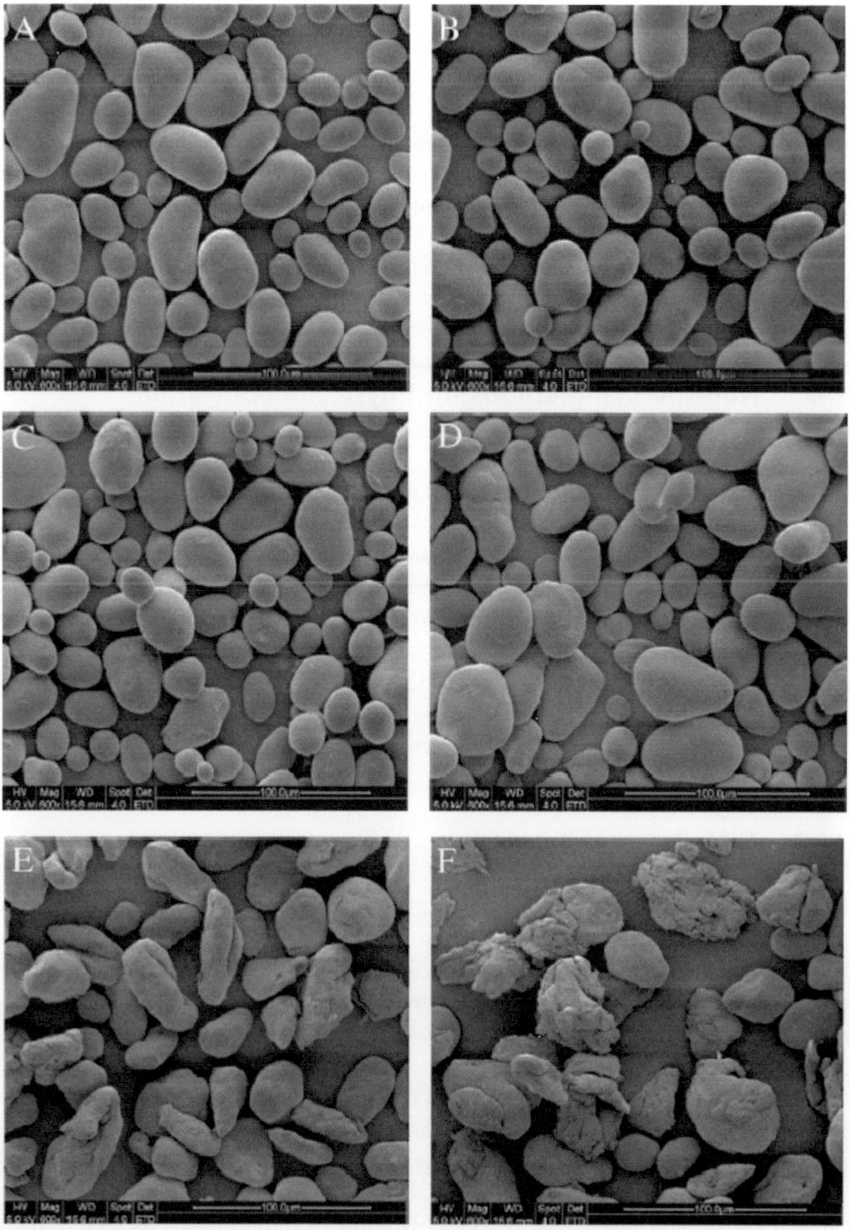

Fig. 3.4 SEM micrographs of (**a**), (**b**) native and (**c**), (**d**), (**e**), and (**f**) microwave-treated potato starch. (Adopted from Xie et al. 2013)

2 min and at a reaction temperature of 100 °C (Biswas et al. 2008). On comparing the microwave-heated enzymatic esterification of starches to that of conventional heating, the microwave-heated (power level 80 W; treatment duration 135 s) esterified starches expressed superior heat stability and improved mechanical properties (Rajan et al. 2008). A similar comparison was attempted on potato starch with thiosemicarbazide, the microwave treatment of starch demonstrated a higher degree of conversion of the reaction mixture with the enhanced thermal stability of resultant modified starch (Siemion et al. 2006). Due to selective and uniform heating, microwave often leads to improved yield at accelerated reaction duration of the modified starches.

3.9 Non-thermal Methods for Physical Modification of Starches

To overcome the limitations of thermal processing, innovative non-thermal processes are being evolved for the physical modification of starches. In recent times, the essential requirement for searching nonpolluting and eco-friendly technological replacements has headed towards investigating nonthermal processes such as ultrasound processing, high-pressure processing, pulsed electric field processing, irradiation, ozone treatment, and cold plasma, as suitable alternatives for conventional chemical methods of starch modification (Pandiselvam et al. 2020). These technologies are often referred to as green technologies as they do not leave any residues in the product and are environment friendly.

3.9.1 High-Pressure Processing

High-pressure processing employs high pressures in the range of 100–1000 MPa under ambient temperature. High pressure destroys the non-covalent bonds which result in structural changes in the starch molecule. This method results in the creation of pre-gelatinized starches at room temperature. High pressures result in the disordered structure of polymers which leads to the gelatinization of starches (Grgić et al. 2019). High-pressure treatment in presence of excess water creates destruction of crystallite structures by the forced infiltration of water molecules into crystalline regions; whereas, the amorphous region shows reversible hydration within the starch granules. The effectiveness of the process depends on the nature of the starch and process parameters such as level of pressure, exposure duration, and starch:moisture ratio. It has been observed that B-type polymorphs exhibit more resistance to high pressures. High pressure treated starches often retain their granular structures in comparison to conventionally heat-treated starches (Chotipratoom et al. 2014). Also, in pressure gelatinization, starch granules remain intact or are only partly disrupted (Pei-Ling et al. 2010). Homogenization at high pressures ranging from 60 to 80 MPa leads to loss of native starch structure, at 100 MPa starch granules fragmented into gel-like arrangements, and increment in their mean diameter was observed (Qiu et al.

2014). High pressure treated waxy and non-waxy rice starch exhibited higher slow digestibility in comparison to thermally gelatinized starches during retrogradation (Tian et al. 2014). Furthermore, high-pressure treated starch gels expressed lower swelling index and lower susceptibility to amylolytic enzymes in comparison to thermally gelatinized starches (Słomińska et al. 2015). High-pressure treated modified starches are suitable to use as a fat substitute, because of the stimulation of fat droplets which are deliberated as microparticles of definite size distribution (Stute et al. 1996). Also, the baking industry requires starches that gelatinize at lower temperatures to save energy; therefore, such starches are better utilized in baked products.

3.9.2 Ultrasonication

Cavitation and radical attack are the two main principles behind the application of ultrasound in starch modification. Ultrasonication affects the starch chains at a molecular level. Cavitation occurs with the formation of gas bubbles that attack the starch granules and cause their explosion. The energy generated during this process increases the temperature and pressure which alters the starch structural arrangement affecting the physicochemical and functional characteristics. Also, the limited breakdown of a water molecule into $^{\bullet}OH$ and $^{\bullet}H$ radicals initiates the structural alterations in the starch molecule and are recognized as a radical attack (Jamalabadi et al. 2019). Ultrasonication alters the starches by creating cracks and pores in the granular arrangement of a starch molecule which promotes the effectiveness of the reactions. The native starches exhibit irregular shapes with smooth surfaces. Low-intensity ultrasound creates cracks and holes on the surface of the granule. This consequence was more noticeable in large granules (Karwasra et al. 2020). With an increase in the intensity, the magnitude of structural alteration increases due to the quicker formation of bubbles and collapse during cavitation (Yang et al. 2020). Largely it has been observed that amorphous regions are more sensitive towards ultrasonication in comparison to crystalline regions. Based on experimental parameters, such as frequency and intensity of ultrasounds, time of exposure, water and temperature of the starch-water system, botanical source of starch, and composition of gas in the atmosphere, ultrasonication demonstrated increased water solubility, decreased gelatinization or pasting viscosity, swelling parameters/retrogradation (Grgić et al. 2019).

3.9.3 Pulsed Electric Field Processing

Pulsed electric field processing employs high-intensity electric pulses (over 10 kV/cm) at a short duration (less than 40 μs). PEF technology has found applications in transforming larger fragments and improving chemical reactions. The benefits of PEF processing comprise constant treatment intensity, continuous processing, low processing temperature, and short duration. The application of PEF in potato starch altered intragranular molecular structure, which resulted in altered physicochemical properties and some new characteristics and functions in treated starch (Han et al. 2009a).

3.9.4 Ozone Processing

Alternative to chemical modification of starches, ozone could be an important replacement for conventional methods. The unreacted ozone is disintegrated into oxygen; therefore, no residues in the final product promise food safety and make it a greener option. Ozone is a strong oxidizing agent with a high electrochemical potential (+2.075 V). The modification of starches by oxidation is accomplished in two steps. Firstly, the hydroxyl groups at positions C-2, C-3, and C-6 are oxidized to carbonyl groups, with subsequent depolymerization due to the disruption of α-(1,4)-glycosidic linkages (Chan et al. 2011). Oxidation principally takes place in the amorphous regions of the starch molecule and is limited in the crystalline region as depicted by the appearance of Maltese crosses. Ozonation alters the shape to rough and fibrous (Çatal and Ibanoğlu 2012). Pores naturally present on the surface of starch granules in corn support the diffusion of ozone into the starch granules. The effectiveness of ozonation depends on several parameters, such as concentration of ozone, exposure duration, temperature of treatment, moisture content, and type of starch. Starches expressed low carbonyl and high carboxyl content after ozone treatment (Chan et al. 2009). Increased carboxylic groups increase the swelling in corn and whole-grain flour. The negative charges lead to repulsion which results in increased swelling power (Obadi et al. 2018). Ozonation of wheat and corn samples leads to increased gelatinization values due to low molecular weight fractions generated by oxidative cleavage of starches (Çatal and Ibanoğlu 2012, 2014). The network created by the reassociation of low molecular weight compounds requires high temperatures to disrupt. The increase in gelatinization temperature was also attributed to the depolymerization of amylopectin chains during ozonation. Conversely, a reduction in pasting temperature, peak viscosity, and breakdown viscosity was observed in wheat starch owing to the fragmentation and weakening of the starch network, limited breakdown of glycosidic bonds, and degradation of the amorphous region of the starch molecule, respectively. The reassociation of the starch molecule was limited by the replacement of carboxyl groups with hydroxyl groups as displayed by the reduced final viscosity (Çatal and Ibanoğlu 2014).

3.9.5 Cold Plasma

There is a fourth state of the matter known as plasma. Plasma delivers a combination of types that includes photons, electrons, free radicals, excited and non-excited molecules, positive and negative ions, and carries a net neutral charge (Thirumdas et al. 2017). Plasma is of two types based on thermodynamic equilibrium's characterization: thermal and non-thermal/cold plasma. The plasma treatment of starches has modified the structural and functional properties of starch. The efficacy of the process depends on the type of starch, treatment duration, plasma generation, and energy input. Plasma can be produced using different sources such as thermal, electric/magnetic fields, radio frequencies, and microwaves. The energy created by these processes increases the thermodynamic energy of the electrons and progresses particle collision, causing generation of plasma. Due to its comparatively simple design, dielectric barrier discharge (DBD) and jet plasma are generally desired techniques in the food processing area. Starch modification is accomplished by the interaction of plasma species with starch molecules. Starch modification with plasma involves different principles which include cross-linking, depolymerization, and plasma etching. In cross-linking, hydroxyl radicals created by the decomposition of a water molecule and oxygen gas generated due to plasma establishes C–O–C linkage between the two polymeric chains (C–OH). In the depolymerization process, the collision of plasma ions causes the depolymerization of amylose and amylopectin fractions of a starch molecule. Plasma etching results in the etching of surface material by physical sputtering or chemical means which in turn improves the hydrophilic properties of starches by increasing the penetration of water into the starch granule (Bie et al. 2016). The degree of surface etching after plasma treatment was directly proportional to the treatment duration in the case of normal and waxy maize starch granules (Zhou et al. 2018). Also, an increase in corn starch granule's brightness was observed using confocal laser scanning microscopy upon plasma treatment. This effect might be attributed to the penetration of surface pinhole channels by plasma. Due to the depolymerization effect of plasma on starches, waxy maize starch expressed lower gelatinization parameters (Zhou et al. 2018) and a lower crystallinity was observed in rice starch (Thirumdas et al. 2017). On the other hand, plasma treatment resulted in an increase in the enthalpy of gelatinization (ΔH) in waxy rice, buckwheat, sorghum, quinoa, and wheat which might be credited to the cross-linking of amylopectin chains and the formation of longer double helices from the untwisted ends of amylopectin (Gao et al. 2019). The decline in pasting temperature in maize starch is related to the weakening of the starch structure while a reduction in peak viscosity was related to the formation of starch–lipid complexes. Plasma-induced cross-linking contributes to improved starch stability and reduced breakdown values. The reordering by leached amylose molecules results in a drop in the setback and final viscosity (Gao et al. 2019). In contrast, rice starch upon plasma treatment expressed an upsurge in peak viscosity, breakdown, and final viscosity. The weakening of bond strength and introduction of water molecules due to the disruption of hydrogen bonding could be the reason for an increase in pasting viscosity and setback and final viscosity, respectively (Thirumdas et al. 2017).

3.10 Chemical Modification of Starch

Chemical modifications are the most widely used method for the modification of starches. Chemical modification of starches can be achieved by different mechanisms namely oxidation, esterification, etherification, and cross-linking.

Under appropriate reaction conditions, starch molecules can undergo hydrolysis, oxidation, esterification, and etherification reactions to create products with desired organoleptic, textural, mechanical, and thermoplastic characteristics for food and non-food applications.

3.10.1 Oxidation

Oxidation is the most commonly used chemical method for the modification of starches. In the process of oxidation, primary and secondary hydroxyl groups of the glucose units (at the C-2, C-3, and C-6 positions) oxidizes and form aldehyde or carboxyl groups. The Oxidation reaction may lead to partial depolymerization of polymeric chains due to the loosening of intermolecular bonds (a-1,4-glycosidic linkages) (Tomasik and Schilling 2004). The commonly used oxidants are air or oxygen (in the presence of catalysts such as transition metal ions), inorganic peroxides, organic peroxides, nitrogenous compounds, organic oxidants, and metal compounds. Out of these, the most commonly used oxidants are hydrogen peroxide, sodium hypochlorite, and sodium periodate owing to concerns regarding the safety and generation of non toxic residues. The extent of starch oxidation is determined by the amount of carbonyl (CO) and carboxyl (COOH) groups and the level of depolymerization. The effectiveness of this method depends on the type of the oxidant, reaction conditions, and nature of the starch. The modified oxidized starches exhibit improved solubility in water, lower viscosity, and reduced affinity to retrograde. The major limitation of the chemical oxidation process is the emission of toxic residual products. However, the oxidized starches are more frequently used as additives by the food processing industry. The oxidants are selected based on high levels of active oxygen (47%) and with high oxidation potential (1.77 V), for effective and efficient utilization of the starch modification process. Different oxidants have their limitations and advantages depending on the type of substrate, for example, hydrogen peroxide has limited reactivity for organic functional groups and does not exhibit oxidizing properties in the existence of electrophilic compounds where it acts as a nucleophile (Arts et al. 1997).

The nature of starch and protein content of the matrix plays an important role during the starch oxidation reaction with sodium hypochlorite as proteins tend to oxidize first during the oxidation reaction. The sodium hypochlorite oxidation of starches initiates in a gelatinized polymer solution where the main hydroxyl moieties are oxidized into either aldehyde or carboxyl groups with a conversion efficiency of only 1 per 35–50 glucose units. The modified starches oxidized by NaClO exhibit enhanced thermal stability and improved resistance to amylase activity. Such types of starches have the capability of complexing calcium ions while exhibiting poly-electrolyte properties (Pietrzyk et al. 2006). These types of starches are better suited for applications as a stabilizer and thickener in ketchup, sauces, puddings, etc. and are designated with an e code of E1404. Oxidation of starch with sodium periodate results in the formation of aldehyde due to cleavage of C_2 and C_3 carbons bonds. The resultant dialdehyde starches are broadly utilized in the food industry and in the manufacture of decomposable plastics (Zhang et al. 2007). In order to attain desired functional properties (lower viscosity and better water solubility), oxidation of starches is preferred before gelatinization due to cleavage of both intramolecular and intermolecular hydrogen bonds (Zhang et al. 2009).

3.10.2 Substitution

In the process of substitution, the starch molecules are substituted by large groups via etherification or esterification which incorporates abnormalities to amylose and amylopectin units and later obstructs the affinity to realign and retrograde by these fragments. The functional properties of such modified starches depend on the botanical origin of starch, reaction conditions, nature of the substituent, degree of substitution, and distribution of the substituent in the starch molecule. The functional characteristics of substituted starches are modulated by the degree of substitution. At a higher degree of substitution, increased hydration and lower gelatinization temperatures were reported due to weak interaction between starch and starch polymer (Ai and Jane 2015; López et al. 2010).

Esterification of starch involves the conversion of hydroxyl groups into hydrophobic ester groups which improves the thermoplastic nature of the starches. The exposure of starch to esterifying agents allows modification of starch with the synthesis of starch esters which produces enhanced molecular interfacial function by inhibiting the linearity of molecular chains which in turn prevents the formation of an intermolecular or intramolecular association. Organic and inorganic acids and their derivatives such as acid anhydrides, oxychlorides, and chlorides are commonly used esterifying agents. Starch acetates esterified with acetic acid or acetic anhydride are commonly used as stabilizers in processed food products such as salad dressings, mayonnaise, and ice creams (Volkert et al. 2010). Starch esters of phosphoric acid are attractively used in various food products for their cold-water solubility and high water-binding capacity (Lewandowicz et al. 1999). Starch phosphate esters without metal ions are widely used as a thickener in the food industry. In the case of distarch phosphates, the amount of cross-linking affects its capability to swell. The higher swelling ability in esterified starches is indicative of enhanced stability to heat and mechanical impact. Low levels of cross-linking create starches with high swelling ability. Also, distarch phosphates display improved freezing and thawing stability and do not possess the tendency to retrograde (Seker et al. 2003; Landerito and Wang 2005).

Etherification of starch is commonly accomplished in the presence of an alkaline catalyst by following reactants: alkyl halides, acrylonitrile, or alkylene oxides. Amylose was altered to a larger magnitude in comparison to amylopectin in hydroxypropylated maize and potato starches (Kavitha 1998). Hydroxypropylated starches showed increased paste viscosity, improved clarity, and enhanced freeze-thaw and refrigerated-storage stabilities (Xu and Seib 1997). These starches are often used as an additive in instant soups, sauces, ready-to-eat desserts, confectionery fillings canned fruits and jams, and frozen foods.

3.10.3 Cross-linking

In the process of cross-linking, intra- and inter-hydrogen bonding in starch granules are replaced by stronger, more permanent, covalent bonds (Wurzburg 1986). The commonly used food-grade chemicals for cross-linking are sodium trimetaphosphate, monosodium phosphate, sodium tripolyphosphate, epichlorohydrin, phosphoryl chloride, a mixture of adipic acid, acetic anhydride, and vinyl chloride. The functional group introduced governs the functional characteristics of cross-linked starches. Depending on the reaction conditions, the cross-linking agent can be selected as different reagent works under different conditions. For example, different reagents have different optimum reacting conditions: $POCl_3$ works better at pH 4 in the presence of a neutral salt (Hirsch and Kokini 2002), and sodium trimetaphosphate works efficiently at high temperature in semi-dry conditions and at moderate temperature in hydrated conditions (Wongsagonsup et al. 2014). With the increased number of crosslinks, the modified starch displays additional resistance to gelatinization. Distarch phosphates and distarch adipates are the most commonly used cross-linked starches. Cross-linked starches offer acid, heat, and shear stability. Cross-linked potato starch paste exhibited enhanced viscosity and resistance to breakdown during prolonged cooking time, increased acid content, or severe agitation (Singh et al. 2016).

3.11 Enzymatic Modification

Enzymatic modification of starches can be accomplished by two main approaches; the first approach is via in vivo suppression or over expression of the enzymes involved in starch biosynthesis and degradation in the transgenic plants, and secondly through in vitro modification with the use of carbohydrate hydrolyzing enzymes from microorganisms. In vitro enzymatic modification of starches is a greener approach to create starches with modified structural attributes and physicochemical characteristics for various food uses and may function as an appropriate alternative method due to greater selectivity and moderate reaction conditions. The catalytic action of enzymes helps in the creation of starches with novel structures due to alterations in molecular mass, branch chain length distribution, and amylose/amylopectin ratio as a result of enzymatic hydrolysis. Enzymatic degradation (with amylase) was found effective for both crystalline and amorphous regions; however, amorphous regions of starch granule are more susceptible to enzymatic action in comparison to the crystalline region (Wang et al. 1995). Subsequently, the altered starch structures created by enzymatic hydrolysis are not able to reassociate. The most suitable applications of enzymatically modified starches in the food industry include freeze-thaw stable gels and in delaying retrogradation during storage. Based on the catalytic reaction, substrate specificity, and sequence similarity, starch-acting enzymes have been segregated into glycosyl hydrolases and glycosyltransferase. Starch modifying enzymes are classified by Coutinho and Henrissat (1999) into several families (CAZY website: http://afmb.curs-mrs.fr/CAZY). Family 13, 57, and

77 comprise a majority of the scientifically significant enzymes for the modification of starches (Park et al. 2008). Family 13 (GH13) consists of α-amylase, pullulanase, isoamylase, glucan branching enzyme, and cyclodextrin-glycosyltransferase, while Family 77 (GH77) comprises amylomaltase and 4-α-glucanotransferase.

Maltogenic amylase has expressed a hydrolytic effect on amylose but the action is very limited in the case of amylopectin. The reason for this action can be described by the geometry of the enzyme's active site, limiting the molecular size and shape of the substrate. Enzymatic debranching can be accomplished by 4-α-glucanotransferase that disproportionate the side chains of glucan, which eventually alters the side chain length (Kim et al. 1999). Enzymatic modifications have been used to modify the textural and functional properties of starches.

Maize starch was studied for structural alterations and in vitro starch digestibility after treatment with β-amylase and transglucosidase. Dual-enzymes treatment resulted in starches with low molecular weight due to cleavage of α-1,4 linkage and transformation of nonreducing D-glucosyl residues of maltose into α-1,6 branch linkage. The increased amount of α-1,6 linkage imparted slow digestibility to starches (33.5%) after double enzymatic hydrolysis for 6 h (Miao et al. 2014). Porous starches with pores of various sizes were developed by the action of amyloglucosidase, α-amylase, cyclodextrin-glycosyltransferase, and branching enzymes (Benavent-Gil and Rosell 2017). Large-sized pores were contributed by amyloglucosidase, while, cyclodextrin-glycosyltransferase contributed smaller sized pores. The porous starches find application in the adsorption of volatile and other molecules. In another finding, a modified granular starch surface with pores was achieved by treatment of tapioca starch with α-amylase, amyloglucosidase, and 2.2 M HCl in various combinations and alone. The resultant modified starch sample exhibited an increase in their specific surface area, volume of pores, and average diameter (Prompiputtanapon et al. 2020). The application of enzymes to catalyze starch acylation is a suitable alternative to the chemical acylation process. The starches developed by this method may deliver as polymers with internal plasticizers. In recent times, acylated starch derivatives have been effectively manufactured as starch nanoparticles dispersed in organic medium with immobilized lipase and aqueous starch gels with lipase and dispersed fatty acids (Alissandratos and Halling 2012).

3.12 Dual Modification of Starches

Singly modified starches witnessed some limitations for their potential in industrial applications. For example, cross-linked starches do not exhibit freeze-thaw stability and characteristics for slowing down the retrogradation process and succinylated starches exhibit poor properties in response to shear and high temperature (Ackar et al. 2015). Therefore, there is a necessity for double/dual modification to alleviate these shortcomings; cross-linking followed by succinylation or the opposite practice, i.e., succinylation of the cross-linked starches can be accomplished to overcome this limitation. Dual modification refers to the modification of starches by two methods which is further classified into homo-dual modification and hetero-dual modification based on the type of method used for alteration. For example, if starch is modified by two methods of the same category, like, physical-physical/chemical-chemical/enzymatic-enzymatic, then it is referred to as homo-dual modification; however, if starch is modified by different combination of methods from different categories (enzymatic-physical, enzymatic-chemical, physical-chemical, etc.), they are classified as a hetero-dual modification. For effective modification of starches by dual modification, the parameters that need to be considered include the nature of starch, order of applied treatments, the modifying agent, reaction conditions, and the medium of reaction. Due to the uniqueness of starch structures, each starch behaves differently with the same order of modification methods and reaction conditions. For instance, legume starches have distinctively high amylose contents and limited swelling of starch granules while root and tuber starches are well-known for their large-sized starch granules with high swelling ability; on the other hand, cereal starches are characterized by their smaller starch granules. If starches of different botanical origins are exposed to similar dual modification treatments, the deviations in the functional characteristics of the specific double modified starches are attributable to, (1) different morphologies of the starch granules; (2) varied ratio of amylose/amylopectin; (3) compositional variations in the residual proteins and lipids; (4) different polymorphism of the starches, A-, B-, or C-type; (5) variation in relative crystallinity; and (6) complexity of amylopectin structure (Ashogbon 2021). However, there are some examples where dual-modified starches do not display superior properties to natural or single modified starches; therefore, careful selection of treatment methods is required and their manufacture may result in excessive expenditure on time and expensive chemicals.

Dual modification of starches involves the preparation of starches by application of the first treatment for subsequent modification by the second treatment. The first treatment prepares the starch by altering the surface of starches or bond strength which makes it suitable for treatment by the second method. The desirable functional characteristics of the dual-modified starches are governed by the second modification treatment while the first alteration improves it. This concept is explained by a few examples; extrusion treatment results in depolymerization of starches which reduces molecular weight and gelatinization parameters due to the creation of smaller starch fraction. These changes during extrusion treatment prepare the starch for easy structural alteration during the second subsequent heat moisture treatment. Another example includes the sonication of starches to achieve structural alteration by the creation of fractures, pores, and cracks to facilitate penetration of the acetyl group during subsequent acetylation (Ashogbon 2021).

The effect of dual treatments of microwave followed by autoclave and microwave followed by hot air oven on properties of taro starch (25% moisture, w/w) was evaluated. A loss of physical integrity of the starch granules was perceived in both types of dual-modified starches. An increased amylose content, swelling, and solubility were observed in both samples. The peak viscosity, holding and final viscosities of dual-modified starches modified by microwave followed by hot air oven were found to be higher. Both the starches expressed better freeze-thaw stabilities than that of native starch (Deka and Sit 2016). Effect of heat moisture treatment with high hydrostatic pressure evaluated for its effects on the properties of potato starch. The dual-modified starches exhibited increased peak viscosities, setbacks, and final viscosities compared to the samples processed with a single method (HMT). The process of dual modification also increased the transition temperatures, swelling power, and altered the relative crystallinity. Dual modification (moisture dependent) displayed a noteworthy capability to modify starches with diverse characteristics especially low glycemic index properties (Colussi et al. 2019). The structural alterations caused by dual modification (hydrolysis and succinylation) in corn starch were studied by Basilio-Cortés et al. (2019). The dual modification resulted in a 44% increment in reaction efficiency and a 45% rise in the degree of substitution as compared to starches modified with succinylation alone. Acid hydrolysis displayed exo-erosion and whitish points on the surface of corn starch granules because of the accumulation of succinyl groups. Acid hydrolysis decreased the peak viscosity by 3–3.5-fold. Also, a notable decline in pasting temperature and peak time to 20 °C and 100 s, respectively, was observed. Thermal properties were also affected by the dual modification, enthalpy of gelatinization showed a reduction while an increased range of gelatinization temperature (6 °C) was observed. FTIR spectra specified the effect of dual modification on the crystallinity of starch, whereas Raman spectra indicated disruption of short-range molecular order in the starch. The dual effect of ozone and dry heating treatment on cassava starch was studied by Lima et al. (2021). The dual alteration stimulated the formation of fissures on the starch granule surface. The alteration significantly affected the amorphous region and displayed a greater extent of starch oxidation. These resultant starches expressed varied functional properties. However, the order of treatments was crucial in deciding the hydrogel properties. Dry heating treatment with subsequent ozone treatment caused stronger gels formation; whereas, ozone treatment before dry heat treatment caused the formation of weaker gels. Lentil starch when subjected to dual modification with ultrasonication and irradiation (5 kGy) showed a decline in pasting properties and amylose content. A substantial reduction in syneresis was perceived with dual treatments after 120 h storage. FTIR examination demonstrated a decline in the strengths of O–H, C–H, and O–C stretches and CH_2 bending with dual treatments. The starch also exhibited a decline in lightness as depicted by lower hunter "L" values following dual modification (Majeed et al. 2017).

Foxtail millet starch modified by annealing and ultrasonication in both the orders (annealing and ultrasonication or ultrasonication and annealing) were characterized by Babu et al. (2019). Ultrasonication before annealing significantly improved the resistant starches level and exhibited improved acid resistance, shear stability, freeze-thaw stability, and gel texture. However, sonication after annealing raised the final viscosity as compared to the dual modification in reverse order. This effect might be attributable to the cavitation caused by the action of sonic waves. Electric field prior to the ultrasonication demonstrated the greatest required characteristics such as better light transmission, improved water absorption capacity, greater solubility (77.50%), increased swelling power (50.48%), and higher levels of resistant starch (79.98%) in potato starch as compared to native, single, and dual-modified starches in different orders (Cao and Gao 2020). The effect of hydroxypropylation and cross-linking at varied levels and different sequences on the properties of taro starch was investigated by Hazarika and Sit (2015). A decline in amylose content followed by dual modification was observed. The dual modification resulted in starches with improved swelling ability, solubility, and clarity with an increased level of hydroxypropylation and showed a decline in these properties with increased levels of cross-linking. The order of treatment considerably affected the freeze-thaw stability of the dual-modified starches. Also, increased viscosities were observed with double alteration treatments. The resultant starch can be suitably used for applications requiring high viscosity and freeze-thaw stability. Native and modified tapioca starch was studied for morphological alterations due to enzymatic using scanning electron microscopy. Native starch granules showed a round shape with truncated ends and smooth surfaces. The surfaces of α-amylase, amyloglucosidase, and the combination of these two enzymes after 6 h of hydrolysis showed damaged surfaces with fractures or surface erosion compared to 1 h of hydrolysis. These findings suggest that α-amylase can hydrolyze superficially and cannot penetrate more deeply due to endo-hydrolysis. However, samples treated with amyloglucosidase demonstrated only single holes after 1 h of hydrolysis and the combination of two enzymes after 6 h of hydrolysis revealed a larger single hole (Fig. 3.5). These results indicated the synergistic effect of α-amylase and amyloglucosidase for endo- and exo-corrosions, respectively, leading to numerous cracks on the surface (Prompiputtanapon et al. 2020).

Some other examples of dual-modified starches include cross-linking/phosphorylation (Iacovou et al. 2017); autoclaving/retrogradation (Ashwar et al. 2016); branching enzyme (BE)/transglucosidase (Guo et al. 2019); HMT/citric acid esterification (Li et al. 2019); MW/acetylation (Zhao et al. 2018); and debranching/HPT (Hu et al. 2019), etc.

3.13 Analyzing Gelatinization of Starch

Starch occurs in the form of granules that are present within the complex food matrix making it difficult to analyze. Therefore, different approaches such as physical, chemical, and enzymatic are required to assess the structural alteration during gelatinization (Table 3.4).

3.13.1 Differential Scanning Colorimetry

Differential scanning calorimetry (DSC) is widely used to study the thermal and gelatinization behavior, glass transition temperature, and crystallization of native and modified starches. The thermal behavior of starches is much more complex because some physicochemical changes take place during the heating of starches. To analyze the thermal properties of starch, it is required to form starch-water

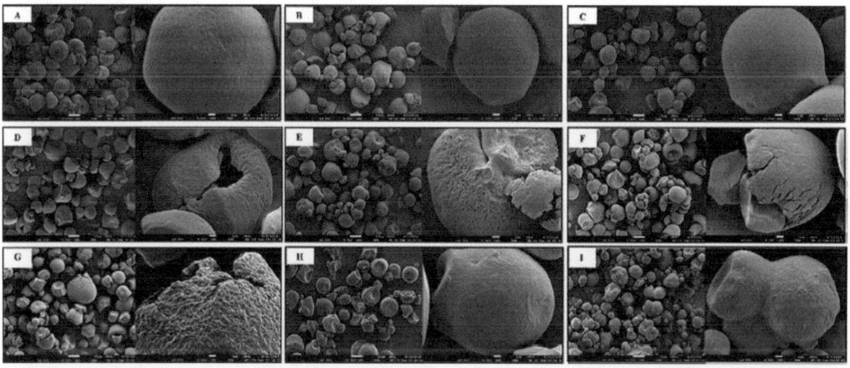

Fig. 3.5 Scanning electron microscopy of granular morphology at 1000× (left) and 6000× (right) in each frame of (**a**) native tapioca starch, (**b**) αA for 1 h, (**c**) αA for 6 h, (**d**) AM for 1 h, (**e**) AM for 6 h, (**f**) αA + AM for 1 h, (**g**) αA + AM for 6 h, (**h**) AS for 1 h, and (**i**) AS for 6 h. *αA* starch treated with α-amylase, *AM* starch treated with amyloglucosidase, *αA + AM* starch treated with α-amylase and amyloglucosidase, *AS* acid hydrolysis. (Adopted from Prompiputtanapon et al. 2020)

formulations in a sealed sample pan for the complete retention of water without rupturing or leakage of water from the formulation during heating. Thermal characterization of starch determined by DSC includes transition temperatures T_o (onset), T_p (Peak), T_c (completion), shapes of curves, and enthalpies (ΔH) which are affected by sample preparation, sample size, type of pan, measurement conditions, and starch-to-water ratio. These factors are carefully considered during the experiments to get stable results because sometimes multiple transitions and instability of water contained in starch make it very difficult to study the thermal behavior of starch using DSC. The DSC experiments are conducted in two steps. In the first heat treatment step, the sealed sample pans are kept for 30 min at a specific temperature in DSC equipment. After heating, samples are promptly cooled and maintained at 25 °C for 5 min. In the second scanning step, treated samples were scanned from 20 to 120 °C at 10 °C per min. The instrument software is used to control the experimental conditions. A typical DSC endotherm is presented in Fig. 3.6. The degree of starch gelatinization (DSG) was calculated as the ratio of enthalpic transition difference between the untreated and treated samples to the enthalpy of the untreated sample as follows:

$$\text{DSG}\ (\%) = [(\Delta H_S - \Delta H_T)/\Delta H_S] \times 100$$

where ΔH_S and ΔH_T are denoted for enthalpies of untreated sample and treated sample which are scanned under the same experimental conditions.

The DSC instrument has a high performance and is sensitive enough to detect the very weak and critical transitions associated with starches having outstanding resolution. It provides reliable and reproducible results with a power compensation strategy for rapid cooling and heating. Besides this, it also gives unparalleled isothermal performance for the isothermal properties of starches. This method

Table 3.4 Methods for measurement of starch gelatinization, retrogradation, and modification

Type of method	Techniques	Principle	Parameters measured	References
Physical method	Turbidity syneresis	Determine the rate and extent of starch retrogradation	Changes in density distribution of gelatinized starch paste	Wang et al. (2006), Ambigaipalan et al. (2013)
	Electrical conductivity method	Electrical conductivity of suspension is measured during starch gelatinization	The points at which a steep increase in electrical conductivity is started and ended are denoted as T_i and T_f	Chaiwanichsiri et al. (2001)
Chemical method	Blue value determination	Determine the concentration of starch from the intensity of the blue iodine complex	Measures absorbance of a starch-iodine complex in aqueous solution. The blue color of starch-iodine complex and retrogradation in starch pastes are principally related with amylose, it is likely to use the vanishing of the blue color to observe the retrogradability of aqueous starch pastes	Jacobson et al. (1997)
Enzymatic method	Enzymatic digestion	During the gelatinization process, the starch granules become solubilized and consequently susceptible to enzyme attack	It measures the degree of starch gelatinization. The degree of starch gelatinization is directly proportional to the degree of enzymatic hydrolysis	Liu and Han (2012)
Thermal analysis	Differential scanning colorimetry (DSA)	Study the thermal and gelatinization behavior, glass transition temperature, and crystallization of starches	Transition temperatures (T_o (onset), T_p (Peak), T_c (completion)), enthalpy change (ΔH) of crystallite melting	Wootton and Bamunuarachchi (1980)

	Different thermal analysis (DTA)	Temperature difference (ΔT) between the test sample and an inert reference sample under controlled conditions of heating or cooling is recorded as a function of temperature or time	Heat absorbed or emitted by a chemical system is determined	Pigłowska et al. (2020)
	Thermogravimetric analysis (TGA)	Measures the mass of a sample as it is heated, cooled, or held at a constant temperature in a defined atmosphere	The weight of the sample is plotted against temperature or time to illustrate thermal transitions in the material—such as loss of solvent and plasticizers in polymers, water of hydration in inorganic materials, and, finally, decomposition of the material	Tian et al. (2011), Liu et al. (2013)
Rheological analysis	Rapid visco analysis (RVA)	Measures pasting viscosities during programmed heating and cooling of a starch suspension	The increase in viscosity is represented in a Brabender plot or profile. The viscosity of starch is generally measured by using a rapid visco Analyser or amylograph	Jane et al. (1999), Hoover and Manuel (1995)
	Texture profile analysis (TPA)	Measures textural properties of starch pastes and gels	Hardness, cohesiveness, adhesiveness, elasticity, and brittleness	Rosenthal (2010)
Spectroscopic analysis	Fourier-transform infrared (FTIR) spectroscopy	Describe the organization and structure of starch at various water contents	The ratio of the heights of the bands at 1047 cm^{-1} and 1022 cm^{-1} is used to express the amount of ordered crystalline to amorphous starches	Van Soest et al. (1995), Xie et al. (2006)
	Near infrared (NIR) spectroscopy	Molecular changes of starch gel during gelatinization and retrogradation	Calibration models for the measurement of thermal and retrogradation properties were built from the spectra	Mariotti et al. (2009), Bao et al. (2007)

(continued)

Table 3.4 (continued)

Type of method	Techniques	Principle	Parameters measured	References
	RAMAN	Based on the discrete vibrational transitions that occur in the ground electronic state of molecules, which correspond to various stretching and bending deformation modes of individual chemical bonds	Measures internal vibrations of molecules	Li-Chan (1996)
	Nuclear magnetic resonance (NMR) spectroscopy	Based on the magnetic properties of some nuclei of starch such as ^1H and ^{13}C. It can be used for the quantification of structural features in starch	^1H NMR analyzes the mobility of starch polymer chains ^{13}C-NMR is widely used to elucidate the granular structure of starch	Gidley (1992)
	Electron paramagnetic resonance (EPR) or Electron spin resonance spectroscopy (ESR)	Study chemical species that have one or more unpaired electrons, such as organic and inorganic free radicals or inorganic complexes possessing a transition metal ion	Study radical processes induced by heating of starch in the temperature range 150–250 °C	Krupska et al. (2012)
X-ray diffraction	Wide-angle X-Ray diffraction (WAXD)	X-ray diffraction patterns and diffraction intensities are altered when starch gelatinizes that provide an indication of relative starch crystallinity Used for determination of short-range information (less than 1 nm) such as crystallinity in starch granules	Determines the relative crystallinity of a starch sample based on peak areas (X_{rA}) or peak heights (X_{rH})	Van Soest et al. (1995), Jenkins et al. (1994)
	Small-angle X-Ray scattering (SAXS)	Usually used for the determination of longer range information (1–100 nm) such as micelle structure	Repetitive crystalline and amorphous lamellae	Jenkins and Donald (1998), Suzuki et al. (1997)

Microscopic	Scanning electron microscopy (SEM)	The resolution of SEM allows precise size determination of the granules	Provide qualitative information on molecular order and architecture of starch granules	Li et al. (2006)
	Confocal laser scanning microscopy (CLSM)	The light source is replaced by a laser and a pinhole in the back focal plane is given which improves the depth of focus	Used for in situ visualization of starch granules during gelatinization. CLSM is used to study the internal structure of dense samples in three dimensions	Dürrenberger et al. (2001)
	Polarized light microscopy (PLM)	A light microscope with heating stage attachment is connected to a camera and a computer is used to observe swelling of starch granules upon heating. The photo-element is placed within the hot stage of the microscope provided with a polarizing filter. The images of the heated samples were continuously taken and saved in the computer, and then analyzed by software	Observe the degree and duration of swelling, integrity, size, and loss of birefringence of starch granules undergoing gelatinization	Chatakanonda et al. (2000)
Chromatographic methods	Gel permeation chromatography (GPC)	Size distributions based on elucidation of the macromolecular structure of starch	Determines size and molecular weight distribution of starch	Liu et al. (2013)

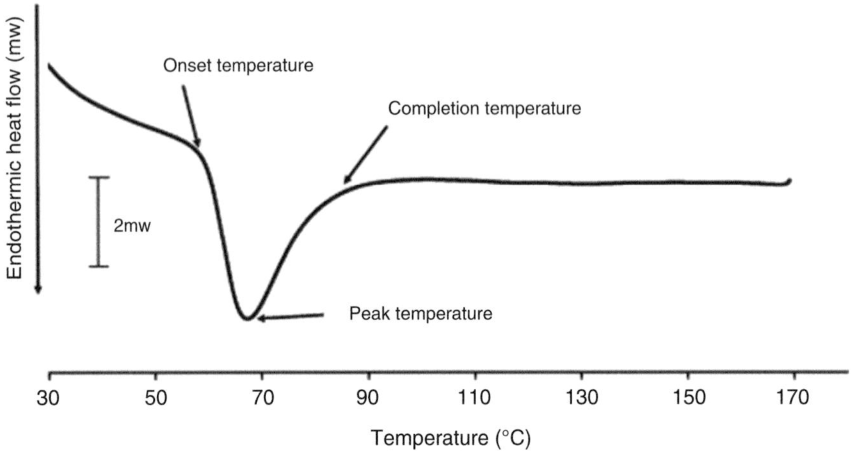

Fig. 3.6 A typical DSC thermogram of starch-water system

provides information on starch gelatinization in a detailed manner and allows determining the enthalpy and the temperature range in which starch gelatinization occurs. This technique is not only quantifying the behavior of starch during the process of gelatinization but also provides quantitative results (Wootton and Bamunuarachchi 1979). The DSC results show an endothermic peak in the temperature region between 54 and 73 °C for different starches known as the gelatinization temperature range (Yu and Christie 2001). The different type of starches expresses their own characteristics transition during heating. The heating temperature and intensity of transitions are important indicators of the gelatinization process. DSC also provides evidence associated with cooking, textural, and digestive properties of the starches (Sinchina 2000).

3.13.2 Enzymatic Digestion Method

The enzymatic method is based on the principle that during the gelatinization process the starch granules become solubilized and consequently susceptible to enzyme attack. The degree of starch gelatinization (DSG) is directly proportional to the degree of enzymatic hydrolysis. The enzymatic method is used to measure the degree of starch gelatinization more precisely, in most of the laboratories where differential scanning calorimetry (DSC) equipment is not available (Di Paola et al. 2003; Zhu et al. 2016). In the enzymatic digestion method, amyloglucosidase also known as glucoamylase enzyme was used to hydrolyze the starch into glucose molecules which are measured colorimetrically (Liu and Han 2012). As the DSG is expressed as percent gelatinized starch relative to the total starch, it is determined in two parallel experiments, one for gelatinized starch and the other for total starch. Therefore, for accurate DSG determination, both tests are equally reliable. The gelatinized starch was measured well by the enzymatic method because in freshly made wet samples gelatinized starch is already fully solubilized. However, due to thermal processing, starchy foods or feeds are frequently cooled or dried, during which some of the gelatinized starch retrogrades progressively into semi-crystalline aggregates differ in form from native starch granules (Tako et al. 2014). Therefore, the gelatinized starch is fully resolubilized before enzymatic analysis (Liu and Han 2012). In some developing methods, researchers used dry products for measuring gelatinized starch while others ignore the requirement of fully resolubilization of gelatinized starch by pretreatments. These methods are tedious and error-prone. However, the starch in most products is partially gelatinized. Therefore, these products contain both types of starch, i.e., gelatinized and native starch. Hence, an appropriate pretreatment is required to enable complete resolubilization of gelatinized starch but minimal solubilization of native starch in the same sample to determine the gelatinized starch more accurately. The total starch determination method also requires chemical and thermal pretreatments to obtain fully solubilized starch samples. Several chemical reagents were used at different concentrations to obtain the fully solubilized starch (Zhu et al. 2016; Liu and Han 2012; Marconi et al. 2004). The thermal treatments also vary with temperature, duration, and mode of application (Zhu et al. 2016). To measure the accurate DSG, total starch measurement is equally important in the same sample. The enzymatic digestion method used to measure the DSG is complex and required not only measuring the contents of both gelatinized and total starch but also determining a correction factor for native starch for sophisticated DSG calculations (Liu and Liu 2020). The enzymatic method used for the measurement of DSG is time-consuming (Di Paola et al. 2003).

3.13.3 Polarized Light Microscopy

The polarized light microscope is used to study the quantitative and qualitative properties of starch granules with a high degree of sensitivity. The starch granules swell, hydrate, and leach amylose upon heating and during continuous heating, the starch granules burst at the end. Upon cooling, a starch paste and gel are formed in which granule leftovers or remnants are embedded within the molecular network. The granule remnants contribute to the properties of starch paste or gel, and, therefore, their visualization is important using microscopic techniques. Starch has a crystal structure that produces birefringent light in presence of polarized light at 500 nm wavelengths (Waigh et al. 2000). For this purpose, a light microscope with a heating stage attachment is connected to a camera and a computer is used to observe swelling of starch granules upon heating. The photo-element is placed within the hot stage of the microscope provided with a polarizing filter. The images of the heated samples were continuously taken and saved in the computer and then analyzed by software. The microscope with a hot stage permits one to observe the degree and duration of swelling, integrity, size, and loss of birefringence of starch granules undergoing gelatinization. The loss of birefringence is measured by observing the decrease in the intensity of polarized light during the heating of the starch. This technique provides a means for rapid screening of starch and initial gelatinization temperatures (Leszczynski and Wroclaw 1987). The rupture temperature and the birefringence endpoint temperature of starch granules were recorded. These experiments are conducted under both normal light and polarized light modes to find the swelling and loss of birefringence (Chatakanonda et al. 2000). The DSG is obtained through the conversion of digital images by using professional image analysis software on the target area of concern. DSG is defined as the percentage of integral optical density (IOD) values that are decreased at a specific temperature from the initial stage of starch. The IOD value of each digital image is calculated by the image analysis software. DSG based on the IOD value is calculated as follows:

$$\text{Background correction } C = A - B \tag{3.1}$$

$$\text{DSG } (\%) = (1 - C/C_0) \times 100 \tag{3.2}$$

where, A and B are used to denote the original IOD value which is calculated from the original digital image and the background IOD value which is calculated when birefringence disappears completely from the original digital image, respectively. C is used to denote the real IOD value of birefringence light resulting from the specific crystal structure of starch in the digital image while C_0 gives the initial real IOD value of birefringence light derived from the specific crystal structure of starch in the initial digital image.

During the whole gelatinization process, the starch granules represent three kinds of states: totally gelatinized (loss of birefringence light), partially gelatinized (decrease of light intensity and area), and ungelatinized (without a change in light intensity) because every starch granule is not gelatinized simultaneously (Li et al. 2013). The IOD method used to measure DSG can dynamically determine starch crystallization and accurately reflects the state of the starch gelatinization process. This method measures two parameters: optical density and area of birefringence. Hence, it provides the DSG of the starch granule especially together with the partially gelatinized zone that can be detected and quantitatively analyzed (Li et al. 2013).

3.13.4 Confocal Laser Scanning Microscopy

Confocal laser scanning microscopy (CLSM), advanced light microscopy, is a powerful tool used for in situ visualization of starch granules during the gelatinization process (Dürrenberger et al. 2001). In this microscopy, the light source is replaced by a laser and a pinhole in the back focal plane is given which improves the depth of focus. CLSM is used to study the internal structure of dense samples in three dimensions. This microscopy detects in-focus regions only, the out-of-focus parts appeared in black in the image. It offers the opportunity to analyze the structure and surface of samples in the epi-reflection mode. In this mode, the laser light which is reflected from the sample is collected as a signal (Van de Velde et al. 2002). The gelatinization behavior of starch granules is studied by monitoring the swelling and expansion of the granules with time. The CLSM imaging technique is suitable for detecting the surface irregularities on the granule surface. This microscopy technique is used to take images of a single focal plane of the sample. This helps in visualizing cross-sections of starch granules without sectioning techniques (Ohtani et al. 2000). To study the gelatinization process, the starch suspensions are heated for a different time, followed by rapid cooling on melting ice and visualized by CSLM. Samples can be visualized in the hydrated state; therefore, the gelatinization process can be followed in situ by CLSM.

3.13.5 Electron Microscopy

Electron microscopy is mainly used to study structural characteristics such as starch granule ultrastructure and granule morphology. The microscopic techniques including scanning electron microscopy (SEM) and transmission electron microscopy (TEM) provide only qualitative information, a higher level of molecular order, and architecture of starch granules. The resolution of SEM allows precise size determination of the granules. However, it cannot provide the differentiation between starch and other particles (Lindeboom et al. 2004; Li et al. 2006). The major limitation of electron microscopy is the sample preparation which includes drying and metal coating, which limits the visualization of starches in their original environment. The imaging of gelatinized starch or starch pastes is much more complicated due to the loss of structure (Van de Velde et al. 2002).

3.13.6 X-Ray Diffraction Method

X-ray diffraction method is used to determine the molecular organization of starch granules. During the process of starch gelatinization, X-ray diffraction patterns and diffraction intensities are altered when starch gelatinizes; this indicates relative starch crystallinity (Varriano-Marston et al. 1980). The procedure to determine the relative crystallinity was given by Van Soest et al. (1995). The X-ray diffraction patterns of different starches such as cereals, legumes, and tubers showed different spectra. This is mainly due to their different microscopic structures that result in different physical properties such as gelatinization temperature (Singh et al. 2003; Betancur-Ancona et al. 2004). The X-ray diffractograms are recorded with a diffractometer and the scattered X-ray radiations are recorded by a proportional moving detector over a 4° and 40° (2θ) angular range. A Matlab script is used to calculate the total area below the characteristic peak that is denoted by A_t, the total area below the peak minus the area below the baseline that is denoted by A_c, the height of the peak denoted by H_t, and the total height of the characteristic peak minus the height of the baseline denoted by H_c at the diffraction angle at 2θ. The ratios $R_A = A_c/A_t$ and $R_H = H_c/H_t$ are calculated for native starch and completely gelatinized starch. The relative crystallinity of a starch sample based on peak areas (X_{rA}) or peak heights (X_{rH}) is determined with the following equations:

$$X_{rA} = (R_A)_s/(R_A)_n; X_{rH} = (R_H)_s/(R_H)_n$$

where, the subscript "n" and "s" is used to denote the native and gelatinized starch, respectively. The calculated relative crystallinity is further used to measure the degree of starch gelatinization (DSG) by using the formula as follows:

$$\mathrm{DSG} = (1 - X_r)$$

where X_r may be either peak areas (X_{rA}) or peak heights (X_{rH}).

3.13.7 Nuclear Magnetic Resonance Spectroscopy

Nuclear Magnetic Resonance (NMR) spectroscopy method is based on the magnetic properties of some nuclei of starch such as ^{1}H and ^{13}C. It can be used for the quantification of structural features in starch. NMR is a useful technique used to measure the molecular order of individual helices at a short-distance range. The X-ray diffraction method monitors crystal structure and measures the relative crystallinity in the starch samples but it is only sensitive to long-range orders and not suitable for fewer crystalline samples. The technique of solid-state ^{13}C-NMR is widely used to elucidate the granular structure of starch. The determination of relaxation times (T_1 and T_2) permits the solid to be separated from the liquid, and, therefore, the movement and microenvironment of water molecules in the starch granule can be determined. The double helix content of starch is determined based on information about the chemical shift (value) and the peak area. The double helix content of starch is found in the range of 40–50% which is greater than starch crystallinity which is in the range of 15–45%. It indicates that granular starch has a significant amount of noncrystalline double helices (Gidley 1992). Besides this, it also provides the degree of branching of starches and amylopectins (Gidley 1985).

3.13.8 Fourier-Transform Infrared Spectroscopy

Fourier-transform infrared (FTIR) spectroscopy is used to describe the organization and structure of starch in various water contents. It is a fast and direct method based on infrared spectroscopy. It is also sensitive to the changes in a molecular structure at short-range order. It provides C–C, C–O, C–H stretching, and C–OH bending in the region of 1300–800 cm^{-1}. The absorbance band observed at 1047 cm^{-1} is characteristic of the amount of ordered or crystalline starches while the band at 1022 cm^{-1} is typical for amorphous starch (Van Soest et al. 1995). Therefore, the ratio of the heights of the bands at 1047 cm^{-1} and 1022 cm^{-1} is used to express the amount of ordered crystalline to amorphous starches (Van Soest et al. 1995; Xie et al. 2006).

The major limiting factor of this method is the overlapping, poorly resolved bands in the spectrum of starch.

3.13.9 Viscosity Method

The viscosity method of starch gelatinization is based on the swelling power and water solubility index (WSI) of starches. The swelling power of starch is defined as the water-holding capacity of starch at different temperatures. It is calculated by the ratio of wet sediment to the initial weight of dry starch at a certain temperature during gelatinization. During the process of gelatinization or heating, the starch-water suspension shows increased viscosity due to increased swelling and water solubility index. Thus, monitoring the viscosity of the starch-water complex can also be used to follow the gelatinization process. The increase in viscosity is represented in a Brabender plot or profile. Physical analytical approaches are generally applied for monitoring the effect of heat on starch. Rapid visco analyzer (RVA), viscograph, and amylograph are some of the analytical methods based on empirical rheological systems that are used to study temperature-dependent viscosity changes with an excess of water (Jane et al. 1999; Hoover and Manuel 1995). Rapid visco analyzer measures the changes in viscosity with varied shear, heating, and/or cooling rate. In RVA, viscosity measurement starts at a temperature below gelatinization, i.e., 30 °C. After attainment of gelatinization temperature, excessive heat partially ruptures the starch granule and increases the viscosity. A typical RVA curve interprets pasting temperature (Tp) which describes the onset temperature of gelatinization and swelling of starch granules, peak viscosity represents the maximum amount of gelatinization, enzymatic and sheer destruction of starch granules, hot paste viscosity is the minimum viscosity of starch paste, breakdown depicts the loss in viscosity, setback illustrates the paste hardening, and final viscosity indicates viscosity after cooling to 50 °C (Fig. 3.7). The changes in viscosity of potato starch during heating from 20 to 70 °C were analyzed during the gelatinization (Chaiwanichsiri et al. 2001). They observed that the viscosity of starch-water suspension is not changed until the temperature reached 64 °C. At the temperature of 64 °C, a sharp increase in the viscosity is recorded due to gelatinization. At this temperature, the free ions and polymers are leached out from the starch granule which is responsible for a sharp increase in viscosity of the starch-water suspension.

In contrast, in an amylograph the temperature sensor is placed directly in the suspension. Amylograph also measures the gelatinization temperature and records the viscosity.

3.13.10 Electrical Conductivity Method

The electrical conductivity measurement found a very effective method to monitor the process of starch gelatinization. During the gelatinization process of potato starch, the electrical conductivity of suspension is measured at the temperature

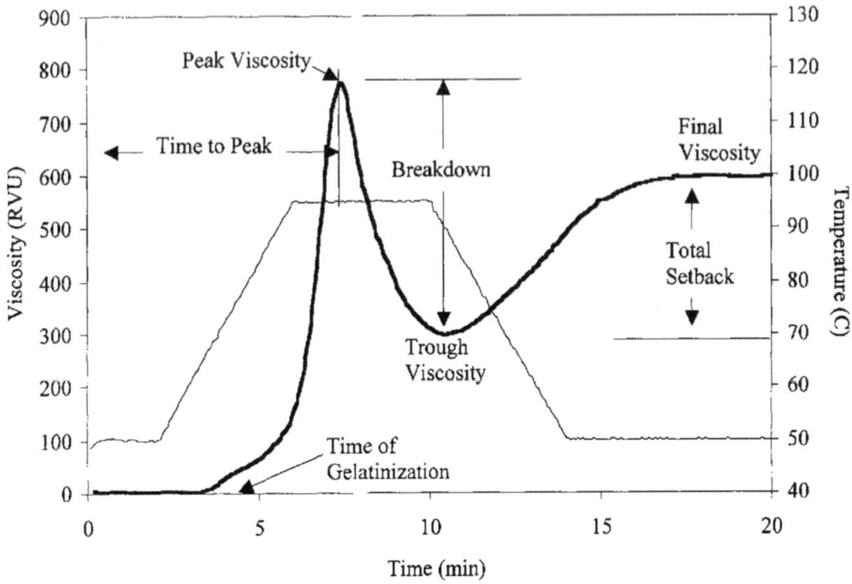

Fig. 3.7 A typical Rapid Viscogram

range from 20 to 95 °C in the frequency range of 200 kHz–20 MHz (Chaiwanichsiri et al. 2001). They found that the frequency has no substantial effect on the electrical conductivity while temperature has a serious effect on electrical conductivity. They reported that the electrical conductivity gradient of the starch-water complex is increased gradually and linearly with an increase in temperature from 20 to 62 °C. After that, the gradient is increased more rapidly and ended at 72 °C. The points at which a steep increase in electrical conductivity is started and ended are denoted as T_i and T_f, respectively. A similar pattern of electrical conductivity gradient is observed for sweet potato, wheat, maize, rice, mung bean, tapioca, and arrowroot starches. The increase in electrical conductivity below T_i is related to ion transfer or molecular diffusion of ions in the background solution because at this temperature starch granules began to swell and the ions inside the granules started to release. As the temperature reached T_f, starch granules are swelled enough to break down and the ions are also completely released into the suspension.

3.14 Functions and Application of Starch in Processed Food Products

Starches deliver an extensive range of functions in processed food products due to their compositional differences, structural variations, and physicochemical nature. Modified starches are being used as a natural additive to acquire desired texture and stability attributes and designing innovative food products. Modified starches can be used to improve the functional characteristics in various food products, for example, improving the cooking properties, decreasing the retrogradation tendency, improving the freeze-thaw stability of paste, decreasing the paste and/or gel syneresis properties, increasing the paste and/or gel clarity, increasing the paste and/or gel textures, improving the film-forming, improving the adhesion properties, and improving the hydrophobic group (for emulsion stabilization). A combination of different native and/or modified starches is commercially added to processed food products to achieve the desired functional characteristics.

Cross-linking improves the tolerance to heat, acid, and shear, such type of modification allows shorter processing time due to faster heat penetration. These starches are most suited for their viscofier and texture improving properties in ambient stable products, bottled sauces and gravies, sterilized soups and sauces, bakery, and dairy products. Cross-linked starches display additional resistance to baking temperatures of 160–230 °C therefore can be used in a variety of baked products such as cakes, breads, and biscuits (Korma et al. 2016; Hung and Morita 2004). Chemically substituted starches (starch acetates) express exceptional chill and freeze-thaw stability and are easy to cook in high solid matrices. Such types of starches exhibit improved paste clarity, increased degree of swelling or hydration capacity, and lower gelatinization temperature reduced retrogradation tendency. Refrigerated and frozen foods and high solid pastes and toppings are the most suitable medium for such types of starches (Mason 2009; Ackar et al. 2015). Hydroxypropylated starch expresses enhanced clarity, better viscosity, reduced syneresis, and good freeze-thaw stability and is used in products such as gravies, dips, sauces, fruit fillings, and puddings. Acid hydrolyzed and enzymatically hydrolyzed starches are better suited for enhanced textural properties. These starches exhibit lower paste viscosity under cold and hot food matrices. Jams, jellies, pastilles, mayonnaise, and salad dressings are the products for applications of such starches (Mason 2009). Oxidized starches improve the adhesion properties in coatings and create soft stable gels at higher levels. Battered meat, poultry, and fish products and confectionery products are suitable matrices for such types of starches (Korma et al. 2016). Enzymatically converted starches are suitably used for fat mimetics, flavor carriers, and dry mix fillers. Lipophilic substitution of starches improves the hydrophobicity of starches to make them appropriate for usage in emulsions for better stability. Therefore, such types of starches are appropriately utilized in salad dressings, beverages, and flavor encapsulating agents. Pre-gelatinized starches are precooked starch that gives properties of cold-water thickening. Instant soups, sauces, dressing, puddings, desserts, bakery mixes, breakfast mixes, and pasta products utilized such types of starches due to their thickening properties at low temperatures. Thermally treated starches show improved tolerance to heat, acid, and shear. Thermally stable starches find application in ambient stable products, bottled sauces, sterilized soups, and sauces (Taggart 2004; Abbas et al. 2010; Egharevba 2019).

3.15 Conclusion

The concept of starch gelatinization is explained by several theories and underlying mechanisms to comprehend the sequential structural changes. With the progression of time, several theories based on different principles explained the process of starch gelatinization. Also, the concept of starch modification is age-old; however, the area still attracts significant attention of researchers and food processors due to the introduction of innovative and advanced processing techniques such as non-thermal processing technologies. Although the effect of different non-thermal technologies has been studied by various researchers, additional exploration is essential to establish detailed information on the properties required for a particular application. It is essential to design and optimize the process parameters for these technologies to ensure the efficiency of the process. Understanding the mechanism of action, the effect of various process parameters, and various interactions need to be prioritized. It is important to ensure the repeatability and reproducibility of results in order to expand the industrial value of starches modified by these advanced methods. The scale-up of such processes also needs extensive input to achieve commercial success. Secondly, the limitations of starches modified by a single alteration technique can be overcome by the synergistic effect of dual modification. Appropriate selection of methods and order of treatment in dual modification with suitable process design and controlled parameters may provide an array of structural alterations and desired functional properties. A detailed comprehension of energy necessities, process economics, dual technologies, product safety parameters, and end-user approval are obligatory to integrate the successful results on a commercial basis.

References

Abbas KA, Khalil SK, Hussin ASM (2010) Modified starches and their usages in selected food products: a review study. J Agric Sci 2(2):90–100

Ackar D, Babic J, Jozinovic A et al (2015) Starch modification by organic acids and their derivatives: a review. Molecules 20(10):19554–19570

Ai Y, Jane J (2015) Gelatinization and rheological properties of starch. Starke 67(3–4):213–224

Alissandratos A, Halling PJ (2012) Enzymatic acylation of starch. Bioresour Technol 115:41–47

Ambigaipalan P, Hoover R, Donner E et al (2013) Retrogradation characteristics of pulse starches. Food Res Int 54:203–212

Arts SJHF, Mombarg EJM, van Bekkum H et al (1997) Hydrogen peroxide and oxygen in catalytic oxidation of carbohydrates and related compounds. Synthesis 6:597–613

Ashogbon AO (2021) Dual modification of various starches: synthesis, properties and applications. Food Chem 342:128325

Ashwar BA, Gani A, Wani IA et al (2016) Production of resistant starch from rice by dual autoclaving-retrogradation treatment: in vitro digestibility, thermal and structural characterization. Food Hydrocoll 56:108–117

Babu AS, Mohan RJ, Parimalavalli R (2019) Effect of single and dual-modifications on stability and structural characteristics of foxtail millet starch. Food Chem 271:457–465

Bao J, Shen Y, Jin L (2007) Determination of thermal and retrogradation properties of rice starch using near-infrared spectroscopy. J Cereal Sci 46(1):75–81

Basilio-Cortés UA, González-Cruz L, Velazquez G et al (2019) Effect of dual modification on the spectroscopic, calorimetric, viscosimetric and morphological characteristics of corn starch. Polymers (Basel) 11(2):333
Becker A, Hill SE, Mitchell JR (2001) Relevance of amylose-lipid complexes to the behaviour of thermally processed starches. Starke 53:121–130
Belitz HD, Grosch W, Schieberle P et al (2009) Food chemistry. Springer, Berlin
BeMiller JN (2011) Pasting, paste, and gel properties of starch–hydrocolloid combinations. Carbohydr Polym 86:386–423
Benavent-Gil Y, Rosell CM (2017) Morphological and physicochemical characterization of porous starches obtained from different botanical sources and amylolytic enzymes. Int J Biol Macromol 103:587–595
Benczedi D, Tomka I, Escher F (1998) Thermodynamics of amorphous starch–water systems. 2. Concentration fluctuations. Macromolecules 31:3062–3074
Berski W, Ptaszek A, Ptaszek P et al (2011) Pasting and rheological properties of oat starch and its derivatives. Carbohydr Polym 83(2):665–671
Betancur-Ancona D, Gallegos-Tintore S, ChelGuerrero L (2004) Wet-fractionation of Phaseoluslunatus seeds: partial characterization of starch and protein. J Sci Food Agric 84: 1193–1201
Bie P, Pu H, Zhang B et al (2016) Structural characteristics and rheological properties of plasma-treated starch. Innov Food Sci Emerg Technol 34:196–204
Biswas A, Shogren RL, Selling G et al (2008) Rapid and environmentally friendly preparation of starch esters. Carbohydr Polym 74(1):137–141
Bonazzi C, Dumoulin E, Raoult-Wack A et al (1996) Food drying and dewatering. Dry Technol 14: 2135–2170
Brasoveanu M, Nemtanu MR (2014) Behaviour of starch exposed to microwave radiation treatment. Starke 66(1–2):3–14
Brummer T, Meuser F, Lengerich BV et al (2002) Effect of extrusion cooking on molecular parameters of corn starch. Starke 54:1–8
Cao M, Gao Q (2020) Effect of dual modification with ultrasonic and electric field on potato starch. Int J Biol Macromol 150:637–643
Çatal H, Ibanoğlu S (2012) Ozonation of corn and potato starch in aqueous solution: effects on the thermal, pasting, and structural properties. Int J Food Sci Technol 47(9):1958–1963
Çatal H, Ibanoğlu S (2014) Effect of aqueous ozonation on the pasting, flow, and gelatinization properties of wheat starch. LWT-Food Sci Technol 59(1):577–582
Chaiwanichsiri S, Ohnishi S, Suzuki T et al (2001) Measurement of electrical conductivity, differential scanning calorimetry and viscosity of starch and flour suspensions during gelatinization process. J Sci Food Agric 81:1586–1591
Chan HT, Bhat R, Karim AA (2009) Physicochemical and functional properties of ozone-oxidized starch. J Agric Food Chem 57(13):5965–5970
Chan HT, Leh CP, Bhat R et al (2011) Molecular structure, rheological and thermal characteristics of ozone-oxidized starch. Food Chem 126(3):1019–1024
Chang SM, Liu LC (1991) Retrogradation of rice starches studied by differential scanning calorimetry and influence of sugars, NaCl, and lipids. J Food Sci 56:564–566
Chatakanonda P, Varavinit S, Chinachot P (2000) Relationship of gelatinization and recrystallization of cross-linked rice to glass transition temperature. Cereal Chem 77(3):315–319
Chen X, He X, Huang Q (2014) Effects of hydrothermal pretreatment on subsequent octenylsuccinic anhydride (OSA) modification of cornstarch. Carbohydr Polym 101:493–498
Chotipratoom S, Choi SH, Choi HW et al (2014) High Hydrostatic Pressure (HHP)-assisted starch modification: acid hydrolysis, hydroxypropylation, acetylation, cross-linking and cationization. J Appl Glycosci 61(2):31–34
Chung HJ, Liu Q, Hoover R (2009) The impact of annealing and heat–moisture treatments on rapidly digestible, slowly digestible and resistant starch levels in native and gelatinized corn, pea and lentil starches. Carbohydr Polym 75:436–447

Collado LS, Corke H (1999) Heat-moisture treatment effects on sweet potato starches differing in amylose content. Food Chem 65:339–346

Collar C, Rosell CM (2013) Bakery and confectioneries. In: Chandrasekaran M (ed) Valorization of by products from plant-based food processing industries. CRC Press, Taylor and Francis Group, Boca Raton, FL, pp 554–582

Colussi R, Kringel D, da Kaur L et al (2019) Dual modification of potato starch: effects of heat-moisture and high pressure treatments on starch structure and functionalities. Food Chem 318: 126475

Cooke D, Gidley MJ (1992) Loss of crystalline and molecular order during starch gelatinisation: origin of the enthalpic transition. Carbohydr Res 227:103–112

Coutinho PM, Henrissat B (1999) Carbohydrate-active enzymes: an integrated database approach. In: Gilbert HJ, Davies GJ, Henrissat B, Svensson B (eds) Recent advances in carbohydrate bioengineering. Royal Society of Chemistry, Cambridge, pp 3–12

Deka D, Sit N (2016) Dual modification of taro starch by microwave and other heat moisture treatments. Int J Biol Macromol 92:416–422

Di Paola RD, Asis R, Aldao MAJ (2003) Evaluation of the degree of starch gelatinization by a new enzymatic method. Starke 55:403–409

Donald AM, Waigh TA, Jenkins PJ et al (1997) Internal structure of starch granules revealed by scattering studies. In: Frazier PJ, Donald AM, Richmond P (eds) Starch: structure and functionality. The Royal Society of Chemistry, Cambridge, pp 172–179

Dürrenberger MB, Handschin S, Conde-Petit B et al (2001) Visualization of food structure by confocal laser scanning microscopy (CLSM). LWT-Food Sci Technol 34(1):11–17

Egharevba HO (2019) Chemical properties of starch and its application in the food industry. In: Emeje M (ed) Chemical properties of starch. Intech Open, London. https://doi.org/10.5772/intechopen.87777

Fadda C, Sanguinetti AM, Del Caro A et al (2014) Bread staling: updating the view. Compr Rev Food Sci Food Saf 13:473–492

Fu Z, Chen J, Luo SJ et al (2015) Effect of food additives on starch retrogradation: a review. Starke 67:69–78

Gao S, Liu H, Sun L et al (2019) The effects of dielectric barrier discharge plasma on physicochemical and digestion properties of starch. Int J Biol Macromol 138:819–830

Germani R, Ciacco CF, Rodriguez-Amaya DB (1983) Effect of sugars, lipids and type of starch on the mode and kinetics of retrogradation of concentrated corn starch gels. Starke 35:377–381

Gidley M (1985) Quantification of the structural features of starch polysaccharides by NMR spectroscopy. Carbohydr Res 139:85–93

Gidley M (1992) Nuclear magnetic resonance analysis of cereal carbohydrates. In: Alexander RJ, Zobel HF (eds) Developments in carbohydrate chemistry. The American Association of Cereal Chemists, St. Paul, MN, pp 163–192

Goodfellow BJ, Wilson RH (1990) A Fourier transform IR study of the gelation of amylase and amylopectin. Biopolymers 30:1183–1189

Gray VM, Schoch TJ (1962) Effect of surfactants and fatty adjuncts on the swelling and solubilisation of granular starches. Starke 14:239

Greenwood CT (1979) Observations on the structure of the starch granule. In: Blanshard JMV, Mitchell JR (eds) Polysaccharides in food. Butterworths and Co. (Publishers) Ltd, London, pp 129–138

Grgić I, Ačkar Đ, Barišić V et al (2019) Non-thermal methods for starch modification - a review. J Food Process Preserv 43(12):1–13

Gudmundsson M, Eliasson AC (1990) Retrogradation of amylopectin and the effects of amylose and added surfactants/emulsifiers. Carbohydr Polym 13:295–315

Guo L, Deng Y, Lu L et al (2019) Synergistic effects of branching enzyme and transglucosidase on the modification of potato starch granules. Int J Biol Macromol 130:499–507

Han Z, Zeng XA, Yu SJ et al (2009a) Effects of pulsed electric fields (PEF) treatment on physicochemical properties of potato starch. Innov Food Sci Emerg Technol 10(4):481–485

Han Z, Zeng XA, Zhang BS et al (2009b) Effects of pulsed electric fields (PEF) treatment on the properties of corn starch. J Food Eng 93(3):318–323

Hazarika BJ, Sit N (2015) Effect of dual modification with hydroxypropylation and cross-linking on physicochemical properties of taro starch. Carbohydr Polym 140:269–278

Hirsch JB, Kokini JL (2002) Understanding the mechanism of cross-linking agents ($pocl_3$, stmp, and epi) through swelling behavior and pasting properties of cross-linked waxy maize starches. Cereal Chem 79(1):102–107

Hodge S, Osman M (1976) Food chemistry. In: Fennema OR (ed) Principles of food sciencc. Part 1. Mercel Dekker, New York, pp 102–114

Hoover R (2001) Composition, molecular structure, and physicochemical properties of tuber and root starches: a review. Carbohydr Polym 45:253–267

Hoover R, Manuel H (1995) A comparative study of the physicochemical properties of starches from two lentil cultivars. Food Chem 53:275–284

Hoover R, Manuel H (1996a) Effect of heat–moisture treatment on the structure and physicochemical properties of legume starches. Food Res Int 29:731–750

Hoover R, Manuel H (1996b) The effect of heat–moisture treatment on the structure and physicochemical properties of normal maize, waxy maize, dull waxy maize and amylomaize V starches. J Cereal Sci 23:153–162

Hoover R, Vasanthan T (1994a) The effect of annealing on the physicochemical properties of wheat, oat, potato and lentil starches. J Food Biochem 17:303–325

Hoover R, Vasanthan T (1994b) Effect of heat–moisture treatment on the structure and physicochemical properties of cereal, legume, and tuber starches. Carbohydr Res 252:33–53

Hoover R, Hughes T, Chung HJ et al (2010) Composition molecular structure, properties, and modification of pulse starches: a review. Food Res Int 43:399–413

Hu X, Jia X, Zhi C et al (2019) Improving properties of normal maize starch films using dual-modification: combination treatment of debranching and hydroxypropylation. Int J Biol Macromol 130:197–202

Hung PV, Morita N (2004) Dough properties and bread quality of flours supplemented with cross-linked corn starches. Food Res Int 37(5):461–467

Iacovou M, Lim J, Maningat CC et al (2017) In vivo digestibility of cross-linked phosphorylated (RS4) wheat starch in ileostomy subjects. Bioact Carbohydr Diet Fibre 1(12):25–36

Jacobson MR, Obanni M, BeMiller JN (1997) Retrogradation of starches from different botanical sources. Cereal Chem 74:511–518

Jamalabadi M, Saremnezhad S, Bahrami A et al (2019) The influence of bath and probe sonication on the physicochemical and microstructural properties of wheat starch. Food Sci Nutr 7(7): 2427–2435

Jane J, Chen YY, Lee LF et al (1999) Effects of amylopectin branch chain length and amylose content on the gelatinization and pasting properties of starch. Cereal Chem 76:629–637

Jayakody L, Hoover R (2008) Effect of annealing on the molecular structure and physicochemical properties of starches from different botanical origins - a review. Carbohydr Polym 74(3): 691–703

Jenkins PJ, Donald AM (1998) Gelatinisation of starch: a combined SAXS/WAXS/DSC and SANS study. Carbohydr Res 308:133–147

Jenkins PJ, Cameron RE, Donald AM (1994) In situ simultaneous small and wide angle X-ray scattering: a new technique to study starch gelatinization. J Polym Sci B Polym Phys 32(8): 1579–1583

Karim AA, Norziah MH, Seow CC (2000) Methods for the study of starch retrogradation. Food Chem 71:9–36

Karwasra BL, Kaur M, Gill BS (2020) Impact of ultrasonication on functional and structural properties of Indian wheat (*Triticum aestivum* L.) cultivar starches. Int J Biol Macromol 164: 1858–1866

Kaur B, Ariffin F, Bhat R et al (2012) Progress in starch modification in the last decade. Food Hydrocoll 26(2):398–404

Kavitha R (1998) Characterization of hydroxypropylated potato starch. Carbohydr Polym 37(2): 115–121

Kim JS, Cha SS, Kim HJ et al (1999) Crystal structure of a maltogenic amylase: provides insights into a catalytic versatility. J Biol Chem 274:26279–26286

Korma SA, Alahmad K, Niazi S et al (2016) Chemically modified starch and utilization in food stuffs. Int J Nutr Food Sci 5(4):264–272

Krupska A, Wieckowski AB, Slominska L et al (2012) Influence of heating time and pressure treatment of potato starch on the generation of radicals: EPR studies. Carbohydr Polym 89:54–60

Landerito NA, Wang YJ (2005) Preparation and properties of starch phosphates using waxy, common, and high-amylose corn starches. II. Reactive extrusion method. Cereal Chem 82(3): 271–276

Lee KH, Kim HJ, Woo KS et al (2016) Evaluation of cold plasma treatments for improved microbial and physicochemical qualities of brown rice. LWT-Food Sci Technol 73:442–447

Leloup VM, Colonna P, Ring SG et al (1992) Microstructure of amylose gels. Carbohydr Polym 18: 189–197

Leszczynski W, Wroclaw (1987) New methods for determination of starch gelatinization temperatures. Starke 39(11):375–378

Lewandowicz G, Fornal J, Walkowski A et al (1998) Effect of microwave radiation on physico-chemical properties and structure of potato and tapioca starches. Carbohydr Polym 34(4): 213–220

Lewandowicz G, Walkowski A, Gawęcki J (1999) Starch phosphates—technological and nutritional characteristics. Food Ind 53(3):34–36

Li JH, Guiltinan MJ, Thompson DB (2006) The use of laser differential interference contrast microscopy for the characterization of starch granule ring structure. Starke 58:1–5

Li Q, Xie Q, Yu S et al (2013) New approach to study starch gelatinization applying a combination of hot-stage light microscopy and differential scanning calorimetry. J Agric Food Chem 61(6): 1212–1218

Li W, Cao F, Fan J et al (2014) Physically modified common buckwheat starch and their physicochemical and structural properties. Food Hydrocoll 40:237–244

Li MN, Xie Y, Chen HQ et al (2019) Effects of heat-moisture treatment after citric acid esterification on structural properties and digestibility of wheat starch, A- and B-type starch granules. Food Chem 272:523–529

Li-Chan ECY (1996) The applications of Raman spectroscopy in food science. Trends Food Sci Technol 7:361–370

Lian XJ, Zhu W, Wen Y et al (2013) Effects of soy protein hydrolysates on maize starch retrogradation by IR spectra and ESI-MS analysis. Int J Biol Macromol 59:143–150

Lima DC, Maniglia BC, Matta Junior MD et al (2021) Dual-process of starch modification: combining ozone and dry heating treatments to modify cassava starch structure and functionality. Int J Biol Macromol 167:894–905

Lindeboom N, Chang PR, Tyler RT (2004) Analytical, biochemical and physicochemical aspects of starch granule size, with emphasis on small granule starches: a review. Starke 56:89–99

Liu K, Han J (2012) Enzymatic method for measuring starch gelatinization in dry products in situ. J Agric Food Chem 60:4212–4221

Liu K, Liu Q (2020) Enzymatic determination of total starch and degree of starch gelatinization in various products. Food Hydrocoll 103:105639

Liu D, Parker ML, Wellner N et al (2013) Structural variability between starch granules in wild type and in ae high-amylose mutant maize kernels. Carbohydr Polym 97:458–468

Liu J, Yang R, Yang F (2015) Effect of the starch source on the performance of cationic starches having similar degree of substitution for papermaking using deinked pulp. Bioresources 10(1): 922–931

López OV, Zaritzkya NE, Garcia MA (2010) Physicochemical characterization of chemically modified corn starches related to rheological behavior, retrogradation and film forming capacity. J Food Eng 100(1):160–168

Łukasiewicz M, Bednarz S, Ptaszek A (2011) Environmental friendly polysaccharide modification—microwave-assisted oxidation of starch. Starke 63(5):268–273

Lund D, Lorenz KJ (1984) Influence of time, temperature, moisture, ingredients and processing conditions on starch gelatinization. Crit Rev Food Sci Nutr 20(4):37–41

Lv S, Gu J, Cao J et al (2015) Effect of annealing on the thermal properties of poly (lactic acid)/ starch blends. Int J Biol Macromol 74:297–303

Majeed T, Wani IA, Hussain PR (2017) Effect of dual modification of sonication and γ-irradiation on physicochemical and functional properties of lentil (*Lens culinaris* L.) starch. Int J Biol Macromol 101:358–365

Marconi E, Messia MC, Palleschi G et al (2004) A maltose biosensor for determining gelatinized starch in processed cereal foods. Cereal Chem 81:6–9

Mariotti M, Sinelli N, Catenacci F et al (2009) Retrogradation behavior of milled and brown rice pastes during ageing. J Cereal Sci 49:171–177

Mason WR (2009) Starch use in foods. In: BeMiller JN (ed) Starch chemistry and technology, 3rd edn. Academic, San Diego, CA, pp 745–795

Miao M, Xiong S, Jiang B et al (2014) Dual-enzymatic modification of maize starch for increasing slow digestion property. Food Hydrocoll 38:180–185

Morrison WR, Tester RF, Snape CE et al (1993) Swelling and gelatinization of cereal starches. IV. Some effects of lipid–complexed amylose and free amylose in waxy and normal barley starches. Cereal Chem 70:385–391

Muzimbaranda C, Tomasik P (1994) Microwaves in physical and chemical modifications of starch. Starke 46(12):469–474

Obadi M, Zhu KX, Peng W et al (2018) Effects of ozone treatment on the physicochemical and functional properties of whole grain flour. J Cereal Sci 81:127–132

Ohtani T, Yoshino T, Hagiwara S et al (2000) High resolution imaging of starch granule structure using atomic force microscopy. Starke 52:150–153

Pałasínski M, Fortuna T, Juszczak L et al (2000) Change in some physico-chemical properties of starch granules induced by heating and microwave radiation. Polish J Food Nutr Sci 9(1):17–22

Pandiselvam R, Kaavya R, Jayanath Y et al (2020) Ozone as a novel emerging technology for the dissipation of pesticide residues in foods–a review. Trends Food Sci Technol 97:38–54

Park KH, Park JH, Lee SY et al (2008) Enzymatic modification of starch for food industry. In: Park KH (ed) Carbohydrate-active enzymes: structure, function and applications. CRC Press, Inc., Boca Raton, FL

Pei-Ling L, Xiao-Song H, Qun S (2010) Effect of high hydrostatic pressure on starches: a review. Starke 62(12):615–628

Pietrzyk S, Fortuna T, Sowa M (2006) Influence of catalysts on effectiveness of starch oxidation process and its physicochemical properties. Food Sci Technol Qual 2(47):69–81

Pigłowska M, Kurc B, Rymaniak L et al (2020) Kinetics and thermodynamics of thermal degradation of different starches and estimation the OH group and H_2O content on the surface by TG/ DTG-DTA. Polymers 12:357

Pinto VZ, Vanier NL, Klein B et al (2012) Physicochemical, crystallinity, pasting and thermal properties of heat-moisture-treated pinhao starch. Starke 64:855–863

Prompiputtanapon K, Sorndech W, Tongta S (2020) Surface modification of tapioca starch by using the chemical and enzymatic method. Starke 72(3–4):1–6

Pukkahuta C, Shobsngob S, Varavinit S (2007) Effect of osmotic pressure on starch: new method of physical modification of starch. Starke 58:78–90

Qiu S, Li Y, Chen H et al (2014) Effects of high-pressure homogenization on thermal and electrical properties of wheat starch. J Food Eng 128:53–59

Raghunathan R, Pandiselvam R, Kothakota A et al (2021) The application of emerging non-thermal technologies for the modification of cereal starches. LWT-Food Sci Technol 138:110795

Rajan A, Sudha JD, Abraham TE (2008) Enzymatic modification of cassava starch by fungal lipase. Ind Crop Prod 27(1):50–59
Ratnayake WS, Otani C et al (2009) DSC enthalpic transitions during starch gelatinization in excess water, dilute sodium chloride, and dilute sucrose solutions. J Sci Food Agric 89(12):2156–2164
Ribota PD, Colombo A, Leon AE, Anon MC (2012) Effects of soy protein on physical and rheological properties of wheat starch. Starke 59:614–623
Roos Y, Karel M (1991) Water and molecular weight effects on glass transitions in amorphous carbohydrates and carbohydrate solutions. J Food Sci 56:1676–1681
Rosenthal AJ (2010) Texture profile analysis-how important are the parameters. J Texture Stud 41: 672–684
Sarko A, Wu HCH (1978) The crystal structures of A, B, and C polymorphs of amylose and starch. Starke 30:73–78
Schirmer M, Jekle M, Becker T (2015) Starch gelatinization and its complexity for analysis. Starke 67(1–2):30–41
Seker M, Sadikoglu H, Ozdemir M et al (2003) Phosphorus binding to starch during extrusion in both single and twin-screw extruders with and without a mixing element. J Food Eng 59(4): 355–360
Shi YC, Seib PA (1992) The structure of four waxy starches related to gelatinization and retrogradation. Carbohydr Res 227:131–145
Siemion P, Kapusniak J, Kozioł JJ (2006) Solid state reaction of starch with thiosemicarbazide. Carbohydr Polym 66(1):104–109
Sinchina WJ (2000) Use of DSC for the characterization of starches. Perkin Elmer Instruments, USA
Singh N, Singh J, Kaur L et al (2003) Morphological, thermal and rheological properties of starches from different botanical sources. Food Chem 81:219–231
Singh J, Colussi R, McCarthy OJ (2016) Potato starch and its modification. In: Singh J, Kaur L (eds) Advances in potato chemistry and technology, 2nd edn. Academic, San Diego, CA, pp 195–247
Słomińska L, Zielonka R, Jarosławski L et al (2015) High pressure impact on changes in potato starch granules. Polish J Chem Technol 17(4):65–73
Smits ALM, Kruiskamp PH, vanSoest JJG et al (2003) The influence of various small plasticisers and malto-oligosaccharides on the retrogradation of (partly) gelatinised starch. Carbohydr Polym 51:417–424
Spies RD, Hoseney RC (1982) Effect of sugars on starch gelatinization. Cereal Chem 59(2): 128–131
Stute R, Klingler RW, Boguslawski S (1996) Effects of high pressures treatment on starches. Starke 48(11–12):399–408
Suzuki T, Chiha A, Yano T (1997) Interpretation of small angle X-ray scattering from starch on the basis of fractals. Carbohydr Polym 34:357–363
Taggart P (2004) Starch as an ingredient: manufacture and implication. In: Elliason AC (ed) Starch in food. CRC Press, Cambridge, pp 363–392
Tako M, Tamaki Y, Teruya T et al (2014) The principles of starch gelatinization and retrogradation. Food Nutr Sci 5:280–291
Tang MC, Copeland L (2007) Analysis of complexes between lipids and wheat starch. Carbohydr Polym 67(1):80–85
Tester RF, Debon SJJ (2000) Annealing of starch: a review. Int J Biol Macromol 27:1–12
Tester RF, Karkalas J, Qi X (2004) Starch - composition, fine structure and architecture. J Cereal Sci 39(2):151–165
Thirumdas R, Kadam D, Annapure US (2017) Cold plasma: an alternative technology for the starch modification. Food Biophys 12(1):129–139
Tian YQ, Yang N, Li Y et al (2010) Potential interaction between β-cyclodextrin and amylose-lipid complex in retrograded rice starch. Carbohydr Polym 80:581–584
Tian Y, Li Y, Xu X et al (2011) Starch retrogradation studied by thermo-gravimetric analysis (TGA). Carbohydr Polym 84(3):1165–1168

Tian Y, Li D, Zhao J (2014) Effect of high hydrostatic pressure (HHP) on slowly digestible properties of rice starches. Food Chem 152:225–229
Tomasik P, Schilling CH (2004) Chemical modification of starch. Adv Carbohydr Chem Biochem 59:175–403
Ulbrich M, Natan C, Flöter E (2014) Acid modification of wheat, potato, and pea starch applying gentle conditions-impacts on starch properties. Starke 66(9–10):903–913
Van de Velde F, van Riel J, Tromp RH (2002) Visualisation of starch granule morphologies using confocal scanning laser microscopy (CSLM). J Sci Food Agric 82(13):1528–1536
Van Soest JJG, De Wit D, Tournois H et al (1994) Retrogradation of potato starch as studied by Fourier-transform infrared-spectroscopy. Starke 46(12):453–457
Van Soest JJG, Tournois H, de Wit D (1995) Short-range structure in (partially) crystalline potato starch determined with attenuated total reflectance Fourier-transform IR spectroscopy. Carbohydr Res 279:201–214
Varriano-Marston E, Ke V, Huang G et al (1980) Comparison of methods to determine starch gelatinization in bakery foods. Cereal Chem 57(4):242–248
Vermeylen R, Goderis B, Delcour JA (2006) An X-ray study of hydrothermally treated potato starch. Carbohydr Polym 64:364–375
Volkert B, Lehmann A, Greco T et al (2010) A comparison of different synthesis routes for starch acetates and the resulting mechanical properties. Carbohydr Polym 79(3):571–577
Waigh TA, Gidley MJ, Komanshek BU et al (2000) The phase transformations in starch during gelatinisation: a liquid crystalline approach. Carbohydr Res 328:165–176
Wang SJ, Copeland L (2013) Molecular disassembly of starch granules during gelatinization and its effect on starch digestibility: a review. Food Funct 4:1564–1580
Wang WJ, Powell AD, Oates CG (1995) Pattern of enzyme hydrolysis in raw sago starch: effects of processing history. Carbohydr Polym 26:91–97
Wang SJ, Gao WY, Jia W et al (2006) Crystallography, morphology and thermal properties of starches from four different medicinal plants of Fritillaria species. Food Chem 96:591–596
Wongsagonsup R, Pujchakarn T, Jitrakbumrung S (2014) Effect of cross-linking on physicochemical properties of tapioca starch and its application in soup product. Carbohydr Polym 101:656–665
Wootton M, Bamunuarachchi A (1979) Application of differential scanning calorimetry to starch gelatinization. I. Commercial native and modified starches. Starke 31(6):201–204
Wootton M, Bamunuarachchi A (1980) Application of differential scanning calorimetry to starch gelatinization. III. Effect of sucrose and sodium chloride. Starke 32(4):126–129
Wurzburg OB (1986) Converted starches. In: Wurzburg OB (ed) Modified starches: properties and uses. CRC Press, Boca Raton, FL
Xie X, Liu Q, Cui SW (2006) Studies on the granular structure of resistant starches (type 4) from normal, high amylose and waxy corn starch citrates. Food Res Int 39:332–341
Xie Y, Yan M, Yuan S et al (2013) Effect of microwave treatment on the physicochemical properties of potato starch granules. Chem Cent J 7:113
Xu A, Seib PA (1997) Determination of the level and position of substitution in hydroxypropylated starch by high-resolution 1H-NMR spectroscopy of alpha-limit dextrins. J Cereal Sci 25(1): 17–26
Yan H, Zhengbiao GU (2010) Morphology of modified starches prepared by different methods. Food Res Int 43:767–772
Yang W, Kong X, Zheng Y et al (2020) Controlled ultrasound treatments modify the morphology and physical properties of rice starch rather than the fine structure. Ultrason Sonochem 59: 104709
Yu L, Christie G (2001) Measurement of starch thermal transitions using differential scanning calorimetry. Carbohydr Polym 46(2):179–184
Zhang SD, Zhang YR, Zhu J et al (2007) Modified corn starches with improved comprehensive properties for preparing thermoplastics. Starke 59(6):258–268

Zhang SD, Zhang YR, Wang XL et al (2009) High carbonyl content oxidized starch prepared by hydrogen peroxide and its thermoplastic application. Starke 61(11):646–655

Zhao K, Saleh ASM, Li B et al (2018) Effects of convention and microwave pretreatment acetylation on structural and physicochemical properties of wheat starch. Int J Food Sci Technol 53(11):2515–2524

Zhou Y, Yan Y, Shi M et al (2018) Effect of an atmospheric pressure plasma jet on the structure and physicochemical properties of waxy and normal maize starch. Polymers 11(1):8

Zhu L, Jones C, Guo Q et al (2016) An evaluation of total starch and starch gelatinization methodologies in pelleted animal feed. J Anim Sci 94:1501–1507

Zia-ud-Din XH, Fei P (2017) Physical and chemical modification of starches: a review. Crit Rev Food Sci Nutr 57(12):2691–2705

Browning Reactions in Foods

4

Prerna Nath, Neha Pandey, Mahesh Samota, Kalyani Sharma, Sakharam Kale, Pankaj Kannaujia, Swati Sethi, and O. P. Chauhan

4.1 Introduction

Browning reactions are a dominant phenomenon that occurs during the processing of the food, its storage, and value addition which represent a stimulating research area for the inferences in food constancy and technology, also in health and nutrition. Browning is the most critical phenomenon that occurs in food during its processing and storage. Generally, browning reactions give rise to undesired changes in the sensory attributes, along with the reduction in the market value of various foods; however, browning reactions may also be advantageous as these provide essential color and flavor to some products such as tea, cocoa, coffee, and fried and baked food products (Whitaker and Lee 1995). They can involve different composites and proceed through various chemical reactions. The enzymatic browning outcome is loss of functional, nutrients, and organoleptic attributes in foods such as softening, blackening, and off-flavor changes. The enzymatic browning has been considered as

P. Nath (✉) · M. Samota · S. Kale · P. Kannaujia
HCP Division, ICAR-Central Institute of Post-Harvest Engineering and Technology, Abohar, Punjab, India
e-mail: prerna.nath@icar.gov.in

N. Pandey
ICAR-Indian Institute of Soybean Research, Indore, Madhya Pradesh, India

K. Sharma
Department of Agricultural Processing and Food Engineering, Indira Gandhi Krishi Vishwavidyalaya, Raipur, India

S. Sethi
FGOP Division, ICAR-Central Institute of Post-Harvest Engineering and Technology, Ludhiana, Punjab, India

O. P. Chauhan
Defence Food Research Laboratory, DRDO, Mysore, India

O. P. Chauhan (ed.), *Advances in Food Chemistry*,
https://doi.org/10.1007/978-981-19-4796-4_4

a serious problem leading to economic losses of fruits like pears, grapes, apples, bananas, etc., and vegetables like mushrooms, lettuce, potatoes, etc. (Whitaker and Lee 1995). The enzymatic browning is due to oxidation reaction, which is another critical reason for food spoilage after microbiological infection (Ioannou and Ghoul 2013). A predominant color reaction occurred among food products (vegetables and fruits) because of the interface of enzymes like PPOs, phenolic compounds, and oxygen. The major groups of pathways leading to browning are enzymatic phenol oxidation and the so-called non-enzymatic browning. The latter is an ideal for heat treatments and consists of a varied number of reactions such as Maillard reaction, chemical oxidation of phenols, maderisation, and caramelization. The impact of these different reactions on the overall antioxidant capacity of foods has been studied barely. Enzymatic or oxidative pathways of polyphenols are commonly responsible for a loss in their antioxidant properties. However, modern observations suggest that partially oxidized polyphenols can exhibit higher antioxidant activity than that of non-oxidized phenols.

In the case of the Maillard reaction, high antioxidant capacity was frequently linked to the formation of brown melanoidins (brown-coloured polymers known as melanoidins are formed). Although in its early stages Maillard reaction leads to the formation of well-known Amadori and Heyn's products (Ames 1988; Rizzi 1994), few literatures are available on the chemical structure of the hundreds of brown products which are formed by a series of sequential and parallel reactions including reductions, aldol condensation, and oxidations in fruits and vegetables. Several variables can be selected for the prevalent mechanism of the overall response and its rate, leading to the establishment of different chemical species that are expected to exert various antioxidant properties.

The antioxidant capabilities of Maillard reaction products (MRPs) are highly influenced by the system's physical and chemical parameters and processing conditions. It is also worth noting that polyphenols, ascorbic acid, and other carbonyl complexes can participate in the Maillard reaction, even if they are generated via oxidative processes. Their role in the production of heat-induced antioxidants is still unknown. Although non-enzymatic browning may not necessarily occur in a progressive manner during storage and subsequent food processing, color changes caused by the desired or undesirable progression of the reaction are likely to be linked to the product's total antioxidant activity. Because food color can be analyzed using quick and non-destructive methods, the discovery of a link between color and antioxidant qualities should help optimize processing conditions not only in terms of sensory attributes but also in terms of antioxidant properties.

Fruits and vegetables provide health advantages to consumers through their high contents of vitamins, fibers, and antioxidant composites. On the other hand, several changes take place during harvesting, preparation (in case of fresh-cut fruits), as well as storage of these commodities. These modifications result in a considerable reduction in microbiological and antioxidant properties (Lindley 1998). As a result, the food processing industry's primary priority is food preservation by preventing oxidation during processing and storage. Indeed, behind microbiological contamination, oxidation is the second most important cause of food loss. Enzymatic browning is the most common oxidative process. Peroxidase (POD) and polyphenol oxidase (PPO) are two oxidoreductase enzymes involved in this process (PPO). The first is a slow hydroxylation of monophenols to diphenols that results in colored molecules. Second, diphenols undergo quick oxidation to quinines, which results in colorful compounds (Queiroz et al. 2008). The substrates and enzymes involved in these reactions are found in the cells' vacuoles and cytoplasm; the reactions can only take place if they are combined and oxygen is present. As a result, all events such as loss of firmness, cutting, and shock, trigger the browning reactions, which further result in nutritional value, flavor, and aroma losses or modifications (Toivonen and Brummell 2008).

As a result, protecting fruits and vegetables against oxidation either in storage or processing has become a top priority in the food industry. Cooling, heating, freezing, packaging, coating, spray-drying, employing chelating, reducing agents, and using antioxidants to prevent enzymatic action, restricting the substrates, and bleaching the pigments are all methods that can be used to avoid enzymatic browning in vegetables and fruits. The efficiency of the anti-browning approach, on the other hand, is dependent on several elements, including concentration, cultivar, and their interactions with other factors, such as pH and application system, among others (Ghidelli et al. 2013).

The fresh-cut fruit and vegetable industry suffers from enzymatic browning, which compromises quality. In various foods, such as tea, coffee, and cocoa, it is useful for the development of color and flavor. Enzymatic browning occurs in a range of fruits and vegetables during processing and storage, like apples, pears, bananas, peaches, lettuce, and potatoes. A number of strategies have been developed to combat this issue. These methods either deactivate polyphenol oxidase (PPO) or avoid contact between the enzyme and its substrate by adding antioxidants or maintaining the structural integrity of the meals. We propose to gather and present recent breakthroughs in all strategies used to avoid enzymatic browning in fruits and vegetables throughout the previous decades in this book chapter. Chemical therapies will be offered by stressing each molecule's main action (antioxidants, acidifying, agents of firmness, or chelating agents). Then, by using modern techniques, physical processes (blanching, freezing, and product environment alteration) will be modernized.

4.2 Browning Reactions

Browning refers to the process of food getting brown as a result of internal chemical reactions. Browning is an interesting area for research to study its implications in food stability, nutrition, and technology. Sensory aspects such as color, texture, flavor, and nutritive value are regarded as essential acceptability criteria in food. Several enzymatic and non-enzymatic browning reactions are responsible for causing browning in the food (Manzocco et al. 2001). The browning reactions in food can be divided generally into two types categories, i.e., enzymatic or nonenzymatic. Phenol oxidative browning of fresh fruit or vegetables is known as enzymatic browning. Artificial heating of foods promotes non-enzymatic browning, which results in color and flavor changes.

4.2.1 Enzymatic Browning

"Enzymatic browning" is considered to be oxidation of phenolic compounds into colored polymers (melanins), changing from intermediary shades of pink, blue, or red to mostly black or brown. Enzymatic browning of fruits and vegetables occurs by exposure to air after processing operations such as trimming, cutting, peeling, pulping, juicing, dehydration, mechanical damage during transportation, and thawing of frozen or cold-stored foods. The enzymes responsible for these reactions mainly fall into two categories of oxidoreductase enzymes: polyphenoloxidase (PPO) and peroxidase (POD). Polyphenol oxidase (PPO) is a common term for the group of enzymes that catalyze phenolic compounds' oxidation to produce brown color on the cut or bruised surfaces of fruits and vegetables. Schoenbein discovered the PPO first time in mushroom (1856).

4.2.1.1 Factors Affecting Enzymatic Browning

4.2.1.1.1 Substrate

Fruits and vegetables are considered to be the best source of vitamins and antioxidant compounds such as phenolic compounds, carotenoids, and vitamins (Table 4.1). The antioxidant compounds present in fruits and vegetables balance oxidative stress in the body which have various health benefits like anti-cancerous, anti-inflamatory, immunity booster, prevention of atherosclerosis, ocular disease, cardiovascular disease, diabetes, obesity, neurogenerative disease, rheumatoid arthritis, and proper functioning of liver, kidney, digestive system (Kaur and Kapoor 2001; Hajhashemi et al. 2010; Wilson et al. 2017).

However, during harvesting, preparation, processing, and storage of these fruits many changes occur for the antioxidant compounds. These changes cause a significant loss of the microbiological and antioxidant qualities which directly affects the sale and price of fruits and vegetables (Lindley 1998). The oxidation of cut fruits and vegetable causes severe loss in all sensory qualities including appearance, texture, and flavor. Oxidation is considered to be a very important cause of food deterioration besides microbiological contamination.

Various natural substrates (phenolic compounds) are associated with enzymatic browning. Some of the important phenols are cinnamic acid and benzoic acid derivatives, flavonoids, catechol derivatives, tannins, and lignins. Chlorogenic and tyrosine acid are the main browning substrates in potatoes, while 3,4-dihydroxyphenylethylamine is the main browning substrate in bananas. Enzymatic browning is mediated by benzoic acid derivatives (protocatechuic acid and gallic acid) and cinnamic acid derivatives (caffeic acid and *p*-coumaric acid). About 60–70% of the phenolic chemicals extracted from carrots are chlorogenic acid (5′-caffeoylquinic). It is involved in the creation of blue-black hues in potatoes

Table 4.1 Class, source, and health benefits of antioxidants present in fruits and vegetables

Class	Health benefits	Sources	Reference
Benzoic acid	Antimicrobial	Prunes, apples, plums, cloudberries, strawberries, apricots, cranberries	Chipley (2005)
Acetophenones	Antibacterial, fungicide, sleep-inducing	Apple, apricot, banana, cauliflower	Salas et al. (2010)
Phenyl acetic acid	Antimicrobial properties	Guava, wines, papaya, raspberry, strawberry, potato, tomato, mango, passion fruit	Cook (2019)
Cinnamic acid	Immunity booster	Grapes, citrus fruits, cabbage, spinach, celery	Bhuyan and Basu (2017)
Phenyl propenes Chromones	Anti-cancerous Anti-inflammatory Antibronchitis, antimicrobial	*Aloe*, *Aquilaria*, *Cassia*, *Hypericum* and *Polygonum*	Semwal et al. (2020)
Coumarins	Antiobesity	Strawberry, black currant, apricot, and cherry	Loncar et al. (2020)
Xanthones	Antioxidant, antidiabetic, anti-Alzheimer, antiobesity, anticancer, and anti-inflammatory	Mangosteen	El-Seedi et al. (2020)
Stilbenes Resveratrol Viniferin	Antioxidant, anti-inflammatory, and antiproliferative effects	All types of berry fruits, passion fruits, and tomatoes	Khawand, Courtois et al. (2018)
Anthraquinones	Anticancer, anti-inflammatory, diuretic, antiarthritic, antifungal, antibacterial, and antimalarial	Rhubarb, noni, *Aloe vera*, cabbage lettuce, beans, and peas	Fouillaud et al. (2018)
Flavonoids Flavonols Anthocyanins Flavanones	Anti-cancerous, reduced risk of CVD, antidiabetic, Anti-inflammatory	Berries, cherries, grapes, plum, apricots, pomegranates, apples, cabbage, onion, radish, citrus fruits, parsley	D'Archivio et al. (2007)
Hydrolyzable Tannins Gallic Acid derivatives	Anticancer, anti-angiogenic, antioxidant, anti-inflammatory, and anti-ulcerative	Mangos, guavas, pimento, pomegranates, plums, apricots, peaches, berries, grapes, and muscadine grapes	Amarowicz and Janiak (2019)
Lignins	Antioxidant, anticancerous	Carrot, spinach, kiwi, curly kale, radish, asparagus, rhubarb, pear, apple, small radish, and kohlrabi	Bunzel et al. (2005)

after cooking in the presence of iron residues, in addition to its appearance in pears and apples. In tomatoes, glucose derivatives of ferulic acid, caffeic acid, and *p*-coumaric acid were found in large quantities.

In the case of enzymatic browning, flavonoids are the primary substrates. Lichens, bacteria, vascular plants, and fungus all have them. Flavonols, for example, are a subclass of this large family. The main flavonoids involved in enzymatic browning in vegetables include flavan-3-ols, isoflavones, flavones, isoflavanones, and anthocyanidins. Flavan-3-ols are the principal plant polyphenols found in grapes and tea. They are not pigmented, but during enzymatic oxidation, they quickly convert to yellow and brown chemicals. Approximately 75% of catechins undergo enzymatic oxidation and polymerization events during the fermentation process of tea leaves in the creation of black tea. Flavan-3-ols, commonly known as epicatechins, are abundant in apples. Epicatechins and catechins have a 5:1 ratio in terms of amount.

Flavonols are plentiful in onion layers in the form of quercetin glycosides. Some flavonols like apigenin and luteolin glycosides can be found in carrots. Flavanones like naringin, or naringenin glycoside, are responsible for the bitter taste of some grapefruits. A compound of alike structure, hesperidin is available in oranges.

Anthocyanidins are brightly colored plant pigments that range in hue from red to purple to blue based on their structure. It is overly sensitive to pH fluctuations, turning blue to red as the pH dips and vice versa. The pericarp of fresh aubergines is colored by delphinidin glycosides. Delphinidin, cyanidin, peonidin, malvidin, pelargonidin, and petunidin glycosides are important flavonoids in ripe fruit, both in terms of amount and quality (figs, blueberries, grapes, strawberries, cherries, blackcurrants, raspberries, and tropical fruits). Pears, apples, oranges, and bananas also contain them in smaller levels.

Likewise, tannins which are responsible for cell wall structure and astringency are substrates for enzymatic browning. Hydrolyzable tannins (digallic, gallic, luteic, and ellagic) and condensed (or catechin) tannins are the two main types of tannins. The ability of tannins to bind polysaccharides and proteins via hydrophobic interactions and/or hydrogen bonds and precipitate them is their main physicochemical feature. Cloudiness in beverages is mainly caused by the formation of complexes between tannins and polysaccharides or proteins (beer and wine). Tannins are also responsible for the astringency (dryness) sensation that develops in the mouth. The pigments generated by the interaction of pyrogallic tannins with residues of iron cause the blackening that sometimes occurs in chestnut purees. Lignins, which provide stiffness to some plant tissues and contribute to enzymatic browning, are polyphenolic polymers.

4.2.1.1.2 Oxidation Causing Enzymes

PPO and peroxidase are two enzymes that contribute to browning reactions to some extent. In the presence of oxygen, PPO attacks phenols, whereas peroxidase produces hydrogen peroxide. PPOs are copper-containing metalloproteins that play a key role in the catalytic mechanism of enzymes. Their optimum pH for enzymatic activity is between 5 and 7, and as pH falls, so does the activity.

4.2.1.1.3 Types of PPO

Enzyme Nomenclature as per substrate affinity

1. Monophenol monooxygenase
2. Diphenol oxygen oxidoreductase
3. Laccase

PPO catalyzes two processes. The first is the gradual hydroxylation of monophenols to diphenols, which yields colorless compounds. The second reaction is the fast conversion of diphenols to quinines, which produces a colored (brown) chemical (Queiroz et al. 2008). Reaction substrates are found in vacuoles, whereas enzymes are present in the cytoplasm; reactions occur when both react in the presence of oxygen. As a result of tissue damage (cutting, shock, loss of stiffness), browning reactions begin, resulting in flavor, odor, and nutritional value losses or alterations (Toivonen and Brummell 2008). As a result, consumers do not prefer browned fruits and vegetables.

Some fresh fruits and vegetables, such as strawberries, potatoes, grapes, apples, bananas, apricots, peaches, and lettuce, suffer significant economic losses as a result of PPO's effect. Due to enzyme-induced browning, up to 50% of the fresh tropical fruits are of no use. In the fruit beverage sector, enzymatic browning is a major issue. PPO activity in plants is desirable in the processing of some dried fruits like black raisins, black figs, zapote, dates, and prunes, the drying of tobacco, the preparation of cider, fermented cocoa and coffee beans, and the fermentation of tea.

The chemistry of enzymatic browning is explained below:

1. The two Cu(I) groups of deoxy PPO (deoxy) are bound to O_2 to give oxy PPO.
2. The two Cu(II) groups of oxy PPO then bind to the oxygen atom of the two hydroxyl groups of catechol to form the O_2*catechol PPO complex.
3. The catechol is oxidized to *o*-benzoquinone and the enzyme is reduced to met PPO. Another molecule of catechol binds to met PPO, is oxidized to *o*-benzoquinone, and the enzyme is reduced to deoxy PPO, completing the cycle.
4. Met PPO must be reduced by reducing the compound, to give deoxy PPO. This time Deoxy PPO binds with O_2 and gives oxy PPO, the monophenol is bound to one of the Cu(II) groups to give the O_2*monophenol-PPO complex.
5. Subsequently, the *o*-position of the monophenol is hydroxylated by an oxygen atom of the O_2 to give catechol, which then dissociates to give deoxy PPO, to complete the cycle. It is to be noted that the first cycle of hydroxylation of a monophenol starts at the Met PPO; all following cycles begin with deoxy PPO.
6. Certain phenols and O_2 are hydroxylated in the *o*-position adjacent and further oxidized to *o*-benzoquinones and then nonenzymatically to melanins (brown pigments).

Browning of fruits and vegetables caused by PPO can be prevented by inactivating the enzyme with heat or lowering the pH to units below the pH optimum, or by adding PPO inhibitors, excluding or removing one or both of the substrates (O_2 and phenols), or adding compounds that inhibit melanin formation.

4.2.1.1.4 Oxygen

Fruits and vegetables have outer covering or skins that exclude oxygen as long as there is no damage to the skins. Exclusion of oxygen is crucial for the inhibition of oxidation reaction and PPO activity, as oxygen is important for enzymatic browning (Ingraham 1955). Controlled atmospheric storage, packaging techniques, and other methods can be used to artificially exclude or lower the content of O_2 in fruits and vegetables.

4.2.1.1.5 Physicochemical Conditions and Existence of Natural Inhibitors

The temperature has a direct impact on enzymatic reactions in general, as well as the color alteration of plant products in particular. The optimal temperature range for PPOs is 25–35 °C, which accounts for both increased catalysis rate and enzyme deactivation as temperature rises. Furthermore, as the temperature rises, non-enzymatic condensation and oxidation processes become more prevalent. As a result, the ideal temperature for enzymatic browning is between 35 and 40 °C. A temperature drop below these values slows the reaction rate but does not completely stop it. Enzymatic browning is preferred at temperatures below those that cause chilling harm. As a result, the best storage temperature for plant products to avoid browning is the temperature at which the highest reduction in enzyme activity occurs while causing no injury to the structure of fruits and vegetable tissues.

Control of Enzymatic Browning

Various approaches have been developed to avoid browning. Inactivation of polyphenol oxidase (PPO), prevention of enzyme–substrate interaction, and various methods of limiting oxygen accessibility are among the preventive approaches. Controlling enzymatic browning in fruits and vegetables, as well as juices and wines, necessitates a chemical understanding of the types and concentrations of oxidative substrates present in the plant, as well as the presence and concentration of reducing compounds like sulfhydryl compounds, ascorbic acid, and O_2 accessibility. Some PPOs hydroxylate monophenols produce *o*-dihydroxyphenols, which are then oxidized to *o*-benzoquinones via enzymes. The *o*-benzoquinones, which are yellowish, are very reactive and unstable. Additional non-enzymatic interactions with O_2 result in more reactions and polymerization of melanin.

Researchers tried various methods for the prevention of enzymatic browning in postharvest storage and processing of fruits and vegetables. In this chapter, we have tried to cover recent trends in the prevention of enzymatic browning (Table 4.2).

4.2.1.1.6 Chemical Treatments

To control the oxidative browning and associated changes in fruits and vegetables, various chemical treatments are used by researchers. As per their mode of action, the used chemical agents are categorized as antioxidant agent, chelating agent, firmness agent, and acidifying agent. Microbial growth and enzymatic browning are also inhibited by sulfites. As a result, they are commonly used as an antibacterial agent in the storage of wine, dried fruit, and juice concentrates.

Table 4.2 Different methods of prevention of enzymatic browning tried by researchers for various fruits and vegetables

Method of prevention	Product	Process followed	Results	References
Chemical	Apple	[Phytic acid, 0.08%], R.T.	~99% inhibition of PPO	Du et al. (2012)
		[Ascorbic acid, 0.3 Mm], 10 min	Browning decreased	Grimm et al. (2012)
		[1% ascorbic acid + 0.1% calcium chloride, pH 3.5], 4 °C/5 min	Prevention of apple texture after UV irradiation	Gomez et al. (2011)
		[Sodium chloride (0.03), acidified sodium chlorite (0.03%), citric acid (2%), calcium chloride (2%)], RT/1 min	Sodium chlorite was found most effective	Luo et al. (2011)
		[Sodium chloride and/or calcium propionate at concentrations between 0% and 2%], 5 min	Combination is effective	Guan and Fan (2010)
	Kiwi	[2% ascorbic acid + 2% calcium chloride], RT/2 min	Effective in the delay of softening and browning of tissues	Antunes et al. (2010)
	Pear	[1-Methylcyclopropene (300 nL/L), 0 °C/24 h] Then [2% ascorbic acid + 0.01% 4-hexylresorcenol + 1% calcium chloride, 4 °C/15 min]	Softening and browning delayed	Arias et al. (2009)
	Eggplant	Calcium ascorbate/citrate (0.4%), 60 °C/1 min	Calcium ascorbate was best for the inactivation of PPO	Barbagallo et al. (2012)
	Artichoke	Ascorbic acid, citric acid, ethanol, cysteine and their combination, sodium chloride, 4-hexyresorcinol, RT/1 min	Cysteine was found most effective in preventing browning	Amodio et al. (2011)
	Longan fruit	0.01% sodium chorite, RT/10 min	Treatment reduce browning and enzyme activity	Khunpon et al. (2011)
		1.5 N HCl then rinsing, RT/20 min	Pericarp browning delayed	Apai (2010)
	Potato	1% sodium acid sulfate + 1% citric acid and 1% ascorbic acid, RT	PPO activity and browning reduced	Calder et al. (2011)

(continued)

Table 4.2 (continued)

Method of prevention	Product	Process followed	Results	References
	Mushroom	2,2′-(hydroxynitrosohydrazino)-bisethnamine, (0.5, 1, 2 Mm)	1 mM concentration maintained firmness and delayed browning	Jiang et al. (2011)
	Chestnut	0.5 μM Nitric acid, 10 min	Delay in browning and decrease in enzymatic activity	Shi et al. (2011)
Blanching	Plum	Water, ascorbic acid (400 ppm), 80 °C, 40 s	Blanching necessary for enzyme inactivation	Gonzalez-Cebrino et al. (2012)
	Red beet	Microwave, 250–450 W, immersed in water for 5 min	90% inactivation of enzyme activity	Latorre et al. (2012)
	Watercress	Thermosonication, 86 °C, 30 s	90% inactivation of enzyme activity, loss of microstructure	Cruz et al. (2011)
	Aonla	• Water • Potassium metabisulphite (0.3%) 80 °C for 3 min	KMS prevents the leaching of nutrients	Gupta et al. (2011)
	Carrot, cauliflower, spinach	Water Steam	Water blanching causes nutritional losses as compared with steam blanching	Mazzeo et al. (2011)
	Indian gooseberry	Hot water, 100 °C for 3 min	Blanching affects all chemical properties and preserves color	Prajapaty et al. (2011)
Modified atmosphere	Apple	High-pressure (150 MPa) Argon treatment	Delayed enzymatic browning	Wu et al. (2012)
	Apple/ mushroom	Different oxygen concentration	21% O_2/79% Ar best combination	O'Beirne et al. (2011)
	Mushroom	Modified Atmosphere: 100% N_2/20% O_2-80% N_2/50 to 100% O_2	80% O_2 best treatment	Wang et al. (2011)
Coating agents	Pomegranate	Starch + glycerol (2:1) Seed oil (300/600 ppm) 15 min at room temperature	Significant delay in browning with a starch coating of 300 ppm seed oil	Oz and Ulukanli (2012)

		Putrescine + carnauba wax treatment Cold storage at 2 °C Exposure at 20 °C for 3 days	Delay of chilling injury, and browning during storage Decrease of respiration and ethylene evolution rate	Barman et al. (2011)
	Melon	Osmotic dehydratation of fresh-cut melon in 40° Brix sucrose solution having 0.5% calcium lactate solution + Coating with 1% pectin	Enhancement in shelf life, improvement in firmness by calcium lactate Increase in soluble solid content of the product, improvement in sensory acceptance	Ferrari et al. (2011)
	Apple	Konjac glucomannan + pineapple fruit extract at diverse concentrations in distilled water	Delayed enzymatic browning. The best result is achieved with pineapple fruit extract (1:1)	Supapvanich et al. (2012)
	Mushroom	Aloe vera gel + Gum tragacanth + calcium chloride and citric acid	Combination of both is more effective	Mohebbi et al. (2012)
Combination	Plum	Blanching High pressure	Combination with high-pressure treatment increases the efficiency of blanching to inactivate PPO	Gonzalez-Cebrino et al. (2012)
	Litchi	Dipping (sodium hypochlorite, potassium metabisulfite, hydrochloric acid, and ascorbic acid) + radiation	The best treatment to prevent enzymatic browning is combination in successive manner: sodium hypochloride (0.2%, 4 min, 52 °C), potassium metabisulfate (3%, 30 min, 26 °C), and hydrochloric acid (0.25 N) containing ascorbic acid (2%, 10 min, 26 °C) followed by gamma irradiation	Kumar et al. (2012)
	Apple	Heat treatment Coating	Combination of two treatments is better to avoid color degradation	Shao et al. (2012)
	Broccoli	Coating heat treatment	Delay in browning due to heat treatment and stability in storage due to coating	Ansorena et al. (2011)
	Apple	Dipping Ultrasonics	Combination of these two methods inactivates enzymes	Jang and Moon (2011)

Treatment with Antioxidant Agents

Antioxidants as the name suggest quench surplus oxygen thus preventing the initiation of browning. The effectiveness of antioxidants depends on the food environment such as water activity (a_w), temperature, pH, and oxygen availability. Ascorbic acid E300, hexylresorcinol E586, glutathione, erythorbic acid E315, Cysteine E920, and cysteine hydrochloride E920 are mentioned as major antioxidants in the literature (Oms-Oliu et al. 2006; Arias et al. 2007). The ascorbic acid is the most frequently and extensively used antioxidant. It prolongs the onset of enzymatic browning by decreasing *o*-quinones to colorless phenolic derivatives. It may too work by reducing copper in PPO enzyme.

Treatment with Chelating Agents

PPO has active copper ions which are responsible for its activity (Du et al. 2012). Thus, the substance which may bind copper ions reduces its enzymatic activity. These substances are known as chelators like citric acid E330, kojic acid, and EDTA E385.

Treatment with Agents of Firmness

Firming agents are usually calcium salts and are utilized in the strengthening of plant cell walls, preventing loss of structure (Khunpon et al. 2011). The main firming agents may include calcium propionate E282 (antifungal), calcium lactate E327 (antimicrobial), calcium chloride (E509), and calcium ascorbate (E302).

Treatment with Acidifying Agents

In general, the optimum pH for enzymatic activity of PPO is between 4 and 7. The fruits have natural lower pH or acidic environment; hence, with the addition of acidifying agents, PPO may inactivate or have reduced activity below pH 3 (Grimm et al. 2012).

Citric acid E330 (also chelating), ascorbic acid E300 (also antioxidant), erythorbic acid E315, and glutathione are acidifying agents to name a few. However, PPO in some fruits like apples upholds about 40% of its maximum activity at pH 3. Thus, acidification alone may not be the best way to reduce enzymatic browning without severely detrimenting food quality. The combinations of various treatments are generally used to get the best result.

4.2.1.1.7 Control by Physical Processes

Temperature is critical in physical processes like blanching and freezing. In reality, polyphenol oxidase (PPO) is extremely sensitive to temperature changes, particularly high temperatures. According to Özel and colleagues (2010), blanching plum fruits at temperatures above 80 °C inactivates PPO, whereas freezing reduces the amount of water available (water activity 0.3) in foods for enzymatic reactions, resulting in lower PPO activity (Lavelli and Caronni 2010).

Blanching

One of the most common methods of heat treatment is blanching. Blanching treatments are classified based on the type of heating medium utilized, such as boiling water, steam, microwave energy, and so on. Blanching time varies depending on the type of blanching, the type and size of the product, and other factors. It is frequently used before the canning, retort processing, dehydration, freezing, and freeze-drying processes. This procedure deactivates the enzymatic systems that cause sensory and nutritional changes, limiting oxidation. Furthermore, the technique enhances the color of plants for better presentation. The activity of polyphenol oxidase fluctuates with temperature; as the temperature rises, the enzyme's relative activity decreases (Özel et al. 2010). Blanching has various disadvantages, such as altering the hardness of the treated product and occasionally imparting a cooked flavor. It also causes nutrient losses, gruel loss, and a reduction in product weight. As a result, the time-temperature combination must be ideal to reduce nutritional and structural losses.

Blanching in Water

Hot-water blanching has the benefit of uniform heat treatment of the foods. On the contrary, the drawback of hot-water blanching is the leaching of some soluble substances and low energy yield (Mazzeo et al. 2011). In order to overcome this shortcoming, chemical agents are added to prevent nutritional losses (Gonzalez-Cebrino et al. 2012; Gupta et al. 2011).

Steam Blanching

Steam blanching is utilized for very fragmented foodstuffs. The release of soluble chemicals is reduced with this technique. To combat with problem of homogeneity of steam blanching and nutritional losses in hot-water blanching, a process using superheated steam (SHS) and a spray of microdrops of hot water (WMD) is applied (Sotome et al. 2009). Changes in color and texture were decreased by the combined treatment.

Microwave Blanching

Because of its effectiveness in inactivating PPO, microwave heating can be used as a blanching procedure (Matsui et al. 2008). Microwave blanching preserves the product's nutritional value (Ramesh et al. 2002). Vina et al. (2007) experimented with microwave blanching mixed with hot-water blanching for Brussels sprouts. To minimize the problems of non-uniform heating in this process, microwave blanching is usually combined with other techniques. Recent research has focused on the development of novel physical blanching technologies such as ohmic heating (Icier et al. 2006) and thermosonication (Cruz et al. 2011). Blanching in combination with other techniques such as coating, dipping, and the use of chemical agents has also been investigated.

Freezing

Lowering water activity reduces enzymatic activity by reducing substrate and reaction product mobility, which might have an inhibitory effect on the enzyme in some instances. It is studied by Lavelli and Caronni (2010) that a water activity below 0.3 in apples caused the inactivation of PPO. Freezing tends to slow reaction rate, and thus is used as one of the physical approach to stop browning reactions in plant tissues. Freezing does not inactivate enzyme permanently, and thus if the product needs to be thawed then food quality is every so often changed and an enzymatic reaction is initiated inside the product rapidly. Consequently, freezing may be used to enhance shelf life of the product, but in association with other methods like blanching (Gossinger et al. 2009).

Conservation in Modified Atmosphere

Oxygen is essential for oxidation reaction along with PPO activity; therefore, to inhibit the browning reactions, change in oxygen level of storage and packaging environment is advised (Ingraham 1955). Some of the studies are mentioned in tabular form (Table 4.2). Recent studies showed Argon or NO_2 have better efficiency in preventing browning without much quality loss (Rocculi et al. 2005; O'Beirne et al. 2011).

Coating

Fruits are routinely coated with coating agents to extend their shelf life during storage. It entails putting a layer of any edible coating material on the fruit or vegetable's surface. It reduces moisture and fragrance loss by obstructing gas transfer and provides a sheen to the product throughout storage (Olivas and Barbosa-Canovas 2005). By changing the atmosphere of coated fruits by separation from the environment, the coating agents postpone enzymatic browning. As per literature, the application of a gel-based coating are better than bath immersion coating method (Oms-Oliu et al. 2010) contributed by selective perviousness of the gel coatings to gaseous diffusion. Chitosan, gum Alginate, or gum Carrageenan are generally used coating materials. Although coating alone cannot prevent enzymatic browning, many coating methods are in use in combination with dipping by the addition of anti-browning agents, preservatives, and firming agents to coating agents.

Combination of Chemical and Physical Approach

The combination of several methods have been utilized by several researchers to avoid phenol oxidation. These studies have been presented in Table 4.2. A number of strategies were combined to increase the protection of vegetables against oxidation. To avoid enzymatic browning and firmness loss, dipping is usually paired with physical methods (modified environment, chemical treatment, coating, and blanching). The use of a combination of dipping and blanching or coating has been shown to prevent browning and food quality loss over time.

Alternative Methods to Substitute Thermal Methods
Thermal techniques are found to be very effective to prevent enzymatic browning; however, they lead to modifications in some of the product parameters such as texture and taste. Thus, identifying non-conventional methods is a crucial challenge in this area. Various methods studied include pulsed electric fields (PEF), irradiation, high-pressure processing (HPP), and ultrasonication. The main objective of these alternative methods is inactivating the browning enzymes using different principles such as light, pressure, or electricity. The most used method is high hydrostatic pressure (HHP); however, it is more effective in inactivating the microorganisms than totally inactivating the enzymes that are more resistant. All these methods are effective to some extent but it is rather difficult to replace thermal methods.

4.2.2 Non-enzymatic Browning

Artificial heating of foods leads to non-enzymatic browning which causes change in color and flavor. Non-enzymatic browning comprises of a group of chemical reactions that take place during the thermal processing and/or storage of foods. Brown substances are formed as a result of it, which have an impact on the sensory quality of foods. As a result, non-enzymatic browning is preferred for coffee and cocoa bean roasting, grilling, bread and pastry items, beer, balsamic vinegar, whiskey, and even some cheeses. Non-enzymatic browning is unpleasant if the process is uncontrolled and impacts the sensory properties of food. It causes some sterilized milk and fruit juices to have an unpleasant flavor, as well as discoloration in dried items (milk powders). Furthermore, it may result in nutritional changes such as the degradation of important amino acids like lysine, a reduction in protein digestibility, the creation of potentially hazardous chemicals, or a change in food's antioxidant capacity. As a result, achieving desired sensory characteristics for final goods while keeping their nutritional value and storage stability is one of the issues in the food processing business when dealing with non-enzymatic browning.

4.2.2.1 Types of Non-enzymatic Browning (NEB)

1. Maillard reaction
2. Caramelization
3. Ascorbic acid browning

The most common type of NEB is Maillard reaction which is named after Louis-Camille Maillard (1878–1936). It comprises of reactions involving interaction of an amino group and a carbonyl group. This includes all the reactions, viz., cyclization, dehydration, rearrangement, retro-aldolization, condensation, and isomerization. All these reactions, on one hand, produce aromatic and low-weight volatile molecules and, on other hand, they produce high molecular weight brown-nitrogenous polymers known as melanoidins.

The compounds participating in the Maillard reaction must have a free carbonyl group and free amino group like lysine and sometimes the side chain of arginine. In a similar way, the N-terminal amino group of proteins, ammonium, and certain primary or secondary amines may also contribute to the Maillard reaction to some extent.

Traditionally, the carbonyl group is the reducing group of sugar. Thus, monosaccharides, reducing disaccharides, or reducing oligosaccharides participate in the Maillard reaction. Although reaction compounds like aldehydes, ketones, polyphenols, ascorbic acid, and organic acids also have one or more carbonyl groups that may participate in the Maillard reaction.

4.3 Mechanism of Browning Reactions

Browning is the most crucial event that occurs in harvested produce at the time of processing and storage. In most cases, browning reactions cause unacceptable changes in taste and smell, which also reduce their market value. But in the case of dried foods (raisins, prunes, dates, and figs), coffee, tea, and cocoa, browning reactions are advantageous as browning reactions in these products impart their characteristic color and flavor (Whitaker and Lee 1995). Two types of reactions that mainly cause browning are enzymatic phenol oxidation and non-enzymatic browning (Manzocco et al. 2001).

4.3.1 Enzymatic Browning

It is among the most crucial reactions that have an effect on food products. Enzyme polyphenols oxidase (PPO) acts as a catalyst in enzymatic browning. It catalyzes the oxidation of phenolic components to quinines, which at last polymerize to colored melanoidins (Marshall et al. 2000). Although, the process of enzymatic browning depends on several factors such as concentration of PPO and phenolic constituents in food products, temperature, pH, and oxygen availability, the intensity of the browning reaction is affected by the concentration of phenolic constituents and enzyme activity (Zawistowski 1991). In the process of enzymatic browning, polyphenols act as a substrate for the PPO enzyme. Polyphenols are found in plastids, while enzymes are in the cytoplasm. Due to mechanical damage at the time of handling or processing of fruits and vegetables, tissues become wounded, plastids are ruptured, and PPO starts reacting with phenolic compounds (Mayer and Harel 1979).

In the first step, PPO catalyzes the hydroxylation of the monophenols, resulting in the formation of diphenols, following second step of oxidation of diphenols into *ortho*-quinones (in the presence of O_2). The whole reaction is catalyzed by PPO, containing two copper moieties at its catalytic site. In the next step, non-enzymatic polymerization of the quinones occurs resulting in the formation of melanins, which are unsolvable, complex, brown-colored compounds with high molecular weight (Queiroz et al. 2008; Peñalver et al. 2005). The extent of color variation of pigments depends on the source of origin of phenolic compounds and environmental factors affecting the oxidation process throughout pigment development (Nicolas et al. 1994). PPOs originated from plants are able to oxidize a number of polyphenolic compounds. Subsequently, molecular oxygen associated with nonenzymatic browning is accountable for additional reactions which consequence in the development of compounds including 5,6-quinone from tyrosine. Also, ortho-quinones covalently react with other polyphenols and give rise to components with red, yellow, green, blue, and black color (Mathies and Whitaker 1984).

4.3.2 Non-enzymatic Browning

4.3.2.1 Maillard Reaction

Non-enzymatic browning is the most complicated reaction of browning. The involvement of a large number of food elements via several pathways makes the process of the Maillard reaction more complicated (Olano and Martínez-Castro 1996). The Maillard reaction is a nonenzymatic browning reaction that occurs when a free amino group (from proteins, peptides, or amino acids) reacts with the carbonyl group of reducing sugar. The Maillard reaction is among the major causes which decline in the quality of food during processing. This quality degradation is due to decreased palatability of proteins, less accessibility to amino acids, and destruction of essential amino acids (Malec et al. 2002). The Maillard reaction entails a series of complex changes that produce several volatile and nonvolatile components. There are three stages to the Maillard reaction: early, middle, and advanced.

The abridgment of fundamental amino groups of distinct amino acids, peptides, or proteins, as well as the carbonyl group of aldose sugar, occurs in the early stages. A water molecule is liberated in this reaction, resulting in the production of a Schiff's base and the Amadori rearrangement, which leads to the Amadori product. This Amadori product is a comparatively stable intermediary product (Feather et al. 1995). Instead of Amadori, Heyns compounds are formed when ketose is the free sugar reacting with amino groups (Machiels and Istasse 2002). Also, Amadori compounds act as a precursor for several other compounds responsible for typical aroma and flavor. The formation of Amadori compounds takes place before sensory changes occur. Thus, estimation of Amadori compounds can be used as a measure for quick observation of quality changes triggered by the Maillard reaction (Olano and Martínez-Castro 1996).

In the next stage, the intermediate stage, the disintegration of Amadori compounds or any compound associated with Schiff's base takes place. This breakdown results in the development of reactive intermediate products (3-deoxyglucosone) and volatiles responsible for flavor. This intermediate component assists in a more rapid cross-linking of proteins as compared to glucose and further breakdown results in the formation of two components: 5-hydroxymethyl-2-furaldehyde (Feather et al. 1995).

In the final stage, the formation of nitrogen-containing brown polymers and copolymers takes place and these polymers are referred to as melanoidins (Badoud

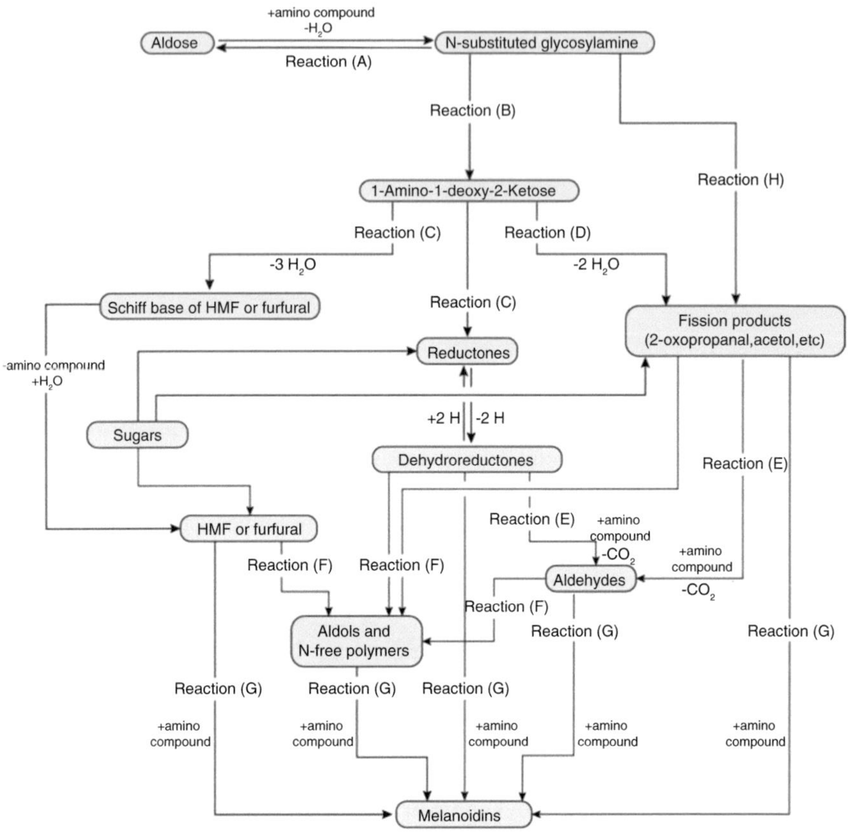

Fig. 4.1 Maillard reaction. (Source: Koubaa et al. 2019)

et al. 1995). Melanoidins are low molecular weight colored compounds which are capable of cross-linking with proteins through α-amino groups of lysine or arginine and give rise to colored compound with high molecular weight known as melanoidins. It is also reported that melanoidins are polymers comprised of numerous units of furans or pyrroles formed during the advanced steps of the Maillard reaction and connected by polycondensation reactions (Martins and van Boekel 2003).

The Maillard reaction can be further explained as different series of multifaceted reactions. They can broadly be divided into four phases (Fig. 4.1):

1. The reaction between amino group and the carbonyl group; (reaction A).
2. Rearrangement of Amadori and Heyns; (reaction B).
3. Ketosamines decomposition (intermediate reactions): sugar degradation (reaction C) and fragmentation (reaction D), degradation of amino acids (reaction E).

4. Polymerization reactions: condensation of aldols (reaction F), condensation of aldehyde-amine, and formation of heterocyclic nitrogen compounds (reaction G).

4.3.2.1.1 Condensation

The nucleophilic interaction of an amino group and a reducing sugar's carbonyl group initiates the Maillard reaction. In partially dehydrated conditions, the reaction is promoted in basic media. In an aqueous solution, the condensation product is highly unstable, losing water molecules and forming a glycosylamine, commonly known as "Schiff's base." Flavor development, physical, and structural change are unlikely to occur during this period, and most are reversible. Amadori or Heyns rearrangement occurs swiftly in glycosylamines generated from amino acids.

4.3.2.1.2 Amadori/Hyens Rearrangement

Schiff's base may produce additional stable byproducts upon irreversible isomerization, such as ketosamines (Amadori products), if the first carbohydrate substrate is an aldose, or aldosamines (Heyns products) if the initial carbohydrate substrate is a ketose. An amine group from another amino acid or protein side chain can attack the carbonyl group of Amadori or Heyns products, causing a fresh nucleophilic attack. Fission can create both carbonyls and amines, which can lead to unexpected compounds due to further rearrangement.

4.3.2.1.3 Degradation of Ketosamines

Ketosamines can further undergo a series of transformations such as enolization, dehydration, oxidation, retro-aldolization, cyclization, and decarboxylation. There are however three main reactions:

1. Formation of highly reactive carbonyl compounds.
2. Strecker degradation.
3. Fragmentation of carbohydrate units (retro-aldolization).

Formation of Highly Reactive Carbonyl Compounds

Whether ketosamines will undergo 1,2 or 2,3 enolization or not, will depend on the pH of the environment. Enolization at position 1,2 initiates the major degradation route of ketosamines to reductones in food products. Rearranging the molecules results in the loss of a water molecule, resulting in the formation of a 2,3 double bond and deamination. At a pH of roughly 5.5, i.e., in an acidic environment, these various processes are favored. Dicarbonyl compounds can lose a water molecule when heated in an acidic medium, resulting in an unsaturated dicarbonyl that can then be cyclized to produce furfuraldehyde molecules, including hydroxymethylfurfural (HMF). These are the most common lactone compounds. In pasteurized fruit juices, the 5-HMF content is employed as an indicator of heat treatment severity. The amount of furfuraldehyde produced decreases in the presence of free amino acids, favoring pyrroles as hetrocycles.

With 2,3 enolization, ketosamines may undergo a distinct breakdown path, culminating in the production of 1-methyl-2, and 3-dicarbonyl (reductones). Under basic pH circumstances or near-neutral pH, this reaction is preferred. These reductones decompose into diverse compounds of mono- and dicarbonyl, i.e, furanone, maltol, and isomaltol.

4.3.2.1.4 Strecker Degradation

Strecker degradation is a significant basis of aroma and flavor molecules. It leads to deprivation of α-amino acids and discharge of carbon dioxide gas, ammonia, and an aldehyde that are responsible for aroma. Reductones are generated as a result of ketosamines reacting with the amino group of free amino acids via a nucleophilic assault on the carbonyl group to produce aromatic aldehydes. The aromatic characteristics of aldehydes generated are determined by the amino acids involved in their synthesis. Strecker degradation of cysteine, for example, results in the synthesis of mercaptoaldehyde, which leads to the formation of thiazoles, thiazolines, thiophenes, trithiolanes, and other sulfur compounds with strong odors. Pyrazines are active fragrant compounds that can be found in roasted foods like grilled meat or fish.

Fragmentation Reactions

Continued heating of Amadori rearrangement molecules can result in direct fragmentation of the amino complex, resulting in several low molecular weight chemicals such as carbonyls and amines. Parallel to enolization reactions, ketosamines and derivatives can also go through retro-aldolization forming small molecules such as formaldehyde, pyruvic acid (slightly caramel-like), formic acid, diacetyl (butter-like), glyoxal, glyceraldehyde, pyruvaldehyde (pungent stinging odor), and acetic acid. Fragmentation chemical reactions take place under neutral or alkaline conditions.

Polymerization Reactions

This reaction leads to the formation melanoidins which are high molecular weight brown nitrogen compounds. Melanoidins contain repeating groups like amine, ether, furan, carbonyl, alcohol, ester, indole, pyrrole, amide, etc. Brown color is an indication of the Maillard reaction; the higher is the polymerization more is the brown color intensity.

Factors influencing the rate of the Maillard reaction at various stages are,

1. Substrates composition
2. Time–temperature combination during heat treatment or storage
3. Water activity
4. pH
5. Presence of inhibitors or activators

The nature and number of colorful and aromatic molecules generated, and the quality of the food product, are affected by adjusting the rate of reaction.

Substrates

The amount and kind of amino acids (lysine) and reducing sugars available for the Maillard reaction in food systems must be considered. When it comes to sugar reduction, the size of the sugar is crucial. Hexoses (glucose, fructose, and mannose) respond faster than pentoses (ribose), which are more reactive than disaccharides (lactose and maltose).

Molecules with carbonyl groups, such as certain lipid oxidation products, phenolic compounds, or any other carbonyl-containing molecule, may affect the balance of aromatic and colorful chemicals created during food preparation or preservation. Similarly, the nature of amino acids also play a great role in reactivity during Maillard browning. Lysine is the most reactive amino acid, followed by arginine, especially in the early phases of the Maillard reaction, due to its strong basic character. Biological agents (enzymes and microbes) can enhance the amount of reducing sugar and stimulate the Maillard process in disaccharides and polysaccharides (milk caramel, bread, beer, etc.). The accessibility of reactive amino acids is influenced by various parameters such as pH and the state of denaturation of the protein.

Temperature and Heating Time

The temperature-time relationship applied to the product has a significant impact on the rate of the Maillard reaction. The rate constant of the browning reactions (Q10, Volume 2) is multiplied by a factor ranging from 2 to 8 when the temperature is raised by 10 °C. Browning is thus accelerated by a rise in temperature and slows down at lower temperatures, but it can still occur at temperatures below 0 °C. The flavor and color created in food products differ depending on the time and temperature combination utilized. Flavor flaws connected to the Maillard process can emerge when stored at room temperature. Caramel or cooked flavors prevail at an intermediate temperature (100 °C), whereas grilled or roasted flavors take over at higher temperatures (150 °C) due to the development of pyrazines. Several heterocyclic flavor compounds have been identified from roasted foods, including oxazoles, thiazoles, quinoxalines, cyclopentapyrazines, and pyrazines, all of which add to the roasted flavor (Hoskin and Dimick 1984).

pH

Because each chemical involved in browning has its optimum pH, which ranges between 6 and 9, the effects of pH are complicated. In addition, sugars' hemiacetal structure is more stable in a dilute acid medium. The cooked meat flavor is best in the pH range of 4.0–5.5; pyrazines are generated in large numbers at pH values over 5, and browning intensity increases with pH. There is steady acid production in the media during the Maillard process, particularly formic acid and acetic acid. The Maillard reaction is inhibited by feedback as a result of their creation.

Water Activity

The Maillard reaction is significantly influenced by water activity. Browning occurs at a maximum when water activity is between 0.5 and 0.8. Because water is a result of the carbonyl–amine condensation reaction, the browning rate decreases at high water activity due to the dilution of the reactants. The delay in the diffusion rate of molecules toward each other represents a decrease in the browning rate at low water activity. Dried foods are thus more resistant to browning than fresh foods, especially below the glass transition temperature.

Presence of Inhibitors and Activators

Metal agents such as copper or iron, as well as sulfites, influence the Maillard process. The reaction is enhanced in the presence of copper. However, in the Maillard reaction, sulfites react with carbonyl compounds to generate sulfonates, which have low reactivity.

4.3.2.2 Caramelization

Another reaction of nonenzymatic browning is caramelization in which several degradative compounds are produced from the caramelization of carbohydrates, without the involvement of amino acids (Ajandouz et al. 2001; Ajandouz and Puigserver 1999). When the food with high sugar content (such as jelly, jams, and fruit juices) is processed with string heat, caramelization takes place (Kroh 1994). The end products of the caramelization process are not always acceptable. It is beneficial for developing caramel-like flavor but it can also lead to the development of mutagenic compounds (Tomasik et al. 1989). Acids and alkalis are responsible for the initiation of caramelization (Namiki 1988) and several compounds formed during caramelization are the same as those formed during the Maillard browning.

The first step of caramelization of reducing carbohydrates is the opening of the hemiacetal ring which under different acidic and alkali conditions go through the process of enolization and results in the formation of isomeric carbohydrates (Kroh 1994). When the reaction is catalyzed by acids, few isomeric carbohydrates are produced. Under acidic conditions, dehydration takes place which results in the formation of two furaldehyde compounds: 5-(hydroxymethyl)-2-furaldehyde (HMF) from hexoses and 2-furaldehyde from pentoses (Fennema 1976). In alkali media, the process of dehydration becomes slower as compared to neutral or acidic media. Under alkali conditions, disintegration results in the formation of acetol, acetoin, and diacetyl which under aerobic conditions form formic, acetic, and other organic acids through oxidative fission. Upon reaction of these products, brown polymers and flavor compounds are formed (Olano and Martínez-Castro 1996). Caramelization is a type of nonenzymatic browning reaction of sugars during high-temperature treatments of foods providing a caramel-like flavor. They can be found in the production of classic sucrose syrups and caramels, which are employed in the confectionery and pastry sectors. These reactions are different from Maillard reactions since there is no need for amino group involvement in the reaction. From the standpoint of food quality, caramel flavor and color are significant, and they can be manipulated to attain desired attributes. However, because of the potential for hazardous chemicals to develop, the process should be carefully monitored to ensure that food safety is not jeopardized.

The reaction progresses differently with different temperature range and the reaction may be accelerated in acidic or basic media. Colors and aromas are mostly produced by deoxyosuloses, O-heterocyclic and carbocyclic intermediates, and low molecular weight sugar fragments, and depend on the type of sugar used (mono-, oligo-, or polysaccharide). (cyclotene) and hydroxymethylfuranone are responsible for the distinctive caramel fragrances (furaneol). Depending on the reaction conditions during Caramelization, various aromatic compounds like hydroxymethylfurfural (HMF) and hydroxyacetylfuran (HAF), hydroxydimethylfuranone (HDF), dihydroxydimethylfuranone (DDF) are formed. There is also production of pyranones like 'maltol' from disaccharides and 'hydroxymaltol' from monosaccharides (Quintas et al. 2007).

4.3.2.3 Ascorbic Acid Browning

In its oxidized state, ascorbic acid (or vitamin C), which is found in a variety of fruits and vegetables, has a molecular structure that is comparable to reductones. Some fruit juices and fruit juice concentrates lose some of their vitamin C activity during Strecker degradation. At least in the early stages of degradation, vitamin C can be decomposed in the absence of nitrogen molecules, resulting in furfural or furoic acid. Brown pigments are formed as a result of the interaction of nitrogen compounds. Nonenzymatic browning was mostly caused by carbonyl compounds created from L-ascorbic acid degradation in freshly produced commercial orange juice, aseptically packaged in Tetrapack® cartons. Carbonylamine reactions have a small contribution, as evidenced by the steady total sugar content of damaged samples (Roig et al. 1999).

Depending on the medium conditions, this reaction happens in the presence or absence of oxygen (pH, presence of catalysts). Ascorbic acid, which is added to fruit juices to reduce enzymatic browning and compensate for vitamin C loss, has the unintended consequence of enhancing nonenzymatic browning. Hence, L-ascorbic acid should always be added in proportion to the level of oxygen present when used as an antioxidant.

4.3.2.4 Lipid Browning

The sensory and nutritional qualities of various foods are influenced by lipids, which chemically alter during processing and storage. Products made from vegetables are subjected to lipid browning due to their low lipid content. Usually, lipid oxidation causes unacceptable flavors and aroma (Nawar 1996) but in few cases, it can lead to some desirable results such as typical cheese or fried-food aromas (Nawar 1985).

Oxidation of lipids can take place through both enzymatic and nonenzymatic reactions. But the non-enzymatic reaction is the main reason behind the oxidation of lipids. Non-enzymatic lipid oxidation proceeds through the free radical mechanism. In this mechanism, hydroperoxides are the preliminary products, which are not stable in nature. Due to the unstable nature of these products, several dendritic reactions takes place through separate reaction pathways and several products are formed (Gardner 1989). On the other hand, enzymatic lipid oxidation is a sequential process that involves lipolytic enzymes. When lipolytic enzymes react with lipids, polyunsaturated fatty acids are produced. These fatty acids are then subjected to oxidation by lipoxygenase or cyclooxygenase and the formation of endoperoxides or hydroperoxides, respectively takes place. These two compounds undergo a number of reactions to form long-chain fatty acids (Gardner 1995).

4.4 Prevention and Control of Non-enzymatic Browning

4.4.1 Removal of Substrates

Before drying certain products, sugar can be eliminated using biological treatments, such as in egg whites, where glucose is first converted or eliminated through enzymatic or fermentation treatment. Temperatures below 10 °C cause the release of reducing sugars, which are involved in the Maillard reaction. However, storage at temperatures over 10 °C (for example, 20 °C for 2 weeks) moves the equilibrium away from reducing sugars and toward starch synthesis. When possible, reducing sugars should be substituted throughout the formulation process. Otherwise, the various ingredients should be heated or preheated separately before final blending. Browning processes are also limited or prevented by altering the amino groups of proteins. The amino group of the lysine side chain is converted to an amide group by the reaction of proteins with transglutaminase, and the enzymatic degradation of asparagine to aspartic acid by asparaginase greatly lowers the incidence of acrylamide in fried items heated at high temperatures.

4.4.2 Physical-Chemical Factors

Foods should not be subjected to severe heat treatments and should be stored at moderate temperatures due to the formation of brown pigment at high temperature. In addition, temperature influences the nature of the products formed. Some alternative treatments like microwave heating can control the development of browning. Residence time in the zone of aw (0.5–0.8) is critical during dehydration and concentration processes. Lowering the pH can in some cases slow down browning, but that is not suitable for all types of products like milk products.

4.4.3 Addition of Inhibitors

Sulfites are the most potent inhibitors of nonenzymatic browning. Carbonyl molecules (reducing sugars, aldehydes, ketones), Schiff's bases, or unsaturated carbonyl compounds compete in the process, resulting in extremely stable sulfonates.

Sulfites prolong the induction period and delay the development of pigments by binding to the most reactive non-enzymatic browning intermediates. Sulfites' role in the structure of melanoidins has yet to be investigated.

4.4.4 Nonconventional Methods

4.4.4.1 Ohmic Heating

When an electric current travels through a substance, the process is known as ohmic heating (Koubaa et al. 2016b; Hashemi et al. 2017).

The food sample is not in direct contact with hot surfaces during ohmic heating, and the procedure allows for low maintenance and excellent energy conversion efficiency (Pereira et al. 2010). When compared to sterilizing with hot water, this method has demonstrated its effectiveness in minimizing non-enzymatic browning by inactivating peroxidases in peas after a short treatment period (Icier et al. 2006). According to a recent study, by utilizing ohmic heating, researchers were able to reduce the amount of furan in sterilized vegetable and vegetable/meat baby foods by 70–90%. (Hradecky et al. 2017). This result could be attributed to ohmic heating's faster heating time, which leads to less furan precursor degradation. Furthermore, when comparing conventionally sterilized samples to those treated with ohmic heating, the analysis of chemicals generated during Maillard reactions and fatty acid oxidation was much greater in conventionally sterilized samples.

4.4.4.2 Pulsed Electric Fields

The mechanism behind PEF treatment is to apply high voltage (typically 50 kV/cm) pulses to foodstuffs put between two high voltage electrodes over short periods (ms to ms) (Koubaa et al. 2016a). Because PEF is a non-thermal technique, it minimizes heat-induced chemical changes. PEF treatment, as opposed to thermal treatment, was found to decrease the development of HMF in various fruit juices (Aguilo-Aguay et al. 2009). Results from this study on potato and asparagus showed that PEF-treated samples had lower Maillard reaction substrate (glucose) compared to non-treated control samples (Lund and Ray 2017; Janositz et al. 2011a, b).

4.4.4.3 High Hydrostatic Pressure

In 1990, high hydrostatic pressure (HHP) products were introduced to the Japanese market for the first time. Applying pressure to the food product in the range of 400–600 MPa for a few seconds to several minutes is the principle (Oey et al. 2016). Under HHP, Maillard browning was studied in lysine/arginine-sugar model systems (Ma et al. 2017). Results indicated different effects of HHP on lysine-sugar and arginine-sugar models. In the lysine-sugar model, intermediate and final stages were controlled, whereas a decrease in the lysine-glucose model and increase in the arginine-sugar model was observed in the degradation rate of Amadori compounds. There have been reports of accelerated Maillard reaction in HHP treated wines by researchers (Tian et al. 2016; Santos et al. 2013, 2015). Color change and proteolysis were seen in reconstituted skim milk treated with either high-pressure thermal processing (HPTP) or standard thermal processing (Devi et al. 2015).

4.5 Formation of Mutagens During Browning Reactions

Application of heat produces deleterious products through the Maillard reaction (MR). These products can cause cytotoxic, carcinogenic, and mutagenic consequences (Capuano and Fogliano 2011). The nature of these compounds varies according to the food products such as in products made from cereal, potato, and coffee bean produces acrylamide, furans, and furfurals while meat products produce heterocyclic amines (HAs).

4.5.1 Acrylamide

Presently, 5-hydroxymethylfurfural (HMF), furan, and acrylamide cover the major area of research as these are highly toxic in nature and extensively found in food products (Anese and Suman 2013; Capuano and Fogliano 2011). International Agency for Research on Cancer categorized acrylamide as carcinogenic to humans (IARC 1994). Researchers reported traces of acrylamide in several food products including coffee (Banchero et al. 2013), biscuits (Graf et al. 2006), bread (Keramat et al. 2011), baby food (Mojska et al. 2012), ripe olives (Casado et al. 2010), potato crisps (Mestdagh et al. 2008a; Ou et al. 2008), and potato strips (Mestdagh et al. 2008a; Zeng et al. 2009, 2010).

At the time of the intermediate stage of the Maillard reaction, acrylamide is produced by the breakdown of asparagines, which takes place in the occurrence of reducing sugars (Gokmen 2015; Stadler et al. 2002, 2004). The formation of acrylamide occurs when food products are exposed to high temperatures as in the case of frying and baking. High temperature and low water content are appropriate conditions for acrylamide formation. Especially food products with high carbohydrate produces more amount of acrylamide. The surface of the food products is more prone to acrylamide and other toxic compound formation as compared to the central portion as the temperature remains high at the food surface (Gokmen 2015). No regulations are imposed on the levels of acrylamide in foodstuffs, except European Food Safety Authority (EFSA) recommendations (The European Commission 2013).

4.5.2 Furan and HMF (5-Hydroxymethylfurfural)

For flavor development, Furan and HMF are of the most significance. Before knowing about the toxic nature of HMF, it was used to estimate the quality of thermally processed food products (Anese and Suman 2013). A large number of food products contain HMF, such as jams, biscuits, toasts, coffee, honey, orange and apple juice, chocolate, and breakfast cereals (Teixido et al. 2011).

Production of HMF takes place via two pathways: the Maillard reaction (MR) and sugar dehydration (caramelization) which results in the distribution of HMF in food products in a broad manner. In the Maillard reaction, HMF is produced due to the heating of reducing hexoses with amino acids or proteins. Under favorable temperature conditions, HMF is produced due to the breakdown of Amadori compounds in the intermediate stage of the Maillard reaction. Another pathway that produces HMF is caramelization under acidic conditions. The later pathway requires higher temperatures as compared to the Maillard reaction for the production of HMF (Gokmen and Morales 2014). Ketoses give rise to higher amounts of HMF as compared to aldoses and the amount of HMF produced varies with the temperature and the concentration of acid catalyst (Perez Locas and Yaylayan 2008).

Another compound furan, potentially carcinogenic to humans (IARC 1995), is found in canned and jarred food products (Kim et al. 2010), infant food (Jestoi et al. 2009), etc. Two main routes are responsible for the production of furan in food products: thermal-oxidative breakdown of ascorbic acid and polyunsaturated fatty acids and the Maillard reaction (Gül Akıllıoğlu et al. 2015). At the time of food processing and cooking, breakdown and reorganization of sugars take place during the Maillard reaction as a result of which furan is formed (Crews and Castle 2007). Furan and HMF are extensively found in food products due to the presence of several pathways which are responsible for their development in food products (Anese and Suman 2013). Due to the carcinogenic effects of furan and HMF, research is being carried out to decrease the development of these compounds in food products (Friedman and Levin 2008; Medeiros et al. 2012). The European Commission has specified the concentration of furans in food products (The European Commission 2014).

4.5.3 Heterocyclic Amines

According to IARC, heterocyclic amines can cause several types of cancer (IARC 1993). PhIP, IQ, IQx, MeIQ and DiMeIQx, Harman, and norharman are some of the frequently found heterocyclic amines in food products (Puangsombat et al. 2012). Production of heterocyclic amines takes place in the muscles of foods under extreme conditions of temperature, for example, pan-frying and grilling of meat or fish at high temperatures or roasting of meat or fish for a long period (Gokmen 2015; Rahman et al. 2014). Several factors affect the production of HAs including temperature, duration, and procedure of cooking, acid content, fat content, the amino acid content of meat, and presence of inhibitors, enhancers, and precursors (Rahman et al. 2014; Vitaglione and Fogliano 2004). Many researchers reported that the Maillard reaction could be responsible for the formation of HAs (Skog et al. 1998; Vitaglione and Fogliano 2004).

When reaction among amino acids, creatine/creatinine, and hexoses takes place, IQ HAs are produced (Skog et al. 1998; Vitaglione and Fogliano 2004). The amino-imidazole part of IQ is formed by creatine by the process of cyclization and the rest of the part of the IQ molecule is produced by breakdown compounds of Strecker (Skog et al. 1998). Breakdown compounds of Strecker include pyridine or pyrazine which are formed during the Maillard reaction between hexoses and amino acids. Linkage of these two parts takes place through Aldol condensation reaction by producing Strecker aldehyde. The reaction between creatine and different amino acids including phenylalanine, leucine, isoleucine, and tyrosine leads to the development of PhIP. In the liquid model approach, glucose affects the production of PhIP from phenylalanine and creatine (Skog et al. 1998). HAs are produced on the external surface of food products (Puangsombat et al. 2012). HAs are toxic components, still, there is no regulation on the use of HAs.

4.6 Preventive and Protective Methods

4.6.1 Choice of the Foodstuff

To check the Maillard reaction, it is extremely important to choose raw material based on its nutritional profile. In potatoes, variety, agronomic factors, and storage environment are some of the variables that affect the quantity of acrylamide in the final product, which might be due to the difference in amino acids constitution, mainly asparagine and reducing sugars of potatoes (Bent et al. 2012; Knutsen et al. 2009). These variabilities in amino acid composition and reducing sugars cause less or more optimum circumstances for the Maillard reaction (Halford et al. 2012), such as in plant-based food products, the quantity of asparagine and reducing sugars act as a criterion for the development of acrylamide: the lesser the quantity of precursor (i.e., asparagines and reducing sugars), the lesser the acrylamide formation (Halford et al. 2012; Loaëc et al. 2014). Nutrients provided during crop production also affect the nutritional status of the plant by modifying the sugars to free amino acids proportion (Elmore et al. 2010). For example, nitrogen in chicory enhances the free asparagine content (Loaëc et al. 2014) and deficiency of sulfur, reduces the asparagine content in potato tubers while increasing asparagine content in wheat grain (Elmore et al. 2008, 2010). Except acrylamide formation, alteration in amino acid and sugar levels can cause changes in the volatile constitution, such as enhanced levels of amino acids results in the formation of volatile compounds such as Strecker aldehydes and their condensation products, while an enhanced level of sugars causes more formation products derived through sugar breakdown (Elmore et al. 2010). Also, the harvesting stage of production affects the amount of the Maillard reaction precursors. Bananas harvested at full maturity, contain a high amount of reducing sugars, which causes high acrylamide formation (Daniali et al. 2013).

In the case of HAs, the quantity is affected by the type of meat (beef, pork, or chicken) and the sample composition (such as chicken with or without skin). pH, moisture content, fat, and protein concentration of the meat samples can alter the amount and type of HAs formed. For example, to reduce the consumption of HAs, chicken must be cooked with skin and then take out the skin before consumption. Throughout the cooking duration, the skin remains in direct contact with the cooking surface, which leads to the development of HAs in the skin. Also, due to higher levels of fat in the skin, HAs are formed because lipids are more efficacious in the transmission of heat through conduction into the food product (Puangsombat et al. 2012). If chicken is directly cooked without skin, HAs formation will directly take place on the chicken meat which is consumed. The storage environment (temperature and period of storage) of raw materials can also affect the quantity of the precursor of the Maillard reaction. For example, potatoes stored at 9 °C enhance the levels of reducing sugars which is an optimum condition for the Maillard reaction (Wicklund et al. 2006). Some researchers have also reported that the initial amount of phenolic compounds in potatoes functions as natural inhibitors of acrylamide throughout the thermal process (Zhu et al. 2010). Thus, the selection of raw material affects the subsequent Maillard reaction. The use of low sugar cultivars of potatoes can be used to restrict the production of acrylamide in potato products (Palermo et al. 2016).

4.6.2 Processing of the Foodstuff

Raw material can be treated either by physical or chemical methods to reduce its reactivity for the Maillard reaction. For example, when raw substances are treated with water baths of different compositions, a proportion of sugars and amino acids are released into these baths. This reduces the number of reactive species required for the Maillard reaction (Viklund et al. 2010). Hot-water treatment (blanching) of potato tubers reduces the acrylamide content of tubers and the final acrylamide content is determined by the duration and temperature of the treatment. In potato tubers, minimum acrylamide levels were reported when blanching was performed at low temperature for a longer duration (at 50 °C for 70 min or at 70 °C for 40 min) (Pedreschi et al. 2004, 2006). Several methods of blanching can be used which include steam, boiling water, and ultrasound treatments. In all the blanching treatments, the highest vitamin content of the food products and the acceptable nutrient amount must be retained along with minimum consequences of the Maillard reaction (Gamboa-Santos et al. 2014; Leong and Oey 2012). By immersing potatoes in water, acidic solutions, NaCl solution, or taurine for different durations, notably reduces the quantity of acrylamide in potato chips and potato strips (Pedreschi et al. 2004, 2006, 2007; Shin et al. 2010). Acidification of the bleaching solution reduces its pH causing adverse conditions for the Maillard reaction and also causes leaching of free asparagines and reducing sugars from the uppermost layers of food products (Pedreschi et al. 2004). The free amino group of taurine can compete with amino groups of other amino acids, resulting in decreased acrylamide content produced in a potato chips model (Shin et al. 2010). To restrict the production of HAs in meat products, a solution named marinade is used which inhibits direct exposure of the food product to the heating surface (Hasnol et al. 2014). But the effectiveness of marinade solution depends on the type of HAs produced (Gibis and Weiss 2012), the formulation of marinade solution, and the duration of marinating (Hasnol et al. 2014; Quelhas et al. 2010). But the use of marinade can alter the odor and aroma of the meat (Gibis and Weiss 2012; Quelhas et al. 2010). In the production of HAs, sugar also acts as a precursor, although the addition of a small amount of sugar to cooked ground meat results in less HAs production. The mechanism restricting the formation of HAs by carbohydrates is not completely understood. It is assumed that either sugar or a few MRPs formed might combine with creatine or creatinine to compete with the HAs formation reactions (Shin and Ustunol 2004). To restrict the production of HAs, honey is the most effective among all the sugars (Hasnol et al. 2014; Shin and Ustunol 2004).

4.6.3 Impact of the Formulation

The formulation is comprised of the most appropriate components of the recipe which results in the production of a food product with required properties in terms of flavor, color, and nutritional value.

4.6.3.1 Choice of the Ingredients

The process of browning and production of unfavorable components can be checked by selecting food ingredients correctly. Substitution of few components with another can affect the Maillard reaction, as in the case of bread formulation in which white flour is substituted with whole-grain flour of maize, oat, soy, wheat bran, rye, or wheat, which results in enhanced nutrient content in bread but can cause the formation of MRPs. Bread made from wheat bran and maize contains a low amount of acrylamide (Serpen et al. 2012). The high amount of water in biscuits helps in the movement of acrylamide precursors which ultimately enhances the production of acrylamide (Anese et al. 2011a), while the high concentration of fat in biscuits decreases the production of acrylamide (Anese et al. 2011a, b). A high concentration of ammonium hydrogen carbonate (baking agent) during gingerbread making results in the formation of more acrylamide, increase in pH, and dark color development in gingerbread (Amrein et al. 2004). Ammonium hydrogen carbonate can be replaced with sodium hydrogen carbonate as the latter can decrease the acrylamide content of foodstuffs by 30% (Amrein et al. 2004; Graf et al. 2006).

4.6.3.2 Role of Salts

Salts can decrease acrylamide content in foodstuffs but their functioning varies with their nature and quantity. Salts are uncomplicated to use and inexpensive. NaCl affects the Maillard reaction, but its end results are disputable. For example, during bread-making, if a small amount of salt is added, it restricts the activity of enzymes resulting in less acrylamide, while more amount of salt enhances the acrylamide formation by restricting the growth of yeast (Claus et al. 2008). Also when NaCl is mixed in olive juice, it results in a 30% reduction of acrylamide content (Casado et al. 2010), but NaCl does not show any decreasing impact on acrylamide content in potato powder (Mestdagh et al. 2008b). A high concentration of NaCl restricts the Maillard reaction (Kwak and Lim 2004), whereas HMF formation does not affect much by Na^+ (Gökmen and Şenyuva 2007). Thus, the occurrence of NaCl alters the circumstances enhancing or restricting the Maillard reaction (Troise and Fogliano 2013). Phytic acid can effectively reduce the non-enzymatic browning in apple juices that arises at the time of storage but increases acrylamide content in Maillard models (Du et al. 2012; Wang et al. 2013). $CaCl_2$ and $MgCl_2$ enhance acrylamide content in ripe olives, while reducing acrylamide content in fried potato strips and a blend of potato flour: semolina (Wang et al. 2013; Mulla et al. 2011; Casado et al. 2010; Ou et al. 2008).

4.6.3.3 Addition of Sulfite-Containing Compounds

Sulfite-containing compounds can inhibit the Maillard reaction, especially in fruits and vegetables (Ahmed et al. 2010; Nursten 2005). Currently, it has also been reported that sulfites are efficient in reducing the formation of acrylamide (Yuan et al. 2011) by restricting the intermediate products which lead to the formation of acrylamide (Casado et al. 2010). Several approaches can be used for sulfuring, such as burning of sulfite-containing compounds, SO_2 gas, and immersion in a solution of $Na_2S_2O_5$, which restrict the Maillard reaction (Coşkun et al. 2013). Sulfite-containing compounds are also able to check the process of enzymatic browning (Roig et al. 1999). Although sulfites are beneficial as they can inhibit the processes of browning, their addition to food products is controlled as these compounds can cause allergies and can interfere in the digestion of carbohydrates (Banon et al. 2007).

4.6.3.4 Addition of Competitive Amino Acid Group

Incorporating proteins or amino acids can check the formation of some specific compounds, majorly by two types of reaction: competitiveness between incorporated amino acid and asparagine throughout the Maillard reaction and bond formation between nucleophilic group of incorporated amino acid and the carbon-carbon double bond of acrylamide (Brathen et al. 2005; Salazar et al. 2012a, b; Claus et al. 2008; Mustafa et al. 2009). Amino acids containing sulfur compound (cysteine and reduced glutathione) are more effective in decreasing acrylamide production in ripe olive (Casado et al. 2010), while cysteine alone can successfully restrict the process of browning and production of acrylamide in potato (Mestdagh et al. 2008b). But the use of sulfur-containing amino acids can develop undesirable flavors in food products as they contain sulfur (Casado et al. 2010; Mestdagh et al. 2008b). Natural compounds containing a high amount of amino acids can also restrict the Maillard reaction. These natural compounds include amaranth flour and amaranth protein extract. Amaranth flour extract when used in fried tortilla chips, cookies, and baked tortilla chips did not affect acrylamide content, while amaranth protein extract did affect acrylamide content. Amaranth protein contains a high amount of lysine and cysteine which causes competition between asparagine and amino acid residues (Salazar et al. 2012a, b).

4.6.3.5 Addition of Microbial Organisms and Enzymes

Asparagines act as a substrate for asparaginase and causes hydrolysis of the amide group of asparagine and as a result of this process, aspartic acid and ammonia are formed. This reaction occurs in products of cereal and potatoes to eliminate asparagines, the precursor of acrylamide (Ciesarová et al. 2006; Capuano et al. 2009; Xu et al. 2016). Asparaginase is naturally found in *Aspergillus*, *Erwinia*, *E. coli*, animals, and some plants but it does not occur in humans (Ciesarová et al. 2006). To enhance the efficiency of asparaginase, it could be combined with other approaches, such as hot-water treatment of potato slices prior to asparaginase application, to decrease the acrylamide content in potato chips (Pedreschi et al. 2011). Asparaginase application does not affect the aroma and flavor of the food product (Anese et al. 2011b; Palermo et al. 2016). The quantity of yeast used for the fermentation process during bread-making also reduces the acrylamide content by reducing the number of asparagines (Huang et al. 2008). Temperature and pH are the two major factors that affect the activities of the microbes (Xu et al. 2016).

4.6.3.6 Addition of Organic Acid

Organic acids lowers the pH and cause undesirable conditions for the Maillard reaction (Mestdagh et al. 2008b). Tartaric, citric, lactic, and acetic acid are used as additives in the production of biscuits, gingerbread, and a potato model system reduces the amount of acrylamide. But the quantity of acid added to the food products is restricted by the alteration of organoleptic properties that can take place (Amrein et al. 2004; Graf et al. 2006; Mestdagh et al. 2008b). The addition of citric acid along with glycine or lysine decreases acrylamide content in a potato powder (Mestdagh et al. 2008b).

4.6.3.7 Addition of Phenolic Compounds

The effect of several polyphenols has been studied to restrict the Maillard reaction, giving positive, negative, and no results at all (Salazar et al. 2012a, b). Phenolic acids including cinnamic acid, ferulic, hydroxybenzoic acids, vanillic, and syringic have shown hardly any effect on reduction of browning but enhanced the anti-browning capacity of citric acid (Kwak and Lim 2005); however coumaric acid, cinnamic acid, gallic acid, caffeic acid, ferulic acid catechin, and epicatechin has no effect on acrylamide formation (Bassama et al. 2010). Ferulic acid has an effect on the formation of fluorescent AGEs, MRPs, and melanoidins. It reduces AGEs production by about 90%, MRPs by 10%, and melanoidins by 28% (Silvan et al. 2011).

4.6.3.8 Addition of Vitamins

As vitamins are antioxidants in nature, they can effectively decrease AGEs. Among vitamins, ascorbic acid is the most researched vitamin which takes part in the Maillard reaction. On contrary, ascorbic acid induces the process of non-enzymatic browning and is also a source of furans (Limacher et al. 2007; Nursten 2005). However, few reports are available on the properties of ascorbic acid which can reduce acrylamide formation and browning (Gokmen 2015; Zeng et al. 2009; Yuan et al. 2011). Reduced acrylamide content due to the action of ascorbic acid might be described by a reaction of asparagine with the carbonyl groups of the oxidation and/or breakdown products of ascorbic acid (Gokmen 2015). In addition, several B vitamins can successfully inhibit the Maillard reaction. In model systems, vitamins B6 (pyridoxamine and pyridoxine) and B7 (biotin) can effectively prevent the development of acrylamide. When added to fried potato strips, vitamin B3 (nicotinic acid) is beneficial (Yuan et al. 2011). In model systems and beef patties, vitamins B6, B7, B3, and ascorbic acid inhibit the synthesis of HAs (Wong et al. 2012).

4.6.3.9 Addition of Hydrocolloids

Hydrocolloids can alter the functional properties of the solvent used in food systems. Hydrocolloids can act as stabilizers, thickeners, emulsifiers, bulking agents, and gelling agents (Viekbe et al. 2014) in the formulation. Alginic acid and pectin are hydrocolloids that are responsible for a reduced quantity of acrylamide in fried potato strips if these are in adequate amounts. On the other hand, carob gum, carrageenan, hydroxypropyl distarch, phosphate, and xanthan gum are hydrocolloids responsible for increasing the quantity of acrylamide (Zeng et al. 2010).

4.6.3.10 Addition of Protein-Rich Components

Protein-rich compounds (such as cod fillet) cause a significant reduction in the acrylamide content of samples as in the case of potatoes. Reduction in acrylamide content is a result of competitive reactions and breakdown processes including proteins (Rydberg et al. 2003).

4.7 Conclusion

Enzymatic browning is the cause of fruit and vegetable degradation. The PPO response can be found in a wide variety of fruits and vegetables. It is responsible for enzymatic browning in fresh fruits and vegetable crops during processing procedures, such as slicing, chopping, and so on, resulting in cell damage. Chemical processes for avoiding enzymatic browning include the use of specific agents such as ascorbic acid, sodium bisulfite, chelating agents, or the addition of antioxidants and agents for control of undesired browning. Physical approaches for avoiding enzymatic browning include cooling, heating, blanching, freezing, and changing product atmospheres such as MAP. However, these actions could be used in various combinations for additional efficacy in improving them for different classes and cultivars, but modest practices, such as the use of edible coating films that vigorously obstruct both browning and textural corrosion, can also be used to elucidate some difficulties. Antisense RNA technology is being used to develop new strategies for preventing enzymatic browning. Peeling, cutting, grating, and other processing processes alter the structural integrity of fruits and vegetables, resulting in quality issues such as browning, off-flavor, and texture degradation. Browning of fruits and vegetables is caused by the enzymatic oxidation of phenolic chemicals, which results in a loss of quality qualities in terms of appearance or nutritional value, as well as the transformation of food products into harmful substances.

The development of anti-browning agents in the food industry is critical to maintaining the quality of fruits and vegetable yields. Efficacy and cost-effectiveness have traditionally been important aspects to consider when designing anti-browning treatments. Nonetheless, current anti-browning agent developments must match customer demands for health benefits, natural sources, and long-term sustainability. Food elements like onions, pineapples, lemons, grapes, and wine, as well as several other dietary components, have been examined for their anti-browning capabilities. Some of them have biological functions, while also inhibiting PPOs. Furthermore, research is being conducted to determine the anti-browning actions of food waste and byproducts. Critical enzymatic browning in foods is a source of major concern that must be investigated using active enzyme inhibitors. Components capable of preventing enzymatic browning in food harvests by interfering with tyrosinase-mediated processes or lowering the proportion of *o*-quinones to *o*-diphenols have been thoroughly recognized and manipulated. The safety of an inhibitor that will be employed in food processing is paramount. There is a continuing need for research into modified inhibitors derived from natural sources to ensure that they do not have any negative side effects. Flavors, colorants, anti-browning agents, spice nutrients, and antibacterial components can all be protected using edible coatings. Coatings extend the shelf life of food and reduce the possibility of germs flourishing on food surfaces. Though precise research on freshly cut fruits is lacking, and their industrial use is still in the works, a new development in edible coatings is on the horizon, with the primary goal of adding and/or organizing the release of active ingredients using nano-technological responses such as multilayered systems and nano-encapsulation. Nanoscale nutrients, additives, and nano-sized delivery systems for bioactive components are now used to enhance the nutritional properties of food. The use of edible coverings in conjunction with nano- and micro-encapsulation of active composites allows for the control of active component release under specific conditions, keeping them away from heat, moisture, and other harsh conditions and increasing their stability and feasibility. However, the use of silver nanoparticles (Ag NPs) could be a future alternative to damaging and expensive enzymatic browning inhibitors, antibacterial agents, and antioxidants. Enzymatic browning was significantly reduced by Ag NPs, with an enzymatic browning reduction index of 8.4. In recent research, chitosan nano-encapsulation has been proven to be beneficial in increasing the PPO inhibitory activity of ascorbyl palmitate.

References

Aguilo-Aguay I, Soliva-Fortuny R, Martin-Belloso O (2009) Avoiding non-enzymatic browning by high-intensity pulsed electric fields in strawberry, tomato and watermelon juices. J Food Eng 92: 37–43

Ahmed M, Akter MS, Eun JB (2010) Peeling, drying temperatures, and sulphite treatment affect physicochemical properties and nutritional quality of sweet potato flour. Food Chem 121:112–118

Ajandouz EH, Puigserver A (1999) Non-enzymatic browning reaction of essential amino acids: effect of pH on caramelization and Maillard reaction kinetics. J Agric Food Chem 47(5): 1786–1793

Ajandouz EH, Tchiakpe LS, Dalle Ore F et al (2001) Effects of pH on caramelization and Maillard reaction kinetics in fructose-lysine model systems. J Agric Food Chem 66(7):926–931

Amarowicz R, Janiak M (2019) Hydrolysable tannins. In: Encyclopedia of food chemistry. Elsevier, Amsterdam, pp 337–343

Ames JM (1988) The Maillard Browning Reaction – an update. Chem Ind 5:558–561

Amodio ML, Cabezas-Serrano AB, Peri G et al (2011) Post-cutting quality changes of fresh-cut artichokes treated with different anti-browning agents as evaluated by image analysis. Postharvest Biol Technol 62(2):213–220

Amrein TM, Schönbachler B, Escher F et al (2004) Acrylamide in gingerbread: critical factors for formation and possible ways of reduction. J Agric Food Chem 52:4282–4288
Anese M, Suman M (2013) Mitigation strategies of furan and 5-hydroxymethylfurfural in food. Food Res Int 51:257–264
Anese M, Quarta B, Frias J (2011a) Modelling the effect of asparaginase in reducing acrylamide formation in biscuits. Food Chem 126:435–440
Anese M, Quarta B, Peloux L et al (2011b) Effect of formulation on the capacity of l-asparaginase to minimize acrylamide formation in short dough biscuit. Food Res Int 44:2837–2842
Ansorena MR, Marcovich NE, Roura SI (2011) Impact of edible coatings and mild heat shocks on quality of minimally processed broccoli (Brassica oleracea L.) during refrigerated storage. Postharvest Biol Technol 59(1):53–63
Antunes MDC, Dandlen S, Cavaco AM et al (2010) Effects of postharvest application of 1-MCP and post-cutting dip treatment on the quality and nutritional properties of fresh-cut kiwifruit. J Agric Food Chem 58(10):6173–6181
Apai W (2010) Effects of fruit dipping in hydrochloric acid then rinsing in water on fruit decay and browning of longan fruit. Crop Protect 29(10):1184–1189
Arias E, Gonzalez J, Oria R (2007) Ascorbic acid and 4-hexylresorcinol effects on pear PPO and PPO catalyzed browning reaction. J Food Sci 72(8):C422–C429
Arias E, Lopez-Buesa P, Oria R et al (2009) Extension of fresh-cut "Blanquilla" pear (Pyrus communis L.) shelf-life by 1-MCP treatment after harvest. Postharvest Biol Technol 54(1): 53–58
Badoud R, Fay LB, Hunston F et al (1995) Periodate oxidative degradation of Amadori compounds. Formation of N-Carboxymethyllysine and N-Carboxymethylamino acids as markers of the early Maillard reaction. In: Lee TC, Kim HJ (eds) Chemical markers for processed and stored foods, ACS Symposium series, vol 631. American Chemical Society, Chicago, IL, pp 208–220
Banchero M, Pellegrino G, Manna L (2013) Supercritical fluid extraction as a potential mitigation strategy for the reduction of acrylamide level in coffee. J Food Eng 115:292–297
Banon S, Diaz P, Rodrígues M et al (2007) Ascorbate, green tea and grape seed extracts increase the shelf life of low sulphite beet patties. Meat Sci 77:626–633
Barbagallo RN, Chisari M, Caputa G (2012) Effects of calcium citrate and ascorbate as inhibitors of browning and softening in minimally processed 'Birgah' eggplants. Postharvest Biol Technol 73:107–114
Barman K, Asrey R, Pal RK (2011) Putrescine and carnauba wax pretreatments alleviate chilling injury, enhance shelf life and preserve pomegranate fruit quality during cold storage. Sci Hortic 130(4):795–800
Bassama J, Brat P, Bohuon P et al (2010) Study of acrylamide mitigation in model system: effect of pure phenolic compounds. Food Chem 123:558–562
Bent GA, Maragh P, Dasgupta T (2012) Acrylamide in Carribean foods-residual levels and their relation to reducing sugar and asparagine content. Food Chem 133:451–457
Bhuyan DJ, Basu A (2017) Phenolic compounds potential health benefits and toxicity, Utilisation of bioactive compounds from agricultural and food production waste. CRC Press, Boca Raton, FL, p 414
Brathen EB, Kita A, Knutsen SH et al (2005) Addition of glycine reduces the content of acrylamide in cereal and potato products. J Agric Food Chem 53:3259–3264
Bunzel M, Seiler A, Steinhart H (2005) characterization of dietary fiber lignins from fruits and vegetables using the DFRC method. J Agric Food Chem 53(24):9553–9559. https://doi.org/10.1021/jf0520037
Calder BL, Skonberg DI, Davis-Dentici K et al (2011) The effectiveness of ozone and acidulant treatments in extending the refrigerated shelf life of fresh-cut potatoes. J Food Sci 76(8):S492–S498
Capuano E, Fogliano V (2011) Acrylamide and 5-hydroxymethylfurfural (HMF): a review on metabolism, toxicity, occurrence in food and mitigation strategies. LWT 44:793–810

Capuano E, Ferrigno A et al (2009) Effect of flour type on Maillard reaction and acrylamide formation during toasting of bread crisp model systems and mitigation strategies. Int Food Res J 42:1295–1302

Casado FJ, Sanchez AH, Montaño A (2010) Reduction of acrylamide content of ripe olives by selected additives. Food Chem 119:161–166

Chipley JR (2005) Sodium benzoate and benzoic acid. In: Michael Davidson P, Sofos JN, Branen AL (eds) Antimicrobials in food, 3rd edn. CRC Press, Boca Raton, FL, pp 11–48

Ciesarová Z, Kiss E, Boegl P (2006) Impact of L-asparaginase on acrylamide content in potato products. J Food Nutr Res 45:141–146

Claus A, Mongili M, Weisz G et al (2008) Impact of formulation and technological factors on the acrylamide content of wheat bread and bread rolls. J Cereal Sci 47:546–554

Cook S (2019) An historical review of phenylacetic acid. Plant Cell Physiol 60(2):243–254

Coşkun AL, Turkyilmaz M, Aksu OT et al (2013) Effects of various sulphuring methods and storage temperatures on the physical and chemical quality of dried apricots. Food Chem 141: 3670–3680

Crews C, Castle L (2007) A review of the occurrence, formation and analysis of furan in heat processed foods. Trends Food Sci Technol 18:365–372

Cruz RMS, Vieira MC, Fonseca SC et al (2011) Impact of thermal blanching and thermosonication treatments on watercress (Nasturtium officinale) quality: thermosonication process optimisation and microstructure evaluation. Food Bioprocess Technol 4(7):1197–1204

D'Archivio M, Filesi C, Di Benedetto R, Gargiulo R, Giovannini C, Masella R (2007) Polyphenols, dietary sources and bioavailability. Ann Ist Super Sanità 43(4):348–361

Daniali G, Jinap S, Hanifah NL et al (2013) The effect of maturity stages of banana on the formation of acrylamide in banana fritters. Food Control 32:386–391

Devi AF, Buckow R, Singh T et al (2015) Colour change and proteolysis of skim milk during high pressure thermal-processing. J Food Eng 147:102–110

Du YJ, Dou SQ, Wu SJ (2012) Efficacy of phytic acid as an inhibitor of enzymatic and non-enzymatic browning in apple juice. Food Chem 135:580–582

Elmore JS, Parker JK, Halford NG (2008) Effects of plant sulfur nutrition on acrylamide and aroma compounds in cooked wheat. J Agric Food Chem 56:6173–6179

Elmore JS, Dodson AT, Muttucumaru NG et al (2010) Effects of sulphur nutrition during potato cultivation on the formation of acrylamide and aroma compounds during cooking. Food Chem 122:753–760

El-Seedi HR, Salem MA, Khattab OM, El-Wahed AA, El-Kersh DM, Shaden Khalifa AM, Saeed A, Abdel-Daim MM, Hajrah NH, Alajlani MM, Halabi MF, Jassbi AR, Musharraf SG, Farag MA (2020) Dietary xanthones. In: Handbook of dietary phytochemicals. Springer, New York, pp 1–20

Feather MS, Mossine V, Hirsch J (1995) The use of aminoguanidine to trap and measure decarbonyl intermediates produced during the Maillard reaction. In: Lee TC, Kim HJ (eds) Chemical markers for processed and stored foods. ACS Symposium series 631. ACS, Chicago, IL, pp 24–31

Fennema OR (1976) In: Fennema OR (ed) Principles of food science. Marcel Dekker, New York

Ferrari CC, Sarantopoulos CIGL, Carmello-Guerreiro SM et al (2011) Effect of osmotic dehydration and pectin edible coatings on quality and shelf life of fresh-cut melon. Food Bioprocess Technol 6:80–91

Fouillaud M, Caro Y, Venkatachalam M et al (2018) Anthraquinones, Chapter 9. In: Phenolic compounds in food characterization and analysis. CRC Press, Boca Raton, FL, pp 130–170

Friedman M, Levin CE (2008) Review of methods for the reduction of dietary content and toxicity of acrylamide. J Agric Food Chem 56:6113–6140

Gamboa-Santos J, Megias-Perez R, Soria AC et al (2014) Impact of processing conditions on the kinetic of vitamin C degradation and 2-furoylmethyl amino acid formation in dried strawberries. Food Chem 153:164–170

Gardner HW (1989) Oxygen radical chemistry of polyunsaturated fatty acids. Free Radic 7(1): 65–86

Gardner HW (1995) Biological roles and biochemistry of the lipoxygenase pathway. HortScience 30(2):197–205

Ghidelli C, Rojas-Argudo C, Mateos M et al (2013) Effect of antioxidants in controlling enzymatic browning of minimally processed persimmon 'Rojo Brillante'. Postharvest Biol Technol 86: 487–493

Gibis M, Weiss J (2012) Antioxidant capacity and inhibitory effect of grape seed and rosemary extract in marinades on the formation of heterocyclic amines in fried beef patties. Food Chem 134:766–774

Gokmen V (2015) Acrylamide in food: analysis, content and potential health effects, 1st edn. Academic, New York

Gokmen V, Morales FJ (2014) Hydroxymethylfurfural. In: Encyclopedia of food safety, vol 2. Elsevier, Amsterdam, pp 404–408

Gökmen V, Şenyuva HZ (2007) Effects of some cations on the formation of acrylamide and furfurals in glucose-asparagine model system. Eur Food Res Technol 225:815–820

Gomez PL, Garcia-Loredo A, Salvatori DM et al (2011) Viscoelasticity, texture and ultrastructure of cut apple as affected by sequential anti-browning and ultraviolet-C light treatments. J Food Eng 107(2):214–225

Gonzalez-Cebrino F, Garcia-Parra J, Contador R et al (2012) Effect of high-pressure processing and thermal treatment on quality attributes and nutritional compounds of "Songold" plum puree. J Food Sci 77(8):C866–C873

Gossinger M, Moritz S, Hermes M et al (2009) Effects of processing parameters on colour stability of strawberry nectar from puree. J Food Eng 90(2):171–178

Graf M, Amrein TM, Graf S et al (2006) Reducing the acrylamide content of a semi-finished biscuit on industrial scale. LWT 39:724–728

Grimm E, Khanal BP, Winkler A et al (2012) Structural and physiological changes associated with the skin spot disorder in apple. Postharvest Biol Technol 64(1):111–118

Guan WQ, Fan XT (2010) Combination of sodium chlorite and calcium propionate reduces enzymatic browning and microbial population of fresh-cut "Granny Smith" apples. J Food Sci 75(2):M72–M77

Gül Akıllıoğlu H, Savaş Bahçeci K, Gökmen V (2015) Investigation and kinetic evaluation of furan formation in tomato paste and pulp during heating. Food Res Int 78:224–230

Gupta RK, Kumar P, Sharma A et al (2011) Color kinetics of aonla shreds with amalgamated blanching during drying. Int J Food Prop 14(6):1232–1240

Hajhashemi V, Vaseghi G, Pourfarzam M et al (2010) Are antioxidants helpful for disease prevention. Res Pharm Sci 5(1):1–8

Halford NG, Muttucumaru NG, Powers S et al (2012) Concentrations of free amino acids and sugars in nine potato varieties: effects of storage and relationship with acrylamide formation. J Agric Food Chem 60:12044–12055

Hashemi SMB, Nikmaram N, Esteghlal SJ et al (2017) Efficiency of ohmic assisted hydrodistillation for the extraction of essential oil from oregano (Origanum vulgare subsp. viride) spices. Innov Food Sci Emerg Technol 41:172–178

Hasnol NDS, Jinap S, Sanny M et al (2014) Effect of different types of sugars in a marinating formulation on the formation of heterocyclic amines in grilled chicken. Food Chem 145:514–521

Hoskin JC, Dimick PS (1984) Role of nonenzymatic browning during the processing of chocolate-a review. Process Biochem 19(3):92–104

Hradecky J, Kludska E, Belkova B et al (2017) Ohmic heating: a promising technology to reduce furan formation in sterilized vegetable and vegetable/meat baby foods. Innov Food Sci Emerg Technol 43:1–6

Huang W, Yu S, Zou Q et al (2008) Effects of frying conditions and yeast fermentation on the acrylamide content in you-tiao, a traditional Chinese, fried, twisted dough-roll. Int Food Res J 41:918–923

IARC (International Agency for Research on Cancer) (1993) Heterocyclic aromatic amines, IARC monographs on the evaluation of carcinogenic risks to humans. Some naturally occurring substances: food items and constituents, heterocyclic aromatic amines and mycotoxins, vol 56. IARC, Lyon, France, pp 165–244

IARC (International Agency for Research on Cancer) (1994) Acrylamide, IARC monographs on the evaluation of carcinogenic risks to humans. Some industrial chemicals, vol 60. IARC, Lyon, France, pp 389–434

IARC (International Agency for Research on Cancer) (1995) Furan, IARC monographs on the evaluation of carcinogenic risks to humans. Dry cleaning, some chlorinated solvents and other industrial chemicals, vol 63. IARC, France, Lyon, pp 393–408

Icier F, Yildiz H, Baysal T (2006) Peroxidase inactivation and colour changes during ohmic blanching of pea puree. J Food Eng 74(3):424–429

Ingraham LL (1955) Reaction-inactivation of polyphenol oxydase: catachol and oxygen dependence. J Am Chem Soc 77(10):2875–2876

Ioannou I, Ghoul M (2013) Prevention of enzymatic browning in fruit and vegetables. Eur Sci J 9(30):310–340. ISSN 1857-7431 311

Jang JH, Moon KD (2011) Inhibition of polyphenol oxidase and peroxidase activities on fresh-cut apple by simultaneous treatment of ultrasound and ascorbic acid. Food Chem 124(2):444–449

Janositz A, Noack AK, Knorr D (2011a) Pulsed electric fields and their impact on the diffusion characteristics of potato slices. LWT 44:1939–1945

Janositz A, Semrau J, Knorr D (2011b) Impact of PEF treatment on quality parameters of white asparagus (Asparagus officinalis L.). Innov Food Sci Emerg Technol 12:269–274

Jestoi M, Järvinen T, Järvenpää E et al (2009) Furan in the baby-food samples purchased from the Finnish markets-determination with SPME-GC–MS. Food Chem 117:522–528

Jiang T, Zheng X, Li J et al (2011) Integrated application of nitric oxide and modified atmosphere packaging to improve quality retention of button mushroom (Agaricus bisporus). Food Chem 126(4):1693–1699

Kaur C, Kapoor HC (2001) Antioxidants in fruits and vegetables-the millennium's health. Int J Food Sci Technol 36:703–725

Keramat J, Le Bail A, Prost C (2011) Acrylamide in baking products: a review article. Food Bioprocess Technol 4:530–543

Khawand, Courtois A, Valls J, Richard T, Krisa S (2018) A review of dietary stilbenes: sources and bioavailability. Toni El Phytochem Rev 17:1007–1029

Khunpon B, Uthaibutra J, Faiyue B et al (2011) Reduction of enzymatic browning of harvested 'Daw' longan exocarp by sodium chlorite. ScienceAsia 37(3):234–239

Kim TK, Kim S, Lee KG (2010) Analysis of furan in heat-processed foods consumed in Korea using solid phase microextraction-gas chromatography/mass spectrometry (SPME-GC/MS). Food Chem 123:1328–1333

Knutsen SH, Dimitrijevic S, Molteberg EL et al (2009) The influence of variety, agronomical factors and storage on the potential for acrylamide formation in potatoes grown in Norway. J Food Sci Technol 42:550–556

Koubaa M, Barba FJ, Grimi N et al (2016a) Recovery of colorants from red prickly pear peels and pulps enhanced by pulsed electric field and ultrasound. Innov Food Sci Emerg Technol 37(Part C):336–344

Koubaa M, Roselló-Soto E, Barba-Orellana S et al (2016b) Novel thermal technologies and fermentation. In: Ojha KS, Tiwari BK (eds) Novel food fermentation technologies. Springer, New York, pp 155–163

Koubaa M, Roohinejad S, Mungure T et al (2019) Effect of emerging processing technologies on Maillard reactions. In: Reference module in food science, Food Chem. Elsevier, Amsterdam, pp 76–82

Kroh LW (1994) Caramelization in foods and beverages. Food Chem 51(4):373–379

Kumar S, Mishra BB, Saxena S et al (2012) Inhibition of pericarp browning and shelf life extension of litchi by combination dip treatment and radiation processing. Food Chem 131(4):1223–1232

Kwak EJ, Lim SI (2004) The effect of sugar, amino acid, metal ion, and NaCl on model Maillard reaction under pH control. J Amino Acids 27:85–90

Kwak EJ, Lim SI (2005) Inhibition of browning by antibrowning agents and phenolic acids or cinnamic acid in the glucose-lysine model. J Sci Food Agric 85:1337–1342

Latorre ME, Bonelli PR, Rojas AM et al (2012) Microwave inactivation of red beet (Beta vulgaris L. var. conditiva) peroxidase and polyphenoloxidase and the effect of radiation on vegetable tissue quality. J Food Eng 109(4):676–684

Lavelli V, Caronni P (2010) Polyphenol oxidase activity and implications on the quality of intermediate moisture and dried apples. European Food Res Technol 231(1):93–100. https://doi.org/10.1007/s00217-010-1256-0

Leong SY, Oey I (2012) Effect of endogenous ascorbic acid oxidase activity and stability on vitamin C in carrots (Daucus carota subsp. sativus) during thermal treatment. Food Chem 134: 2075–2085

Limacher A, Kerler J, Conde-Petit B et al (2007) Formation of furan and methylfuran from ascorbic acid in model systems and food. Food Addit Contam 24:122–135

Lindley MG (1998) The impact of food processing on antioxidants in vegetable oils, fruits and vegetables. Trends Food Sci Technol 9(8–9):336–340

Loaëc G, Niquet-Léridon C, Henry N et al (2014) Effects of variety, agronomic factors, and drying on the amount of free asparagines and crude protein in chicory. Correlation with the acrylamide formation during roasting. Int Food Res J 63:299–305

Loncar M, Jakovljević M, Šubarić D et al (2020) Coumarins in food and methods of their determination. Foods 9(5):645

Lund MN, Ray CA (2017) Control of Maillard reactions in foods: strategies and chemical mechanisms. J Agric Food Chem 65:4537–4552

Luo YG, Lu SM, Zhou B et al (2011) Dual effectiveness of sodium chlorite for enzymatic browning inhibition and microbial inactivation on fresh-cut apples. LWT 44(7):1621–1625

Ma XJ, Gao JY, Tong P et al (2017) Tracking the behavior of Maillard browning in lysine/arginine–sugar model systems under high hydrostatic pressure. J Sci Food Agric 97:5168–5175

Machiels D, Istasse L (2002) Maillard reaction: importance and applications in food chemistry. Ann Med Vet 146(6):347–352

Malec LS, Gonzales ASP, Naranjo GB et al (2002) Influence of water activity and storage temperature on lysine availability of a milk-like system. Int Food Res J 35(9):849–853

Manzocco L, Calligaris S, Mastrocola D et al (2001) Review of non-enzymatic browning and antioxidant capacity in processed foods. Trends Food Sci Technol 11:340–346

Marshall MR, Kim J, Wei CI (2000) Enzymatic browning in fruits, vegetables and seafoods. FAO, Rome, Italy. http://www.fao.org/waicent/faoinfo/agricult/ags/Agsi/ENZYMEFINAL/COPYRIGH.HTM

Martins SIFS, van Boekel MAJS (2003) Melanoidins extinction coefficient in the glucose/glycine Maillard reaction. Food Chem 83(1):135–142

Mathies G, Whitaker JR (1984) Modifications of proteins by polyphenol oxidase and peroxidase and their products. J Food Biochem 8:137

Matsui KN, Gut JAW, de Oliveira PV et al (2008) Inactivation kinetics of polyphenol oxidase and peroxidase in green coconut water by microwave processing. J Food Eng 88(2):169–176

Mayer AM, Harel E (1979) Polyphenol oxidases in plants. Phytochemistry 18(2):193–215

Mazzeo T, N'Dri D, Chiavaro E et al (2011) Effect of two cooking procedures on phytochemical compounds, total antioxidant capacity and colour of selected frozen vegetables. Food Chem 128(3):627–633

Medeiros VR, Mestdagh F, De Meulenaer B (2012) Acrylamide formation in fried potato products-present and future, a critical review on mitigation strategies. Food Chem 133:1138–1154

Mestdagh F, De Wilde T, Fraselle S et al (2008a) Optimization of the blanching process to reduce acrylamide in fried potatoes. J Food Sci Technol 41:1648–1654

Mestdagh F, Maertens J, Cucu T et al (2008b) Impact of additives to lower the formation of acrylamide in a potato model system through pH reduction and other mechanisms. Food Chem 107:26–31

Mohebbi M, Ansarifar E, Hasanpour N et al (2012) Suitability of aloe vera and gum tragacanth as edible coatings for extending the shelf life of button mushroom. Food Bioprocess Technol 5(8): 3193–3202

Mojska H, Gielecińska I, Stoś K (2012) Determination of acrylamide level in commercial baby foods and an assessment of infant dietary exposure. Food Chem Toxicol 50:2722–2728

Mulla MZ, Bharadwaj VR, Annapure US et al (2011) Effect of formulation and processing parameters on acrylamide formation: a case study on extrusion of blends of potato flour and semolina. LWT 44:1643–1648

Mustafa A, Fink M, Kamal-Eldin A et al (2009) Interaction effects of fermentation time and added asparagine and glycine on acrylamide content in yeast-leavened bread. Food Chem 112:767–774

Namiki M (1988) Chemistry of Maillard reactions: recent studies on the browning reaction. Mechanism and the development of antioxidant and mutagens. Adv Food Nutr Res 32:116–170

Nawar WW (1985) Lipids. In: Fennema OR (ed) Food chemistry. Marcel Dekker, New York, pp 139–244

Nawar WW (1996) Lipids. In: Fennema OR (ed) Food chemistry, 3rd edn. Marcel Dekker, New York, pp 225–319

Nicolas JJ, Richard-Forget FC, Goupy PM et al (1994) Enzymatic browning reactions in apple and apple products. Crit Rev Food Sci Nutr 34:109–157

Nursten H (2005) The Maillard reaction-chemistry, biochemistry and implications. The Royal Society of Chemistry, Cambridge, UK

O'Beirne D, Murphy E, Eidhin DN (2011) Effects of argon enriched low-oxygen atmospheres and of high-oxygen atmospheres on the kinetics of polyphenoloxidase (PPO). J Food Sci 76(1): E73–E77

Oey I, Faridnia F, Roohinejad S, Leong SY, Lee P, Kethireddy V (2016) High hydrostatic pressure processing at high temperature: impact on quality attributes and future outlook from consumers' perspective. In: Amit J (ed) Food processing technologies: impact on product attributes. CRC Press, Boca Raton, FL

Olano A, Martínez-Castro I (1996) Nonenzymatic browning. In: Nollet ML (ed) Handbook of food analysis, vol 2. Marcel Dekker, New York, pp 1683–1721

Olivas GI, Barbosa-Canovas GV (2005) Edible coatings for fresh-cut fruits. Crit Rev Food Sci Nutr 45(7–8):657–670

Oms-Oliu G, Aguilo-Aguayo I, Martin-Belloso O (2006) Inhibition of browning on fresh-cut pear wedges by natural compounds. J Food Sci 71(3):S216–S224

Oms-Oliu G, Gemma González LA, Varela P, Martin-Belloso O, Soliva-Fortuny R, Munuera IP, Fiszman S (2010) Recent approaches using chemical treatments to preserve quality of fresh-cut fruit: a review. Postharvest Biol Technol 57(3):139–148. https://doi.org/10.1016/j.postharvbio.2010.04.001

Ou S, Lin Q, Zhang Y et al (2008) Reduction of acrylamide formation by selected agents in fried potato crisps on industrial scale. Innov Food Sci Emerg Technol 9:116–121

Oz AT, Ulukanli Z (2012) Application of edible starch-based coating including glycerol plus oleum nigella on arils from long-stored whole pomegranate fruits. J Food Process Preserv 36(1):81–95

Özel A, Colak A, Arslan O et al (2010) Purification and characterisation of a polyphenol oxidase from Boletus erythropus and investigation of its catalytic efficiency in selected organic solvents. Food Chem 119(3):1044–1049

Palermo M, Gökmen V, De Meulenaer B et al (2016) Acrylamide mitigation strategies: critical appraisal of the Food Drink Europe toolbox. Food Funct 7:2516–2525

Pedreschi F, Kaack K, Granby K (2004) Reduction of acrylamide formation in potato slices during frying. LWT 37:679–685
Pedreschi F, Kaack K, Granby K (2006) Acrylamide content and color development in fried potato strips. Food Res Int 39:40–46
Pedreschi F, Bustos O, Mery D et al (2007) Color kinetics and acrylamide formation in NaCl soaked potato chips. J Food Eng 79:989–997
Pedreschi F, Mariotti S, Granby K, Risum J (2011) Acrylamide reduction in potato chips by using commercial asparaginase in combination with conventional blanching. LWT 44:1473–1476
Peñalver MJ, Fenoll LG, Rodríguez-López JN et al (2005) Reaction mechanism to explain the high kinetic autoactivation of tyrosinase. J Mol Catal B Enzym 33:35–42
Pereira RN, Souza BWS, Cerqueira M, Vicente A (2010) Effects of electric fields on protein unfolding and aggregation: influence on edible films formation. Biomacromolecules 11(11). https://doi.org/10.1021/bm100681a
Perez Locas C, Yaylayan VA (2008) Isotope labeling studies on the formation of 5 (hydroxymethyl)-2-furaldehyde (HMF) from sucrose by pyrolysis-GC/MS. J Agric Food Chem 56:6717–6723
Prajapaty VK, Nema PK, Rathore SS (2011) Effect of pretreatment and drying methods on quality of value-added dried aonla (Emblica officinalis Gaertn) shreds. J Food Sci Technol 48(1):45–52
Puangsombat K, Gadgil P, Houser TA et al (2012) Occurrence of heterocyclic amines in cooked meat products. Meat Sci 90:739–746
Queiroz C, Lopes MLM, Fialho E et al (2008) Polyphenol oxidase: characteristics and mechanisms of browning control. Food Rev Int 24(4):361–375
Quelhas I, Petisca C, Viegas O et al (2010) Effect of green tea marinades on the formation of heterocyclic aromatic amines and sensory quality of pan-fried beef. Food Chem 122:98–104
Quintas M, Guimarães C, Baylina J et al (2007) Multiresponse modelling of the caramelisation reaction. Innov Food Sci Emerg Technol 8:306–315
Rahman U, Sahar A, Khan MI et al (2014) Production of heterocyclic aromatic amines in meat: chemistry, health risks and inhibition: a review. LWT 59:229–233
Ramesh MN, Wolf W, Tevini D et al (2002) Microwave blanching of vegetables. J Food Sci 67(1): 390–398
Rizzi GP (1994) The Maillard reaction in food. In: Labuza TP, Reineccius GA, Morrier VM, O'Brien J, Baynes JW (eds) Maillard reaction in chemistry food and health. The Royal Society of Chemistry Publisher, Cambridge, pp 11–19
Rocculi P, Romani S, Dalla Rosa M (2005) Effect of MAP with argon and nitrous oxide on quality maintenance of minimally processed kiwifruit. Postharvest Biol Technol 35(3):319–328
Roig MG, Bell JF, Rivera ZS et al (1999) Studies on the occurrence of non-enzymatic browning during storage of citrus juice. Int Food Res J 32:609–619
Rydberg P, Eriksson S, Tareke E et al (2003) Investigations of factors that influence the acrylamide content of heated foodstuffs. J Agric Food Chem 51:7012–7018
Salas PG, Soto AM, Carretero AS et al (2010) 1,2 Phenolic-compound-extraction systems for fruit and vegetable samples. J Mol Struct 15:8813–8826
Salazar R, Arámbula-Villa G, Hidalgo FJ et al (2012a) Mitigating effect of piquin pepper (Capsicum annuum L. var. Aviculare) oleoresin on acrylamide formation in potato and tortilla chips. LWT 48:261–267
Salazar R, Arámbula-Villa G, Vázquez-Landaverde PA et al (2012b) Mitigating effect of amaranth (Amarantus hypochondriacus) protein on acrylamide formation in foods. Food Chem 135: 2293–2298
Santos MC, Nunes C, Rocha MAM et al (2013) Impact of high pressure treatments on the physicochemical properties of a sulphur dioxide-free white wine during bottle storage: evidence for Maillard reaction acceleration. Innov Food Sci Emerg Technol 20:51–58
Santos MC, Nunes C, Rocha MAM et al (2015) High pressure treatments accelerate changes in volatile composition of sulphur dioxide-free wine during bottle storage. Food Chem 188:406–414

Semwal RB, Semwal DK, Combrinck S et al (2020) Health benefits of chromones: common ingredients of our daily diet. Phytochem Rev 19:761–785
Serpen A, Gökmen V, Mogol BA (2012) Effects of different grain mixtures on Maillard reaction products and total antioxidant capacities of breads. J Food Compost Anal 26:160–168
Shao XF, Tu K, Tu S et al (2012) A combination of heat treatment and chitosan coating delays ripening and reduces decay in "Gala" apple fruit. J Food Qual 35(2):83–92
Shi J, Li J, Zhu S et al (2011) Browning inhibition on fresh-cut chestnut kernel by exogenous nitric oxide. Int J Food Sci Technol 46(5):944–950
Shin HS, Ustunol Z (2004) Influence of honey-containing marinades on heterocyclic aromatic amine formation and overall mutagenicity in fried beef steak and chicken breast. J Food Sci 69: 147–153
Shin DC, Kim CT, Lee WJ et al (2010) Reduction of acrylamide by taurine in aqueous and potato chipmodel systems. Int Food Res J 43:1356–1360
Silvan JM, Assar SH, Srey C et al (2011) Control of the Maillard reaction by ferulic acid. Food Chem 128:208–213
Skog KI, Johansson MAE, Jägerstad MI (1998) Carcinogenic heterocyclic amines in model systems and cooked foods: a review on formation, occurrence and intake. Food Chem Toxicol 36:879–896
Sotome I, Takenaka M, Koseki S et al (2009) Blanching of potato with superheated steam and hot water spray. LWT 42(6):1035–1040
Stadler RH, Blank I, Varga N et al (2002) Food chemistry: acrylamide from Maillard reaction products. Nature 419:449–450
Stadler RH, Robert F, Riedeker S et al (2004) In-depth mechanistic study on the formation of acrylamide and other vinylogous compounds by the Maillard reaction. J Agric Food Chem 52: 5550–5558
Supapvanich S, Prathaan P, Tepsorn R (2012) Browning inhibition in fresh-cut rose apple fruit cv. Taaptimjaan using konjac glucomannan coating incorporated with pineapple fruit extract. Postharvest Biol Technol 73:46–49
Teixido E, Núñez O, Santos FJ et al (2011) 5-Hydroxymethylfurfural content in foodstuffs determined by micellar electrokinetic chromatography. Food Chem 126:1902–1908
The European Commission (2013) Commission recommendation 2013/647/EU on investigations into the levels of acrylamide in food. Official Journal of the European Union
The European Commission (2014) Commission recommendation 2014/663/EU of 11 September 2014 amending the Annex to recommendation 2013/711/EU on the reduction of the presence of dioxins, furans and PCBs in feed and food. Official Journal of the European Union
Tian Y, Huang J, Xie T et al (2016) Oenological characteristics, amino acids and volatile profiles of Hongqu rice wines during pottery storage: effects of high hydrostatic pressure processing. Food Chem 203:456–464
Toivonen PMA, Brummell DA (2008) Biochemical bases of appearance and texture changes in fresh-cut fruit and vegetables. Postharvest Biol Technol 48(1):1–14
Tomasik P, Palasinski M, Wiejak S (1989) The thermal decomposition of carbohydrates. Part I: The decomposition of mono-, di- and oligosaccharides. Adv Carbohydr Chem Biochem 47:203–270
Troise AD, Fogliano V (2013) Reactants encapsulation and Maillard reaction. Trends Food Sci Technol 33:63–74
Viekbe C, Al-Assaf S, Phillips GO (2014) Food hydrocolloids and health claims. Bioact Carbohydr Diet Fibre 4:101–114
Viklund GÅI, Olsson KM, Sjöholm I et al (2010) Acrylamide in crisps: effect of blanching studied on long-term stored potato clones. J Food Compost Anal 23:194
Vina SZ, Olivera DF, Marani CM et al (2007) Quality of Brussels sprouts (Brassica oleracea L. gemmifera DC) as affected by blanching method. J Food Eng 80(1):218–225
Vitaglione P, Fogliano V (2004) Use of antioxidants to minimize the human health risk associated to mutagenic/carcinogenic heterocyclic amines in food. J Chromatogr A 802:189–199

Wang CT, Wang CT, Cao YP et al (2011) Effect of modified atmosphere packaging (MAP) with low and superatmospheric oxygen on the quality and antioxidant enzyme system of golden needle mushrooms (Flammulina velutipes) during postharvest storage. Eur Food Res 232(5): 851–860

Wang H, Zhou Y, Ma J et al (2013) The effects of phytic acid on the Maillard reaction and the formation of acrylamide. Food Chem 141:18–22

Whitaker JR, Lee CY (1995) Recent advances in chemistry of enzymatic browning. In: Lee CY, Whitaker JR (eds) Enzymatic browning and its prevention, ACS symposium series. American Chemical Society, Washington, DC, pp 2–7

Wicklund T, Østlie H, Lothe O et al (2006) Acrylamide in potato crisp-the effect of raw material and processing. LWT 39:571–575

Wilson DW, Nash P, Buttar HS et al (2017) The role of food antioxidants, benefits of functional foods, and influence of feeding habits on the health of the older person: an overview. Antioxidants 6(4):81

Wong D, Cheng KW, Wang M (2012) Inhibition of heterocyclic amine formation by water-soluble vitamins in Maillard reaction model systems and beef patties. Food Chem 133:760–766

Wu ZS, Zhang M, Wang S (2012) Effects of high pressure argon treatments on the quality of fresh-cut apples at cold storage. Food Control 23(1):120–127

Xu F, Oruna-Concha MJ, Elmore JS (2016) The use of asparaginase to reduce acrylamide levels in cooked food. Food Chem 210:163–171

Yuan Y, Shu C, Zhou B et al (2011) Impact of selected additives on acrylamide formation in asparagine/sugar Maillard model systems. Int Food Res J 44:449–455

Zawistowski J (1991) Polyphenol oxidase. Oxidative enzymes in foods. Elsevier Applied Science Ltd, London, pp 217–273

Zeng X, Cheng KW, Jiang Y et al (2009) Inhibition of acrylamide formation by vitamins in model reactions and fried potato strips. Food Chem 116:34–39

Zeng X, Cheng KW, Du Y et al (2010) Activities of hydrocolloids as inhibitors of acrylamide formation in model systems and fried potato strips. Food Chem 121:424–428

Zhu F, Cai YZ, Ke J et al (2010) Compositions of phenolic compounds, amino acids and reducing sugars in commercial potato varieties and their effects on acrylamide formation. J Sci Food Agric 90:2254–2262

Protein Chemistry and Gelation

5

Taranjeet Kaur, Ajay Pal, Sunil Kumar, and Vinod Saharan

5.1 Protein

The word "protein" comes from the Greek word "proteios," which means "principal" or "first." Proteins are vital for biological systems because of their unique functioning, which allows them to accomplish the greatest functional and structural aspects. They are the ultimate multifunctional macromolecules of amino acids linked together by peptide bonds (Fig. 5.1). Carbon, hydrogen, oxygen, and nitrogen are the primary elements of proteins, with sulfur as a minor component. Proteins have a wide range of composition and structure, with hydrophilic and hydrophobic, structured and unstructured, charged (positive and negative) and uncharged regions all present in a single molecule (Loveday 2019). Proteins of different origins have distinct nutritional values determined by their essential amino acids and digestibility. Proteins are vital in diets because of their nutritional value and ability to harmonize structural alterations. These are the most critical functional component of foods since they can be used as gelling, foaming, or emulsifying agents, affecting the textural features (Vaclavik and Christian 2008). Water holding capacity, solubility, swelling, gelling capacity, and lowering the interfacial tension are the essential functional qualities of proteins in culinary applications. Proteins are dynamic and prospective food components because they are sensitive to changes in physicochemical

T. Kaur · A. Pal (✉)
Department of Biochemistry, College of Basic Sciences and Humanities, CCS Haryana Agricultural University, Hisar, Haryana, India

S. Kumar
Indian Institute of Wheat and Barley Research, Karnal, India

V. Saharan
Department of Molecular Biology and Biotechnology, Rajasthan College of Agriculture, Maharana Pratap University of Agriculture and Technology, Udaipur, Rajasthan, India

O. P. Chauhan (ed.), *Advances in Food Chemistry*,
https://doi.org/10.1007/978-981-19-4796-4_5

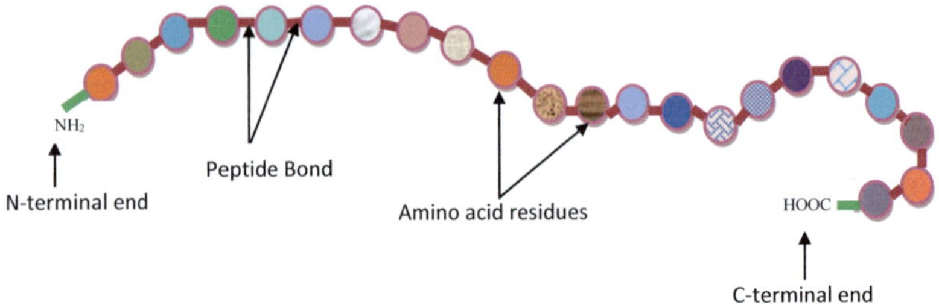

Fig. 5.1 Polypeptide chain

conditions such as pH, ionic strength, and temperature. These entitics are subjected to various structural changes due to multiple types of processing (Mills et al. 2009).

5.2 Amino Acids

An amine (NH_2) and carboxyl (COOH) group and a side chain often denoted as the R group makes up this significant family of organic molecules. Amino acids' side chains vary in size, charge, and reactivity and are unique to each amino acid. Proteins are amino acids linked together by a peptide bond or a CO-NH bridge (Loveday 2019). A peptide bond is formed when the α-carboxylic acid moiety of one amino acid reacts with the α-amino group of another amino acid. A dipeptide, tripeptide, tetrapeptide, and oligopeptide are formed when two, three, four, or a few amino acids are combined. The polypeptide is made up of between 10 and 100 amino acids. Proteins are polypeptides with more than 100 amino acids and a molecular mass greater than 10 kDa. Because the H of the α-NH_2 group and the OH of the α-COOH group is eliminated while creating a peptide bond, amino acid units composing the polymer are called amino acid residues. The amino-terminal (N-terminal) end of the polypeptide chain has a free α-amino group, whereas the carboxy-terminal (C-terminal) end has a free α-carboxylic acid group (Fig. 5.2).

The N-terminal amino acid is printed on the left-hand side when displaying the protein sequence. A protein's biosynthetic route also starts production at the N-terminal end. Although there are only 20/22 amino acids, nature contains a wide variety of proteins generated by altering the sequence of these amino acids. All other amino acids, except for proline, are α-amino acids because their carboxyl and amino groups are covalently connected to α-carbon atoms. The side chains of amino acids determine the qualities they exhibit. They can be polar, nonpolar, aliphatic, hydrophilic, hydrophobic, acidic, basic, or aromatic (Khan et al. 2017; Vasudevan et al. 2017). As a result, amino acids are polar, nonpolar, hydrophilic, hydrophobic, acidic, basic, aliphatic, and aromatic. Proteins rely on amino acids for their structure, activities, and biological value. The biological relevance of amino acids varies from essential to partially essential to nonessential. Nonessential amino acids can be produced by the human body and hence do not need to be taken from

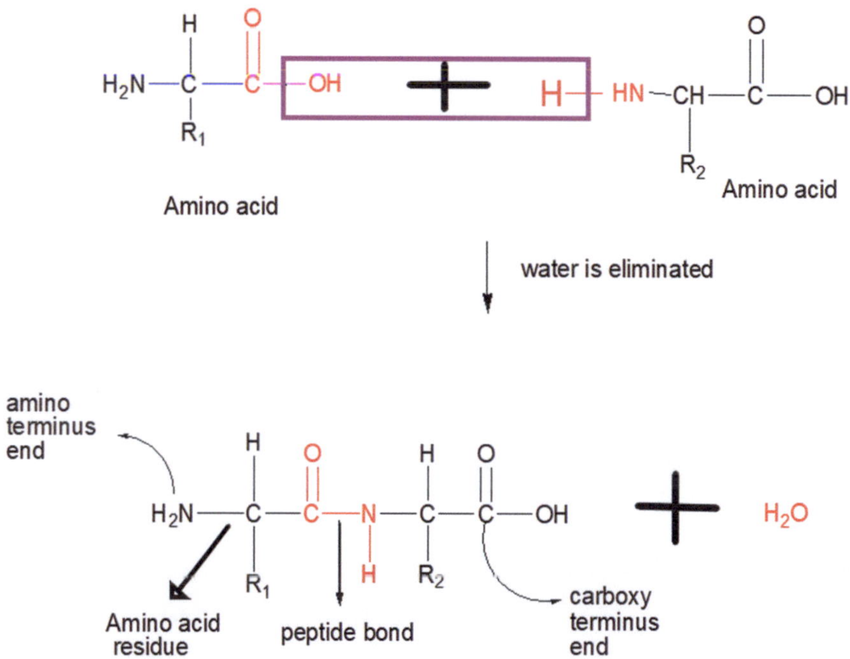

Fig. 5.2 Formation of peptide bond

Table 5.1 List of essential and nonessential amino acids

Essential amino acids	Nonessential amino acids
Histidine	Alanine
Isoleucine	Arginine
Leucine	Asparagine
Lysine	Aspartate
Methionine	Cysteine
Phenylalanine	Glutamine
Threonine	Glycine
Tryptophan	Glutamate
Valine	Proline
Cysteine (in infants)	Serine
Tyrosine (in infants)	Tyrosine

the diet. Because they cannot be made, the remaining amino acids are necessary or indispensable and must be obtained through diet (Table 5.1). The ability of the tissue to grow, heal, and maintain itself will be harmed if certain amino acids are missing.

5.2.1 Classification of Amino Acids

Based on the properties of the side chains, amino acids are divided into six classes.

5.2.1.1 Neutral Amino Acids with Nonpolar Aliphatic Chain (Six Amino Acids)

Glycine is the simplest and smallest amino acid with only one hydrogen atom on its side chain. Alanine with a methyl group as a side chain is more water-soluble than those with hydrophobic chains. The nonpolar side chains of valine, leucine, and isoleucine are chemically unreactive. The amino acid proline is hydrophobic by nature. Its α-carbon and nitrogen are included in a pyrrolidine and result in conformational rigidity.

5.2.1.2 Neutral Aliphatic Amino Acids with Nonionizable Polar Chain (Two Amino Acids)

Serine and threonine are polar due to their primary and secondary alcohol group, respectively, very soluble in water and can participate in hydrogen bonding.

5.2.1.3 Aromatic Neutral Amino Acids (Three Amino Acids)

Tryptophan with an indole heterocyclic side chain and phenylalanine with a benzene ring are nonpolar and hydrophobic. Tyrosine possesses a phenolic hydroxyl group that contributes to its polarity.

5.2.1.4 Sulfur-Containing Amino Acids (Two Amino Acids)

Because of the sulfhydryl group, cysteine is polar and very reactive, and it can create a disulfide bond with another cysteine. The hydrophobic nonpolar side chain of methionine, on the other hand, makes it chemically unreactive.

5.2.1.5 Acidic Amino Acids (Dicarboxylic) and Corresponding Amides (Four Amino Acids)

Under physiological conditions, aspartic and glutamic acids have a second carboxyl group in their side chains that are ionized, negatively charged, and highly polar. Asparagine and glutamine are polar derivatives of aspartic and glutamic acid, with an amide functional group at the carbon atom distal to the α-carbon. They do not have any charge on their side chains, yet they can participate in hydrogen bonding.

5.2.1.6 Basic Amino Acids (Three Amino Acids)

Under physiological conditions, lysine and arginine are ionized, positively charged, and highly polar. At pH 7.0, imidazole-containing histidine is poorly protonated.

5.3 Peptide Bond

It is a chemical bond produced between amino acids due to a condensation reaction. Its geometry is planar. The partial double bond character of the peptide bond is conferred by the sharing of electrons between the N and O atoms. Because there is a less steric hindrance between groups linked to α-carbon (Cα), it is predominantly *trans* in configuration. The planarity and rigidity of the peptide bond are because free rotation around the double bond is not possible due to the 180° torsion angle (ω)

between C and N. Around C_α-N and C_α-C, on the other hand, free rotation is permitted. For C_α-N and C_α-C, the designation for rotation or torsion angle is ϕ and ψ, respectively. These angles define the backbone conformation, allowing the polypeptide backbone to fold and adopt a specific conformation. Although an infinite number of ϕ and ψ values are theoretically feasible, proteins only achieve a limited number of them. In a plot known as the Ramachandran plot, the sterically permitted values of torsion angles are displayed.

5.4 Sources of Protein

A wide range of dietary options is available. Foods derived from plants and animals, as well as the highly publicized sports supplement industry and microorganisms, are among them (Fig. 5.3).

5.4.1 Conventional Protein Sources

5.4.1.1 Animal Proteins

Protein is abundant in animal products such as meat, milk, milk products, egg, chicken, and fish. Animal proteins are considered complete because they contain a balanced amount of all essential amino acids. Because of high digestibility, superior amino acid score (AAS), and greater water solubility, they have a higher nutritional quality than plant proteins (Tilman and Clark 2014). However, they are frequently associated with high saturated fat and cholesterol levels. As a result, when consumed in large quantities, it can raise the risk of ailments such as hypertension and cardiac disease.

Some Important Animal Proteins

Milk proteins: Milk is ideal because of its high nutritional value, true digestibility, and high net postprandial protein utilization. The principal protein categories in

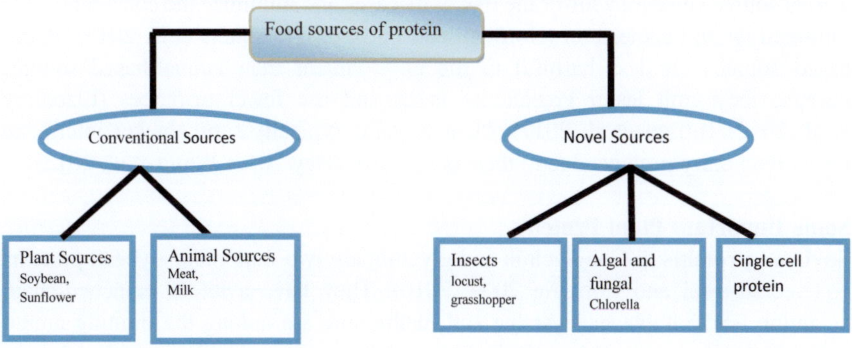

Fig. 5.3 Sources of protein

bovine milk are casein and whey, which constitute 70–80% and 20% of total protein, respectively. Caseins, the solid portion of skim milk, gives milk its white color. Whey proteins are a type of acid-soluble protein that contains a lot of cysteine and branched-chain amino acids. They are generally obtained as by-products of cheese production (Hoffman and Falvo 2004; Sun-Waterhouse et al. 2014). Caseins and whey protein concentrates are commercially available dairy protein preparations.

Meat proteins: Meat has a high protein level but varies depending on age, species, and tissue. It contains all amino acids and "bioactive sequences" produced during proteolysis (Belitz et al. 2009). Diverse forms of meat contain various proteins, most notably myofibrillar, sarcoplasmic, and connective tissue proteins (Sun-Waterhouse et al. 2014).

Egg proteins: A chicken egg's yolk, white and shell are three sections comprising 17.5%, 11.0%, and 3.0% (w/w) protein, respectively. Almost all of the necessary amino acids are found in egg proteins. The protein content of egg yolk is higher than egg white (Strixner and Kulozik 2011). Most proteins are glycoproteins generated by a glycosidic bond between the carbohydrate moiety and the seryl or threonyl residues of the protein or by an amide linkage between the carbohydrate moiety and the asparaginyl residues of the protein. The proteins ovalbumin, ovotransferrin, ovomucoid, ovomucin, and lysozyme are abundant in egg white (Belitz et al. 2009). Eggs are employed because of their excellent functional qualities, such as solubility, foaming, emulsification, gelling, and high protein quality.

5.4.1.2 Plant Proteins

Protein can be found in vegetables, beans, and fruits. When compared to fruits and vegetables, beans have superior protein content. On the other hand, plant proteins lack biological value, net protein utilization, protein digestibility corrected amino acid score (PDCAAS), and protein efficiency ratio compared to animal proteins. This is because they are frequently deficient in essential amino acids. Various vegetables, fruits, grains, and legumes must be consumed to fulfill essential amino acids. When plant-based foods are coupled with essential amino acids, they become a great source since they lower the risk of diseases and minimize the consumption of saturated fat and cholesterol (Hoffman and Falvo 2004; Nehete et al. 2013). Plant-based sources are less harmful to the environment than animal-based sources because they emit lesser greenhouse gases and use fewer resources (Elzoghby et al. 2012; Tarhini et al. 2017). Plant proteins typically form thicker interfacial layers than dairy proteins due to their bigger size (Day 2013; Wong et al. 2012).

Some Important Plant Proteins

Soybean proteins: β-Conglycinin and glycinin are two important soybean proteins (Keerati-u-rai and Corredig 2009, 2010). They have a lot of branched-chain amino acids; however, cysteine and methionine are among the limiting amino acids (Hoffman and Falvo 2004). Soy proteins are complete because they have a well-balanced amino acid profile and are easily digestible. They have a PDCAAS value of 90–99 (Fukushima 2011; Petrusán et al. 2016). It is a good substitute for animal proteins because it gets a 1.0 on the PDCAAS scale, the highest possible value. In addition to their nutritional benefits, soy proteins play an essential function in the food industry due to their ability to gel, emulsify, and hold water (Quintana et al. 2019). Heat treatment causes significant changes in conformation, aggregation, and surface reactivity in glycinin and β-conglycinin. Soybeans are low in fat and cholesterol, with no saturated fat.

Sunflower proteins: Globulins and albumins are the dominant proteins, and glutelins, prolamins, and oleosins are minor fractions (González-Pérez 2015). They are high in branched-chain amino acids, necessary for muscle repair. Sunflower proteins meet all amino acid requirements except lysine, the limiting amino acid (Nehete et al. 2013). Due to its low digestibility, the PDCAAS score is low (0.6). Sunflower proteins are helpful because of their solubility; however, de-phenolized protein isolates have better properties than protein isolates with higher phenolic content (Malik and Saini 2017).

Sorghum proteins: Sorghum contains prolamins, glutelins, albumins, and globulins. Prolamins, also known as kafirins, make up 50–82% of storage proteins in the endosperm. Kafirins have a lot of proline, leucine, glutamic acid, and alanine. Nonetheless, they have lower essential amino acids, such as lysine, tryptophan, and threonine, resulting in poor nutritional quality and utilization (De Mesa-Stonestreet et al. 2010). Sorghum has a low protein digestibility due to protein interactions with tannins and fiber, which function as anti-nutritional agents (Badigannavar et al. 2016).

Canola proteins: Napin and cruciferin are two important protein components that makeup 25% and 65% of total seed proteins, respectively (Perera et al. 2016). These proteins are well-balanced, with a high amount of glutamate, arginine, glutamine, leucine, isoleucine, and low levels of methionine and cysteine. The limiting amino acid is lysine, followed by valine (Ivanova et al. 2016). Because of their possible functional capabilities, such as gelling, foaming, and emulsifying, these proteins are an alternative to other proteins.

Wheat proteins: Wheat is part of breads, pastas, noodles, and cakes, among other processed meals. The most abundant protein components in wheat are gliadin and glutenin. Wheat proteins, unlike animal proteins, lack numerous vitally essential amino acids, particularly lysine, tryptophan, and threonine. They can operate as fat/water binders, offer viscoelasticity, and give wheat a texture suitable for baking (Day 2013).

5.4.2 Novel Protein Sources

Various non-traditional protein sources supply high-quality proteins and are commonly used by particular cultural and community groups.

5.4.2.1 Insect Proteins

Crickets, locusts, and grasshoppers, in particular, are abundant protein sources and are considered significant alternative protein sources. Their protein content varies between 9% and 91% (dry weight). Due to the thick exoskeleton, their digestibility varies greatly (EFSA 2015; Van Huis 2016). The amino acid makeup of different orders and species of insects differs. In many insect proteins, tryptophan and lysine are limiting essential amino acids. Some insect proteins have protein values comparable to meat and plant proteins (Yi et al. 2013). Insect proteins exhibit good foamability, solubility, gelling, and emulsifying ability (Zielińska et al. 2018).

5.4.2.2 Algal and Fungal Proteins

Algae are high in protein and are widely regarded as a nutritional powerhouse (Scieszka and Klewicka 2018). Some species contain protein identical to that found in typical protein sources. *Chlorella* and *Arthrospira platensis* are unicellular microalgae commercially grown in bioreactors or open ponds and have received the GRAS (Generally Regarded as Safe) status. On a dry weight basis, they comprise 51–58% and 21–70% proteins, respectively (Bleakley and Hayes 2017; Teuling et al. 2017). The limiting amino acids in algal protein include methionine, cysteine, lysine, and tryptophan, while arginine and glutamine are abundant (Bleakley and Hayes 2017). Food proteins generated from *Fusarium venenatum* mycelium are known as mycoproteins. These proteins can emulsify, froth, and gel (Ursu et al. 2014). The functional characteristics of these proteins are comparable to those of plant proteins.

5.4.2.3 Single-Cell Protein

Single-cell proteins (SCP) or microbial proteins are derived from unicellular or multicellular microorganisms, primarily fungi, microalgae, and bacteria. Their protein quality varies by species, growth stage, nutrition sources, and environmental conditions (Reihani and Khosravi-Darani 2019; Ursu et al. 2014; Vieira et al. 2018). Microbes may synthesize essential amino acids at levels near the FAO/WHO standard value of 40%, making them high-quality protein sources (FAO/WHO 2007). Microbial proteins can be employed as nourishment or food additives like preservatives, colorants, and texture modifiers to add or improve the functionality of the foods (Ritala et al. 2017).

5.5 Quality Assessment of Proteins

Protein quality is determined by the essential amino acid composition, digestibility and bioavailability, stability, metabolic compatibility, and health advantages. This evaluation is based on several metrics (Fig. 5.4).

Digestibility: The ability of a human or animal to digest and absorb amino acids from dietary protein sources is measured by digestibility. The relationship between food and a person's eating determines digestibility. It is not a set feature

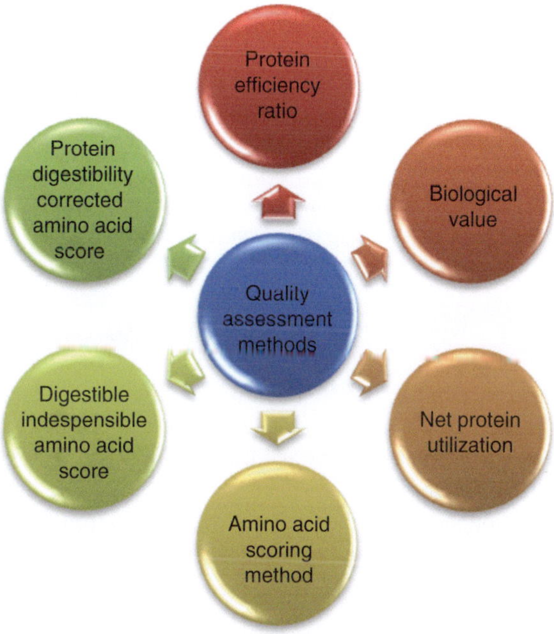

Fig. 5.4 Methods of quality assessment of proteins

of food; instead, it can vary depending on the individual. Protein accessibility to digestive processes varies depending on the physical and chemical environment and the presence of chemicals that can obstruct digestion and absorption (Loveday 2019).

Bioavailability: It is defined as the proportion of total dietary amino acids absorbed in an accessible form that may be used for body protein synthesis and other metabolic pathways that make up the metabolic requirement (FAO 2013).

5.5.1 Protein Efficiency Ratio (PER)

It measures the growth of animals to determine the efficacy and quality of a protein. In this method, rats are fed a test protein and the weight increase is determined in grams per gram of protein consumed. The computed value is then compared to a reference value of 2.7, representing the effect of casein on growth. A value of more than 2.7 suggests a high-quality protein source. This approach however measures rat growth and does not have a high correlation with human growth, and this ratio is rapidly becoming obsolete.

5.5.2 Biological Value (BV)

It is calculated by dividing the nitrogen used in tissue formation by the nitrogen taken from the diet. This product is multiplied by 100 to get the percentage of nitrogen used. However, there are several inherent flaws in this scoring system. It ignores several vital aspects that affect protein digestion and interaction with other foods before absorption. It also assesses a protein's maximum potential quality rather than an estimate based on required levels.

5.5.3 Net Protein Utilization (NPU)

NPU is similar to BV; however, it directly measures the retention of absorbed nitrogen. Its goal is to determine what proportion of amino acids taken are transformed into proteins and used by the body. Protein intake vs. nitrogen excretion is used to measure it. NPU and BV both measure the exact parameters of nitrogen retention; however, the former is determined from nitrogen absorbed, while the latter is calculated from nitrogen ingested (Hoffman and Falvo 2004).

5.5.4 Amino Acid Scoring Method (AAS)

This method estimates the efficiency of meeting the recommended levels of essential amino acids for different age groups based on dietary protein intake recommendations (https://www.nal.usda.gov/fnic/protein-and-amino-acids).

$$\text{Amino acid score} = \frac{\text{mg of amino acid in 1 g of test protein}}{\text{mg of amino acid in reference pattern}} \times 100$$

The lowest score for essential amino acids in a protein or diet of the most limiting amino acid would represent a first approximation of its utilization efficiency by children. It allows correction of protein requirements for dietary protein quality, although it may undervalue the protein quality for adults (Millward 2012). An AAS of 100 indicates that the protein contains all the essential amino acids in the recommended amount, and it does not however account for their bioavailability.

5.5.5 Protein Digestibility Corrected Amino Acid Score (PDCAAS)

It assesses the quality of protein based on human requirements for essential amino acids and their ability to digest them. It assigns a score to proteins based on their amino acid profile and bioavailability. The score compares the concentration of the first limiting amino acid in test protein to the concentration of that amino acid in a reference protein. When a test protein is compared to a standard amino acid profile, it is given a 0 to 1, with 1 indicating maximum amino acid digestibility. A protein's

PDCAAS is computed by multiplying its AAS by its digestibility; values more than 100% are not acceptable and are truncated to 100%. PDCAAS of a protein is examined through the feces. It looks at how much protein is absorbed after passing through the small and large intestines. The small intestine is home to the gut's bacteria that might consume amino acids and peptides that the body never absorbs. Therefore, it may sometimes overestimate the protein bioavailability. This issue has been overcome by digestible indispensable amino acid scores (Petrusán et al. 2016).

$$\text{PDCAAS\%} = \frac{\text{mg of limiting amino acid in 1 g of test protein}}{\text{mg of same amino acid in 1 g of reference protein}} \times \text{Fecal true digestibility} \times 100$$

5.5.6 Digestible Indispensable Amino Acid Score (DIAAS)

Protein digestibility may not necessarily reflect individual dietary indispensable amino acid digestibility, so it is preferable to use a score based on individual nutritional indispensable amino acid digestibility. Because good protein digestibility does not imply bioavailability of all essential amino acids, DIAAS was proposed as a new measure for protein quality (Schaafsma 2000; FAO 2013). The amount of protein absorbed after it has left the small intestine is measured in the ileum content to calculate the DIAAS of a protein.

$$\text{DIAAS\%} = \frac{\text{mg of digestible indispensable amino acid in 1 g of the dietary protein}}{\text{mg of the same dietary indispensable amino acid in 1 g of the reference protein}} \times 100$$

A DIAAS equal to 100% indicates a quality comparable to reference protein (FAO 2013).

5.6 Hierarchy in Protein Structure

There are four different levels of structural organization of a protein (Fig. 5.5).

5.6.1 Primary Structure

It is the simplest and unique protein structure that is determined by genes. The primary structure of a polypeptide chain describes the quantity and sequence of amino acids and the position of disulfide bonds, if any, are present. Peptide bonds keep it together. The amino acid sequence determines the higher level of organization and the specific role of protein. A single amino acid substitution in the linear sequence can have significant biological consequences for a protein's function.

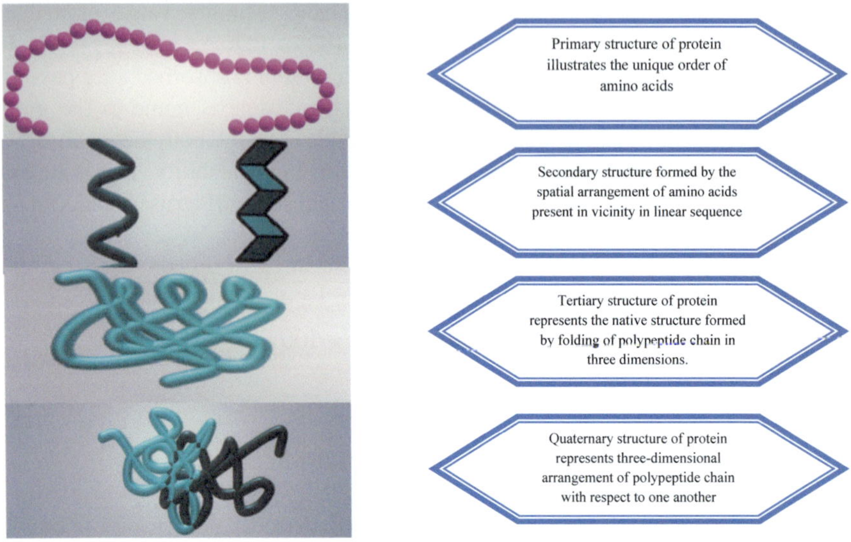

Fig. 5.5 Level of protein organization

Analysis of the basic structure of a protein from multiple species can be used to understand the folding of a protein, its function, evolution, and disease diagnosis (Vasudevan et al. 2017).

5.6.2 Secondary Structure

It is configured between amino acid residues close to one another (3–4 residues apart) in the linear chain. If the overall torsion angles (ψ and φ) in a polypeptide segment are similar, it can achieve a stable secondary structure, and hydrogen bonds mainly stabilize it. The secondary structure of a protein can take many forms. Folding can occur in two ways: helical twisting or packing one polypeptide segment onto another, resulting in helical and β-structures, respectively. Variations in the propensity of amino acids lead to the formation of these structures. Certain amino acids are more likely to form a α-helix, whereas others are more likely to produce a β-structure.

5.6.2.1 Helical Structures

A polypeptide segment can attain various helical structures due to hydrogen bonds between imino hydrogen and carbonyl oxygen.

3_{10}-Helix: It is a rare secondary structural component. By establishing a hydrogen bond, carbonyl oxygen and imino hydrogen of the *n*th and *n* + third residues contribute to the structure of this helix. The number 10 in the helix name refers to atoms in a hydrogen bond loop, and each turn of the helix contains 3 residues.

3.6_{13}-Helix: This helix is formed via hydrogen bonding between the carbonyl oxygen of the *n*th residue and the imino hydrogen of the *n* + fourth residue. Because it was the first helical structure to be discovered, it is also known as the α-helix. It is a spiral structure with a backbone of polypeptide bonds and amino acid side chains extending outwards. The structure is stabilized by hydrogen bonding between the NH and C=O groups of the main chain. Each twist of the helix comprises 3.6 residues that are 1.5 Å apart each and contain 13 atoms in a hydrogen bond loop. α-Helices are common secondary structures and generally right-handed helical coils. Left-handed α-helices are uncommon as the L-configuration of amino acids in proteins prevents them. The values for ψ and φ angles are −47° and −57°, respectively. Amino acids differ in their propensity to form α-helix as amino acids like Met and Leu create α-helix. In contrast, others such as Pro and Gly are helix breakers, disrupting the regularity of helical conformation. Proteins with high α-helix content include hemoglobin and myoglobin.

4_{16}-Helix: The hydrogen bonding between the carbonyl oxygen and imino hydrogen of the *n*th and *n* + fifth residues results in a helical configuration. π-Helix is another name for it. Each turn of the helix with a pitch of 4.6 Å contains four amino acid residues, resulting in an 87° turn in the helix. In one hydrogen bond loop, there are a total of 16 atoms.

5.6.2.2 Beta-Pleated Sheet

This is another common secondary structure that is completely expanded. Hydrogen bonds between the carbonyl oxygen and imino hydrogen of nearby polypeptide segments allow one segment to fold on top of another. The distance between neighboring amino acids is 3.5 Å. β-Pleated sheets are divided into two categories based on the orientation of polypeptide segments: parallel and antiparallel. Some amino acids, such as Val, Ile, Phe, and Tyr, aid in constructing the β-sheet, while others, such as Pro, Gly, and Tyr, break it down. Silk fibroin and chymotrypsin are two examples of proteins rich in β-sheet structures.

Parallel β-sheet: The segments of the polypeptide chain are parallel in orientation, meaning the β-strands run in the same direction. The distance between two amino acids is 6.5 Å, and the ψ and φ angles are +113° and −119°, respectively. This form is less stable because the geometry of individual amino acid forces the hydrogen bond to occur at an angle, making them longer and weaker.

Antiparallel β-sheet: The polypeptide chain segments in this form are oriented antiparallel. Two amino acids in an antiparallel β-sheet are 7 Å apart, and the sheet is defined by ψ and φ angles of +135° and −139°, respectively. Close packing, resulting from stronger hydrogen bonds that lie precisely opposite to one other, makes the antiparallel arrangement more stable.

5.6.2.3 β-Turn

Folding a protein into a spherical form frequently necessitates a reversal in the orientation of the polypeptide chain. The secondary structures β-turn, β-bend, and reverse turn produce a 180° change in the direction of the polypeptide chain. The structure is called a hairpin structure because it resembles a hairpin. Four successive residues define the structure, maintained by a hydrogen bond between the carbonyl and imino hydrogen of the *n*th and n + third residues. Gly, Asn, Pro, and Asp are amino acid residues with a high propensity for forming β-turns, with Pro and Gly being favored in every reverse turn. Pro causes a kink in the polypeptide, and Gly overcomes the steric barrier created by Pro, leaving Asp or Asn as the remaining residues. The hairpin structure is found on the surface of globular proteins and aids in the joining of the β-sheet. β-Turns are classified as type I, II, or III, with the least prevalent type III.

5.6.2.4 Coils or Loops

These are the unordered structures in numerous segments of the polypeptide chain, in addition to the secondary structures discussed above, that have unequal and irregular ψ and ϕ angles (Khan et al. 2017).

5.6.3 Tertiary Structure

It is the three-dimensional structure created by folding all secondary structural parts of a polypeptide chain. Based on the spatial arrangement of all of its constituents, this is the thermodynamically preferred native conformation. Folding the linear sequence of amino acids brings distantly present amino acids in the linear sequence closer to each other in their native configuration. It forms a biologically functional, three-dimensional protein structure from a linear sequence of amino acids. The structure relies on a network of noncovalent hydrogen bonding, hydrophobic interactions, electrostatic interactions, and van der Waals interactions between amino acid side chains. The fundamental driving force behind this process is hydrophobic effects, which result in the segregation of hydrophobic amino acids from water by burying them in the core of the native protein. Most hydrophilic residues are found on the surface, interacting with water. The three-dimensional structure of proteins determines their functional features, allowing them to recognize and interact with various molecules. Because the linear sequence of amino acids depicts the native structure, any alteration renders the protein unusable (Vasudevan et al. 2017).

5.6.4 Quaternary Structure

Few proteins have more than one polypeptide chain, and the quaternary structure is the biologically functional structure generated by the aggregation of these chains in three-dimensional space. Noncovalent interactions like van der Waals forces, hydrogen bonds, hydrophobic bonds, and electrostatic bonds keep it together. It only applies to multi-subunit proteins and denotes how the subunits are oriented and arranged about one another. Each polypeptide chain is referred to as a subunit or monomer, and when dissociated these subunits are physiologically inactive. Identical and nonidentical subunits combine to generate homogeneous and heterogeneous quaternary structures. Homodimer comprises two copies of the same polypeptide chain, whereas heterodimer is made up of two different polypeptide chains (Khan et al. 2017).

5.7 Classification of Proteins

Proteins execute a wide range of functions due to their chemical diversity. Although it is nearly impossible to categorize all proteins, the following is the primary classification based on different criteria (Fig. 5.6).

5.7.1 Based on Function

Proteins are classified into numerous classes based on their functions:

Catalytic proteins: This protein class works as a catalyst in nearly every reaction. They have high catalytic power and can speed up the rate of the reactions in which they are involved. Example: enzymes.

Structural proteins: These provide a structural foundation for cells and play a critical role in stabilizing numerous structures by giving strength. Example: collagen and elastin.

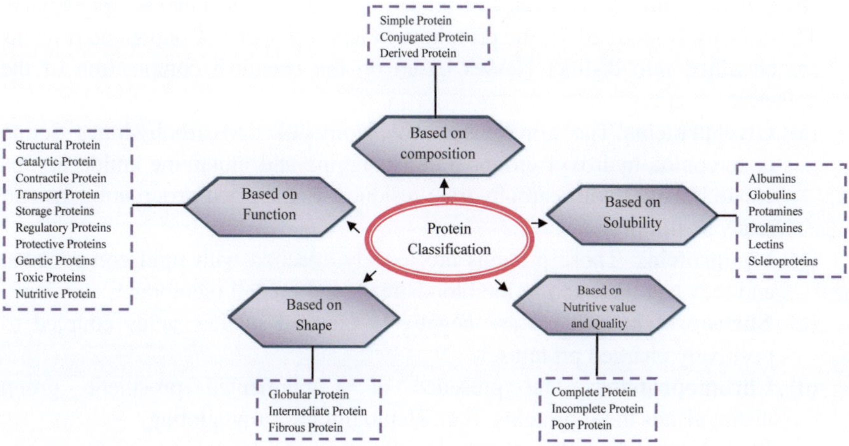

Fig. 5.6 Classification of proteins

Contractile proteins: Contractile proteins, as force generators, aid in movement and mobility. Example: myosin and actin.

Transport proteins: These proteins move life-sustaining chemicals through the circulation and extracellular fluids. Transport proteins, such as hemoglobin, myoglobin, albumin, and transferrin, form channels in the plasma membrane and transport organic and inorganic molecules across the cell membrane and into the cells from one compartment to another.

Regulatory proteins: These proteins are essential in regulating cellular and physiological activity and modulating gene expression. Example: Insulin and growth hormone.

Genetic proteins: These are genetic proteins connected with nucleic acids and offer structural stability. Example: Histone proteins.

Protective proteins: These proteins protect against dangerous chemicals by binding to them and neutralizing or eliminating them from the body. Example: Immunoglobulins, interferons, and clotting factors.

Nutritive proteins: These proteins have a high nutritional value and offer nutrition when taken correctly. Example: Casein and albumin.

Toxic proteins: These are hazardous and can cause tissue damage. Example: Bacterial exotoxins and endotoxins, snake venom nucleic acid degradation enzymes.

5.7.2 Based on Composition

1. **Simple proteins**: These proteins are made up entirely of amino acids and have a simple structural arrangement. Simple proteins are hydrolyzed to produce only constituent amino acids. Example: Albumins and prolamines.
2. **Conjugated proteins**: These proteins are made up of amino acids and a non-amino acid component. Apoenzyme is the former, and cofactor is the latter, either tightly (prosthetic group) or loosely (coenzyme) attached to apoenzyme. The cofactor is required for the protein to function properly. Conjugated proteins are classified into distinct classes based on the chemical composition of the cofactor:
   (a) **Glycoproteins**: These proteins are covalently linked to carbohydrates. Serine or threonine hydroxyl groups and asparagine and glutamine amide groups create linkages with carbohydrate residues. E.g., Blood group antigens and many serum proteins.
   (b) **Lipoproteins**: These proteins are loosely attached with lipid components, and they can be found in the bloodstream and on cell membranes.
   (c) **Nucleoproteins**: These are negatively charged nucleic acids coupled to positively charged proteins.
   (d) **Chromoproteins**: The presence of a pigmented prosthetic group distinguishes these proteins. E.g., Hemoglobin or myoglobin.
   (e) **Phosphoproteins**: Phosphorus is present in these proteins as phosphoric acid, which is esterified to the hydroxyl groups of serine and threonine

residues. Generally, they have a structural or reserve function. E.g., Casein and vitellin.

(f) **Metalloproteins**: Metal ions serve as a cofactor in these proteins. E.g., Hemoglobin (iron), cytochrome (iron), tyrosinase (copper), and carbonic anhydrase (zinc).

3. **Derived proteins**: When acids, alkalies, or enzymes hydrolyze conjugated and simple proteins, they produce degradation or derived proteins. Protein hydrolysis results in smaller and smaller chains as time goes on.

5.7.3 Based on the Shape

Proteins can be classified into the following groups based on their shape:

Globular proteins: They have a compact, spherical or oval form and are made up of polypeptide chains coiled around themselves. These are folded so that hydrophilic amino acids are on the surface and hydrophobic amino acids are buried deep within. Because they contain multiple secondary structures, they have a more complex structure than fibrous proteins. Protamines, globulins, and albumins, for example, are water-soluble.

Fibrous proteins: These proteins are made up of elongated molecules that support the structural features of cells. Because they contain several hydrophobic amino acids, they are water-insoluble. Hydrophobic amino acids on their surfaces make it easier to package them into complex supramolecular structures. These are distinguished by a single secondary structure leading to a basic tertiary structure. For example, collagen, elastin, and keratins have primarily mechanical and structural purposes.

Intermediate proteins: These proteins are short, soluble in water, and have structures that are intermediate to linear and globular structures, e.g., fibrinogen.

5.7.4 Based on Solubility

Albumins: These are the most abundant proteins in nature, soluble in water, and can coagulate with heat. Example: Legumelins and ova-albumin.

Globulins: These proteins are soluble in dilute salt solutions but not in pure water, and they coagulate when heated. These proteins are deficient in methionine. Example: Egg globulin and legumin.

Protamines: These proteins are water and dilute acid-soluble and do not coagulate when heated. Because of the abundance of lysine and arginine, these are extremely basic, making them compatible with acidic proteins. These proteins are devoid of aromatic and sulfur-containing amino acids. E.g., Zinc insulinate and salmine.

Prolamins: These are seed storage proteins, soluble in 70–80% alcohol but insoluble in pure water. These are deficits of lysine and are rich in glutamine and proline. E.g., Zein, hordein, and gliadin.

Lectins: These proteins have a high affinity for sugar moieties, and these can be precipitated using 30–60% ammonium sulfate. E.g., Phytohemagglutinin.

Scleroproteins: These proteins are insoluble in salt solutions, water, and organic solvents and soluble only in hot and strong acids. E.g., Collagen, keratin, cartilage and tendon.

5.7.5 Classification Based on Nutritional Value and Quality

Nutritionally rich proteins: These are complete proteins and are sometimes known as first-class proteins. These proteins, such as casein in milk, contain sufficient amounts of all nine essential amino acids to meet the body's growth and repair needs.

Incomplete proteins: These proteins have all essential amino acids, but one or more are limiting amino acids because they are not present in the proper amount. They cannot support tissue synthesis, but they might support body weight. Methionine deficiency exists in pulse proteins, while lysine deficiency exists in cereal proteins. It is possible to achieve enough growth if both are integrated into the diet.

Poor proteins: Some essential amino acids are absent in these proteins. As a result, these proteins cannot sustain cell growth and repair. They cannot maintain their original body weight to support life; for example, zein derived from corn lacks lysine and tryptophan.

5.8 Functional Role of Proteins as Food Ingredients

The literature on the functions of various dietary proteins is listed in Table 5.2. Water-holding, gelation, solubility, and foaming or emulsifying capacity are all functional properties of proteins. The protein functionality is dictated by the composition of amino acids, the ratio of hydrophilic and hydrophobic amino acids, structure and molecular weight of protein (Del Rosario Moreira et al. 2011).

5.8.1 Surface Properties

Surfaces/interfaces in the food items are abundant in emulsions/foams. Proteins, made up of polar and nonpolar amino acids, can help dispersions to form and stay stable by adsorbing to the interface and forming films with high interfacial elasticities and viscosities (Murray et al. 2011). Other natural biopolymers, such as starch and cellulose, are readily available and low-cost materials for possible food applications. These materials can be particularly useful in stabilizing emulsions and foams with proper hydrophobic alteration (Wege et al. 2008). On the other hand, protein-based particles are likely to be more versatile as stabilizing agents due to their intrinsic amphiphilicity and surface activity (Chen 2009). A multistep

Table 5.2 Summary of literature on the functionality of food proteins

Source	Functionality	Reference
Roe protein isolate from skipjack tuna	Water holding capacity, solubility, buffering capacity, foaming and emulsifying ability, and antioxidant activity	Cha et al. (2020)
Protein isolate and concentrate from defatted walnut flour	Capacity to absorb water and fat, protein solubility, and foaming ability	Mao and Hua (2012)
Protein isolate from Rainbow trout fish	Water and oil holding capacity and emulsifying property	Lone et al. (2015)
Soluble protein extracts from *C. sorokiniana* and *P. tricornutum*	Interfacial activity and unusual high salt stability up to 500 mM	Ebert et al. (2019)
Proteins from *Chlorella vulgaris*	Excellent emulsifying property	Ursu et al. (2014)
Red kidney and navy beans Adzuki bean	High formability and foam stability, emulsifying activity Oil adsorption, emulsifying activity	Sai-Ut et al. (2009)
Protein hydrolysate prepared from *Decapterus maruadsi*	Emulsifier and foaming agent with anti-oxidative property, excellent solubility	Thiansilakul et al. (2007)
Olesins from rapeseed cake	Emulsifier and antioxidant	Östbring et al. (2020)
Hemp seed albumin fraction Globulin fraction of hemp seed	Serve as an excellent ingredient for food formulation Useful for formation of food emulsions	Maloma and Aluko (2015)
Protein isolates from lupin	Good solubility and foaming properties and foam stability	Vogelsang-O'Dwyer et al. (2020)

procedure governs the stabilization of an air-water or oil-water interface. Initially, absorption at the interface necessitates the diffusion of proteins to the interface. The kinetic absorption barrier, determined by parameters such as exposed hydrophobicity and net charge, governs their retention at the contact (Wierenga et al. 2003, 2005). Protein molecules rearrange after effective absorption at the interface to produce a thermodynamically stable yet dynamic monolayer that coats the droplets.

Hydrophobins are proteins that are particularly well suited to emulsions and foams. They lower the surface tension, allowing for smaller air or oil cells to form, and they coat the surface quickly, preventing coalescence. Hydrophobins bond to one another to create an elastic membrane at the contact. They are suitable for stabilizing food foams and emulsions because they tend to self-assemble into films at a hydrophobic/hydrophilic interface. Hydrophobins are the most surface-active protein because of their rapid diffusion to a surface and their highly amphiphilic character (Green et al. 2013) (Fig. 5.7).

5.8.1.1 Emulsifiers

Emulsions are immiscible liquid mixes that are thermodynamically unstable. The dispersed phase of one of the liquids is present in droplets in the continuous phase of the other liquid surrounding the droplets (Esquena 2016). Unless an energy barrier is

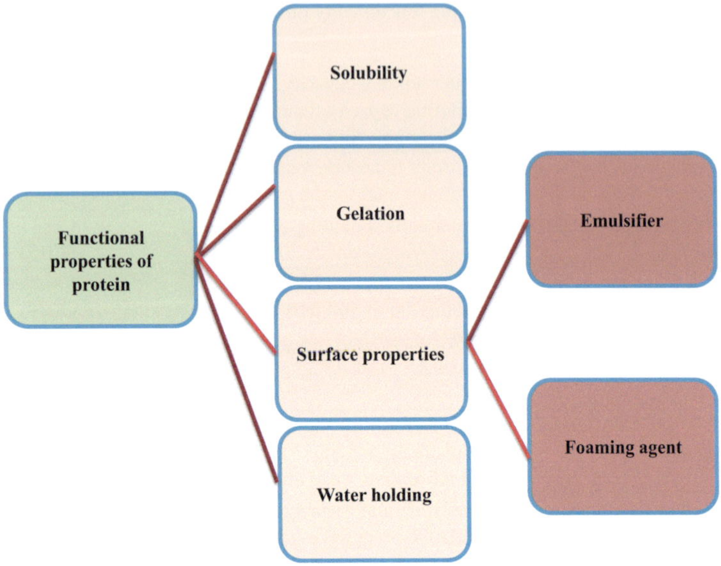

Fig. 5.7 Functional properties of protein

built at the interface to prevent droplet fusion, these phases will separate during storage due to multiple physical destabilizing mechanisms such as creaming, flocculation, coalescence, and Ostwald ripening (Hoffmann and Reger 2014). Emulsions are either oil-in-water (O/W) or water-in-oil (W/O) combinations, as in milk, creams, mayonnaise, and soups, or water-in-oil (W/O) mixtures, as in margarine and butter (Divsalar et al. 2012; Pal 2011). Emulsifiers, which physically separate the continuous phase from the dispersed phase by producing a layer on the surface of the dispersed phase, can be employed to achieve stability. They range in size from tiny molecules to huge macromolecules. Food emulsifiers are amphiphilic compounds with hydrophilic-lyophilic balancing numbers that are attractive to water and oil. Film-forming abilities are particularly important in the emulsifying characteristics of proteins with both hydrophilic and hydrophobic regions (Foegeding and Davis 2011). The quantity of various amino acids in their polypeptide chain, such as polar and nonpolar, has a significant impact on their emulsifying ability. It also regulates emulsion qualities, including chemical and physical consistency (McClements 2015; Papalamprou et al. 2010).

Proteins' emulsifying ability varies depending on their hydrophilicity and hydrophobicity. As seen in a few forms of gelatin, hydrophilic proteins are less effective emulsifiers because they cannot firmly adsorb to the surfaces of oil droplets (Gomez-Guillen et al. 2011). Hydrophobic proteins like gliadin and zein, on the other hand, prefer to mix, resulting in a decrease in surface activity and water solubility (Luo and Wang 2014). Proteins having a balance of hydrophilic and hydrophobic groups make them water-soluble and surface-active, making them suitable emulsifiers.

Small molecular weight emulsifiers diffuse quickly to the interface, allowing them to generate good emulsions, whereas proteins are larger and diffuse slower.

They adsorb and unfurl at the interface through selective contact with both surfaces, stabilizing the oil and water phase. By coating the lipid droplets, they provide an energy barrier to particle association and phase separation, preventing individual droplets from coalescing. Protein-protein intermolecular interactions can lead to more stable emulsions, resulting in highly viscoelastic films that form around the emulsion droplets. They allow emulsified solutions to be included in food products that can be preserved for an extended period (Lam and Nickerson 2013). By enhancing solubility, exposing hidden hydrophobic groups, boosting surface hydrophobicity, and reducing molecular weight, partial hydrolysis has been utilized to improve the emulsifying capabilities of proteins, allowing improved adhesion to the oil-water interface (Tsumura 2009). When two or more emulsifiers interact, they can have synergistic or antagonistic effects on emulsion stability. Low molecular weight emulsifiers have a higher surface activity and are more quickly adsorbed than proteins hence they tend to displace proteins from the oil-water interface in a competitive displacement process.

5.8.1.1.1 Characteristics of Emulsifiers

The development and stability of emulsions are dependent on the capacity of an emulsifier:

1. To immediately adsorb to the droplet surface generated during homogenization.
2. To reduce interfacial tension by a significant amount to promote more droplet disruption.
3. To prevent droplet build-up by introducing repulsive forces and forming a protective layer (McClements 2015).

Emulsion formation: It requires the diffusion of protein to the interface. Charged groups are found on the surface of proteins, which come into contact with water molecules. Water's favorable interaction with surface charge lowers protein molecules' total energy. Hydrophobic groups are removed from contact with the aqueous phase, whereas charged groups increase the number of interactions. There is less possibility for charged groups to interact with solvent at the interface; therefore, protein unfolding is necessary. If the groups closer to the interface are in a flexible region of the protein molecule, the molecule may unfold and expose its buried hydrophobic groups to the surface, allowing them to contact the lipid phase (Cabra et al. 2008). Assume that these groups are in an aqueous environment; in that situation, a rise in total energy and random variations in protein structure lead these groups to return to the molecule's core. The hydrophobic groups are introduced into the lipid phase when a protein unfolds near a lipid. This insertion has very low activation energy and happens on its own. The insertion of a hydrophobic group occurs spontaneously with low activation energy; however, the reaction is not easily reversible. Other regions of the protein molecule approach the surface over time, and, if they are flexible, they may be introduced into the lipid phase. The protein will unfurl at the interface if this

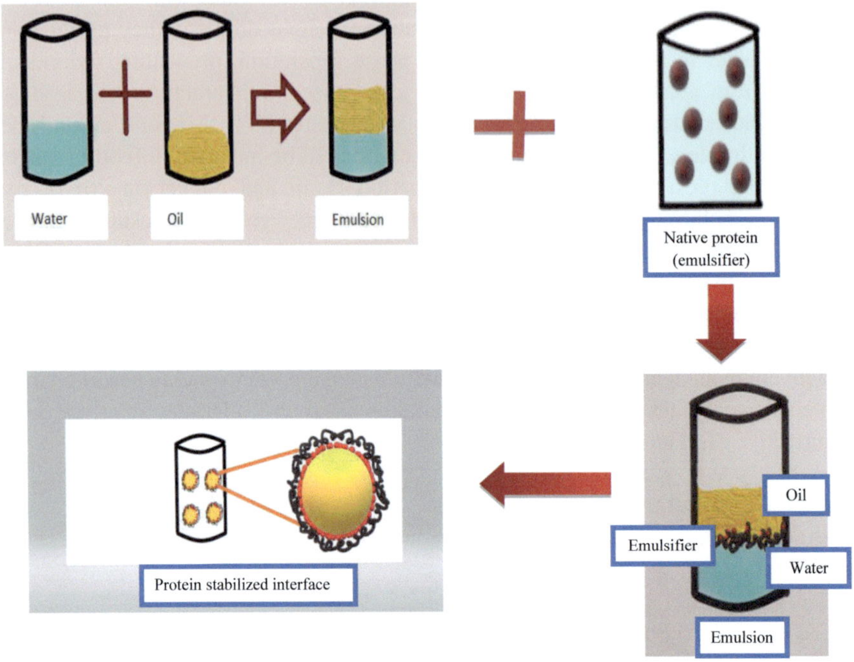

Fig. 5.8 Protein as an emulsifying agent

continues. Protein realignment leads to forming a strong viscoelastic film (Fig. 5.8).

Emulsion stabilization: Physicochemical processes such as gravity separation (creaming and sedimentation), aggregation (flocculation, coalescence, and partial coalescence), Ostwald ripening, phase inversion, and chemical degradation can all cause emulsions to become unstable. Changes in pH, ionic strength, dilution, component interactions, temperature, mechanical forces, and water activity are just some of the circumstances that a product may be exposed to throughout its existence. The emulsifier must provide strong electrostatic forces and steric repulsive forces that can overcome the attractive attraction between oil droplets to avoid droplet aggregation and eventual phase separation.

Droplets can take on a negative or positive charge once the viscoelastic film is produced, depending on whether the emulsion pH is above or below the protein's isoelectric point. The higher the electrostatic repulsion between oil droplets, the more stable the emulsion, but pH levels near the protein's isoelectric point cause droplet flocculation/aggregation, which leads to coalescence and instability. Adsorbed proteins usually form thin, electrically charged interfacial coatings; therefore, electrostatic repulsion is the primary mechanism that prevents droplet flocculation. As a result, protein-coated droplets are more likely to flocculate at pH levels near to their isoelectric points (McClements 2005).

Steric stabilization: The loops or tails of protein segments may radiate from the interface, depending on the size, structure, and conformational freedom of the protein. By forming an osmotic barrier, these prevent droplets from colliding, resulting in steric stability (Tcholakova et al. 2006). Proteins with large molecular weights and extended structures induce steric repulsion and, as a result, inhibit aggregation better. Protein in the continuous phase raises the viscosity of the emulsion and inhibits the mobility and diffusion of oil droplets within the emulsion (Jafari et al. 2012).

The emulsifying capabilities of dietary proteins are usually described as follows:

1. Emulsion activity (maximum interfacial area per unit mass of protein in a stabilized solution).
2. Emulsion capacity (maximum amount of oil that a unit mass of protein can emulsify under certain conditions).
3. Emulsion stability (a protein's ability to produce an emulsion that remains unaltered for a set time at a specified temperature and gravity field) (Amarowicz 2010).

5.8.1.1.2 Parameters for Determining the Efficacy of Emulsifiers

Proteins differ greatly in their ability to form emulsions due to differences in surface hydrophobicity, surface loads, adsorption kinetics, electrical property, and saturation surface pressure (Wierenga et al. 2006). The effectiveness of emulsifiers in lowering the tension of an oil-water contact can provide basic information about their qualities. The surface pressure versus emulsifier concentration profile is calculated after the interfacial tension is predicted to raise the emulsifier level.

Saturation surface pressure: The difference between interfacial tension in the presence and absence of an emulsifier at the interface can be used to calculate surface pressure. This value is related to disrupting oil droplets within a homogenizer, and it represents the efficacy of an emulsifier in lowering interfacial tension once it adsorbs to the droplet surfaces. The surface pressure value rises from zero to a constant value as the emulsifier concentration increases from zero to a level above saturation. The smaller the droplets formed under given homogenization conditions, the higher the value of surface pressure at saturation, provided there is enough emulsifier available and it adsorbs quickly.

Surface activity: The level of surface activity, and hence their inclination to adsorb to an oil-water interface, is regulated by the relative balance of the number of groups such as nonpolar and polar exhibited on their surfaces. The hydrophobic groups on the surface of proteins are responsible for this activity, calculated as the reciprocal of the emulsifier concentration at which the surface pressure reaches 50% of the saturation value. It specifies an emulsifier's affinity for the oil-water interface and is based on its ability to hide the thermodynamically unfavorable interaction at the interphase. Emulsifiers with a high surface activity can adsorb to oil droplets at a lower concentration than emulsifiers with a low surface activity because they have a higher affinity (McClements 2015). As a result, when a group of emulsifiers is studied, those with high surface activity are better at adsorbing to the interface.

Protein load at the surface: It is determined from the gradient of an interfacial tension versus the logarithm of the emulsifier concentration plot and is connected to the quantity of emulsifier required to stabilize a certain amount of interfacial area. A considerable amount of emulsifiers is needed to produce emulsions with increased surface loads.

Adsorption energy: During homogenization, it indicates the scale at which the emulsifier coats the surfaces of droplets. It is critical because interfacial tension, and therefore droplet rupture and coalescence, are influenced by this count. Emulsifiers are designed to move quickly towards the surface of droplets and form a strong bond with them. Natural emulsifiers' rate of adsorption on droplets is determined by their molecular size and surface characteristics. In general, smaller emulsifiers move and absorb more quickly than larger ones (McClements et al. 2017).

5.8.1.1.3 Destabilizing Mechanism in Protein Stabilized Emulsion

The flocculation and aggregation of oil droplets can lead to partial or complete coalescence where electrostatic repulsion is minimized (near the isoelectric point of a protein film or in the presence of high ionic strength) (Tcholakova et al. 2006). In this process, films from adjacent droplets combine by exchanging material, resulting in a bigger droplet. Creaming occurs when larger droplets gravitationally detach and move upwards due to density differences.

Ostwald ripening can also occur in protein-stabilized emulsions, where tiny oil droplets spread through the continuous phase and coalesce with larger ones; however, it is uncommon. Protein surface hydrophobicity and affinity for the interface are affected by changes in environmental circumstances. As a result, a protein with the highest affinity for an interface may supplant another protein there. Because of their small molecular weight, mobility, and intense affinity, surfactant molecules (tweens) frequently displace proteins at the interface. Finally, changes in protein structure and the thermodynamic force for phase separation reduce the stability of an emulsion over time (Tcholakova et al. 2006). Because of protein structural changes over time, proteins generate noncovalent connections with neighboring proteins at the interface. The emulsion's stability suffers due to this conformational change and new noncovalent links. Protein removal from the surface of a fat globule is energy inefficient and does not occur at a significant pace. Temperature changes are a prominent source of emulsion destabilization in food items. Water becomes more ordered as the temperature drops, resulting in a smaller energy difference between hydrophobic groups exposed to the aqueous phase and those buried in the oil phase (Lam and Nickerson 2013).

5.8.1.2 Foaming Agent

Foam is the continual dispersion of gas bubbles in a liquid or semi-solid medium. Air bubbles are separated by a thin liquid film and twisted as polyhedrons due to a large volume of gas in the fluid (Narsimhan and Xiang 2017). In the absence of a surface-

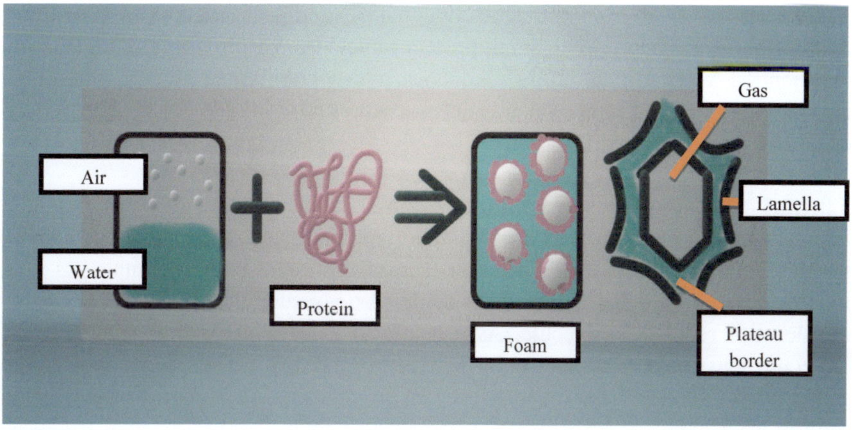

Fig. 5.9 Proteins as a foaming agent

dynamic food emulsifier, foams are highly unstable and collapse due to rupturing of the thin film due to drainage and van der Walls attraction. Because gas has a significant level of solubility in a continuous phase of foam, the high free energy at the gas-liquid interface causes the intrinsic instability of foam. Unless bubbles are embedded in a solid matrix, this results in a continuous diffusive mass movement across bubbles of various sizes under the effect of local Laplace pressure gradients. Due to the larger size and lower density of bubbles in foams than in equivalent emulsion oil droplets, the risk of gravity creaming and rupturing of thin films increases (Goff and Vega 2007). As a result, a surface-active emulsifier that adsorbs onto the air-liquid interface is required to prevent bubble coalescence. A foaming agent, which generates a layer of adsorbed molecules separating the air bubbles from a continuous liquid phase, determines the stability of air bubbles in foam. Foaming factors in foods include proteins and low molecular weight surfactants (LMWS). Surface strain is reduced by emulsifier/protein adsorption, changing the interaction and interfacial rheological properties. Balancing out the thin films between air pockets extends the usability period of froth.

When compared to low molecular weight emulsifiers, protein adsorption is slower; nonetheless, the air pocket size of froths balanced out by a protein is larger than that resolved by an emulsifier. When a protein settles the froth, it is more stable than when an emulsifier is used to level it out. Because protein desorption from the interface is extremely difficult because it necessitates the simultaneous expulsion of all adsorbed portions, it is regarded as irreversible. Creating a dense particle layer at the gas-liquid interface creates a colloidal armor that can slow or stop disruptive bubble coalescence and coarsening processes (Fig. 5.9) (Zhang et al. 2008). As a result, froth is considered a colloidal framework with a large surface area per unit volume. The stability of froths is influenced by surface characteristics and intermolecular interactions (Narsimhan and Xiang 2017).

Because of their high surface-active nature and self-assembling properties, hydrophobins are appropriate small protein molecules for providing foam stability (Tchuenbou-Magaia et al. 2009). Aqueous hydrophobins can swiftly saturate an air-water contact and noncovalently form a vital and dynamic elastic film (Green et al. 2013).

5.8.1.2.1 Characteristics of Foaming Agents

Foamability and foam stability are two important qualities to consider when describing the foaming properties of a solution. The adsorption of particles onto a gas-fluid interface is required for the froth to settle. As a result, the fluid should neither moisten nor repel the molecule. As a result, to exist at a gas-fluid interface, the molecule must have a contact point between 0 and 180°. The energy required to separate the molecule from the interface can provide significant particle adequacy in settling froths.

Foamability: It is the ability to entrap gas (gas incorporated in a specific volume of solution) and is linked to the development and disappearance of bubbles. Whipping, supersaturation of liquid with gas (e.g., champagne), or yeast fermentation are mechanical, chemical, and biological processes for creating bubbles (Drenckhan and Saint-Jalmes 2015). Their molecular size and structure determine the foaming capability of proteins. Ordered, inflexible, and larger proteins decrease the surface strain faster than degraded, undersized, and adaptive proteins; as a result, disordered proteins have greater foaming limits. The primary goal of protein adsorption is to reduce their free energy by exposing their hydrophobic moieties, which are available either superficially or in the center, to the air/water contact (Narsimhan and Xiang 2017).

Increased exposed hydrophobicity of proteins through lipid conjugation and decreased net charge due to repulsive interactions improve the adsorption rate to an interface (Le Floch-Fouéré et al. 2011; Wierenga et al. 2003). According to studies, increasing protein focus increases the rate of protein adsorption at the air-water interface, resulting in a decrease in surface tension and, as a result, an increase in the frothing limit (Lech et al. 2016).

Foam stability: It is decided by the qualities of thin films to endure rupture against the coalescence of two bubbles, and it is the capacity to keep the gas for a particular period. A balance of opposing forces gives thin films their stability (repulsive electrostatic forces and attractive van der Waals forces). The van der Waals interaction is the principal cause of diminishing film bursts without emulsifiers or proteins, resulting in bubble coalescence. Surface charge and electrical twofold layers are conferred in this region by adsorption of charged emulsifiers/proteins at the gas-fluid interface. A depleting film encounters a power of repulsion (or positive free energy of interaction) due to the cover of two electrical twofold layers connected to two essences of the film (Gochev et al. 2014). The degree of protein adsorption at interfaces, influenced by physicochemical parameters like hydrophobicity, size, and secondary conformation, is also a factor in electrostatic repulsion. It has been shown that for froth settled by tiny clusters of particles, film thickness drops quickest in the middle, with the internal film being more slender than the external film, eventually causing it to rupture because film thickness reaches zero in the middle. For lower molecule fixation and higher surface strain, the rate of film seepage is faster. A longer period of realistic usability is shown by a higher initial film thickness and a larger air pocket size. For larger air pocket sizes and higher initial film thickness, the film's lifetime is increased (Narsimhan and Xiang 2017).

5.8.1.2.2 Destabilization Mechanism of Foams

The bubbles in foam have a dodecahedral form, and thin liquid layers divide them. A crisscrossing of three such films forms a plateau boundary. Due to its radius of curvature, the plateau boundary has less pressure than the pressure inside the bubble; however, the pressure inside the bubble is similar to the pressure in film. A radial pressure gradient runs through the film from its center to its perimeter, causing liquid seepage from the thin film to the plateau border (Karakashev and Manev 2015). The principal driving factor is plateau border suction. Because the radius of curvature between microscopic bubbles is small, the film between them drains more quickly. Reduced foam stability is caused by higher air-liquid ratios, temperatures, and ionic strength. It is well known that as the ionic concentration rises, so does film seepage, resulting in the compression of an electrical double layer and a low surface potential. The repelling force is reduced due to this (Kotsmar et al. 2009). Because fat improves film drainage, it acts as a foam destabilizer. The greater the destabilizing effect at the air-water interface, the larger the fat droplet (Indrawati and Narsimhan 2008).

5.8.2 Gelation

By turning fluids into solid structures with elasticity and different textures and juiciness, gelation makes a range of foods. Most dietary proteins capable of generating gels, such as whey proteins, plant globulins, and egg white proteins, are globular proteins with well-defined secondary and tertiary structures (Nicolai 2019). All protein gelation reactions begin when the balance of repulsive and attractive interactions in a solution is shifted, favoring attractive interactions. A structural transition is required to alter the balance of forces that produce protein stability, and it changes the structure from unreactive to reactive, which is needed for intermolecular interactions (Vander Linden and Foegeding 2009). As uncoiled polypeptide segments interact at specific sites to form a three-dimensional cross-linked protein network, gel formation is caused by the partially unfolded protein. In colloidal dispersions, this network can only occupy discrete sections of space, whereas, in gels, it can percolate across the entire space (Mezzenga and Fischer 2013; Nicolai 2019). Large amounts of water are trapped in a gel network, which behaves elastically on a timeframe and is viewed as a solid.

5.8.2.1 Mechanism of Gelation

To promote gelation, various techniques are used, including enzymatic processes, high pressure, acidity, salt or adjuvant, and heating or chilling. Gelation is most commonly induced by heating or cooling, and denaturation and aggregation are also required to create gels. In denaturant cosolvents, proteins can lose their naturally folded tertiary structure, most commonly due to thermal treatment, resulting in unfolding. As a result, a new force is induced between oppositely charged patches in the form of hydrogen bonding, hydrophobic contacts, disulfide bonding, or electrostatic interactions. Because their hydrophobic and sulfhydryl groups are exposed on the surface of the native proteins, amino acids can interact with and bind to other proteins. Because the attractive forces are higher than the repulsive forces, irreversible aggregates form (Fig. 5.10).

Denaturation and aggregation do not co-occur in all proteins. Because not all proteins are transformed into reactive intermediates instantly, a reservoir of native proteins will be gradually absorbed into the network. The rate of protein aggregation is determined by how the protein becomes active after it has been denatured (Weijers et al. 2002). As a result, there is a decreasing population of native proteins and a growing population of aggregates. The rate of aggregation and gelation increases as

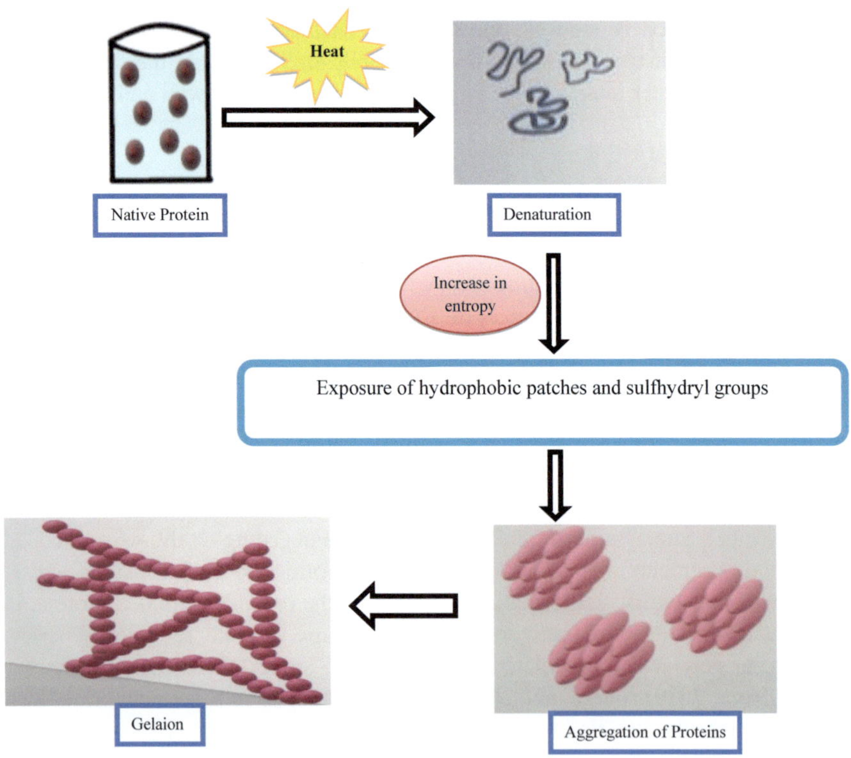

Fig. 5.10 Gelation of protein

the temperature rises (Brodkorb et al. 2016). The net charge on proteins, affected by pH determines the intensity of attraction and repulsive forces between proteins. The stabilization of proteins in solution is due to repulsive forces between proteins in equilibrium with the system. Due to local densification under increased attractive interactions and decreased repulsion, phase separation is common during the gelation of globular proteins. In linear molecules, phase separation may or may not be achievable (Vander Linden and Foegeding 2009). Phase separation will occur if the interactions that contribute to gelation during the early transition are significant. At high concentrations, this will happen via a spinodal decomposition mechanism. Nuclcation phenomena are observed when the chosen concentration is marginally higher than necessary for nucleation type behavior but lower than required for spinodal decomposition.

5.8.3 Water Holding Capacity

The term "water holding capacity" (WHC) is frequently used to characterize the hydration qualities of foods such as animal protein, gels, vegetables, and dietary fibers (Sman et al. 2013). It keeps track of the amount of water used in the processing and water linked to proteins. Water-binding capacity is another name for WHC (WBC). These phrases have slightly different meanings. The former refers to the ability to tie up water in raw meat and the latter refers to the amount of water bonded to meat throughout the processing process (Pospiech and Montowska 2011).

Given that myofibrils account for 85% of the volume of muscle cells, it is apparent that they contain a substantial amount of water. It has 1% water, so tightly bound that it remains intact even when frozen and heated. As a result, it is categorized as bound water. Despite being regularly shuffled with encircling water molecules, this bound water shows minor changes in postmortem muscle. Immobilized water, which accounts for 85% of total water in meat, is secured by bound water through steric processes. This water is challenging to remove, but it can be done by a process that causes protein structural disintegration (Pearce et al. 2011). The remaining half is categorized as free water since it can seep in freely if the necessary circumstances are met. It is found in the sarcoplasmic fluid and is held within and between myofibrils by capillary forces.

5.8.4 Solubility

One of the essential features of proteins chosen for use in food is their solubility. Because of noncovalent interactions, proteins are soluble. Its size, conformation, amino acid content, and proportion of polar and nonpolar amino acids all play a role. It rises when electrostatic repulsion increases and falls as attractive forces between proteins grow. At their isoelectric point (p*I*), proteins are the least soluble because the attractive interactions predominate. The hydrophobic amino acids are buried inside the core of a uniquely folded native protein.

On the other hand, hydrophilic ones are exposed on the surface to form a hydrogen bond with an aqueous medium. When proteins are unfolded however their hydrophobic amino acids are exposed to an aqueous solution, causing them to become unstable. Aggregates are produced by protein-protein interaction between exposed hydrophobic regions to achieve stability, which causes a decrease in solubility (Aryee and Boye 2016).

5.9 Facets of Physical and Chemical Modification of Proteins During Food Processing

The majority of food proteins are ingested after varying degrees of processing. Before eating, foods are processed in various ways to improve functional, nutritional, and sensory qualities. Food processing employs multiple techniques, including heat, high pressure, radiation, high-intensity ultrasound, and biochemical processing (Lepski and Brockmeyer 2013). Different processing processes modify the structure of food proteins in different ways. The literature on protein alteration resulting from processing is summarized in Table 5.3. Physical and chemical processing has a wide range of effects on the structure of food proteins, adding to the already complicated interactions between the structure and function of the protein. Unfolding, aggregation, cross-linking between constituents, and chemical alterations such as oxidation and glycosylation are examples of structural modifications (Gerrard et al. 2012). Protein nutritional quality, bio-accessibility, and bioavailability may be impacted by changes in physical structure (Orlien and Bolumar 2019). Processing, such as heating and enzymatic treatments, arguably has an immense impact on proteins among the numerous unit operations in the food sector. It affects protein digestibility and, as a result, immune system sensitization (Fasolin et al. 2019). Due to the partial denaturation of proteins, mild heat treatment can improve protein digestibility.

Still, extreme heat treatment can cause various chemical reactions, including cross-linking, racemization, and a reduction in digestibility (Damodaran 2007). Proteins make up the physical structure of muscle meals and are responsible for features including water retention, fiber swelling, extractability, gelation, emulsification, and binding. Under varied processing methods, the quality features of muscle meals provided by proteins vary dramatically (Xiong 2014). Muscle meals are frequently heated before ingestion. The modification that most proteins go through throughout processing is depicted in Fig. 5.11. Many of them undergo structural transformation/denaturation during processing or home cooking. Cooking causes side-chain modifications, an increase in surface hydrophobicity, cross-linking with other polypeptides, fragmentation, and aggregation of proteins, changes in the primary structure of proteins, such as protein carbonylation, aromatic residue modification, and the occurrence of the Maillard reaction (Yu et al. 2017). Protein breakdown in thermal processing begins with denaturation and progresses to aggregation, resulting in myofibril shrinkage and water expulsion (Kong et al. 2007). Proteins' native structures are altered, which can have positive and negative

Table 5.3 Summary of literature on changes induced in food proteins as a consequence of processing

Source	Processing method	Modifications induced	References
Bovine sarcoplasmic protein from *M. longissimus dorsi*	Combined pressure (200–600 MPa) and mild temperature (10–30 °C)	Reduction in solubility, WHC of protein, and change in meat color	Marcos et al. (2010)
Tropomyosin Tod p1 from Fresh squids	High hydrostatic pressure	Structural modifications as 53% of α-helix converted into random coils and β-sheet. Free sulfhydryl group dropped. Surface hydrophobicity increased. Digestibility improved. Allergenicity decreased	Jin et al. (2015)
Soy protein isolate and soy protein concentrate	Thermo-mechanical processing by shearing sample at 100–140 °C	Increase in carbonyl content and reduction in surface-exposed hydrophobicity	Duque-Estrada et al. (2019)
Bovine meat	Combination of pH, temperature, and time	Oxidative modification in myosin, methionine, and aromatic amino acid, deamidation, formation of pyroglutamic acid	Deb-Choudhury et al. (2020)
Reconstituted micellar casein concentrates and milk protein concentrates	High-pressure processing (150–450 MPa) at ambient temperature	Disintegration of casein micelle, pressure above 350 MPa induced weak gel network formation and aggregation for micellar casein concentrate, alteration in soluble mineral and protein levels	Cadesky et al. (2017)
Myofibrillar protein from *Hypophthalmichthys molitrix*	High pressure (200–500 MPa) for 10 min at 20 °C	More sulfhydryl groups, hydrophobic regions, and amino acid residues were found, conformational stability of myofibril is reduced, secondary structural changes α-helix destroyed	Qiu et al. (2014)
Hake myofibrils from *Merluccius merluccius*	High hydrostatic pressure	Reduction in α-helix, increase in β-sheet, denaturation, unfolding of protein, increase in the number of available sulfhydryl groups, gelation, thermal denaturation	Cando et al. (2014)
Myofibrillar protein from grass crap	γ-irradiation	Increased surface hydrophobicity, decrease in sulfhydryl groups, emulsifying properties,	Shi et al. (2015)

(continued)

Table 5.3 (continued)

Source	Processing method	Modifications induced	References
		ordered structure of myofibrils, intermolecular cross-linking in myofibrillar protein	
Trichiurushaumela surimi	High pressure (300–500 MPa for 5 min)	Improvement in WHC, gel strength, whiteness, chewiness, cohesiveness, and springiness of gel. Reduction in α-helix and β-turn. Increase in sheet and random coil in protein	Chen et al. (2020)
Trichiurushaumela surimi	High-pressure synergistic heat (90 °C for 30 min)	Increase in gel strength, partial degradation of actin and myosin heavy chain, conversion of partial β-sheet into α-helix	Chen et al. (2020)
Myofibrillar proteins from bighead carp *Hypophthalmichthys nobilis*	Oxidation with ozone	Increase in protein (salt extractable), total sulfhydryl and active sulfhydryl content, Ca^{2+}-ATPase activity, carbonyl content, and surface hydrophobicity	Zhang et al. (2015)
Egg white proteins	Heat and high pressure (400–700 MPa) at 10–60 °C	Formation of moist and creamy foams with small bubble size that have low sensitivity to bubble coalescence	Vander Plancken et al. (2007)
	At pH 7.6	High levels of unfolding and extensive solubility loss, stable foams	
	At pH 8.8	Extreme protein unfolding, voluminous foam obtained	
Whey protein concentrate	Doses of γ-irradiation (0–100 kGy)	Occurrence of Maillard reactions with antioxidant potential, dose-dependent reduction in free amino groups, formation of fluorescent compounds, and brown pigment	Chawla et al. (2009)
Buckwheat seeds and groats	Roasting	Maillard reaction products induced in both, and protein content of groats decreased	Zielinski et al. (2009)
Processed proteins of poultry	Heat	Decreased digestibility, appearance of D-aspartic acid	Bellagamba et al. (2015)

(continued)

Table 5.3 (continued)

Source	Processing method	Modifications induced	References
M. longissimus thoracis	Heat treatment at 100 °C for 10–30 min	Protein denaturation, oxidation, unfolding, formation of protein aggregates, Schiff bases, increase in surface hydrophobicity	Traore et al. (2012)
Lamb Lions	Vacuum packaged and cooked (60–80 °C)	Carbonyl content of protein increased, and the levels of α-aminoadipic acid and γ-glutamic semialdehyde increased	Roldan et al. (2014)
Peanut flour	Roasting	Reduced WHC and oil binding capacity, protein denaturation, decrease in solubility, reduced foaming capacity by half, decreased functionality	Yu et al. (2007)
	Fermentation	Increased WHC and oil binding capacity, forming ability by threefold, increased functionality and solubility	
Whey protein	Heating at charge shielding conditions at temperature 50, 60, 70, and 80 °C for 60 min	Unfolding of protein accompanied by an increase in surface hydrophobicity, formation of microparticles with less surface activity, lower formability, and foam stability	Grossmann et al. (2019)
Myofibrillar proteins from silver crap	Vacuum freeze-drying	Partially denatured protein, increase in -SH group and surface hydrophobicity. Decrease of α-helical structure	Niu et al. (2019)
	Vacuum spray drying	Denaturation of protein, Higher water retention, and emulsifying capacity	
Fish fillets from *Acipenser gueldenstaedtii*	Boiling Steaming Microwaving Roasting Deep-frying	Amino acids' side-chain modifications. Oxidation of aromatic amino acid, lysine. Appearance of α-Aminoadipic acid semialdehyde	Hu et al. (2017)
Rapeseed press protein cake	Heat	Reduced emulsifying capacity	Ostbring et al. (2019)
Egg white from commercial eggs	Boiling for 15 min at 100 °C	Formation of Amyloid fibrils and unfolding of the polypeptide	Monge-Morera et al. (2020)

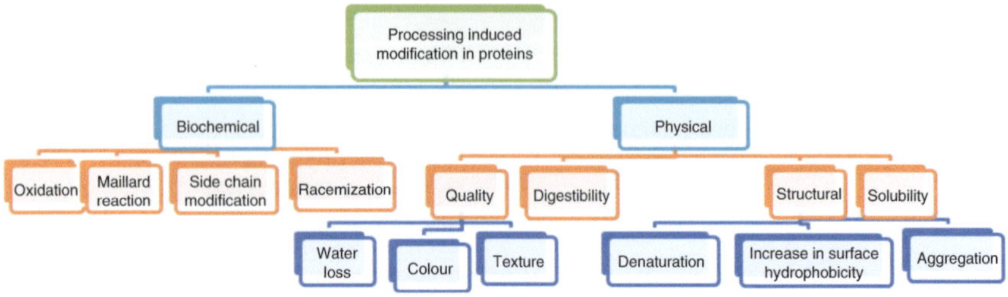

Fig. 5.11 Aspects of physical and chemical changes in protein during processing

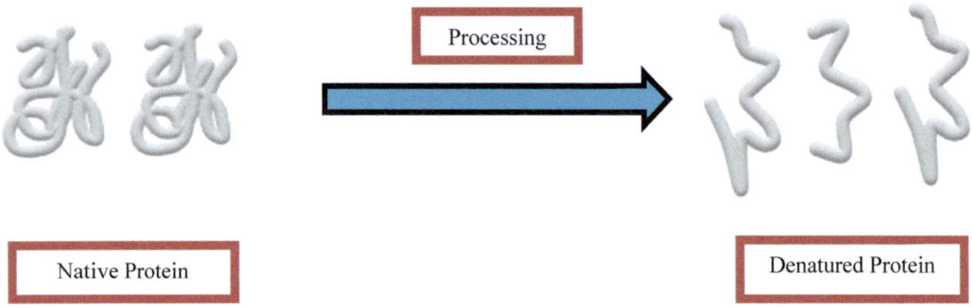

Fig. 5.12 Denaturation of food protein due to processing

consequences on protein digestion and critical amino acid availability (Singh et al. 2014). Heat-induced alterations in proteins' native and primary structures have been linked to various quality issues, including color, texture, and gelation (Sun and Holley 2011; Suman and Joseph 2013).

5.9.1 Structural Changes in Protein

When proteins are subjected to various processing conditions, such as heat, pressure, pH, and so on, their structure changes, making them either more resistant or prone to hydrolysis. Proteins with more disulfide links are less prone to structural alterations than those with less. These proteins are additionally resistant to digestion because the energy required to break these covalent bonds is greater than that needed to break noncovalent bonds. Disulfide bonds can be disrupted depending on the intensity of processing conditions. It is possible to increase aggregation by creating new connections between free thiol residues (Tang et al. 2009).

5.9.1.1 Denaturation

Denaturation is the process of proteins unfolding from their native form to an altered configuration due to a disruption in secondary and tertiary structure induced by a change in interactions (Fig. 5.12). It causes connective tissue to degrade, resulting in changes in features like reduced WHC, a more rigid and compact texture, and so on (Kong et al. 2007; Skipnes et al. 2011). When polypeptide chains are heated, thermal energy is released, enhancing polypeptide motion, destabilizing noncovalent bonds, and denaturing the protein. Hydrophobic contacts, which are principally responsible for stabilizing the unfolded state, are stabilized throughout this phase. The protein's type and structure determine denaturing conditions, and the protein's net charge determines the structural response. Proteins are more digestible because of this denaturation. For most single-domain proteins, Li-Chan and Lacroix (2018) explained that a simple two-state transition (native-denatured) represents the shift from folded, native structure to unfolded, denatured structure. They also realized that the two-state transition model does not always adequately describe the denaturation process. Instead, a stable partially folded state known as "a molten globule" is observed for some proteins under certain conditions. The molten globule state has little or no tertiary structure, yet it has a secondary structural composition and compactness close to the native protein structure. Proteins are refolded to reduce free energy by forming new noncovalent interactions and disulfide and intermolecular non-covalent bonds between two unfolded proteins. However, because new interactions are not representatives of the original conformation, this conformation differs from native (Becker and Yu 2013; Damodaran 2007; Villanea et al. 2016).

5.9.1.2 Increase in Protein Hydrophobicity

Hydrophobic portions of the protein are folded inward in their native conformation and unfolding due to denaturation exposes them outward. The exposed hydrophobic groups and increased surface area of unfolded or partially refolded proteins enhance aggregation risk. The unmasking of hydrophobic groups from the core of proteins increases surface hydrophobicity (He et al. 2014).

5.9.1.3 Aggregation

Internal disulfide bridges and weak interactions within proteins can be destroyed through chemical or physical degradation of proteins caused by denaturing factors such as temperature and pH. It causes an increase in free energy, which causes proteins to refold to reduce their free energy state. As a result of eliminating heat-exposed hydrophobic residues from water and preventing contact with surrounding water via noncovalent linkages and disulfide bridges, aggregation occurs (Becker and Yu 2013; Carbonaro et al. 2012). Aggregates are also produced by cross-linking different proteins due to oxidation and covalent linkages joining various amino acids (Fig. 5.13). This is particularly common at alkaline pH levels (Schwarzenbolz and Henle 2010). Because aggregated and cross-linked proteins are less accessible to digestive enzymes, their digestibility is reduced (Villanea et al. 2016). Because milk protein caseins lack a tertiary structure, they do not undergo usual denaturation and aggregation when heated. They are, nevertheless, vulnerable to chemical changes caused by processing (Pellegrino et al. 2011).

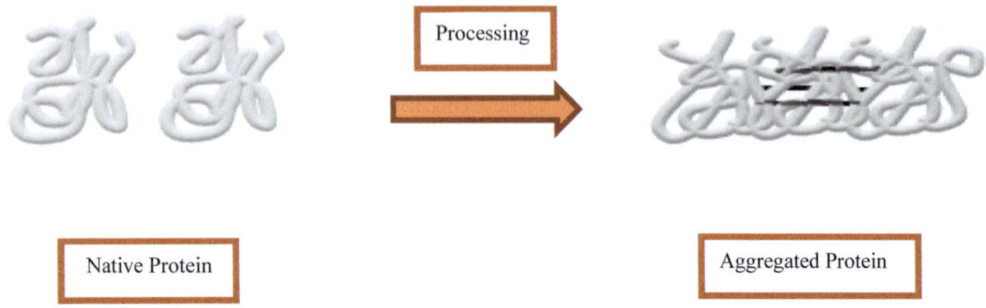

Fig. 5.13 Aspects of physical and chemical changes in protein during processing

5.9.2 Solubility

Most proteins are soluble because their hydrophobic groups are buried inside the native structure's core. Their solubility however reduces as protein aggregates. The protein aggregates are soluble at first due to processing-induced denaturation and reorientation of groups, but they become insoluble as their size exceeds the solubility limit (Shivu et al. 2013). Extreme thermal processing enhances the formation of protein aggregates by causing a change in the linkages that exist in and between proteins (He et al. 2014).

5.9.3 Digestibility

Processing conditions determine whether proteins will be more vulnerable or resistant to enzymatic processes. Vulnerable proteins can result from unfolding or developing a random coil, whereas resistant proteins can result from chemical changes of amino acids or the formation of aggregates. The creation of random coils is caused by collapsing the protein's tertiary and secondary structure, and it accelerates enzyme-catalyzed hydrolysis by revealing previously buried moieties. The formation of random coils indicates successful denaturation of protein, which is positively associated with in vitro digestibility. Thermal inactivation of protease inhibitors enhances vulnerability to enzymatic hydrolysis/digestibility in addition to altering the secondary structure (Carbonaro et al. 2012). Overcooking can reduce digestibility and absorption; even if digestion is unaffected or improved by heat treatment, the modified peptide may not function the same as the native (Wada and Lönnerdal 2014; Oberli et al. 2016; Yu et al. 2017). In general, unfolding proteins makes amino acids more accessible to proteolytic enzymes. On the other hand, aggregation and chemical changes restrict the accessibility of proteolytic enzymes, depending on the precise amino acid being changed.

5.9.4 Chemical Modifications

Denaturing proteins, which result in a random coil with numerous possible conformations, is a common outcome of processing regimens. Food proteins may also be subjected to a range of chemical changes. While a protein's shape changes, it exposes reactive groups that are not there in their natural condition. Sidechains of amino acids can react with each other and with other dietary components. Chemical alterations can take many forms, with reactions including the production of color and flavor compounds being particularly desirable (Fayle and Gerrard 2002). Although chemically modified amino acids may still be nutritionally adequate, their utilization by the body may be hampered by several circumstances. If a protein is not identified by the active site of the gut proteolytic enzymes, it will not be hydrolyzed, and if the protein is successfully hydrolyzed, the modified amino acid or peptide may not be able to pass through the gut wall. If taken through the gut, the body may be unable to convert the modified amino acid to parent amino acid or any other useful molecule. It will be eliminated in urine in such instances (Meade et al. 2005).

5.9.4.1 Racemization

Racemization of amino acid residues occurs under various conditions commonly encountered during food preparation. Proteins containing D-amino acids have a lower nutritional quality than those containing L-amino acids due to the combined impact of proteolytic enzymes' inability to cleave peptide bonds containing a D-amino acid, as well as inadequate absorption.

5.9.4.2 Side-Chain Modification

After heat treatment, a range of side-chain changes mediated by oxidation or the Maillard reaction has been identified in proteins.

5.9.4.2.1 Maillard Reaction

It is a complicated web of nonenzymatic processes involving interactions between reducing sugars and lysine residue to produce sugar-derived brown adducts (Arena et al. 2014). The reaction usually happens at room temperature when reactive amino groups bind with the carbonyl group of reducing sugar. This reaction produces Schiff's base, which is then transformed by Amadori rearrangement to create ketose amine, reductones, furfurals, and other cyclic compounds (Fig. 5.14).

Due to a free amino group at the epsilon carbon, lysine is a highly reactive amino acid and prone to Maillard reaction during processing. The intermediate chemical fructoselysine is formed when transformed, and cross-linking of two different amino acids occurs occasionally. For example, lysinoalanine is formed when alanine and lysine are cross-linked. In contrast, *N*-carboxymethyl lysine (CML) and 5-hydroxymethyl-2-furfural (HMF) are formed when alanine and lysine are cross-linked (Van Rooijen et al. 2014). These derivatives are poorly absorbed and have low biological activity. A fascinating subset of Maillard reaction products is protein cross-linking. The covalently bonded aggregate is less accessible to digestive enzymes, which results in a loss of essential amino acids like lysine (Gerrard et al.

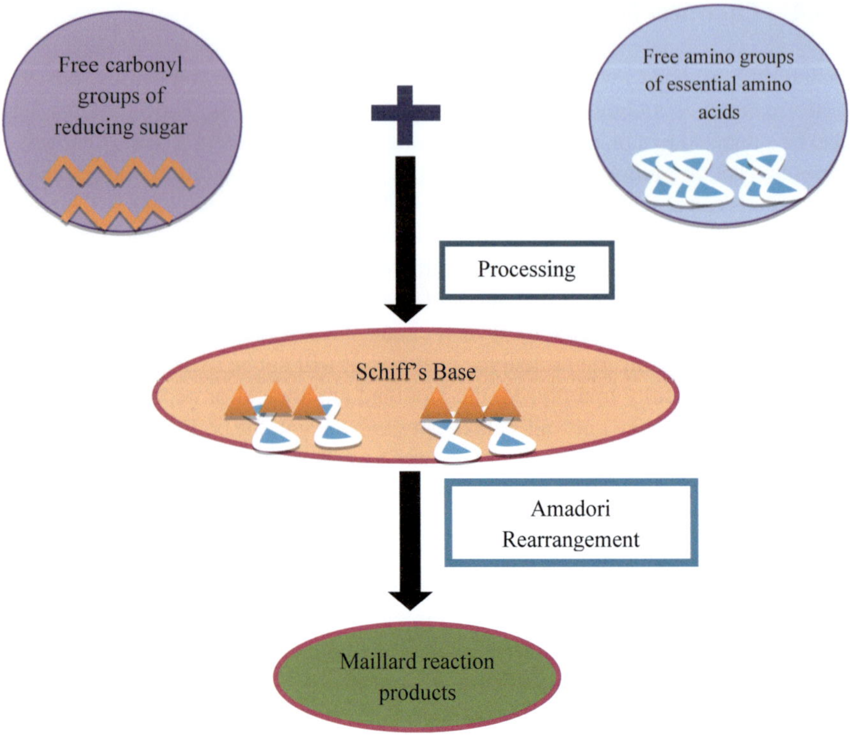

Fig. 5.14 Maillard reaction

2012). Heterocyclic amines (HCAs) are formed during meat cooking by the reaction of creatinine with other small organic molecules or by the thermal degradation of proteins or amino acids (Gibis 2016).

5.9.4.2.2 Oxidation

Proteins are easily damaged by reactive oxygen and nitrogen species (ROS and RNS). Cooking heat promotes protein oxidation by triggering ROS production (Traore et al. 2012). It works by a free radical chain reaction, in which a free radical removes a hydrogen atom from a protein molecule. Carbonylation, disulfide formation, and the production of dityrosine bridges occur when proteins are oxidized (Fig. 5.15) (Cui et al. 2012; Sun et al. 2011). Fragmentation happens when free radicals react with the polypeptide chain's backbone at a specific location (Lund et al. 2011). Because oxidation can cause protein-protein interaction and side-chain modification, sulphydryl groups are lost, aromatic amino acids are modified, essential amino acids are lost, and protein carbonyls are formed. These alterations result in digestibility, nutritional value, solubility, texture, and color changes (Bax et al. 2013).

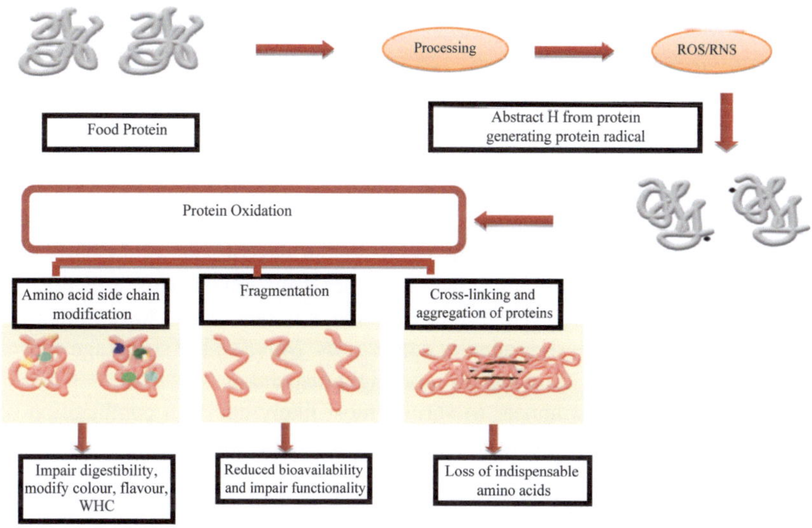

Fig. 5.15 Processing-induced oxidation of proteins

5.9.5 Quality Changes

Protein quality is evaluated by its capacity to perform specific metabolic tasks and is determined by its constituent amino acids, primary structure, and native structure. Protein denaturation is responsible for most of the physical changes during heat processing. Muscle fiber contraction, connective tissue deterioration, and reduced water-binding capacity are caused by protein structural disruption, resulting in a rigid and more compact texture (Kong et al. 2007).

5.9.5.1 Water Loss

The denaturation and shrinking of muscle proteins cause water loss during the cooking process. Many proteins are involved in shrinkage and water ejection, as indicated by longitudinal and transverse shrinkage at different temperatures. As a result, their water holding capacity is reduced (Kondjoyan et al. 2013). Cooking usually results in more water loss in aged meat than unaged meat. Water is necessary because it affects both the structure and quality of muscle. Cooking causes drip loss, making muscle proteins more rigid by increasing stiffness and decreasing suppleness. However, some proteins, such as collagen, can gelatinize and contain water when the heating period is extended. Cooking also affects the juiciness of meat by causing drip loss, which occurs when the juice in the muscle disgorges owing to denaturation and a loss of muscular structure. The juiciness of the meat varies depending on the cooking temperature (Kondjoyan et al. 2013).

5.9.5.2 Color

Color is a physical characteristic that has an impact on customer acceptance. Several mechanisms, including pigment loss, ascorbic acid oxidation, and enzymatic and nonenzymatic browning, affect it during thermal processing (Kong et al. 2007; Ovissipour et al. 2013). Muscle color is determined not just by pigment but also by the muscle structure. As a result, any alteration in the 3-D structural lattice generated by cooking may modify light scattering, transmission, and reflection, influencing color. These structural alterations are induced by a variety of temperature and cooking methods. The increase in lightness during cooking could be due to structural changes in myofibrillar proteins and other structural proteins, affecting the degree of myofibrillar spacing (Hughes et al. 2014). It is clear that as the temperature rises, the structure of the flesh gradually compresses. Muscle fiber shrinkage causes wide gaps between fibers, allowing for more light dispersion.

Furthermore, myofibrils appear to shrink, most likely due to a combination of protein denaturation, destruction of water-binding sites, and increased hydrophobicity. The sarcolemma would be disrupted, resulting in a more dense protein structure (García-Segovia et al. 2007). More light scattering and a lighter look could result from myofibril shrinking and reduced fiber diameter.

5.9.5.3 Texture

The texture is defined as a set of physical characteristics that describe the composition and structure of food as it is perceived in the mouth. The two most common and widely used procedures for determining the mechanical qualities of foods are compression and piercing testing (Kong et al. 2007; Ovissipour et al. 2013). Meat proteins undergo thermal denaturation, which results in significant water loss. Because water is the principal plasticizer in meat, its loss causes the cooked substance to stiffen and harden. Juiciness has an inverse association with water loss from the structure, and there appears to be a positive correlation between juiciness and softness.

5.10 Conclusion

Proteins are amphoteric molecules composed of amino acids linked through peptide bridges. They hold a prominent position in living cells due to their involvement in diverse biological processes. Proteins of different origins provide a variety of activities, including solubility, water retention, gelling, foaming, and emulsification. Their functions are affected by pH, ionic strength, and structure. Denaturation, reduced digestibility, and nutritional value result from changes in proteins at all the structural levels.

References

Amarowicz R (2010) Modification of emulsifying properties of food proteins by enzymatic hydrolysis. Eur J Lipid Sci Technol 112:695–696

Arena S, Salzano AM, Renzone G et al (2014) Non-enzymatic glycation and glycoxidation protein products in foods and diseases: an interconnected, complex scenario fully open to innovative proteomic studies. Mass Spectrom Rev 33:49–77

Aryee ANA, Boye JI (2016) Comparative study of the effects of processing on nutritional, physiochemical and functional properties of lentil. J Food Process Preserv 41(1):e12824

Badigannavar A, Girish G, Ramachandran V et al (2016) Genotypic variation for seed protein and mineral content among post-rainy season-grown sorghum genotypes. Crop J 4:61–67

Bax ML, Sayd T, Aubry L et al (2013) Muscle composition slightly affects in vitro digestion of aged and cooked meat: identification of associated proteomic markers. Food Chem 136:1249–1262

Becker PM, Yu P (2013) What makes protein indigestible from tissue-related, cellular, and molecular aspects. Mol Nutr Food Res 57:1695–1707

Belitz HD, Grosch W, Schierberle P (2009) Eggs. In: Food chemistry. Springer, Berlin, pp 546–561

Bellagamba F, Caprino F, Mentasti T et al (2015) The impact of processing on amino acid racemization and protein quality in processed animal proteins of poultry origin. Ital J Anim Sci 14:238–245

Bleakley S, Hayes M (2017) Algal proteins: extraction, application, and challenges concerning production. Foods 6(5):33

Brodkorb A, Croguennec T, Bouhallab S et al (2016) Heat-induced denaturation, aggregation and gelation of whey proteins. In: Advanced dairy chemistry, Proteins: applied aspects, vol 1B. Springer, Berlin, pp 155–178

Cabra V, Roberto A, Amelia F (2008) Emulsifying properties of proteins. Bol Soc Quím Méx 2:80–89

Cadesky L, Walking-Ribeiro M, Kriner KT et al (2017) Structural changes induced by high-pressure processing in micellar casein and milk protein concentrates. J Dairy Sci 100:7055–7070

Cando D, Moreneo HM, Tovar CA et al (2014) Effect of high pressure and/or temperature over gelation of isolated Hake myofibrils. Food Bioprocess Technol 7:3197–3207

Carbonaro M, Maselli P, Nucara A (2012) Relationship between digestibility and secondary structure of raw and thermally treated legume proteins: a Fourier transform infrared (FT-IR) spectroscopic study. Amino Acids 43:911–921

Cha JW, Yoon IS, Lee GW et al (2020) Food functionalities and bioactivities of protein isolates recovered from skipjack tuna roe by isoelectric solubilisation and precipitation. Food Sci Nutr 8: 1874–1887

Chawla SP, Chander R, Sharma A (2009) Antioxidant properties of Maillard reaction products obtained by gamma-irradiation of whey proteins. Food Chem 116:122–128

Chen L (2009) Protein micro/nanoparticles for controlled neutraceutical delivery in functional foods. In: McClements DJ, Decker EA (eds) Designing functional foods: measuring and controlling food structure breakdown and nutrient absorption. CRC Press, Boca Raton, FL, pp 572–600

Chen Y, Xu A, Yang R et al (2020) Myofibrillar protein structure and gel properties of *Trichiurushaumela surimi* subjected to high pressure or high pressure synergistic heat. Food Bioprocess Technol 13:589–598

Cui X, Xiong YL, Kong B et al (2012) Hydroxyl radical-stressed whey protein isolate: chemical and structural properties. Food Bioprocess Technol 5:2454–2461

Damodaran S (2007) Amino acids, peptides and proteins. In: Damodaran KLPS, Fennema OR (eds) Fennema's food chemistry, vol 4. CRC Press, Boca Raton, FL, pp 217–329

Day L (2013) Proteins from land plants—potential resources for human nutrition and food security. Trends Food Sci Technol 32:25–42

De Mesa-Stonestreet NJ, Alavi S, Bean SR (2010) Sorghum proteins: the concentration, isolation, modification, and food applications of kafirins. J Food Sci 75:R90–R104
Deb-Choudhury S, Haines S, Harland D, Clerens S, Van Koten C, Lee E, Thomas A, Dyer J (2020) Multi-parameter evaluation of the effect of processing conditions on meat protein modification. Heliyon 12(6):e04185
Del Rosario Moreira M, Pereda M, Marcovich NE et al (2011) Antimicrobial effectiveness of bioactive packaging materials from edible chitosan and casein polymers: assessment on carrot, cheese, and salami. J Food Sci 76:M54–M63
Divsalar A, Saboury AA, Mohammad N et al (2012) Characterization and side effect analysis of a newly designed nanoemulsion targeting human serum albumin for drug delivery. Colloids Surf B Biointerfaces 98:80–84
Drenckhan W, Saint-Jalmes A (2015) The science of foaming. Adv Colloid Interf Sci 222:228–259
Duque-Estrada P, Berton-Caarabin CC, Nieuwkoop M et al (2019) Protein oxidation and in vitro gastric digestion of processed soy-based matrices. J Agric Food Chem 67:9591–9600
Ebert S, Grossmann L, Hinrichs J et al (2019) Emulsifying properties of water-soluble proteins extracted from the microalgae *Chlorella sorokiniana* and *Phaeodactylum tricornutum*. Food Funct 10:754–764
Elzoghby AO, Samy WM, Elgindy NA (2012) Protein-based nanocarriers as promising drug and gene delivery systems. J Control Release 161:38–49
Esquena J (2016) Water-in-water (w/w) emulsions. Curr Opin Colloid Interface Sci 25:109–119
European Food Safety Authority Scientific Committee (EFSA) (2015) Scientific opinion on a risk profile related to production and consumption of insects as food and feed. EFSA J 13(10):4257
FAO/WHO (2007) Protein and amino acid requirement in human nutrition. World Health Organization, Geneva. ISBN 92 4 120935 6
Fasolin LH, Pereira RN, Pinheiro AC et al (2019) Emergent food proteins – towards sustainability, health and innovation. Food Res Int 125:108586
Fayle SE, Gerrard JA (2002) The Maillard reaction. Royal Society of Chemistry, Cambridge, UK
Foegeding EA, Davis JP (2011) Food protein functionality: a comprehensive approach. Food Hydrocoll 25:1853–1864
Food and Agricultural Organization of the United Nations (2013) Report of an FAO Expert Consultation. Dietary protein quality evaluation in human nutrition
Fukushima D (2011) Soy proteins. In: Phillips G, Williams P (eds) Handbook of food proteins. Woodhead, Cambridge, UK, pp 210–232
García-Segovia P, Andrés-Bello A, Martínez-Monzó J (2007) Effect of cooking method on mechanical properties, color and structure of beef muscle (*M. pectoralis*). J Food Eng 80: 813–821
Gerrard JA, Lasse M, Cottam J et al (2012) Aspects of physical and chemical alterations to proteins during food processing – some implications for nutrition. Br J Nutr 108:S288–S297
Gibis M (2016) Heterocyclic aromatic amines in cooked meat products: causes, formation, occurrence, and risk assessment. Compr Rev Food Sci Food Saf 15:269–302
Gochev G, Retzlaff I, Exerowa D, Miller R (2014) Electrostatic stabilization of foam films from β-lactoglobulin solutions. Colloids Surf A Physiochem Eng Asp 460:272–279
Goff HD, Vega C (2007) Structure-engineering of ice-cream and foam-based foods. In: McClements DJ (ed) Understanding and controlling the microstructure of complex foods. Cambridge, UK, Woodhead, pp 558–574
Gomez-Guillen MC, Gimenez B, Lopez-Caballero ME et al (2011) Functional and bioactive properties of collagen and gelatin from alternative sources: a review. Food Hydrocoll 25: 1813–1827
González-Pérez S (2015) Sunflower proteins. In: Martínez-Force E, Dunford N, Salas J (eds) Sunflower. Chemistry, production, processing, and utilization. AOCS Press, Urbana, IL, pp 331–393

Green AJ, Littlejohn KA, Hooley P et al (2013) Formation and stability of food foams and aerated emulsions: hydrophobins as novel functional ingredients. Curr Opin Colloid Interface Sci 18: 292–301

Grossmann L, Beicht M, Reichert C et al (2019) Foaming properties of heat-aggregated microparticles from whey proteins. Colloids Surf A Physiochem Eng Asp 579:123572

He R, He HY, Chao D et al (2014) Effects of high pressure and heat treatments on physicochemical and gelation properties of rapeseed protein isolate. Food Bioprocess Technol 7:1344–1353

Hoffman JR, Falvo MJ (2004) Protein – which is best? J Sports Sci Med 3:118–130

Hoffmann H, Reger M (2014) Emulsions with unique properties from proteins as emulsifiers. Adv Colloid Interf Sci 205:94–104

Hu L, Ren S, Shen Q et al (2017) Proteomic study of the effect of different cooking methods on protein oxidation in fish fillets. RSC Adv 7:27496–27505

Hughes JM, Oiseth SK, Purslow PP et al (2014) A structural approach to understanding the interactions between colour, water-holding capacity and tenderness. Meat Sci 98:520–532

Indrawati L, Narsimhan G (2008) Characterization of protein stabilized foam formed in a continuous shear mixing apparatus. J Food Eng 88:456–465

Ivanova P, Chalova V, Uzunova G et al (2016) Biochemical characterization of industrially produced rapeseed meal as a protein source in food industry. Agric Sci Proc 10:55–62

Jafari SM, Beheshti P, Assadpoor E (2012) Rheological behaviour and stability of D-limonene emulsions made by a novel hydrocolloid (Angum gum) compared with Arabic gum. J Food Eng 109:1–8

Jin Y, Deng Y, Qian B et al (2015) Allergenic response to squid (*Todarodes pacificus*) tropomyosin Tod p1 structure modifications induced by high hydrostatic pressure. Food Chem Toxicol 76: 86–93

Karakashev SI, Manev ED (2015) Hydrodynamics of thin liquid films: retrospective and perspectives. Adv Colloid Interf Sci 222:398–412

Keerati-u-rai M, Corredig M (2009) Heat-induced changes in oil-in-water emulsions stabilized with soy protein isolate. Food Hydrocoll 23:2141–2148

Keerati-u-rai M, Corredig M (2010) Heat-induced changes occurring in oil/water emulsions stabilized by soy glycinin and β-conglycinin. J Agric Food Chem 58:9171–9180

Khan RH, Siddiqi MK et al (2017) Protein structure and function. www.austinpublishinggroup.com/ebooks

Kondjoyan A, Oillic S, Portanguen SP et al (2013) Combined heat transfer and kinetic models to predict cooking loss during heat treatment of beef meat. Meat Sci 95:336–344

Kong F, Tang J, Rasco B et al (2007) Kinetics of salmon quality changes during thermal processing. J Food Eng 83:510–520

Kotsmar C, Pradines V, Alahverdjieva VS et al (2009) Thermodynamics, adsorption kinetics and rheology of mixed protein–surfactant interfacial layers. Adv Colloid Interf Sci 150:41–54

Lam RSH, Nickerson MT (2013) Food proteins: a review on their emulsifying properties using a structure function approach. Food Chem 141:975–984

Le Floch-Fouéré C, Pezennec S, Pézolet M, Rioux-Dubé JF et al (2011) Unexpected differences in the behavior of ovotransferrin at the air-water interface at pH 6.5 and 8.0. J Colloid Interface Sci 356:614–623

Lech FJ, Delahaije RJ, Meinders MB et al (2016) Identification of critical concentrations determining foam ability and stability of β-lactoglobulin. Food Hydrocoll 57:46–54

Lepski S, Brockmeyer J (2013) Impact of dietary factors and food processing on food allergy. Mol Nutr Food Res 57:145–152

Li-Chan ECY, Lacroix IME (2018) Properties of proteins in food systems: an introduction. In: Proteins in food processing, Woodhead Publishing series in food science, technology and nutrition. Woodhead, Cambridge, UK, pp 1–25

Lone DA, Wani NA, Wani IA et al (2015) Physico-chemical and functional properties of rainbow trout fish protein isolate. Int Food Res J 22:1112–1116

Loveday SM (2019) Food proteins: technological, nutritional, and sustainability attributes of traditional and emerging proteins. Annu Rev Food Sci Technol 10:311–339

Lund MN, Heinonen M, Baron CP et al (2011) Protein oxidation in muscle foods: a review. Mol Nutr Food Res 55:83–95

Luo Y, Wang Q (2014) Zein-based micro- and nano-particles for drug and nutrient delivery: a review. J Appl Polym Sci 131:40696

Malik MA, Saini CS (2017) Polyphenol removal from sunflower seed and kernel: effect on functional and rheological properties of protein isolates. Food Hydrocoll 63:705–715

Maloma SA, Aluko RE (2015) A comparative study of the structural and functional properties of isolated hemp seed (*Cannabis sativa* L.) albumin and globulin fractions. Food Hydrocoll 43: 743–752

Mao X, Hua Y (2012) Composition, structure and functional properties of protein concentrates and isolates produced from walnut (*Juglans regia* L.). Int J Mol Sci 13:1561–1581

Marcos B, Kerry JP, Mullen AM (2010) High pressure induced changes on sarcoplasmic protein fraction and quality indicators. Meat Sci 85:115–120

McClements DJ (2005) Food emulsions: principles, practice and techniques, 2nd edn. CRC Press Taylor & Francis Group, Boca Raton, FL

McClements DJ (2015) Food emulsions: principles, practices, and techniques. CRC Press, Boca Raton, FL

McClements DJ, Bai L, Chung C (2017) Recent advances in the utilization of natural emulsifiers to form and stabilize emulsions. Annu Rev Food Sci Technol 8:205–236

Meade SJ, Reid EA, Gerrad JA (2005) The impact of processing on the nutritional quality of food proteins. J AOAC Int 88:904–922

Mezzenga R, Fischer P (2013) The self-assembly, aggregation and phase transitions of food protein systems in one, two and three dimensions. Rep Prog Phys 76:046601

Mills ENC, Sancho AI, Rigby NM et al (2009) Impact of food processing on the structural and allergenic properties of food allergens. Mol Nutr Food Res 53:963–969

Millward DJ (2012) Amino acid scoring patterns for protein quality assessment. Br J Nutr 108:S31–S43

Monge-Morera M, Lambrecht MA, Deleu LJ et al (2020) Processing induced changes in food proteins: amyloid formation during boiling of hen white. Biomacromolecules 21:2218–2228

Murray BS, Durga K, Yusoff A et al (2011) Stabilization of foams and emulsions by mixtures of surface active food-grade particles and proteins. Food Hydrocoll 25:627–638

Narsimhan G, Xiang N (2017) Role of proteins on formation, drainage, and stability of liquid food foams. Annu Rev Food Sci Technol 9:45–63

Nehete JY, Bhambar RS, Narkhede MR et al (2013) Natural proteins: sources, isolation, characterization and applications. Pharmacogn Rev 7:107–116

Nicolai T (2019) Gelation of food protein-protein mixtures. Adv Colloid Interf Sci 270:147–164

Niu J, Zhao B, Guo X et al (2019) Effects of vacuum freeze-drying and vacuum spray-drying on biochemical properties and functionalities of myofibrillar proteins from silver crap. J Food Qual 2019:1–8

Oberli M, Lan A, Khodorova N et al (2016) Compared with raw bovine meat, boiling but not grilling, barbecuing, or roasting decreases protein digestibility without any major consequences for intestinal mucosa in rats, although the daily ingestion of bovine meat induces histologic modifications in the colon. J Nutr 146:1506–1513

Orlien V, Bolumar T (2019) Biochemical and nutritional changes during food processing and storage. Foods 8(10):494

Ostbring K, Tullberg C, Burri S et al (2019) Protein recovery from rapeseed press cake: varietal and processing condition effects on yield, emulsifying capacity and antioxidant activity of the protein rich extract. Foods 8(12):627

Ostbring K, Nilsson K, Ahlstrom C et al (2020) Emulsifying and anti-oxidative properties of proteins extracted from industrially cold-pressed rapeseed press cake. Foods 9(5):678

Ovissipour M, Rasco B, Tang J et al (2013) Kinetics of quality changes in whole blue mussel (*Mytilus edulis*) during pasteurization. Food Res Int 53:141–148
Pal R (2011) Rheology of simple and multiple emulsions. Curr Opin Colloid Interface Sci 16:41–60
Papalamprou EM, Doxastakis GI, Kiosseoglou V (2010) Chickpea protein isolates obtained by wet extraction as emulsifying agents. J Sci Food Agric 90:304–313
Pearce KL, Rosenvold K, Andersen HJ et al (2011) Water distribution and mobility in meat during the conversion of muscle to meat and ageing and the impacts on fresh meat quality attributes - a review. Meat Sci 89:111–124
Pellegrino LS, Cattaneo, De Noni I (2011) Nutrition and health: effects of processing on protein quality of milk and milk products. In: Fuquay JW, McSweeney PLH, Fox PF (eds) Encyclopedia of dairy sciences. Academic, Cambridge, MA, pp 1067–1074
Perera SP, McIntosh TC, Wanasundara JP (2016) Structural properties of cruciferin and napin of *Brassica napus* (Canola) show distinct responses to changes in pH and temperature. Plants (Basel) 5(3):36
Petrusán JI, Rawel H, Huschek G (2016) Protein-rich vegetal sources and trends in human nutrition: a review. Curr Top Pept Protein Res 17:1–19
Pospiech E, Montowska M (2011) Technologies to improve water-holding capacity of meat. In: Joo ST (ed) Control of meat quality. Research Signposts, Kerala, India, pp 61–80
Qiu C, Xia W, Jiang Q (2014) Pressure-induced changes of silver carp (*Hypophthalmichthys molitrix*) myofibrillar protein structure. Eur Food Res Technol 238:753–761
Quintana RRB, Mendoza-Wilson AM, Montfort GRC et al (2019) Plant based proteins. In: Proteins: sustainable source, processing and applications. Elsevier, Amsterdam, pp 97–130
Reihani SFS, Khosravi-Darani K (2019) Influencing factors on single-cell protein production by submerged fermentation: a review. Electron J Biotechnol 37:34–40
Ritala A, Hakkinen ST, Toivari M et al (2017) Single cell protein—state-of-the-art, industrial landscape and patents 2001–2016. Front Microbiol 8:2009
Roldan M, Antequera T, Armenteros M et al (2014) Effect of different temperature-time combinations on lipid and protein oxidation of sous-vide cooked lamb loins. Food Chem 149: 129–136
Sai-Ut S, Ketnawa S, Chaiwut P et al (2009) Biochemical and functional properties of proteins from red kidney, navy and adzuki beans. Asian J Food Agro Ind 2:493–504
Schaafsma G (2000) The protein digestibility-corrected amino acid score. J Nutr 130:1865S–1867S
Schwarzenbolz U, Henle T (2010) Non-enzymatic modifications of proteins under high-pressure treatment. High Press Res 30:458–465
Scieszka S, Klewicka E (2018) Algae in food: a general review. Crit Rev Food Sci Nutr 59:1–23
Shi Y, Li RY, Tu ZC et al (2015) Effect of γ-irradiation on the physiochemical properties and structure of fish myofibrillar proteins. Radiat Phys Chem 109:70–72
Shivu B, Seshadri S, Li J et al (2013) Distinct β-sheet structure in protein aggregates determined by ATR–FTIR spectroscopy. Biochemistry 52:5176–5183
Singh TK, Øiseth SK, Lundin L et al (2014) Influence of heat and shear induced protein aggregation on the in vitro digestion rate of whey proteins. Food Funct 5:2686–2698
Skipnes D, Johnson SO, Skara T et al (2011) Optimization of heat processing of farmed Atlantic cod (*Gadus morhua*) muscle with respect of cook loss, water holding capacity, color, and texture. J Aquat Food Prod Technol 20:331–340
Sman VRGM, Paudel E, Voda A et al (2013) Hydration properties of vegetable foods explained by Flory-Rehner theory. Food Res Int 54:804–811
Strixner T, Kulozik U (2011) Egg proteins. In: Phillips GO, Williams PA (eds) Handbook of food proteins. Woodhead, Cambridge, UK, pp 150–209
Suman SP, Joseph P (2013) Myoglobin chemistry and meat color. Annu Rev Food Sci Technol 4: 79–99

Sun XD, Holley RA (2011) Factors influencing gel formation by myofibrillar proteins in muscle foods. Compr Rev Food Sci Food Saf 10:33–51
Sun W, Zhou F, Zhao M et al (2011) Physicochemical changes of myofibrillar proteins during processing of Cantonese sausage in relation to their aggregation behaviour and in vitro digestibility. Food Chem 129:472–478
Sun-Waterhouse D, Zhao M, Waterhouse ING (2014) Protein modification during ingredient preparation and food processing: approaches to improve food processability and nutrition. Food Bioprocess Technol 7:1853–1893
Tang CH, Chen L, Ma CY (2009) Thermal aggregation, amino acid composition and in vitro digestibility of vicilin rich protein isolates from three *Phaseolus* legumes: a comparative study. Food Chem 113:957–963
Tarhini M, Greige-Gerges H, Elaissari A (2017) Protein-based nanoparticles: from preparation to encapsulation of active molecules. Int J Pharm 522:172–197
Tcholakova S, Denkov ND, Ivanov IB et al (2006) Coalescence stability of emulsions containing globular milk proteins. Adv Colloid Interf Sci 123:259–293
Tchuenbou-Magaia FL, Norton IT, Cox PW (2009) Hydrophobins stabilised air filled emulsions for the food industry. Food Hydrocoll 23:1877–1885
Teuling E, Wierenga PA, Schrama JW et al (2017) Comparison of protein extracts from various unicellular green sources. J Agric Food Chem 65:7989–8002
Thiansilakul Y, Benjakul S, Shahidi F (2007) Compositions, functional properties and antioxidative activity of protein hydrolysates prepared from round seed (*Decapterus maruadsi*). Food Chem 103:1385–1394
Tilman D, Clark M (2014) Global diets link environmental sustainability and human health. Nature 515:518–522
Traore S, Aubry L, Gatellier P et al (2012) Effect of heat treatment on protein oxidation in pig meat. Meat Sci 91:14–21
Tsumura K (2009) Improvement of the physicochemical properties of soybean proteins by enzymatic hydrolysis. Food Sci Technol Res 15:381–388
Ursu AV, Marcati A, Sayd T et al (2014) Extraction, fractionation and functional properties of proteins from the microalgae *Chlorella vulgaris*. Bioresour Technol 157:134–139
Vaclavik VA, Christian EW (2008) Essentials of food science, 3rd edn. Springer/LLC, New York, pp 21–33
Van Huis A (2016) Edible insects are the future? Proc Nutr Soc 75:294–305
Van Rooijen C, Bosch G, Wierenga PA et al (2014) The effect of steam pelleting of a dry dog food on the Maillard reaction. Anim Feed Sci Technol 198:238–247
Vander Linden E, Foegeding EA (2009) Gelation: principles, models and applications to proteins. In: Modern biopolymer science. Elsevier, Amsterdam, pp 29–91
Vander Plancken I, Loey AV, Hendrickx ME (2007) Foaming properties of egg white protein affected by heat or high pressure treatment. J Food Eng 78(4):1410–1426
Vasudevan D, Sreekumari S, Vaidyanathan K (2017) Proteins: structure and function. In: Textbook of biochemistry for medical students. Jaypee Brothers Medical Publishers, New Delhi, pp 36–51
Vieira E, Cunha SC, Ferreira IMPLVO (2018) Characterization of a potential bioactive food ingredient from inner cellular content of Brewer's spent yeast. Waste Biomass Valor 10: 3235–3242
Villanea SS, Hendriks WH, Bruininx EMAM et al (2016) Protein structural changes during processing of vegetable feed ingredients used in swine diets: implications for nutritional value. Nutr Res Rev 29:126–141
Vogelsang-O'Dwyer M, Bez J, Petersen IL et al (2020) Techno-functional, nutritional and environmental performance of protein isolates from blue lupin and white lupin. Foods 9:230
Wada Y, Lönnerdal B (2014) Bioactive peptides derived from human milk proteins—mechanisms of action. J Nutr Biochem 25:503–514
Wege HA, Kim S, Paunov VN et al (2008) Long-term stabilization of foams and emulsions with in-situ formed microparticles from hydrophobic cellulose. Langmuir 24:9245–9253

Weijers M, Visschers RW, Nicolai T (2002) Light scattering study of heat-induced aggregation and gelation of ovalbumin. Macromolecules 35:4753–4762

Wierenga PA, Meinders MBJ, Egmond MR et al (2003) Protein exposed hydrophobicity reduces the kinetic barrier for adsorption of ovalbumin to the air-water interface. Langmuir 19:8964–8970

Wierenga PA, Meinders MB, Egmond MR et al (2005) Quantitative description of the relation between protein net charge and protein adsorption to air-water interface. J Phys Chem B 109(35):16946–16952

Wierenga PA, Egmond MR, Voragen AGJ et al (2006) The adsorption and unfolding kinetics determines the folding state of proteins at the air-water interface and thereby the equation of state. J Colloid Interface Sci 299:850–857

Wong BT, Zhai J, Hoffmann SV et al (2012) Conformational changes to deamidated wheat gliadins and β-casein upon adsorption to oil-water emulsion interfaces. Food Hydrocoll 27:91–101

Xiong YL (2014) Chemical and physical characteristics of meat—protein functionality. In: Dikeman M, Devine C (eds) Encyclopedia of meat sciences, 2nd edn. Academic, Oxford, pp 267–273

Yi L, Lakemond CM, Sagis LM et al (2013) Extraction and characterisation of protein fractions from five insect species. Food Chem 141:3341–3348

Yu J, Ahmedna M, Goktepe I (2007) Peanut protein concentrate: production and functional properties as affected by processing. Food Chem 103:121–129

Yu TY, Morton JD, Clerens S et al (2017) Cooking-induced protein modifications in meat. Compr Rev Food Sci Food Saf 16:141–159

Zhang S, Sun D, Dong X, Li C et al (2008) Aqueous foams stabilized with particles and nonionic surfactants. Colloids Surf A Physiochem Eng Asp 324:1–8

Zhang T, Xue Y, Li Z et al (2015) Effects of ozone-induced oxidation on the physiochemical properties of myofibrillar proteins recovered from bighead carp (*Gypophthalmichthys nobilis*). Food Bioprocess Technol 8:181–190

Zielińska E, Karaś M, Baraniak B (2018) Comparison of functional properties of edible insects and protein preparations thereof. LWT Food Sci Technol 9:168–174

Zielinski H, Michalska A, Amigo-Benavent M et al (2009) Changes in protein quality and antioxidative properties of buckwheat seeds and groats induced by roasting. J Agric Food Chem 57:4771–4776

Chemistry of Food Fats, Oils, and Other Lipids

6

Satish Kumar Sharma, Riya Barthwal, Deepa Saini, and Neha Rawat

6.1 Introduction

Tailored items made from fats and oils, margarines, liquid oils, shortenings, and other specialty oils are basic ingredients used by food processors and restaurants, as well as at home for food preparation. Fats and oils have been used by humans for food and a number of other purposes since prehistoric times as they are easy to isolate from their source. Boiling fatty tissues from animals, for example, releases free-floating fats, while oil can be extracted from olives and oilseeds. Because of their special properties, fats and oils have found application. Foods with these ingredients have been found to have more lubricity, texture, taste, and satiety. They have also been discovered to play major role in human nutrition. Fats and oils are the most energy-dense out of the three basic foods (proteins, carbohydrates, and fats), and many of them contain essential fatty acids that the human body cannot produce. Fats and oils are found naturally in a variety of places, each offering unique and distinct content. Hundreds of seeds, nuts, and fruits contain oil, and all animals as well as marine sources produce fat; however, just a few of these sources are economically significant. Glyceryl esters of fatty acids, or triglycerides, make up the majority of edible fats and oils, with certain non-glyceridic part present in minimal or trace amounts. Fats and oils are used conversely, and the terms are typically chosen depending on the physical condition of the substance at room temperature and custom. At room temperature, fats appear solid, while oils appear liquid. In the end, it is the chemical composition of the particular fat or oil that determines its characteristics, which in turn decides its suitability for different processes and applications. Food lipids are an indispensable part of our diets and affect composition, taste, texture, shelf life, and so many other characteristics of foods. Lipids

S. K. Sharma (✉) · R. Barthwal · D. Saini · N. Rawat
Department of Food Science and Technology, GB Pant University of Agriculture and Technology, Pantnagar, Uttarakhand, India

O. P. Chauhan (ed.), *Advances in Food Chemistry*,
https://doi.org/10.1007/978-981-19-4796-4_6

undergo so many changes during processing by application of heat, radiation, physical force, and other methods. Many of such changes are undesirable. Alteration in the chemical composition and properties of lipids is also affected by the presence of enzymes in food, exposure to air, and so many other factors. Excess intake of fats and oils has also been linked to human health. Research on food fats and oils, so far, has primarily focused on two aspects, i.e., reducing undesirable changes in fats and oils during processing and storage, and exploring processing methods to reduce consumption of fats and oils. Food fats are predominantly obtained from plant or animal sources, but, exploring alternative sources of food lipids had also been the focus of researchers in recent years.

6.2 Classification of Lipids

Lipids are the compounds that are soluble in organic solvents and consist of free and esterified fatty acids, sterols, carotenoids, and fat-soluble vitamins (McClements and Decker 2008). Majority of the lipids exist as triglycerides in foods and have glycerol esterified to three fatty acids. Fatty acids are carboxylic acids attached to a long aliphatic chain. There is no solitary, internationally accepted classification system available for lipids. Because of the structure of these components, the names of the compounds fit into certain groups.

The classification based on the types of lipids, i.e., simple, compound, and derived has been presented in Table 6.1.

Another way of classifying the lipids is based on the hydrocarbon chain length, i.e., short chain fatty acids (SCFA, <6 Carbons) like butyric, medium chain fatty acids (MCFA, 6–12 Carbon) such as capric, long chain fatty acids (LCFA, 13–21 Carbons), for example, oleic, very long chain fatty acids (VLCFA, >22 Carbons) like cerotic acid. A triglyceride may comprise a mix of these acids, with most fats having medium to long chain fatty acids.

The dietary fats are also classified based on the presence (unsaturated) or absence (saturated) of double bonds in their structures. The classification of dietary fats based on the presence of saturated and unsaturated fatty acids and cholesterol has been presented in Table 6.2. It can be visualized that most deep-fried fast foods and commercially baked products are largely known to contain bad cholesterol. Largely, plants make the least of the cholesterol and so are the products prepared by using plant fats. On the contrary, animal fats are considered to be the main source of cholesterol in the food. *Trans* fats are found rare in nature; however, the extreme processing conditions may lead to their formation, which remains undesirable though. These are extremely deleterious to human health and take the lives of about 50,000 people every year worldwide. The WHO has formed a guideline for the elimination of industrially produced *trans* fats from products, i.e., ghee, margarine, snack foods, etc., by adopting a REPLACE (RE-Review, P-Promote, L-Legislate, A-Assess, C-Create, E-Enforce) strategy (WHO 2018).

Table 6.1 Classification of fats, oils, and other food lipids

Class	Simple		Compound	Derived	Miscellaneous
Definition	Esters of fatty acids and simple alcohols		Esters of fatty acids and alcohols with other groups	Derived from compound lipids or neutral lipids and possess general properties of lipids. These are composed of hydrocarbon rings and a long hydrocarbon side chain	
Composition	Fatty acids, simple alcohols		Fatty acids, sulfur, phosphorus, carbohydrates, proteins	Hydrocarbon rings and a long hydrocarbon side chain	
Further categories	Fats: esters of fatty acids and glycerol	Waxes: long chain fatty acid ester and long chain alcohols			
Examples	• Groundnut oil • Sunflower oil	• Bees wax • Cutin • Suberin	• Phospholipids • Glycolipids • Sulfolipids • Lipoprotein	• Steroids • Fatty acids • Alcohols • Fat-soluble vitamins	• Aliphatic hydrocarbons • Terpenes

6.3 Sources of Fats and Oils

There are three main sources of lipids used in foods, i.e., plants, animals, and marine sources. Important plant sources include corn, sunflower, soybean, cottonseed, peanut, olive, canola, pumpkin seed, safflower, sesame, bran, palm, linseed, coconut, etc. Animal sources predominantly include butterfat and lard. However, marine oils include whale and salmon, cod liver oil (Aloyor et al. 2009; Ogori 2020). The details of the sources, composition, and application of different fats and oils in the food industry have been given in Table 6.3. It can be visualized that the predominant sources containing saturated fatty acids come from animal fats except coconut from plant sources. The oils containing too much of PUFA may not be suitable for deep-fat frying considering the chances of formation of *trans* fats.

Table 6.2 Types of dietary fats

Type	Saturated	Monounsaturated	Polyunsaturated	*Trans*[a]
Other often used names	Bad fats	Good fats	Good fats	Bad fats
Chemistry	No double bonds	Single double bond	Two or more double bonds	
Effect on blood LDL cholesterol level	Raise LDL	Lower when they replace saturated fats	Even better than Monounsaturated fats to lower LDL cholesterol	Raise LDL and even lower HDL
Sources	• Fatty portion of meat • Full-fat milk, cream, cheese, butter • Mostly baked products (biscuits and pastries) • Most deep-fried fast foods • Coconut and palm oil	• Avocado, nuts (peanuts, hazelnuts, cashews, almonds, and other nut butter) • Margarine spreads (such as canola or olive oil-based choices) • Oils such as peanut, canola, and olive	• Fish and seafood • Polyunsaturated margarine • Vegetable oils (soy oils, sunflower, corn, or safflower) • Nuts (such as walnuts and Brazil nuts) and seeds	Pies, biscuits, pastries, cakes, and in deep-fried meals, milk, cheese, beef, and lamb

[a] *Trans* fats are uncommon—they are only formed in the stomach of sheep and cows, due to which, *trans* fats are naturally present in little amounts in milk, beef, and lamb

Erucic acid present in mustard oil is considered to have toxic effects on human health when consumed in high doses. These considerations, so far, have been made based on some animal trials; however, several reports deny relationship of human consumption of erucic acid and increased heart diseases (FSANZ 2003).

6.4 Factors Affecting Fat Properties

Fats are the ester of fatty acids and glycerol; therefore, the structure of fatty acids esterified to glycerol determines its properties.

Table 6.3 Sources of fats and oils

Name	Composition			Application	Remarks	References
	Saturated FA	MUFA	PUFA			
Butterfat	12–32%	29–32%	2–4%	Used as a table spread, margarine, and in manufacture of dairy products, i.e., cheese, ice cream, milk, whipping and coffee cream, etc.	Usually obtained from cow's milk. Has distinct flavor and yellowish color, which makes it popular	Mishra and Manchanla (2012)
Lard	46.2%	45.2%	11.0%	It is used for cooking, baking, frying, and can also be used for making candles, soaps, and moisturizers	Rendered from the fatty tissues of pig	Ogori (2020)
Tallow	54.9%	40.9%	4.2%	Used in high-temperature cooking such as frying	Obtained mainly from cattle. Firmer and harder at ambient temperature as compared to lard	Ogori (2020)
Fish Oil	15–34%	22–48%	34–41% (EPA/DHA)	Food supplements, nutraceuticals	Composition varies from species to species	Gomes et al. (2019)
Cod liver oil	23.3%	41.3%	34.5% (EPA/DHA)	Food supplements, nutraceuticals	Derived from the liver of cod fish	Gomes et al. (2019)
Soybean oil	15%	25%	61%	Used in mayonnaise, margarine, salad oils, and cooking	α-Linolenic and linolenic acids account for 11% and 89% out of the total essential fatty acids	Ogori (2020)
Palm oil	44–45% (Palmitic)	39–40% (Oleic)	1% (Linoleic)	Cooking/frying oils, margarines. Shortenings, specialty tins and spray dried products	Linoleic acid at low concentration makes it relatively stable to oxidative deterioration	Srilakshmi (2003)
Canola oil	9–17%	23–59%	30–59%	Shortening, margarine cooking, and frying	It is high in tocopherol. Third in production	Kostik et al. (2013)
Sun flower oil	10%	55–75% (Oleic)	15–35% (Linoleic)	Cooking, margarine, salad dressing, but not suitable for frying as having poor oxidation stability	It is the fourth most popular vegetable oils and preferred over soybean, cotton seed, and palm oils	Ogori (2020)

(continued)

Table 6.3 (continued)

Name	Composition			Application	Remarks	References
	Saturated FA	MUFA	PUFA			
Coconut oil	92%	6.0%	1.6%	Frying, production of margarine as milk fat substitute, in infant and sports foods, as nondairy creamers	Being slid it is resistant to oxidative changes. Has a tendency of foaming due to its very low molecular weight; hence, does not mix with other oils easily	Jana et al. (2015)
Palm kernel oil	48% Lauric, 16% myristic	15% Oleic		In margarine, frying oil, filling creams, nondairy whipping creams, nondairy ice creams		Ogori (2020)
Cotton seed oil	21%	16%	55%	Used in shortening, margarine, deep frying, as a salad oil	Cannot be processed to have a high oxidative and flavor stability	Yang et al. (2019a, b)
Groundnut (peanut) oil	10.0%	71.0%	18.0%	Deep frying and cooking oil, shortenings preparation, mayonnaise, and margarines	Highly unsaturated and is therefore prone to rancidity	Jana et al. (2015)
Olive oil	19%	68% (Oleic)	18%		Virgin olive oil has never been deodorized as olive oil flavor is desirable and liked by the consumers	Jana et al. (2015)
Corn oil	15%		High	It is primarily used in corn margarines, cooking/salad oil	Produced as a by-product of corn starch production	Ogori (2020)
Sesame seed oil		41% (Oleic)	41% (Linoleic)	Cooking oil	Highly resistant to oxidation	Ogori (2020)
Rice bran oil	20%	40%	40%	Used as salad oil, cooking and frying oil, in mayonnaise, shortening production, margarine	It is a by-product of rice milling	Ogori (2020)

Flaxseed (linseed) oil	9%	15%	75% 50% (linolenic)	Health food	Imparts rancid flavor in short time therefore it is stored under cold, oxygen, and light-free conditions and is preserved by the addition of an antioxidant	Anna et al. (2015), Ogori (2020)
Safflower oil	9%	28%	62–80% (linoleic)	Used for deep frying that turns toxic when exposed to high heat	More prone to rancidity	Jana et al. (2015), Ogori (2020)
Mustard oil	12%	60% 42% erucic acid, 12% Oleic acid	21% PUFA	Cooking and frying	One of the healthiest edible oils because of its low level of saturated fatty acids (8%)	Ogori (2020)

6.4.1 Chain Length

Predominance of short-chain fatty acids tends to give softer fats with lower melting points as compared to that given by long chain fatty acids (Potter and Hotchkiss 1996).

6.4.2 Unsaturation

Higher degree of unsaturation also leads to softening of the texture of fats. Each triglyceride of the fat has its own melting point; therefore, the softening characteristics are based on the mixtures of these triglycerides. Palm oil has a wide range (27–45 °C) while coconut oil has a narrow range (23–26 °C). Short-chain fatty acids and presence of *cis*-configuration of unsaturated fatty acids reduce melting point. The melting point of saturated fatty acids is being highest and that of the *cis* fatty acids being lowest, while *trans* fatty acids have melting points in between these two. These melting points are responsible for imparting the mouth-feel, shelf life, rheology, and other quality parameters in cookies and other processed foods.

6.4.3 Solid Fat Content

Solid fat content (SFC) is the proportion of solid fat to the entire fat (Ghotra et al. 2002). SFC has a significant effect on the textural and functional properties of food products. Creaming properties and spread ability of fat can be correlated to the proportion of SFC in shortenings. Cookie produced by using higher SFC needs higher breaking force. SFC effects rheology and shelf life of baked products. In shortenings, higher SFC results in insufficient oil volume for satisfactory aeration, whereas lower SFC cannot retain air till the end of mixing. In general, 15–20% SFC is usually recommended for cookie.

6.4.4 Polymorphism

Polymorphism is the ability of any compound to exist in more than one crystalline form. These forms in fats and oils are α (alpha), β' (beta prime), and β (beta), which vary in crystalline structure, free energy, physical, and chemical properties but their chemical composition remains the same. Polymorphism affects the consistency, plasticity, hardness, texture, mouth-feel, and stability of cookies and other baked products. Out of the three polymorphic forms, β (beta) form is the most stable; however β' (beta-prime) is still preferred for its smoothness, softness, and creaming properties. Presence of milk fat, palm oil, and cottonseed oil favor the formation of β' crystal while, peanut oil, coconut oil, and palm kernel oil favor the formation of β crystals. Smaller crystals have a larger surface area than bigger ones, so they can hold a large volume of liquid oil inside crystal network, therefore, small and stiffly knit crystals have better creaming properties (Devi and Khatkar 2016).

6.4.5 Viscosity

Viscosity, which indicates resistance of one part of fluid with the other, is the fundamental factor of its rheology. Oil viscosity has a direct relationship with its chemical properties. Higher the unsaturation, lower the viscosity. Also, an increase in temperature reduces the viscosity of fluids. Every 10 °C temperature, decreases the viscosity by about 30%. Rheology is the deformation and flow of matter in response to applied stress and is determined by the microstructure of the oil and fat crystals in addition to SFC. Spreadability, which is the main rheological property, is affected by the composition and SFC of fats, considered along with shape and size (Tang and Marangoni 2007). Fats act like stiff solids until a disfiguring stress exceeds. For liquid fats, viscosity is sufficient to define its rheology.

6.5 Functional Characteristics of Fats and Oils

6.5.1 Plasticity

This is practically the most important characteristic of fats to be considered for their suitability in the preparation of baked products and is defined as the capacity to retain its shape yet being molded by applying light pressure (Devi and Khatkar 2016). Increase in the degree of unsaturation and temperature of fat, increases its plasticity.

6.5.2 Shortening Power

Mechanical operation and incorporation of edible oil improved shortening power and plasticity of fats. Shortenings obtained their name because they coat and cover the protein molecules in flour and shorten the strands of developed gluten. Unsaturated fatty acids have better covering power than saturated fatty acids (Devi and Khatkar 2016).

6.6 Processing of Raw Materials, Oils, and Fats

6.6.1 Extraction

Vegetable oils are usually obtained by expression or expelling; however, solvents may also be added along with mechanical stress for the extraction of vegetable oils. Expressed vegetable oils, having not undergone any chemical or heat abuse, are also called virgin oils. Virgin fruit oils include olive oil from the Mediterranean region and palm oil from equatorial Africa. Olive oil is monounsaturated oil derived from the expression of olives. The formation of free fatty acids is considered a criterion to categorize virgin oils. Extra virgin oil contains free fatty acids (FFA) (free oleic acid) less than 1% while virgin oil may contain up to 3.3% free oleic acid. Endogenous lipoxygenase and lipase enzymes come into contact with the oil during the extraction process and activate oxidative and hydrolytic reactions, rendering the oil unstable. Hydrolysis increases acidity and oxidation increases peroxide value, both of which contribute to the shortening of oil's shelf life. It is also vital to pick fruit carefully to avoid damaging its skin, to expel the oil as soon as possible after harvesting to reduce contact time, and to store the oil in tight containers away from oxygen and light after decanting.

6.6.2 Rendering

This method is predominantly used for the production of animal fats mainly for soap and shortenings industries. Heat is used to extract fat from animal tissues, allowing it to flow out and be separated from non-fat elements by gravity. This process can be carried out in heated vats or using continuous cooker-extruder technology. Fats obtained in this manner can be fractionated to obtain higher and lower melting phases, or they can be used directly. Tallow (a cattle by-product) and lard (pig fat) are two examples of fat created via rendering.

6.6.3 Butter

This fat is obtained from mammal's milk, by churning it, and separation of fat predominant portion (butter) and the remaining watery portion (buttermilk). At normal temperature, butter contains around 20–25% water and possesses plastic behavior. However, it readily solidifies at lower temperatures and becomes less spreadable as crystals. Commercial butter contains added salt for flavor enhancement and better microbial stability.

6.6.4 Refining

Other than triglycerides, extracted oils contain tiny amounts of free fatty acids, carbohydrates, phospholipids, proteins and their degradation products, pigments, water, (primarily carotenoids and chlorophyll), and oxidized fat products. These compounds need to be removed to provide stability and enhance the quality of oils; thus, they are subjected to refining, which is of two types, i.e., physical and chemical refining. In chemical refining, oil is subjected to neutralization by using alkali. During this process, soapstock is developed which is consist of unused alkali, neutral oil, sodium salts of fatty acids, and other compounds. Removal of soapstock causes loss of neutral oil thereby reducing the yield of oil. Whereas, in physical refining all free fatty acids are removed by steam stripping followed by degumming and oil pretreatment so terminate the development of soapstock; thus, reducing the loss of neutral oil (Tandy and McPherson 1984).

6.6.4.1 Settling and Degumming

Settling is primarily done to remove proteinaceous material, water, phospholipids, and carbohydrates from the fat by heating and then allowing it to stand for some time to withdraw the aqueous phase. In the case of high-phospholipid soybean oil, 2–3% water is added to the oil and the mixture is stirred at 50 °C. Hydrated phospholipids are separated by settling or centrifugation. This is also called as degumming. Sometimes phosphoric acid may also be added instead of water to ensure phospholipid separation.

6.6.4.2 Neutralization

To extract free fatty acids from fats, caustic soda is forcefully combined with heated fat at 60–80 °C and then left to settle for the aqueous phase. Fat is removed from the top layer by settling and centrifugation and the remaining aqueous solution is named as foots or soapstock used for producing soap. Although the primary goal of alkali treatment is to remove free fatty acids, it also removes phospholipids and coloring matter.

6.6.4.3 Bleaching

Color of oils is very important and is sometimes impaired due to the presence of some inherent compounds. Oil is heated to 85 °C and bleached with adsorbents such as Fuller's earth, acid-activated montmorillonite clays, or activated carbons among many others. This eliminates almost all the coloring materials. However, some undesirable odors may develop during such treatments. Therefore, some silica-based and composite adsorbents are being explored which may be cheap and recyclable and do not impart any off-odors to the oils.

6.6.4.4 Deodorization

Some undesirable flavoring substances in oils, formed due to oxidation of oils, are removed by steam or nitrogen steam distillation under reduced pressures, by the process called deodorization. Thermal destruction of non-volatile off-flavoring compounds also happens during this process. Partially, *cis-trans* isomerization of polyunsaturated fatty acids may occur as a result of heating at temperatures as high as 240 °C for extended periods of time, which is an undesired side effect of this procedure. Further, some of the natural protectants of oils, i.e., tocopherols, sterols are also unfortunately removed during this treatment. However, addition of citric acid chelates traces of pro-oxidant metals and provides additional stability to the oils even at reduced tocopherol levels.

Although oil refining improves its oxidative stability due to the removal of pro-oxidants, i.e., chlorophylls, exceptions, i.e., cottonseed oil is also available wherein crude cottonseed oil has higher antioxidant capacity due to the presence of gossypol and tocopherols than its refined counterpart. Therefore, the benefits derived due to refining are immense in terms of color improvement, enhancement of stability, flavor, and removal of acute toxicants, i.e., gossypol present in cottonseed oil and aflatoxins in peanut oil.

6.6.4.5 Storage and Packaging

Because refined oil is oxidizable and may lose quality during storage, it must be handled, transported, stored, and bottled in stainless steel vessels and through stainless steel pipes. Opaque packages (at least yellow color) are preferred to protect from the pro-oxidative action of light.

6.7 Fat Modification

Most edible oils, in their native form, have limited applications in the food industry and therefore need chemical or physical modification, in order to alter their textural properties. In the food industry, three principal modification processes used are fractionation (separation of stearin and olein), interesterification (rearrangement of fatty acids over the triglycerides required for chemical blending of properties of different oils), and hydrogenation (saturation of double bonds in the fat, leading to its solidification and increase in hardness).

6.7.1 Fractionation

Fats are made up of a variety of triglycerides with varying melting temperatures, hardness, and solid fat percentage. To expand the use of edible oils and fats in associated food products, fractionation is required (Abeshima 1998). It is the process of separation of harder and higher melting fats from the softer and liquid fats at higher temperatures. Fat is held to create solid and liquid fractions at a particular temperature. Solid crystals (stearin fraction) are separated from liquid oil (olein fraction) based on their different melting points. Fractionated oils include palm oil, palm kernel oil, butter fat and shea butter, coconut oil, and cocoa butter. Palm oil, on the other hand, can be separated twice to provide three fractions: very high-melting (>50 °C) stearin, moderate melting (about 34 °C), and low-melting olein (Talbot 2009). Dry fractionation, unlike wet fractionation, does not employ a solvent to dissolve the fat (typically acetone). The solution is cooled to crystallize the stearin fraction. Oil fatty acids do not undergo any chemical modifications as a result of fractionation. Saturated fatty acids concentrate in stearin, while unsaturated fatty acids concentrate in olein. Palm oil has been fractionated to generate cooking and frying oils, specialized fats from tropical solid fats, and soft butter from anhydrous milk fat.

6.7.2 Interesterification

The ester bonds between the glycerol and fatty acids are broken and rebuilt when a mixture of oils is held at a high temperature in the presence of a chemical or enzyme catalyst. Because the fatty acid groups do not always reform the same bond, the positions of the fatty acids on the triacylglycerol molecules are randomized. Interesterification does not modify the overall fatty acid content, but it does affect the melting and crystallization functions of fats by changing their location.

When the interesterification process is modified with enzyme catalysts to break the connections between glycerol and fatty acids at 1- and 3-locations only, having left fatty acids esterified at the central 2-position, the structured triglycerides are formed. Interesterification changes the physical properties of oils by rearranging the location of fatty acids within triacylglycerols, the primary component of dietary fat. This feature is ideal for usage in a variety of food applications. Chemical reactions in which an ester is reacted with alcohol (alcoholysis), acid (acidolysis), or another ester (interesterification or ester exchange) to form a new ester are known as *trans*-esterification. Interesterified fats are used to reformulate products to remove fats containing *trans* fatty acids, which are known to be harmful to cardiovascular health when produced during partial hydrogenation. These can also be used to lower the amount of saturated fatty acids (SFA) in the end product (up to 20% in spreads) while still keeping acceptable physical qualities (Berry et al. 2019).

6.7.3 Hydrogenation

Due to the existence of numerous double bonds, vegetable oils are polyunsaturated and liquid in character, making them difficult to handle and prone to oxidation, resulting in a reduced shelf life. By adding hydrogen and converting a double bond into a single bond in the presence of a catalyst, hydrogenation is used to convert unsaturated fatty acids to saturated fatty acids (usually nickel). The hydrogenation reaction can transform liquid oils into solid fats. This process "hardens" margarines and shortenings, turning them solid or semi-solid.

Because *trans* fatty acids have a greater melting temperature than similar *cis* unsaturated acids, the *cis* double bond is sometimes changed to a *trans* double bond. Because of the health risks connected with *trans* fatty acids, the solid modification process of hydrogenation is being largely dismissed and discouraged, either voluntarily or by law.

6.7.4 Effects of Fat Modification

It is feasible to build a large variety of fats and oils with characteristics appropriate to specific uses by combining hydrogenation, fractionation, and interesterification with the easy blending of native and modified oils. Fat modification affects their properties to a great extent.

Fats create texture in food by generating crystalline network structures. The physical qualities of our food product are significantly affected by the modification of fats and oils. Cheeses made from homogenized milk or creams are significantly whiter and opaquer than the non-homogenized ones. Also, the moisture contents of the cheese made from the homogenized products tend to be higher (Rudan et al. 1998).

Fat modification extends the shelf life by making the fats lesser prone to oxidation. Therefore, the processed products wherein such fats are used also have better stability, flavors, and quality. The incidence of rancidity is reduced in such products. The substitution of unsaturated fat for saturated fat in the diet is recommended for the prevention of cardiovascular disease (Huang et al. 2013).

6.8 Role of Fats and Lipids in Different Foods

There are a large number of functional characteristics imparted by fats and oils to different food products, i.e., bakery products, dairy products, confectionery, snack, etc. Some of such functions have been detailed in Table 6.4.

Table 6.4 Functional role of fats and oils

Food	Function	Examples	References
Chocolate	Affects the rheological properties of fluid chocolate, prevention of bloom, melting properties, and flavor release	Milk fat and cocoa butter	Jahurul et al. (2013), Beckett (2009)
Ice cream	Flavor, texture, color and mouth-feel, shape retention, air phase stabilization, and melting resistance	Milk fat and vegetable fat	Su and Lannes (2012), Karaca et al. (2009)
Bread	Improve gas retention in dough, increase dough softness and volume, lubrication, aeration, help heat transfer, provide texture	Butter, commercial fats, solid and liquid margarines and oils	Cauvain (2003), Shahidi (2005), Manzocco et al. (2012), Smith and Johansson (2004)
Cake	Texture, lubrication, softness, air entrapment, heat transfer, structure integrity, and extended shelf life	Vegetable fats	Zhou et al. (2011)
Biscuits	Tenderness, texture, mouth-feel, lubrication, incorporation of air, structural integrity, heat transfer, shelf life	Solid or semi-solid fats at room temperature	O'Brien (2009), Ghotra et al. (2002)

6.9 Snack Foods and Frying

Snacking can be defined as the consumption of any food or drink in between regular meals. Feeling of satiety persisting after eating is one of the key factors in suppression of overconsumption of food which otherwise can cause the problem of obesity and consequently many health problems. Snacking in between meals promotes feeling of satiety and suppresses overconsumption in the next meal (Chapelot 2011; Chaplin and Smith 2011; Piernas and Popkin 2010). Fried snacks have persisted since history in eating habits of people. For the production of snack foods, such as French fries, pakoras, samosas, chips, patties, cutlets, poppers, and various other products, frying is one of the most common methods used. Temperatures for frying can range from 170 to 190 °C (Zaidi 2017). The crisp texture, flavor, aroma, and taste of fried foods are appealing and liked by everyone. Deep frying consists of immersing food in oil and cooking. The fried product that results has a particular structure. The surface area of the outer zone contributes to the initial visual effect. The outer zone is the surface area, which contributes to the initial visual effect. This surface has an even, golden brown color that is presumably the consequence of a browning process that occurs when the carbohydrates and proteins in the product react in the presence of high temperatures (Reda 2004). The degree of browning is determined by the time and temperature of frying, rather than the shortening or source of fat or oils in the fryer, in conjunction with the chemical content of the item being fried. The crisp crust developed on the outer skin of the meal as a result of dehydration during frying is the second component of the outer zone. The moisture content of this layer is reduced to 3% or less during frying, and water pushed out is responsible for the steam generated. During frying, the fat shortening fills the void created by the moisture loss. This absorbed fat has a tenderizing impact on the crust in addition to adding flavor, crispness, and pleasant eating characteristics. The amount of frying fat absorbed by the food varies depending on the product being fried; for example, potato chips with a large surface area and little core absorb about 30–40% fat, whereas French fried potatoes with a smaller surface area and crust area absorb only 7–10% fat compared to the core area (Stevenson et al. 1984).

6.9.1 Changes During Frying

Deep-fat frying is a method of preparing foods that involves immersing them in an edible fluid (fat) at a temperature much above the boiling point of water. This unit operation also involves concurrent heat and mass transfer, resulting in water vapor (bubbles) and oil counter flows at the piece's surface (Bouchon et al. 2003). On the basis of visual observations, moisture data analysis, and temperature profiles, the procedure has been separated into four stages (Farkas et al. 1996). The first stage, initial heating, is defined as the time it takes for the product's surface to reach the boiling point of water. This time is usually brief, and the food loses only a small amount of water. The second stage, surface boiling, is characterized by fast water loss, the beginning of crust development, and a boiling convection regime due to the strong turbulence associated with nucleate boiling. The third stage, the period of falling rate, is the moment when the majority of the moisture is lost. It is the longest of all the stages, and the temperature in the core region reaches the boiling point of the water. The final stage is the bubble termination point, which explains when the product's moisture loss stops (Singh and Mermelstein 1995; Farinu and Baik 2005). Furthermore, frying induces physical and chemical changes in key food components, as well as significant microstructural changes, all of which affect the quality of the fried product and the frying oil.

6.9.2 Changes in Key Structural Components and Composite Structure Development

Microstructural changes in the central part are milder and comparable to those that take place during potato cooking (Bouchon and Aguilera 2001). Starch granules are gelatinized at around 60–70 °C, are quickly swelled by intracellular water, and occupy the entire cell interior. The middle lamella between cells disintegrates and separate cells in a narrow temperature range (60–80 °C), giving a so-called mealy texture and proteins might denature. The crust formation is the outcome of several changes that take place primarily in the outer layer of the product at the cellular and subcellular level, where the temperature exceeds 100 °C. These physiochemical alterations encompass physical damage caused by the cutting of the product and the development of a rough surface with intracellular material release, protein denaturation and subsequent dehydration, starch gelatinization, cellular adhesion breakdown, water evaporation and rapid cell dehydration located in the developing crust, and the self-absorption of oil. With regards to cell integrity, evidence has accumulated since the first histological study by Reeve and Neel (1960) of deep-fat fried potatoes which stated that a significant proportion of inner cells hold their individuality. Cells also seem to shrink all through frying in respect to the outer layers, with no extended deformation, while cell walls become wrinkled and convoluted around dehydrated gelled starch (Van Marle et al. 1992). Rapid dehydration has also been suggested to decrease starch swelling, and, therefore, cell walls do not really break as can sometimes occur throughout ordinary cooking. Using hot-stage video microscopy, the researchers examined the geometrical changes in potato cells and starch granules during real-time heating in oil (Bouchon and Aguilera 2001). It could be seen that changes in area and shape (circularity) began at 70–80 °C, without significant damage to the cell wall. In a study of structural changes of potatoes during frying and another on physical changes during deep-fat frying of tortilla chips also similar observations were assessed (Costa et al. 2000; McDonough et al. 1993). The development of a composite structure: a porous, crispy, moist cooked interior, or core and oily outer layer or crust is majorly due to these structural changes. As a result, it has been demonstrated that there are three distinct microstructures in finished commercial French fries: (1) a thin outer layer (approx. 250 m) developed by remnants of cell walls of damaged cells by cutting; (2) an intermediate layer of shrunken intact cells extending to the evaporation front; and (3) a core with water-saturated gelatinized starch-containing cells (Aguilera and Gloria 1997). In French fries, the crust only extends to the outermost layer (about 1 mm); whereas, in potato chips, it covers the entire product. During deep-fat frying, potato chips are thinly sliced potatoes (typically less than 2 mm thick) that are dehydrated to a final moisture level of 0.02 kg/kg or less (Baumann and Escher 1995). Changes in microstructure are responsible for the textural qualities of fried foods. A chip should be hard and snap easily when bent, making a crunchy sound (Krokida et al. 2001). Starch swelling gelatinization and, the stability of the cell wall and middle lamellae pectic compounds are frequently linked to firmness. These variations are also due to the amount and location of oil in fried foods.

6.9.3 Chemical Reactions of Oil During Deep-Fat Frying

When food is immersed in hot oil in the presence of oxygen, the oil is exposed to three agents that cause changes in its composition: food water (which causes hydrolytic changes), oxygen (which comes into contact with the oil and tends to cause oxidative changes from the outside to the inside of the food), and, finally, high temperature which causes thermal alterations such as isomerization and scission reactions—aldehydes and ketones—developing various degradation products like hydroperoxides and epoxides (Reda 2004; Moretto and Fett 1998).

6.9.3.1 Hydrolysis of Oil

Water, steam, and oxygen activate chemical reactions in the frying oil and food. Water, a weak nucleophile, attacks the ester bond of triacylglycerols, forming di- and monoacylglycerols, glycerol, and free fatty acids (Chung et al. 2004). The value of free fatty acids is used to track frying oil quality. Instead of occurring at the water-oil contact, thermal hydrolysis occurs largely in the oil stage (Lascaray 1949). Hydrolysis is better in oils with short unsaturated fatty acids than in oils with long-saturated fatty acids. Long-saturated fatty acids are less water-soluble than short and unsaturated fatty acids. For hydrolysis, short-chain fats and oils can easily access water from meals (Nawar 1969). Large amounts of water hydrolyze the oil. The oil is hydrolyzed faster by water than by steam (Pokorny 1989). Extensive contact between oil and the aqueous phase of food increases oil hydrolysis. Mono- and diacylglycerols in cotton seed oil increased initially and then plateaued throughout the frying of potato chips at 155–195 °C (Houhoula et al. 2003). The changing of frying oil with fresh oil on a regular basis decreases frying oil hydrolysis. Sodium hydroxide and other alkali used to clean a fryer are added to the oil hydrolysis process. The amount of time spent on frying has no effect on the oil hydrolysis. Free fatty acids and their oxidized constituents provide an off-flavor in deep-fat frying, making the oil less desirable. Oil hydrolysis is sped up by di- and monoacylglycerols, glycerol, and free fatty acids. At 150 °C, glycerol evaporates, and the residual glycerol in the oil encourages the synthesis of free fatty acids through hydrolysis. The maximum free fatty acid concentration for frying oil is recommended to be between 0.05% and 0.08% (Stevenson et al. 1984).

6.9.3.2 Oxidation of Oil

Oil oxidation is a dangerous chain of chemical reactions that includes oxygen and causes oil to lose its consistency. Due to oxidation, oil progressively becomes rancid, resulting in unpleasant flavors and odors. Autoxidation is the reaction of oxygen with unsaturated lipids to produce a lipid hydroperoxide, which then undergoes additional reactions with or without the addition of other molecules (Sun et al. 2011). Oil autoxidation is accelerated by the presence of free fatty acids, mono- and diacylglycerols, metals such as iron, and thermally oxidized substances. Carotenoids, tocopherols, and phospholipids exhibit both antioxidant and pro-oxidant behavior depending on the oil system, while chlorophylls and phenolic substances inhibit oil autoxidation in the dark.

The oxygen reacts with the oil in deep-fat frying (Houhoula et al. 2003). Thermal oxidation occurs at a faster rate than autoxidation. The non-radical singlet state oil does not react with the triplet di-radical oxygen state due to the spin barrier. Ordinary oxygen is a di-radical molecule that can be found in the air. Radiant oxygen requires radical oil to oxidize oil. Oil must be in a radical state to react with radical oxygen in oil oxidation processes. To become a radical, the hydrogen on the carbon of oil with the weakest link will be eliminated first. The energy required to break the carbon-hydrogen bond on the 11th carbon of linoleic acid is 50 kcal/mol (Min and Boff 2002). The beginning, propagation, and termination of thermal oxidation are all part of the mechanism.

Lipid oxidation produces volatile and non-volatile chemicals that interfere with taste and flavor. The volatiles in the frying oil increase at first but progressively decrease during the frying process. Saturated aldehydes C6–C9, enals (e.g., 2-decenal), dienals (e.g., 2,4-heptadienal), and hydrocarbons are significant volatile chemicals to maintain process quality (Bordin et al. 2013). The generation of non-volatile breakdown products is caused by the oxidation and polymerization of unsaturated lipid acids. Aldehydes impact the flavor of deep-fried foods because 2-*trans*-4-*trans*-decadienal correlates to a flavor, while other aldehydes produce off-flavors (Boskou 2002).

The thermal stability of oils is determined by their chemical structure. Oils that are saturated are more stable than oils that are unsaturated (Reda 2004). Unsaturated fatty acids are the primary precursors of the volatile chemicals found in oxidized oils (Morales et al. 1997; Kiritsakis 1998). Linolenic acid is an important unsaturated fatty acid that is rapidly lost during the frying process, altering the balance of saturated and unsaturated fatty acids in the frying oil and increasing the production of off-flavors (Fellows 2009; Solinas et al. 1984). The growth and aggregation of unwanted chemicals occur as a result of long-term use and/or reuse of oil.

These molecules may be attributed to the release and dissolution of particles present in the food or thermo-oxidative reaction products in the oil. All of these factors significantly increase the viscosity of the oil; reduce surface tension between both the food and the oil and raise the supply of food surface oil, enabling the absorption of oil (Dobarganes et al. 2000; Jorge et al. 2005). The kind and quality of oil or fat used determines how much oil or fat is absorbed (Del Re and Jorge 2007). The viscosity of the oil, the temperature of the oil, the type of food, its geometric shape, and the length of the frying process all influence the oil's penetration into food. Triacylglycerols (TAGs) are converted to FFAs and glycerol through lipolysis, a catabolic process. After being released into the bloodstream, FFAs are transmitted and picked up by various tissues to be used for oxidation and eventually ATP generation.

6.9.3.3 Polymerization of Oil

During frying, polymerization takes place, producing a wide range of chemical reactions that lead to the formation of high molecular weight and high polarity compounds. Polymers can be formed by the Diels-Alder reaction from free radicals or triglycerides. Within one fatty acid, cyclic fatty acids can be formed; dimeric fatty acids can be formed between two fatty acids, either between or within triglycerides. As these molecules continue to cross-link, polymers with high molecular weight are obtained. The development of dimers and polymers depends on the kind of oil, temperature of frying, and the number of frying processes. The number of polymers increased as the number of frying and the frying temperature increases. During deep-fat frying, oil rich in linoleic acid is polymerized more easily than that of oil rich in oleic acid. Polymers formed during deep-fat frying are oxygen-rich. The oxidation of oil is being speeded up by oxidized polymer compounds. During deep-fat frying, polymers accelerate further oil degradation, increase oil viscosity, decrease heat transfer, generate foam, and create undesirable color in the food. The high absorption of oil into food is also caused by polymers (Zhang et al. 2012).

6.10 Factors Affecting Oil Degradation

6.10.1 Fresh Oil Replenishment

A high ratio of fresh oil to total (whole) oil provides better frying oil quality. Regular oil replenishment reduces the development of free fatty acids, diacylglycerols, and polar compounds, while increasing the frying life and quality of oils (Zhang et al. 2012).

6.10.2 Frying Time and Temperature

Frying duration increases the concentration of polar molecules, free fatty acids like triacylglycerol dimers and triacylglycerols, dimers, and oxidized polymers. The high frying temperature accelerates the thermal oxidation and polymerization of oils. High frying temperatures reduced peroxide bonding in polymers, but polymers with ether bonding or carbon to carbon bonding were boosted. Because oxygen solubility in oil increases as it cools from frying temperature, intermittent heating and cooling produces more oil degradation than continuous heating and cooling (Choe and Min 2007).

6.10.3 Quality of Frying Oil

The oxidation rate of frying oil increases as the concentration of unsaturated fatty acids in the oil increases. This explains why corn oil is a superior frying oil than canola or soybean oil, which contains significantly more unsaturated fatty acids. Linolenic acid content is important for frying efficiency, flavor quality, and oil stability. Oil with low linolenic acid content produces fewer free fatty acids and fewer polar molecules. Hydrogenation can improve the frying stability of oil. However, hydrogenation produces *trans* fatty acids and metallic flavors, and it does not improve the quality of low linolenic acid oils. Oil filtration is commonly used to minimize the amount of free fatty acids in the oil (Warner 2002).

6.10.4 Moisture Content of Foods

Because of the high moisture content of foods, oil hydrolysis is accelerated during deep-fat frying. More oil is hydrolyzed when the moisture content of the food is high.

6.10.5 Antioxidants

Antioxidants in oils and foods that are naturally present or added affect the quality of the oil during deep-fat frying. Oxidation of oil at room temperature is slowed by butylated hydroxy anisole (BHA), tocopherols, propyl gallate (PG), butylated hydroxytoluene (BHT), and tert-butylhydroquinone (TBHQ). They become less efficient at frying temperature due to losses related to volatilization or breakdown (Zhang et al. 2012).

6.10.6 Dissolved Oxygen Contents in Oil

Flushing carbon dioxide or nitrogen into oil lowers the dissolved oxygen content, which minimizes oil oxidation during deep-fat frying.

6.11 Blending of Oils

The most significant considerations in food technology are the quality, stability, and nutritional properties of oils. There is no pure oil with adequate oxidative stability and strong mechanical and nutritional properties. As a result, vegetable oils are modified in a variety of ways to increase their commercial uses and nutritional content. Combining different formulations and qualities of vegetable fats/oils is one of the simplest ways to create new, unique products with desirable oxidative, textural, and nutritional properties, resulting in improved industrial uses (Hashempour-Baltork et al. 2016). Over the last century, edible oils based on regional agriculture have received more attention due to regional preferences and flavor tastes. Some unusual oils, such as soyabean and palmolein, have been manufactured for the past 20 years, but they have only been approved by the general public in their refined form. Blending will alleviate the burden on real individual oil's geographic tastes thereby aiding in the stabilization of a country's edible oil price. A basic approach for preparing more durable edible oils with a wide variety of ideal fatty acid composition is to mix various amounts of high-oleic sunflower oil (HOSO) with polyunsaturated vegetable oils (Frankel and Huang 1994). The majority of the issues faced are related to fat and oil polymorphism, which often reduces the proportion of a given fat that can be used. Plant processing capability also imposes constraints.

6.12 Advances in Frying Technique

6.12.1 Vacuum Frying

In order to avoid loss of heat-labile components of the product, frying at low temperature and under reduced pressure is suggested and the technology is referred as vacuum frying technology (Kang et al. 2016). Preferably, the operating pressure ought to be lower than 7000 Pa (about 2 in.Hg), resulting in a large reduction in the boiling point of the water, enabling frying to be carried out at temperatures even lower than 90 °C (Dueik and Bouchon 2011). It is important to emphasize that for required structural changes to take place, the operating temperature should not be too low. This technique operates under vacuum conditions; moisture evaporates at a rapid rate thereby reducing nutritional losses due to heat and low oil absorption as compared to other methods. Therefore, the product's organoleptic properties are improved (Da Silva and Moreira 2008). In deep-fried products, this technique has been suggested to reduce the oil content (Ayustaningwarno et al. 2018). Vacuum frying is a variation of traditional frying that uses a low pressure of less than 50 Torr (6.65 kPa) and a low temperature (Banerjee and Sahu 2017). Its advantages include a lower oil and acrylamide content, the preservation of nutritious components, color, flavor, and fewer negative impacts on oil quality (Sobukola et al. 2013). During vacuum frying, low pressure is achieved using either liquid ring or oil sealed vacuum pumps. It is becoming more popular in current times as a result of new research discoveries proving a link between traditional fried food intake and human health risks (Pankaj and Keener 2017). A vacuum frying chamber, a cold condenser, and a vacuum pump make up a vacuum frying system (Diamante et al. 2015). A sealed container outfitted with an oil heater and a frying basket is the vacuum fryer. To condense the vapor on a cool surface and capture the rising steam, the refrigerated condenser is loaded. The vacuum pump facilitates the low frying pressure needed and eliminates noncondensable gases (Diamante et al. 2015). The vacuum frying process has been successfully implemented to process various fruits and vegetables into snack products such as fried chips as well as for other products like fried fish and shellfish (Banerjee and Sahu 2017). Work was done on the optimization of the vacuum frying process for producing fried plantain chips with enhanced features (Akinpelu et al. 2014). Findings demonstrated that 133 °C at 9.91 mmHg and 6 min frying were the favored technological requirements for vacuum fried plantain chips (Akinpelu et al. 2014). Moreover, 94% of acrylamide formation in potato chips is reduced by lower operating temperature during vacuum frying. Vacuum frying is an appropriate method to enhance nutritional quality, strengthen color, and achieve less oil degradation due to the retention of vital phytochemicals and important nutrients, as well as enable less oxidation compared to traditional frying (Banerjee and Sahu 2017). It was found that vacuum fried snacks contain as little as 27% less oil than light-colored atmospheric fried snacks (Sobukola et al. 2013). The technology of vacuum frying has added numerous advantages compared to conventional frying, including improvements in the fried products' sensory and textural properties.

6.12.2 Air Frying

This is a revolutionary method of decreasing oil absorption in fried foods. Foods are delivered into a frying chamber by indirect exposure between an external emulsion of oil droplets in heated air and the food matrix (Shaker 2015). The product is constantly moving to ensure that the contact between the two phases is uniform. Moisture is lost as a result of this and a crust forms (Ghaitaranpour et al. 2018). When compared to traditional deep-fat frying, air-frying typically uses very little or no oil. In terms of oil uptake, air-frying is thought to be superior to traditional deep-fat frying (Andres et al. 2013). Another study comparing the evolution of chromatic parameters during hot air-frying and oil-frying found that air-fried French fries had substantially superior color features, implying that air-frying lowers the Maillard reaction significantly (Heredia et al. 2014). Air-frying is the most expensive and ineffective frying method when compared to the other frying methods. Furthermore, there is a paucity of literature on the use of air fryers. In order to overcome the inadequacies of AF, it may be necessary to incorporate microwave energy or other techniques, such as induction, radio frequency, and other thermal means to further heat the food product (Devi et al. 2020).

6.12.3 Microwave Frying

It is an electromagnetic wave-based quick heating system. The electromagnetic wave ranges from 300 MHz to 300 GHz, with wavelengths ranging from 1 mm to 1 m (Roknul Azam et al. 2019). It has been used in food processing units for drying, baking, pasteurizing, sterilizing, tempering, blanching, and thawing, among other things (Das and Arora 2018). Microwave heating provides the advantage of quick heat treatment throughout the frying process, which reduces processing time while also lowering the overall oil content in the finished fried product (Schiffmann 2017). A comparison of air-frying versus microwave frying of potato chips and french fries has been published by a number of researchers (Oztop et al. 2007; Parikh and Takhar 2016). They compared the fried product's oil uptake as well as other sensory features in their study and discovered that microwave heating produced better results in both cases. The uniform delivery of heat is the most significant challenge for the microwave heating system (Schiffmann 2017). Many writers have proposed combining this technique with other technologies like vacuum frying (Zhang et al. 2016) and ultrasound (Devi et al. 2018) to overcome this barrier and improve the features of fried items.

6.12.4 Radiant Frying

Radiant heating systems involve exposure of heat at a wavelength within a range from 0.78 to 1000 mm through the substance via an infrared source. Electromagnetic waves are being used by the microwave also, but far more thermal efficacy and responses with much quicker heating ability are demonstrated by the infrared radiant heating system (Devi et al. 2020). In many food processing operations, such as pasteurization, drying, sterilization, baking, and blanching, infrared heating has extensive applications. However, there are few frying researches that have used a radiant heating system. A comparative finding indicated that french fries manufactured by radiant heating produce products of greater quality than oven heating or immersion frying (Lloyd et al. 2004). A study in the production of potato chips also demonstrated that healthier chips could be produced by infrared pre-drying treatment (Su et al. 2018a).

6.12.5 Ultrasound-Assisted Frying

Ultrasound is currently frequently used as a nonthermal synergistic method to boost production in a variety of food processing systems (Azam et al. 2020; Majid et al. 2015; Wang et al. 2018). The frequency range of the acoustic waves employed in this treatment ranges from 20 kHz to 10 MHz. For the snack food business, ultrasound has emerged as a revolutionary technique. In a prior study, ultrasound was utilized as a pretreatment to make potato chips (Oladejo et al. 2017). They discovered that, when compared to untreated potato chips, ultrasonic pretreated potato chips exhibited a lower rate of oil uptake. Some other research studies have also used ultrasound treatment with microwave-assisted vacuum frying (Al Faruq et al. 2019; Devi et al. 2018; Islam et al. 2019; Su et al. 2018b). All these researches indicated that the application of ultrasound technology during vacuum frying produces a product with lower oil content and crispy texture along with high production rate efficiency. Better quality products (low and crisper oil content) with a high rate of production efficiency.

6.12.6 Spray Frying

Another kind of novel frying process is spray frying which involves spraying the heated oil towards the sample rather than placing the sample in the oil. The amount of oil sprayed determines the efficacy of this frying technique. In a study comparing the physicochemical and microstructural features of fried rice crackers made with spray- and air-frying procedures, it was discovered that spray frying resulted in a 45.4% lower oil uptake and a better color than air-frying. However, air-fried crackers had better textural properties than those produced from spray frying method (Udomkun and Innawong 2018).

6.13 Transformation of Frying Medium

Modifying frying oil can also be a way to reduce oil absorption during frying. Main factors that affect the oil absorption post-frying are temperature and viscosity (Ghaderi et al. 2018). When frying is done under elevated temperature, viscosity of oil is decreased, and, hence, the oil uptake is reduced (Yang et al. 2019a, b). Reduction in hysteresis and increase in wettability affect heat and mass transfer during frying, movement of oil into the food, and drainage rate during post-frying (Li et al. 2008).

6.14 Fried Foods and Health

The fact that whether a fried food is healthy or unhealthy for a particular person depends upon the following factors:

6.14.1 Age of Individuals

Being in growing stage and more active, children need more fat than adults. Adults, in contrast, require less fat. Oils that are high in saturated fatty acids should not be consumed by those who have cardiovascular disease. The total fat consumption as a percentage of energy must not be less than 15% and must not exceed 30%, as per the World Health Organization (WHO) recommendation. Saturated fat (SFA) intake as percentage of energy should not exceed 10% (7% for cardiac patients) of the total energy intake.

6.14.2 Types of Fats and Oils Used During Frying

Cooking oils containing more saturated fats, such as palm oil, are relatively more stable than oils containing more unsaturated fatty acids, like soybean oil, which can easily break down at high frying temperatures, resulting in the formation of polar compounds. It is therefore suggested to use oils with more saturated fatty acids for frying, provided this is occasionally done. Oils with greater levels of unsaturated fatty acids, on the other hand, are often much healthier, making sure that they are only used once for frying.

6.14.3 Frequency of Consumption

Fried food when consumed occasionally may not be as when consumed regularly in conjugation with processed or junk foods.

6.14.4 Reuse of Frying Medium

Frying oil if used repeatedly for frying is considered a bad practice. Numerous oxidative and thermal reactions are caused by repeated frying, resulting in changes in the oil's physicochemical, nutritional, and sensory properties. These changes include dark color development, increased viscosity and free fatty acid content, refractive index changes, decreased iodine value and surface tension, and an increased foaming tendency. There are also changes in the flavor and stability of compounds present in the oil. During frying, various by-products are generated which are volatile and non-volatile in nature, such as alcohols, free fatty acids, dimers, cyclic compounds, and polymers (Zaidi 2017).

6.15 Dairy Products

Milk is a liquid food which is normal mammary secretion obtained after complete milking of healthy milch animal and free from colostrum. Owing to the presence of different nutritional elements, milk is recognized as a complete food. Major constituents of milk are water, fat, protein, milk sugar, vitamins, and minerals. In view of numerous factors, fat is an important component of milk. Fat in milk may vary from 3.0% to 6.0% and the percentage of milk fat is influenced by feed, breed, age, physical condition of animal, season, stage of lactation, etc. (Boro et al. 2018). It is the main energy-giving component of milk and accounts for many of the physical properties, characteristics of production, and sensory characteristics of milk and milk products.

Fat in milk is present in form of emulsion, i.e., oil-in-water emulsion. Milk fat is predominantly present in spherical droplets with a diameter ranging from <0.2 to >15 μm and bovine milk typically contains more than 1010 fat globules per mL

(Huppertz et al. 2020). Milk fat is among the most complex natural fats, with its triglyceride molecule containing over 400 different fatty acids (Schröder and Vetter 2013; Umar et al. 2018). Milk lipids are largely comprised of triacylglycerols (approximately 95%) and the remaining are diacylglycerols, cholesterol, phospholipids, monoacylglycerols, free fatty acids, and some other components like fat-soluble flavoring components, fat-soluble vitamins, and β-carotene are also present in trace amounts. Among all fatty acids palmitic acid, oleic acid, and stearic acid are most abundantly present in milk fat (Metin and Hartel 2012). Alterations in fatty acid composition of triacylglycerol molecule can lead to modification of the physical properties of fat (MacGibbon 2020). In milk fat following types of fatty acids are present:

Saturated fatty acids: These fatty acids, which range in chain length from 4 to 18 carbon atoms and are unbranched molecules, make up a large portion of the total fatty acids, accounting for 70–75% of total fatty acids. The most significant fatty acids are 16:0 and 18:0. Short-chain and medium-chain fatty acids in milk fat have certain special qualities, as they are absorbed in non-esterified form in the bloodstream and promptly processed in the liver. Furthermore, short-chain fatty acids and, to a lesser extent, medium fatty acids are responsible for triacylglycerols' low-melting point, which keeps milk fat liquid at physiological temperatures (MacGibbon 2020).

Unsaturated fatty acids: Around 18–24% of fatty acids is cis-monoenoic, with oleic acid being the most common *cis*-monounsaturated fatty acid. Milk fat contains a small amount of *cis*-polyenoic acids (Rego et al. 2016; Schwendel et al. 2017). Furthermore, partial bio-hydrogenation of unsaturated dietary lipids in the rumen results in trans fatty acids.

Milk as well as other dairy items are considered as important foods because they are a key source of several critical nutrients in the human diet. The popular dairy products are butter, cheese, yogurt, ice cream, infant formulas, etc., in which fat plays a vital role to maintain their physicochemical quality.

6.15.1 Role of Fat in Butter

Butter is defined as a "product with a milk fat content of not less than 80% but less than 90%, a maximum water content of 16% and a maximum dry non-fat milk content of 2%" (Lee et al. 2018). Fat in milk is present in spherical droplets form with diameter ranging from 0.1 to 10 μm. The core of the fat droplet constitutes majorly of triglycerides and a small amount of monoglycerides, diglycerides, and cholesterol esters, while the outer part is a membrane comprised of proteins, cholesterol, and polar lipids. This outer fat globule membrane is sensitive to chemical, physical, physiological, and enzymatic changes (Lopez et al. 2011). During the production of butter, separation of milk cream and skim milk resulted in disruption of fat globule membrane (Holzmüller et al. 2016). After separation of cream

(35–40% fat), whipping is done. At the initial stage of whipping oil-in-water emulsion is stabilized by adsorption of β-casein and whey proteins, to the air/water interface (Brooker et al. 1986). Further, stabilized foam is produced by aggregation and partial coalescence due to the collision of fat globules. For this process cooling is important otherwise excess release of butter oil will take place (Anderson et al. 2005; Hotrum 2004). During the initial period of storage, liquid fat is captured into solid fat matrix and increases the solid fat content (Rønholt et al. 2014). Studies show that salt used for preservation acts as pro-oxidant (Cui et al. 2018; Méndez-Cid et al. 2017). Free fatty acid released by the action of lipase on triacylglycerols causes rancid flavor. Lipase is heat labile and can be inactivated by heating but microbial lipase produced during low-temperature storage for long period are more heat stable and can remain active (Deeth 2006; Suryavanshi and Ghosh 2010).

6.15.2 Role of Fat in Cream

The fine dispersion of fat globules in the hydrophilic phase causes creaminess, which is mostly dependent on the fat amount. Fat globules in separated cream range in size from 1 to 8 m in diameter. The fat content of milk fat is the fundamental reason for its importance in various types of cream products. Different cream products, such as coffee cream, cultured cream, whipped cream, recombined cream, and cream liqueurs, have features due to the interaction of fat with non-fat components. Fat globules with a diameter of 0.4–0.6 m and a low degree of aggregation are characteristics of high-quality coffee cream. In case of whipping cream indirect heating is done at ≥135 °C for a few seconds to limit the heat-induced change. Size of fat globules is also responsible for whipping properties of unhomogenized cream; therefore, by controlling the homogenization process the physical properties of cream can be controlled. This unit operation is very important for cultured cream production because uniform sized fat globules are must for a better product (Eden et al. 2016).

6.15.3 Role of Fat in Cheese

Milk fat imparts flavor, color, and texture and contributes functional behavior to the cheese. During aging process, fat affects the metabolic activity of microbiota involved in the aging of cheese along with other multiple effects. Lipolysis action releases fatty acids responsible for the aroma of the product, especially short-chain fatty acids such as caproic, butyric, capric, and caprylic (Collins et al. 2003). Both low-fat and full-fat cheese have the same major odorants, but at differing relative concentrations (Drake et al. 2010). Furthermore, low-fat cheese has a waxy look, is more fracturable, hard, and springy, cohesive, less smooth, less meltable, and springy than full-fat cheese, and is more fracturable, hard, and springy than full-fat cheese (Johnson et al. 2009). Low-fat cheese retains its springiness to a greater extent than full-fat cheese throughout storage. The moisture of fat-serum channels in

Mozzarella cheese is absorbed into the protein matrix during preservation. It totally fills the spaces between the fat globules when it grows into new protein matrix material (McMahon et al. 1999; McMahon and Oberg 2017). As a result, fat aids in the retention of moisture within the protein matrix. The presence of many fat-serum channels in Mozzarella cheese creates parallel weak places that aid in the propagation of fractures when the cheese is torn apart, which is how string cheese is made. As a result of the lower fat content, there are fewer fat-serum channels and more protein-protein interactions, resulting in little stringiness (Mulvaney et al. 1997).

6.15.4 Role of Fat in Ice Cream

In frozen dairy, dessert fat provides richness of flavor, smooth texture, foam stabilization, assist in inducing desirable melting properties, and is a good carrier of added flavor compounds (Goff and Hartel 2013). Fat acts as an indicator of the quality and value of ice cream. The triglycerides of milk fat cover a wide range of melting points +40 to −40 °C. The crystallization pattern of milk fat is very complex due to difference in fatty acids and numerous types of triglycerides. The role of fat in ice cream preparation begins with very initial step, i.e., formation of the emulsion. Emulsification produces a large number of small, distinct droplets, which is essential for the creation of structure during dynamic freezing (Koxholt et al. 2001; Hayes et al. 2003; Biasutti et al. 2013). The finished product is smoothed out thanks to the fat. As a result, low-fat blends must modify the ratio of other constituents, primarily protein, polysaccharide stabilizer, and emulsifier, to compensate for the loss of intrinsic smoothness. If too much fat instability occurs in high-fat mixes, the ice cream will taste oily, and a problem known as "does not melt" may result (Goff and Hartel 2013).

6.15.5 Role of Fat in Milk Powder

Even a small amount of fat in milk powder can affect the functionality of the product namely, flowability, lipid oxidation, and rehydration (Mahmoodani et al. 2018). Taking this factor into consideration, it is clear that the requirements of packaging and storage are important in prevention of oxidation. Some researchers have demonstrated the importance of fat in milk powder flowability, stating that more fat results in inferior flow behavior. As a result, WMP is more cohesive than SMP but somewhat less cohesive than cream powder (Fitzpatrick et al. 2004; Kim et al. 2005). High-fat content in milk powder also affects its solubility because of slow wetting and sinking during rehydration process. This is why powders are lecithinated these days, as it lowers the surface tension of the liquid in which the powder is to be dispersed. Dairy powders are dried to extend their shelf life and make transportation easier. When exposed to high ambient temperatures during shipping and storage, surface fat can worsen product degradation (Waldron et al. 2020).

6.16 Bakery Products

Bread and biscuits are the most popular bakery items. Other bakery products include cake, toast/rusk, bun/roll, doughnut, pizza, puff pastry, cookies, muffins, etc. Fats and oils are used in various forms in different food products. Because they influence the rheological properties (fluidity, plasticity, and texture), structure (aeration and lightness), and sensory characteristics (taste, color, odor, creaminess, creaminess, and melting) of the products studied (Martínez-Cervera et al. 2012), it is critical to use the best fat in the right proportions with the other ingredients. In bakery items, such as breads, cakes, and biscuits, fat provides a range of purposes, including gas retention, aeration, lubrication, heat transfer in dough, and desirable texture in the end product (breads), air incorporation, softness, mouth-feel, and structural and sensory qualities (cakes and biscuits).

6.16.1 Fat and Baking

Fat is used to prepare almost all of the bakery products. It is also known as shortening. Fat content varies from product to product and so its effect. Cakes, puff pastry, breads, cookies, scones, and pie crusts are commonly made with fats such as butter, shortenings, or hydrogenated fats.

6.16.2 Fats Used in Bakery Products

6.16.2.1 Lard

It is made entirely of fat and is derived from nutritious hog fat. It is used for shortening bread, pastry, biscuit, and crackers. It is crystalline in nature. Hydrogenated lard treated with antioxidants has a better shelf life.

6.16.2.2 Margarine

It is prepared with hydrogenated soybean oil or other plant-based ingredients. It is made up of 80% vegetable fat and 20% water. It is also known as artificial butter. Margarine and butter have a similar structure but different flavors. Margarine has poor flavor but still it has taken the place of butter in most bakeries today. It is used for making puff pastry.

6.16.2.3 Butter

It is manufactured from churning cream and comprises 80% fat and 20% water. It is considered as an emulsion. It gives better flavor and bloom, even texture, and lighter cakes of good appearance. Diacetyl is responsible for the flavor. Its melting point is low due to which it has poor creaming properties. It is incorporated in the icings for cakes and it may be melted and brushed onto the baked products.

6.16.2.4 Clarified Butter

It is the butter that has been cooked to eliminate the milk particles and give it a clear appearance. Cocoa butter (extracted from cocoa beans), oils (100% fat liquid lipids produced from seeds such as soybean, corn, or cottonseed), and shortening are examples (prepared from hydrogenated soybean oil or other vegetable sources, consisting 100% fat).

6.16.3 Functions of Fat in Baking

6.16.3.1 Tenderizing Agents

Shortenings are fats that shorten the gluten strands in flour. Shortening of gluten imparts tenderness to products. This also enhances the use of other ingredients. Butter and lards contribute to specific organoleptic characteristics to bakery products. Butter is responsible for the flaky texture of puff pastry such as croissants.

6.16.3.2 Lubrication

When preparing the dough, the fat fraction allows the gluten network to expand more easily, resulting in simpler mixing and handling. Fat forms a film on gluten strands which weakens the structure making it tender and gets easily disintegrated by breaking down the gluten structure. This property of tendering the product is called lubrication. Fats act as dough lubricants in breads, assisting with loaf rise and crumb softening.

6.16.3.3 Creaming Ability

When fat is pounded with a paddle for mixing, it absorbs air to a certain level. This is crucial while baking a cake. The lighter the cake, the better will be the creaming ability. Good shortening incorporates about 270% of air when creamed with granulated sugar. This also increases the volume of the product.

6.16.3.4 Stabilizing Ability

Fat provides sufficient strength to the dough and batter to prevent their collapse during baking. Fat distribution and proper aeration result in a well-stabilized structure of the baked product.

6.16.3.5 Plastic Range

This refers to the temperature at which the fatty acid component melts and shortening remains workable and "stretches" without breaking (too hot) or softening (too cold) (too warm). Excellent fat is one that remains "plastic" over a temperature range of 4–32 °C. Dough produced with this fat can be handled interchangeably from the walk-in cooler to the bench in a hot bakeshop.

6.16.3.6 Emulsifying Ability

This determines how much liquid can be incorporated in a cake batter without curdling taking place. Higher the water absorption, higher the emulsifying power. This results in easy mixing of ingredients to give fine batter.

6.16.3.7 Moistening Ability

Fats check the drying out of the dough or cake; hence, maintain the texture as well as mouthfeel of the baked product.

6.16.3.8 Nutrition

Fats are an excellent source of concentrated energy. They are high in fatty acids, which are necessary for good health.

6.16.3.9 Preservative

Fats and oils delay staling (in case of cake) as well as help to keep baked products for a longer period of time. This is a result of fat's interference with starch gelatinization. Biscuits having more fat remain fresh for a longer time as compared to breads with less fat.

6.16.4 FDA Regulations

GRAS (Generally Recognized as Safe) ingredients include fats. However, the FDA has declared that *trans* fats, such as partially hydrogenated oils, are no longer deemed safe since 2015. Saturated fat consumption should be limited to 10% of total daily calories.

6.16.5 Role of Fat in Biscuits

Fat is essential in baked goods because it contributes to tenderness, softness, and overall texture, as well as improving mouth feel, structural integrity, air incorporation, heat transfer, lubrication, and shelf life extension. The shortening used has a big impact on the consistency and eating characteristics of the filler. To make handling of the batter during the manufacturing process easier, the fats used to manufacture cookies should be solid or semi-solid at room temperature. As a result, the amount of saturated fatty acids in the body rises. To be functional, shortenings must have plastic characteristics. Fats are crystalline in nature, and the three fundamental polymorphs are α, β, and β′ (Marangoni et al. 2012). To promote optimal creaming, the fat must be in "crystallized" state (Wilderjans et al. 2013). When compared to non-emulsified shortenings, the use of emulsified bakery shortening aids in the fine dispersion of fat in the batter or dough (Jacob and Leelavathi 2007).

6.16.6 Role of Fat in Cake

Cake has distinct sensory characteristics. Cake batter is considered as an oil-in-water emulsion containing dry ingredients like sugar, milk powder, flour, etc., dissolved in the continuous aqueous phase. Fat has a principal role in cake, both from the sensory as well as technological point of view. Aerated structure of the cake is responsible for its better quality. Aeration is achieved through the incorporation of air during whipping and the formation of bubbles during cooking. The act of fat in the whipping cream process incorporates these air bubbles; these bubbles are trapped inside the continuous phase of the emulsion at room temperature. When this batter is cooked during cooking, the air bubbles travel from the fatty phase to the aqueous phase structure, resulting in a voluminous and foamed structure.

6.16.7 Role of Fat in Bread

Fats play an important role in bread making. Fats improve gas retention in the dough which in turn increases its volume and softness. It also lubricates and helps heat transfer in the dough imparting a desirable texture. The amount of fat to be added is determined by the flour type. Breads are made with a number of shortenings like butter, oils, solid, and liquid margarines. Oil softens the mouth and gives it a moist feel. The air bubbles in the mixing process are involved in the solid fraction, which contributes to the dough and final product structure. In baking processes, fat crystals act as a cell membrane for gas cells, causing them to grow larger. To ensure good bread quality, a particular fraction of solid fat in shortenings is needed. The lubricating activity allows more expansion during fermentation and bread baking, resulting in a smoother final texture and improved dough rise, oven spring, and overall efficiency. Wheat flour includes lipids extracted from membranes, organelles, and spherosomes. Since they form a lipid coating at the gas/liquid interface, lipids or polar lipids control the production mechanism of breads by disrupting the stabilization of gas bubbles in the dough. This favors increased gas retention in the dough.

6.17 Chocolates and Confectionery

Chocolates are basically the suspension of milk, cocoa solids, and sugar in a continuous fat phase. This provides the color, flavor, and form to the final product. White, dark, and milk chocolates have 30.9%, 28.0%, and 30.7% of fat, respectively (Afoakwa et al. 2007; Leite et al. 2013). The two fat forms used in chocolates, i.e., milk fat and cocoa butter provide rheological properties of fluid chocolate, melting properties, flavor release, gleam, and prevention of bloom (Beckett 2009).

6.17.1 Role of Cocoa Butter

Cocoa butter imparts the rheological properties to liquid chocolates. Although cocoa butter commonly consists of palmitic, stearic, oleic, linoleic acids, lauric and myristic acids, their constitution may differ based on the location. In commercial chocolate brands, white chocolates are reported to contain higher contents of lipids than diet chocolates (Reis et al. 2011).

Since cocoa butter is the fat source of high costs, the current trends indicate the production of low-fat chocolates (Medeiros and Lannes 2009; Do et al. 2010). Further, the future would explore technological solutions for creating acceptable, rheological characteristics, taste, nutritional quality, functional characteristics, and stability even with the use of alternative fats (Afoakwa et al. 2007; Jahurul et al. 2013). Palm oil, palm olein, cotton seed oil, palm kernel oil, mango seed fat, soy oil, rapeseed oil, coconut oil, etc. have been explored for this purpose. As far as hardness, mouthfeel, and flavor release are concerned, coconut and palm kernel oil were similar to cocoa butter; however, an entirely different fatty acid composition with high amounts of lauric acids, made them incompatible with cocoa butter (Talbot 2009).

Due to different stable crystal forms and temperatures of melting and crystallization, milk fat and cocoa butter do not mix well during chocolate manufacture. To bring about desirable changes in the crystalline structures of cocoa butter, addition of about 50% milk fat is essential (Haylock and Dodds 2009). We often observe white crystals or powder on the surface of chocolates. This happens when chocolate is kept too warm to separate cocoa butter from the crystallized chocolate mixture which then resolidifies and comes to the surface. This is known as *fat bloom*. If the cocoa butter is replaced with milk fat to the extent that it is above 2.5% in the final product, this is helpful in preventing fat bloom (Sonwai and Rousseau 2010).

6.17.2 Fat Alternatives in Chocolates

With the expanding demand for low-fat or no-fat chocolates and confectionery, new ingredients, perception, skills, and knowledge are being developed. However, reducing fat increases the viscosity of molten chocolates; thus, impairing textural properties as well as taste. Chocolates thus prepared are hard, have poor eating quality, and difficult to swallow. However, replacement of sugars by Sucralose and fat by Benefat (a structured triglyceride), olestra, and caprenin gave a highly acceptable product (Richter and Lannes 2007). Carbohydrate-based substitutes (Dextrins, Avicel-microcrystalline cellulose, Inulin-soluble fiber, Carrageenan) and protein-based fat substitutes (Simplesse) have also been used (Food Safety Network 2014). Gelatin and gums could not be used for satisfaction to replace fat as they impaired the softness and eating quality of chocolates (Amir et al. 2013). However, cocoa butter could be replaced up to 10% level successfully with β-glucan-enrich hydrocolloids. The chocolates thus produced were low caloric, soft, and had better boundary lubrication qualities. Shumacher et al. (2010) also succeeded to develop chocolate with above 70% acceptability by addition of quinoa for producing a protein-rich dark chocolate.

6.17.3 Chocolate Spreads

Chocolate spreads have 30% fat content, are kept at room temperature and are prepared to be spread on sandwiches. In these products, fat is in continuous system, while sugar and other particles are in dispersed form. Its structure is soft giving a creamy sensation, but always has the risk of sweeping out. Its preparation includes mixing in open air; therefore, needs high oxidation stability. As a consequence, functional fat system should be able enough to entrap liquid oil in crystal network so strongly that the product may be stirred without releasing the oil. Due to the desirable properties of hydrogenated oils which form small crystals that can be entrapped easily, such oils are traditionally used in fat spreads. Other systems include high-melting triacylglycerols that crystallize promptly into tiny crystals, entrapping liquid oil in use (Norberg and AarhusKarlshamn 2006).

6.18 Fat Replacers

Fat replacers/substitutes are substances that take over the role of all or a part of the fat but still provide the food with the same texture, flavor, and mouthfeel as that of the original full-fat food. Fat replacements serve two purposes (Fig. 6.1). The amount of fat in food is reduced, and the calorie content of the food is typically reduced. Fat replacers are derived from protein, carbohydrate, and fat-based compounds in the form of surface-active agent, hemicelluloses, bulking agents, maltodextrins, and functional blends (Kew et al. 2020; Ognean et al. 2006).

6.18.1 Protein-Based Fat Replacers

Protein is known for their multifunctional properties in the food system such as water binding, viscosity, flavor and fat binding, emulsification, solubility, gelation, and textural properties. Composition, pH, processing methods, and additives affect the functionality of proteins. Due to these unique properties, protein has the potential to make it appropriate and fit for fat replacement. Besides, mimicking fat properties, the use of protein-based fat replacers also decreases the detrimental consequences of protein interactions in low-fat foods (Yashini et al. 2019; Kinsella and Melachouris 1976). The desired physicochemical properties of proteins to act as fat replacers are isoelectric precipitation, protein-protein complex formation, heat denaturation and coagulation, protein-carbohydrate complex formation, hydrophobic effects, protein, and carbohydrate stabilized emulsion. The benefit of protein-derived fat replacers is in terms of flavor interaction (Karnjanapratum et al. 2017; Nath et al. 2016; Chen et al. 2015; Schirle-Keller et al. 1994).

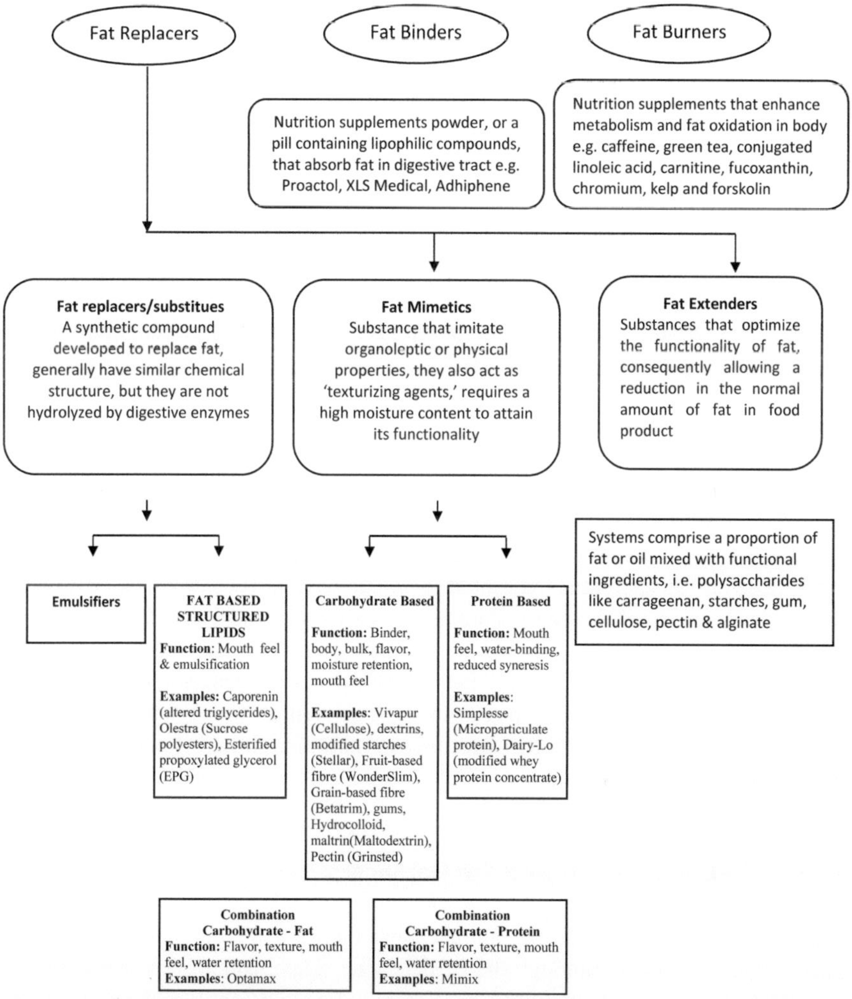

Fig. 6.1 Food fat management strategies

6.18.1.1 Milk Proteins

Casein is found in milk as big spherical micelles which are heat stable up to 140 °C. Its emulsification properties enhance textural properties, amalgamation of fat and water, and sensory properties. This allows casein to be used as a fat substitute in dairy products such as dips, cheeses, spreads, and yogurts (Mulvihill and Ennis 2003). Due to the synthesis of microparticulated whey proteins, whey protein isolate is readily dispersed in meals and has the ability to replace fat completely or partially (Queguiner et al. 1992). A study showed that the incorporation of whey protein concentrates into nonfat goat's milk yogurt enhanced the water retention capacity as well as texture (Zhang et al. 2015). With the inclusion of fermented whey protein concentrate, the yield and quality of cheese were enhanced (Jooyandeh 2009). Further, microparticulate whey protein poses better gelling, emulsifying, and heat stability properties, which makes it a good fat replacer. When used in mayonnaise it results in a complex structure of protein and pectin, which increases the flow behavior index, elasticity index, storage stability and decreases pseudoplastic properties (Sun et al. 2018; Torres et al. 2011).

6.18.1.2 Egg White Protein

Studies showed the potential of egg white protein microbubbles as fat replacers. It was observed that these proteins act as texture modifiers when added to food systems (Rovers et al. 2016). Trailblazer is a fat replacer prepared by a combination of egg white, xanthan gum, and whey protein.

6.18.1.3 Plant Protein

Supro range, Lita A, Lita C, Lita D, and UltraBake NF are some commercially available fat replacers. Traditionally used soy protein is utilized in minced meat as extender. As a result, soy protein isolate is proven to be a better fat replacement and aroma provider, avoiding rancid off-flavors, and lowering cholesterol and moisture levels (Keeton 1994). Zein is hydrophobic protein so it is acceptable as a fat replacer also it shares many of the same features as fats (Wang and Padua 2012). Studies showed that zein-based fat analogs up to 40% gave acceptable results regarding appearance, calorific value, and sensory as well as rheological properties (Yashini et al. 2019). Protein from split pea flour, soy, as well as wheat starch, acts as fat substitutes. Moreover, pea protein was used to minimize shrinking and improve the texture of low-fat hamburgers.

6.18.2 Carbohydrate-Based Fat Replacers

Starch is easily digestible in the human digestive tract. The granular structure of native starch is similar, although the particle size, shape, and amylase-amylopectin ratios vary. Starches with granular sizes similar to fat emulsions have been identified as promising fat replacers (Lindeboom et al. 2004; Malinski et al. 2003). When dispersed singly, in a manner comparable to that of emulsion droplets, starch granules exhibit similar textural and sensory characteristics. As a result, researchers sought to substitute fat in low-fat food products such as cheese, sausage, yogurt, mayonnaise, and frozen desserts with cross-linked starches rather than native starch (Radi et al. 2009). The inclusion of starch as fat replacer increases the water-holding capacity, production, and gel stiffness as well as flow behaviors and sensory quality adjustments. Low-dextrose equivalent maltodextrins can be converted into randomly oriented microgels with a diameter of 1–3 μm to provide better fat replacement outcomes (Chronakis 1997). This maltodextrin microgel was similar to that of crystal-like fat particles therefore it contributed towards the fat-like behaviors and properties (Loret et al. 2004). When compared to full-fat mayonnaise, low-fat mayonnaise made with xanthan gum, guar gum, or citrus fiber resembled the function of oil emulsions and had the same sensory acceptance (Su et al. 2010). Low-fat ice cream had similar compositions, with the combination of guar gum and basil seed gum providing superior creaminess than guar gum alone (Javidi et al. 2016). Carbohydrates, along with other types of dietary ingredients, have given a variety of fat replacements in food items.

6.18.3 Fat-Based Replacers

As the name indicates, fat-based replacers are largely made from fats and have functional, sensorial, and textural characteristics similar to native fats (Hahn 1997; Mela 1996). Fat-based replacers can be categorized as modified fats (which are largely triglycerides) and synthetic fats (Non-triglycerides).

6.18.3.1 SALATRIM: A Modified Fat

Salatrim is the abbreviation for Short and Long chain Acid Triglyceride Molecules. This is a group of triglycerides that comprise a combination of long chain (e.g., stearic acid) and short-chain (e.g., acetic, butyric, or propionic acids) fatty acids that are reconfigured on a glycerol backbone. Salatrim provides very less calories because these short-chain fatty acids are deficient in energy also stearic acid is not absorbed fully. Salatrim is predicted to give 21 kJ/g (5 kcal/g) according to US regulation or 25 kJ/g (6 kcal/g) according to European Union regulation, compared to conventional fat's 38 kJ/g (Lone et al. 2008; Finley et al. 1994a, b). Salatrim, which is commercialized under the trade name Benefat by Cultor Food Science, has been granted GRAS (Generally Recognized as Safe) certification by the FDA.

6.18.3.2 Olestra: A Synthetic Fat

Olestra with a brand name "Olean" does not exist naturally; therefore is created from edible vegetable oils and sucrose. In ordinary fats are triglycerides, i.e., three fatty acids attached to one glycerol. In Olestra structure, glycerol is replaced with sucrose and instead of three fatty acids about 6–8 fatty acids may be attached. As a consequence, it is not hydrolyzed in the stomach; further, its large size prevents it from being absorbed. Therefore, Olestra goes undigested adding zero calories to the meal. The Food and Drug Administration (FDA) authorized Olestra under the brand name "Olean" in 1996 for use in salty snack foods like those now marketed by P&G Food Ingredients.

6.18.3.3 Caprenin

It is a designed fat substitute that lowers the caloric content of food. It is made up of glycerol and fatty acids (behenic acid, capric acid, and caprylic acid) and provides about 4 kcal/g, due to incomplete absorption of these fatty acids. This was a product of Procter and Gamble intended to replace cocoa butter but was withdrawn during the 1990s due to incidences of increased serum cholesterol. But in candies and confectionery coatings it is used to reduce calories.

6.19 Conclusion

With the rising popularity of low-fat and no-fat products among consumers, the demand for newer fats with enhanced properties and fewer health challenges is bound to increase in times to come. Although some success has been achieved in the development of structured fats in recent past, this is an extremely time consuming and cost-intensive process. So, the future strategies shall stay around, adoption of those routes that reduce R&D costs and saves time as well. Breeders and geneticists have to work closely with food processors to explore and develop newer germplasm better suited for today's needs. Preference is also seen towards minimal chemical processing of oils and fats; therefore, better physical refining and improved chemical refining methods need to be explored for providing healthier fats. Controlled and programmed absorption technologies for fats in digestive systems shall also be the focus. We must not forget that nothing in this nature is bad or useless. We must be educated adequately, for its proper use and optimal benefits. The same holds true for fats. Just eradicating fats from our diets is not needed. The need is to adopt controlled and balanced use of the available forms of all fats.

References

Abeshima T (1998) Fractionation of edible oils and fats. J Oleo Sci 47(6):553–561

Afoakwa EO, Paterson A, Fowler M (2007) Factors influencing rheological and textural qualities in chocolate: a review. Trends Food Sci Technol 18:290–298. https://doi.org/10.1016/j.tifs.2007.02.002

Aguilera JM, Gloria H (1997) Determination of oil in fried potato products by differential scanning calorimetry. J Agric Food Chem 45:781–785

Akinpelu OR, Idowu MA, Sobukola OP et al (2014) Optimization of processing conditions for vacuum frying of high quality fried plantain chips using response surface methodology (RSM). Food Sci Biotechnol 23(4):1121–1128

Al Faruq A, Zhang M, Adhikari B (2019) A novel vacuum frying technology of apple slices combined with ultrasound and microwave. Ultrason Sonochem 52:522–529

Aloyor EO, Ozigagu CE, Oboh OI et al (2009) Chromatographic analysis of vegetable oils: a review. Sci Res Essays 4:191–197

Amir IZ, Sharon WXR, Syafiq A (2013) D-Optimal mixture design on melting and textural properties of dark chocolate as affected by cocoa butter substitution with Xanthan gum/Guar gum blends. Int Food Res J 20(4):1991–1995

Anderson M, Brooker BE, Needs EC (2005) The role of proteins in the stabilization/destabilization of dairy foams. In: food emulsions and foams. Woodhead, Cambridge, pp 100–109

Andres A, Arguelles A, Castello ML et al (2013) Mass transfer and volume changes in French fries during air frying. Food Bioprocess Technol 6(8):1917–1924

Anna L, Jacek Z, Magdalena D et al (2015) Communication fatty acid profile and biological activities of linseed and rapeseed oils. Molecules 20:22872–22880. https://doi.org/10.3390/molecules201219887

Ayustaningwarno F, Dekker M, Fogliano V et al (2018) Effect of vacuum frying on quality attributes of fruits. Food Eng Rev 10(3):154–164

Azam SR, Ma H, Xu B et al (2020) Efficacy of ultrasound treatment in the removal of pesticide residues from fresh vegetables: a review. Trends Food Sci Technol 97:417–432

Banerjee S, Sahu CK (2017) A short review on vacuum frying-a promising technology for healthier and better fried foods. J Nutr Health 1:56–59
Baumann B, Escher F (1995) Mass and heat transfer during deep-fat frying of potato slices rate of drying and oil uptake. LWT Food Sci Technol 28:395–403
Beckett ST (2009) Chocolate flow properties. In: Beckett ST (ed) Industrial chocolate manufacture and use, 4th edn. Blackwell, Oxford, UK, pp 224–246
Berry SE, Bruce JH, Steenson S et al (2019) Interesterified fats: what are they and why are they used? A briefing report from the roundtable on interesterified fats in foods. Wiley, London, pp 363–380
Biasutti M, Venir E, Marino M et al (2013) Effects of high pressure homogenisation of ice cream mix on the physical and structural properties of ice cream. Int Dairy J 32(1):40–45
Bordin K, TomiheKunitake M, Kazue AK et al (2013) Changes in food caused by deep fat frying-a review. Arch Latinoam Nutr 63(1):5–13
Boro P, Debnath J, Kumar DT et al (2018) Milk composition and factors affecting it in dairy Buffaloes: a review. J Entomol Zool 6:340–343
Boskou D (2002) Frying fats. In: Sokorski ZE, Kolakowska A (eds) Chemical and functional properties of food lipids. CRC Press, Boca Raton, FL, p 385
Bouchon P, Aguilera JM (2001) Microstructural analysis of frying potatoes. Int J Food Sci Technol 36:669–676
Bouchon PB, Aguilera JM, Pyle DL (2003) Structure oil-absorption relationships during deep-fat frying. J Food Sci 68(9):2711–2716
Brooker BE, Anderson M, Andrews AT (1986) The development of structure in whipped cream. Food Struct 5(2):12
Cauvain SP (2003) Bread making: an overview. In: Stanley P (ed) Bread making improving quality. CRC Press, Boca Raton, FL. https://doi.org/10.1201/9780203495001
Chapelot D (2011) The role of snacking in energy balance: a biobehavioral approach. J Nutr 141(1): 158–162
Chaplin K, Smith AP (2011) Definitions and perceptions of snacking. Curr Top Nutraceut Res 9(1/2):53
Chen L, Wang P, Kang ZL et al (2015) Effect of soybean oil emulsified and unemulsified with chicken plasma protein on the physicochemical properties of frankfurters. CYTA J Food 13(3): 445–455
Choe E, Min DB (2007) Chemistry of deep-fat frying oils. J Food Sci 72(5):77–86
Chronakis IS (1997) Structural–functional and water-holding studies of biopolymers in low fat content spreads. LWT Food Sci Technol 30(1):36–44
Chung J, Lee J, Choe E (2004) Oxidative stability of soybean and sesame oil mixture during frying of flour dough. J Food Sci 69(7):574–578
Collins YF, McSweeney PL, Wilkinson MG (2003) Lipolysis and free fatty acid catabolism in cheese: a review of current knowledge. Int Dairy J 13(11):841–866
Costa RM, Oliveira F, Boutcheva G (2000) Structural changes and shrinkage of potato during frying. Int J Food Sci Technol 36:11–25
Cui L, Fan J, Sun Y et al (2018) The prooxidant activity of salts on the lipid oxidation of lecithin-stabilized oil-in-water emulsions. Food Chem 252:28–32
Da Silva PF, Moreira RG (2008) Vacuum frying of high-quality fruit and vegetable-based snacks. LWT Food Sci Technol 41(10):1758–1767
Das I, Arora A (2018) Alternate microwave and convective hot air application for rapid mushroom drying. J Food Eng 223:208–219
Deeth HC (2006) Lipoprotein lipase and lipolysis in milk. Int Dairy J 16(6):555–562
Del Re PV, Jorge N (2007) Behavior of sunflower oil, soybeans and corn for frying of breaded meat product prefried frozen. Cienc Agrotec 31:1774–1779
Devi A, Khatkar BS (2016) Physicochemical, rheological and functional properties of fats and oils in relation to cookie quality: a review. J Food Sci Technol 53(10):3633–3641

Devi S, Zhang M, Law CL (2018) Effect of ultrasound and microwave assisted vacuum frying on mushroom (Agaricus bisporus) chips quality. Food Biosci 25:111–117
Devi S, Zhang M, Ju R et al (2020) Recent development of innovative methods for efficient frying technology. Crit Rev Food Sci Nutr 61:1–16
Diamante LM, Shi S, Hellmann A et al (2015) Vacuum frying foods: products, process and optimization. Int Food Res J 22(1):15–22
Do TAL, Mitchell JR, Wolf B et al (2010) Use of ethylcellulose polymers as stabilizer in fat-based food suspensions examined on the example of model reduced-fat chocolate. React Funct Polym 70(10):856–862. https://doi.org/10.1016/j.reactfunctpolym.2010.07.012
Dobarganes MC, Velasco J, Dieffenbacher A (2000) Determination of polar compounds, polymerized and oxidized triacylglycerols, and diacylglycerols in oils and fats. Pure Appl Chem 72(8):1563–1575
Drake MA, Miracle RE, McMahon DJ (2010) Impact of fat reduction on flavor and flavor chemistry of Cheddar cheeses. J Dairy Sci 93(11):5069–5081
Dueik V, Bouchon P (2011) Development of healthy low-fat snacks: understanding the mechanisms of quality changes during atmospheric and vacuum frying. Food Rev Int 27(4): 408–432
Eden J, Dejmek P, Lofgren R et al (2016) Native milk fat globule size and its influence on whipping properties. Int Dairy J 61:176–181
Farinu A, Baik OD (2005) Deep fat frying of foods—transport phenomena. Food Rev Int 21(4): 389–410
Farkas BE, Singh RP, Rumsey TR (1996) Modeling heat and mass transfer in immersion frying: model development. J Food Eng 29:211–226
Fellows PJ (2009) Food processing technology: principles and practice. Elsevier, Amsterdam
Finley JW, Klemann LP, Leveille GA et al (1994a) Caloric availability of salatrim in rats and humans. J Agric Food Chem 42:495–499
Finley JW, Leveill GA, Dixon RM et al (1994b) Clinical assessment of salatrim, a reduced-calorie triacylglycerol. J Agric Food Chem 42:581–596
Fitzpatrick JJ, Iqbal T, Delaney C et al (2004) Effect of powder properties and storage conditions on the flowability of milk powders with different fat contents. J Food Eng 64(4):435–444
Food Safety Network (2014) Providing reliable information to help keep food safe & healthful. University of Guelph, Guelph. https://www.uoguelph.ca/foodsafetynetwork/fat-substitutes
Frankel EN, Huang SW (1994) Improving the oxidative stability of polyunsaturated vegetable oils by blending with high-oleic sunflower oil. J Am Oil Chem Soc 71(3):255–259
FSANZ (2003) Erucic acid in food – a toxicological review and risk assessment. Food Standards Australia New Zealand. Technical report series No. 21. ISBN 0-642-34526-0. ISSN 1448-3017
Ghaderi A, Dehghannya J, Ghanbarzadeh B (2018) Momentum, heat and mass transfer enhancement during deep-fat frying process of potato strips: influence of convective oil temperature. Int J Therm Sci 134:485–499
Ghaitaranpour A, Koocheki A, Mohebbi M et al (2018) Effect of deep fat and hot air frying on doughnuts physical properties and kinetic of crust formation. J Cereal Sci 83:25–31
Ghotra BS, Dyal SD, Narine SS (2002) Lipid shortenings: a review. Food Res Int 35:1015–1048. https://doi.org/10.1016/S0963-9969(02)00163-1
Goff HD, Hartel RW (2013) Cleaning and sanitizing for microbiological quality and safety. In: Ice Cream. Springer, Boston, MA, pp 379–401
Gomes R, Martins S, Afonso C et al (2019) Comparison of fish and oil supplements for a better understanding of the role of fat level and other food constituents in determining bioaccessibility. Food Sci Nutr 7:1179–1189
Hahn NI (1997) Replacing fat with food technology - a brief review of new fat replacement ingredients. J Am Diet Assoc 97:15–16
Hashempour-Baltork F, Torbati M, Azadmard-Damirchi S et al (2016) Vegetable oil blending: a review of physicochemical, nutritional and health effects. Trends Food Sci Technol 57:52–58

Hayes MG, Lefrancois AC, Waldron DS et al (2003) Influence of high pressure homogenisation on some characteristics of ice cream. Milchwissenschaft 58(9–10):519–523
Haylock SJ, Dodds TM (2009) Ingredients from milk. In: Beckett ST (ed) Industrial chocolate manufacture and use, 4th edn. Blackwell, Reino Unido
Heredia A, Castello ML, Arguelles A et al (2014) Evolution of mechanical and optical properties of French fries obtained by hot air-frying. LWT Food Sci Technol 57(2):755–760
Holzmüller W, Muller M, Himbert D et al (2016) Impact of cream washing on fat globules and milk fat globule membrane proteins. Int Dairy J 59:52–61
Hotrum NE (2004) Emulsion droplet spreading at air/water interfaces: mechanisms and relevance to the whipping of cream. Wageningen University and Research, Wageningen
Houhoula DP, Oreopoulou V, Tzia C (2003) The effect of process time and temperature on the accumulation of polar compounds in cottonseed oil during deep-fat frying. J Sci Food Agric 83(4):314–319
Huang X, Lindholm B, Stenvinkel P et al (2013) Dietary fat modification in patients with chronic kidney disease: n-3 fatty acids and beyond. J Nephrol 26(6):960–974
Huppertz T, Uniacke-Lowe T, Kelly AL (2020) Physical chemistry of milk fat globules. Adv Dairy Chem 2:133–167
Islam M, Zhang M, Fan D (2019) Ultrasonically enhanced low-temperature microwave-assisted vacuum frying of edamame: effects on dehydration kinetics and improved quality attributes. Dry Technol 37(16):2087–2104
Jacob J, Leelavathi K (2007) Effect of fat-type on cookie dough and cookie quality. J Food Eng 79: 299–305. https://doi.org/10.1016/j.jfoodeng.2006.01.058
Jahurul MHA, Zaidul ISM, Norulaini NAN et al (2013) Cocoa butter fats and possibilities of substitution in food products concerning cocoa varieties, alternative sources, extraction methods, composition, and characteristics. J Food Eng 117:467–476. https://doi.org/10.1016/j.jfoodeng.2012.09.024
Jana O, Ladislava M, Jarmila VA et al (2015) Fatty acids composition of vegetable oils and its contribution to dietary energy intake and dependence of cardiovascular mortality on dietary intake of fatty acids. Int J Mol Sci 16:12871–12890. https://doi.org/10.3390/ijms160612871
Javidi F, Razavi SM, Behrouzian F et al (2016) The influence of basil seed gum, guar gum and their blend on the rheological, physical and sensory properties of low fat ice cream. Food Hydrocoll 52:625–633
Johnson ME, Kapoor R, McMahon DJ et al (2009) Reduction of sodium and fat levels in natural and processed cheeses: scientific and technological aspects. Compr Rev Food Sci Food Saf 8(3): 252–268
Jooyandeh H (2009) Effect of fermented whey protein concentrate on texture of Iranian white cheese. J Texture Stud 40(5):497–510
Jorge N, Soares BBP, Lunardi VM et al (2005) Physico-chemical alterations of sunflower, corn and soybean oils in deep fat frying. Quim Nova 28(6):947–951
Kang Q, Chen Y, Gu Z (2016) The method of fat content reduction of the vacuum fried purple sweet potato chips. Zhongguo Shipin Xuebao 16(11):174–181
Karaca OB, Guven M, Yasar K et al (2009) The functional, rheological and sensoru characteristics of ice creams with various fat replacers. Int J Dairy Technol 62:93–99. https://doi.org/10.1111/j.1471-0307.2008.00456.x
Karnjanapratum S, Sinthusamran S, Sae-leaw T et al (2017) Characteristics and gel properties of gelatin from skin of Asian bullfrog (*Rana tigerina*). Food Biophys 12(3):289–298
Keeton JT (1994) Low-fat meat products—technological problems with processing. Meat Sci 36(1–2):261–276
Kew B, Holmes M, Stieger M et al (2020) Review on fat replacement using protein-based microparticulated powders or microgels: a textural perspective. Trends Food Sci Technol 106: 457–468
Kim EHJ, Chen XD, Pearce D (2005) Effect of surface composition on the flowability of industrial spray-dried dairy powders. Colloids Surf B Biointerfaces 46(3):182–187

Kinsella JE, Melachouris N (1976) Functional properties of proteins in foods: a survey. Crit Rev Food Sci Nutr 7(3):219–280. https://doi.org/10.1080/10408397609527208
Kiritsakis AK (1998) Flavor components of olive oil—a review. J Am Oil Chem Soc 75(6): 673–681
Kostik V, Memeti S, Bauer B (2013) Fatty acid composition of edible oils and fats. J Hyg Eng Des 4:112–116
Koxholt MM, Eisenmann B, Hinrichs J (2001) Effect of the fat globule sizes on the meltdown of ice cream. J Dairy Sci 84(1):31–37
Krokida MK, Oreopoulou V, Maroulis ZB et al (2001) Effect of pre-treatment on viscoelastic behavior of potato strips. J Food Eng 50:11–17
Lascaray L (1949) Mechanism of fat splitting. Ind Eng Chem Res 41(4):786–790
Lee CL, Liao HL, Lee WC et al (2018) Standards and labeling of milk fat and spread products in different countries. J Food Drug Anal 26(2):469–480
Leite PB, Lannes SCS, Rodrigues AM et al (2013) Rheological study of chocolates made with different cocoa (Theobroma cacao L.) varieties. Braz J Food Technol 16(3):192–197. https://doi.org/10.1590/S1981-67232013005000024
Li Y, Ngadi M, Oluka S (2008) Quality changes in mixtures of hydrogenated and non-hydrogenated oils during frying. J Sci Food Agric 88(9):1518–1523
Lindeboom N, Chang PR, Tyler RT (2004) Analytical, biochemical and physicochemical aspects of starch granule size, with emphasis on small granule starches: a review. Starch-Starke 56(3–4): 89–99
Lloyd BJ, Farkas BE, Keener KM (2004) Quality comparison of French fry style potatoes produced by oven heating, immersion frying and controlled dynamic radiant heating. J Food Process Preserv 28(6):460–472
Lone BS, Heidi TC, Maja TA et al (2008) The effect of salatrim, a low-calorie modified triacylglycerol, on appetite and energy intake. Am J Clin Nutr 87(5):1163–1169. https://doi.org/10.1093/ajcn/87.5.1163
Lopez C, Briard-Bion V, Menard O et al (2011) Fat globules selected from whole milk according to their size: different compositions and structure of the biomembrane, revealing sphingomyelin-rich domains. Food Chem 125(2):355–368
Loret C, Meunier V, Frith WJ et al (2004) Rheological characterisation of the gelation behaviour of maltodextrin aqueous solutions. Carbohydr Polym 57(2):153–163
MacGibbon AKH (2020) Composition and structure of bovine milk lipids. Adv Dairy Chem 2:1–32
Mahmoodani F, Perera CO, Abernethy G et al (2018) Lipid oxidation and vitamin D3 degradation in simulated whole milk powder as influenced by processing and storage. Food Chem 261:149–156
Majid I, Nayik GA, Nanda V (2015) Ultrasonication and food technology: a review. Cogent Food Agric. https://doi.org/10.1080/23311932.2015.1071022
Malinski E, Daniel JR, Zhang XX et al (2003) Isolation of small starch granules and determination of their fat mimic characteristics. Cereal Chem 80(1):1–4
Manzocco L, Calligaris S, Pieve S et al (2012) Effect of monoglyceride-oil-water gels on white bread properties. Food Res Int 49(2):778–782. https://doi.org/10.1016/j.foodres.2012.09.011
Marangoni AG, Acevedo N, Maleky F et al (2012) Structure and functionality of edible fats. R Soc Chem 8:1275–1300
Martínez-Cervera S, Sanz T, Salvador A et al (2012) Rheological, textural and sensorial properties of low-sucrose muffins reformulated with sucralose/polydextrose. LWT Food Sci Technol 45(2):213–220. https://doi.org/10.1016/j.lwt.2011.08.001
McClements DJ, Decker EA (2008) Lipids. In: Srinivasan D, Parkin KL, Fennema OR (eds) Fennema's food chemistry, 4th edn. CRC Press/Taylor & Francis, Boca Raton, FL, pp 155–216
McDonough C, Gomez MH, Lee JK et al (1993) Environmental scanning electron microscopy evaluation of tortilla chip microstructure during deep-fat frying. J Food Sci 58:199–203

McMahon DJ, Oberg CJ (2017) Pasta-filata cheese. In: Cheese technology and major cheese groups. Cheese: chemistry physics and microbiology, vol 2. Elsevier, Amsterdam, pp 1041–1068
McMahon DJ, Fife RL, Oberg CJ (1999) Water partitioning in mozzarella cheese and its relationship to cheese meltability. J Dairy Sci 82(7):1361–1369
Medeiros ML, Lannes SCS (2009) Chemical evaluation of cocoa substitutes and study of formulated chocolate drink powder. Food Sci Technol 29(2):247–253. https://doi.org/10.1590/S0101-20612009000200002
Mela DJ (1996) Implications of fat replacement for food choice and energy balance. Chem Ind 9: 329–332
Méndez-Cid FJ, Centeno JA, Martínez S et al (2017) Changes in the chemical and physical characteristics of cow's milk butter during storage: effects of temperature and addition of salt. J Food Compos Anal 63:121–132
Metin S, Hartel RW (2012) Milk fat and cocoa butter. In: Cocoa butter and related compounds. Academic, San Diego, pp 365–392
Min DB, Boff JM (2002) Lipid oxidation of edible oil. In: Food lipids: chemistry, nutrition, and biotechnology. Marcel Dekker, New York, pp 335–364
Mishra S, Manchanla SC (2012) Cooking oils for heart health. Eur J Prev Cardiol 1:123–131
Morales MT, Rios JJ, Aparicio R (1997) Changes in the volatile composition of virgin olive oil during oxidation: flavors and off-flavors. J Agric Food Chem 45(7):2666–2673
Moretto E, Fett R (1998) Technology of vegetable oils and fats. Varela, Sao Paulo, Brazil, p 150
Mulvaney S, Barbano DM, Yun JJ (1997) Systems analysis of the plasticization and extrusion processing of Mozzarella cheese. J Dairy Sci 80(11):3030–3039
Mulvihill DM, Ennis MP (2003) Functional milk proteins: production and utilization. In: Advanced dairy chemistry—1 Proteins. Springer, Boston, MA, pp 1175–1228
Nath PM, Kumar V, Praveen PK, Ganguly S (2016) Effect of chicken skin, soy protein and olive oil on quality characteristics of chicken nuggets. Int J Environ Sci Technol 5(3):1574–1585
Nawar WW (1969) Thermal degradation of lipids. J Agric Food Chem 17(1):18–21
Norberg S, AarhusKarlshamn AAK (2006) Chocolate and confectionery fats. In: Gunstone FD (ed) Modifying lipids for use in food, Part III. Applications of modified lipids in food. Woodhead, Cambridge, UK, pp 487–516
O'Brien RD (2009) Fats and oils, 3th edn. CRC Press, Boca Raton, FL
Ognean CF, Darie N, Ognean M (2006) Fat replacers: review. J Agroaliment Process Technol 12(2):433–442
Ogori AF (2020) Extraction and constituents of fats and oils. J Food Sci Nutr 6:060
Oladejo AO, Ma H, Qu W et al (2017) Effects of ultrasound pretreatments on the kinetics of moisture loss and oil uptake during deep fat frying of sweet potato (Ipomea batatas). Innov Food Sci Emerg Technol 43:7–17
Oztop MH, Sahin S, Sumnu G (2007) Optimization of microwave frying of potato slices by using Taguchi technique. J Food Eng 79(1):83–91
Pankaj S, Keener KM (2017) A review and research trends in alternate frying technologies. Curr Opin Food Sci 16:74–79
Parikh A, Takhar PS (2016) Comparison of microwave and conventional frying on quality attributes and fat content of potatoes. J Food Sci 81(11):2743–2755
Piernas C, Popkin BM (2010) Trends in snacking among US children. Health Aff 29(3):398–404
Pokorny J (1989) Flavor chemistry of deep fat frying in oil. In: Flavor chemistry of lipids foods. The American Oil Chemists' Society, Champaign, IL, pp 113–155
Potter NN, Hotchkiss JH (1996) Food science, 5th edn. Chapman and Hall Inc., New York, pp 359–379
Queguiner C, Dumay E, Salou-Cavalier C et al (1992) Microcoagulation of a whey protein isolate by extrusion cooking at acid pH. J Food Sci 57(3):610–616
Radi M, Niakousari M, Amiri S (2009) Physicochemical, textural and sensory properties of low-fat yogurt produced by using modified wheat starch as a fat replacer. J Appl Sci 9(11):2194–2197

Reda SY (2004) Comparative study of vegetable oils subjected to the thermal stress. Master Dissertation

Reeve R, Neel E (1960) Microscopic structure of potato chips. Am Potato J 37:45–52

Rego OA, Cabrita AR, Rosa HJ et al (2016) Changes in milk production and milk fatty acid composition of cows switched from pasture to a total mixed ration diet and back to pasture. Ital J Anim Sci 15(1):76–86

Reis MG, Coelho NRA, Castro EMO (2011) Estudo do teor de lipideosem chocolates. Estudos 38(3):519–532

Richter M, Lannes SCS (2007) Chocolate confectionery for special purposes: chemical and sensorial evaluation. Ciência Tecnol Alime 27(1):193–200. https://doi.org/10.1590/S0101-20612007000100034

Roknul Azam SM, Zhang M, Law CL et al (2019) Effects of drying methods on quality attributes of peach (Prunus persica) leather. Dry Technol 37(3):341–351

Rønholt S, Kirkensgaard JJK, Mortensen K et al (2014) Effect of cream cooling rate and water content on butter microstructure during four weeks of storage. Food Hydrocoll 34:169–176

Rovers TA, Sala G, Van der Linden E et al (2016) Potential of microbubbles as fat replacer: effect on rheological, tribological and sensorial properties of model food systems. J Texture Stud 47(3):220–230

Rudan MA, Barbano DM, Guo MR et al (1998) Effect of the modification of fat particle size by homogenization on composition, proteolysis, functionality, and appearance of reduced fat Mozzarella cheese. J Dairy Sci 81(8):2065–2076

Schiffmann R (2017) Microwave-assisted frying. In: The microwave processing of foods. Woodhead, Cambridge, UK, pp 142–151

Schirle-Keller JP, Reineccius GA, Hatchwell LC (1994) Flavor interactions with fat replacers: effect of oil level. J Food Sci 59(4):813–815

Schröder M, Vetter W (2013) Detection of 430 fatty acid methyl esters from a transesterified butter sample. J Am Oil Chem Soc 90:771–790

Schwendel BH, Wester TJ, Morel PC et al (2017) Pasture feeding conventional cows removes differences between organic and conventionally produced milk. Food Chem 229:805–813

Shahidi F (2005) Bailey's industrial oil and fat products, 6th edn. Wiley, Hoboken, NJ

Shaker MA (2015) Comparison between traditional deep-fat frying and air-frying for production of healthy fried potato strips. Int Food Res J 22(4):1557–1563

Shumacher AB, Brandeli A, Macedo FC et al (2010) Chemical and sensory evaluation of dark chocolate with addition of quinoa (*Chenopodium quinoa* Willd.). J Food Sci Technol 47(2): 202–206

Singh RP, Mermelstein N (1995) Heat and mass transfer in foods during deep-fat frying: engineering aspects of deep-fat frying of foods. Food Technol (Chicago) 49(4):134–137

Smith PR, Johansson J (2004) Influences of the proportion of solid fat in a shortening on loaf volume and staling of bread. J Food Process Preserv 28:359–367. https://doi.org/10.1111/j.1745-4549.2004.23079.x

Sobukola OP, Dueik V, Munoz L et al (2013) Comparison of vacuum and atmospheric deep-fat frying of wheat starch and gluten-based snacks. Food Sci Biotechnol 22:177–182

Solinas M, Angerosa F, Camera L (1984) Oxidation progress in plant oils during frying. Determination of volatile components by HRGC and HPLC. Ann Ist Super Sanita 10:137–154

Sonwai S, Rousseau D (2010) Controlling fat bloom formation in chocolate: impact of milk fat on microstructure and fat phase crystallization. Food Chem 119:286–297. https://doi.org/10.1016/j.foodchem.2009.06.031

Srilakshmi B (2003) Food science. New Age International Publishers, Hyderabad, India

Stevenson SG, Vaisey-Genser M, Eskin NA (1984) Quality control in the use of deep frying oils. J Am Oil Chem Soc 61(6):1102–1108

Su F, Lannes SCS (2012) Rheological evaluation of the structure of ice cream mixes varying fat base. Appl Rheol 22(6):63871

Su HP, Lien CP, Lee TA et al (2010) Development of low-fat mayonnaise containing polysaccharide gums as functional ingredients. J Sci Food Agric 90(5):806–812
Su Y, Zhang M, Adhikari B et al (2018a) Improving the energy efficiency and the quality of fried products using a novel vacuum frying assisted by combined ultrasound and microwave technology. Innov Food Sci Emerg Technol 50:148–159
Su Y, Zhang M, Zhang W et al (2018b) Ultrasonic microwave-assisted vacuum frying technique as a novel frying method for potato chips at low frying temperature. Food Bioprod Process 108:95–104
Sun YE, Wang WD, Chen HW et al (2011) Autoxidation of unsaturated lipids in food emulsion. Crit Rev Food Sci Nutr 51(5):453–466
Sun C, Liu R, Liang B et al (2018) Microparticulated whey protein-pectin complex: a texture-controllable gel for low-fat mayonnaise. Food Res Int 108:151–160
Suryavanshi MV, Ghosh JS (2010) Spoilage of white unsalted butter by psychrophilic lipolysis of *Pseudomonas aeruginosa* NCIM 2036. Br J Dairy Sci 1(1):26–29
Talbot G (2009) Vegetable fats. In: Beckett ST (ed) Industrial chocolate manufacture and use, 4th edn. Blackwell, Reino Unido
Tandy DC, McPherson WJ (1984) Physical refining of edible oil. J Am Oil Chem Soc 61(7): 1253–1258
Tang D, Marangoni AG (2007) Modeling the rheological properties and structure of colloidal fat crystal networks. Trends Food Sci Technol 18(9):474–483
Torres IC, Janhøj T, Mikkelsen BO et al (2011) Effect of microparticulated whey protein with varying content of denatured protein on the rheological and sensory characteristics of low-fat yoghurt. Int Dairy J 21(9):645–655
Udomkun P, Innawong B (2018) Effect of pre-treatment processes on physicochemical aspects of vacuum-fried banana chips. J Food Process Preserv 42(8):e13687
Umar KM, Fahimul HM, Rauf A (2018) Effect of temperature on milk fats of cow, buffalo, and goat used for frying local food products. Food Qual Saf 2(1):51–57
Van Marle JT, Clerkx ACM, Boekestein A (1992) Cryo-scanning electron microscopy investigation of the texture of cooked potatoes. Food Struct 11(3):3
Waldron DS, Hoffmann W, Buchheim W et al (2020) Role of milk fat in dairy products. In: McSweeney PLH, Fox PF, O'Mahony JA (eds) Advanced dairy chemistry, vol 2. Springer, Cham, pp 245–305
Wang Y, Padua GW (2012) Nanoscale characterization of zein self-assembly. Langmuir 28(5): 2429–2435
Wang L, Xu B, Wei B et al (2018) Low frequency ultrasound pretreatment of carrot slices: effect on the moisture migration and quality attributes by intermediate-wave infrared radiation drying. Ultrason Sonochem 40:619–628
Warner K (2002) Chemistry of frying fats. In: Akoh CC, Min DB (eds) Food lipids: chemistry, nutrition, and biotechnology, 2nd edn. Marcel Dekker, New York
WHO (2018) WHO plan to eliminate industrially produced trans fatty acids from global food supply. WHO News. https://www.who.int/news/item/14-05-2018-who-plan-to-eliminate-industrially-produced-trans-fatty-acids-from-global-food-supply. Accessed 05 Mar 2020
Wilderjans E, Luyts A, Brijs K et al (2013) Ingredient functionality in batter type cake making. Trends Food Sci Technol 30:6–15. https://doi.org/10.1016/j.tifs.2013.01.001
Yang A, Qi M, Wang X et al (2019a) Refined cottonseed oil as a replacement for soybean oil in broiler diet. Food Sci Nutr 7:1027–1034
Yang Y, Wang L, Li Y et al (2019b) Investigation the molecular degradation, starch-lipid complexes formation and pasting properties of wheat starch in instant noodles during deep-frying treatment. Food Chem 283:287–293
Yashini M, Sunil CK, Sahana S et al (2019) Protein-based fat replacers-are view of recent advances. Food Rev Int. https://doi.org/10.1080/87559129.2019.1701007
Zaidi A (2017) Can we re-use cooking oil, once it is utilized for frying foods? FSSAI. https://foodsmart.fssai.gov.in/reusingoil.html. Accessed 25 Feb 2021

Zhang Q, Saleh AS, Chen J et al (2012) Chemical alterations taken place during deep-fat frying based on certain reaction products: a review. Chem Phys Lipids 165(6):662–681
Zhang T, McCarthy J, Wang G et al (2015) Physiochemical properties, microstructure, and probiotic survivability of nonfat goats' milk yogurt using heat-treated whey protein concentrate as fat replacer. J Food Sci 80(4):788–794
Zhang T, Li J, Ding Z et al (2016) Effects of initial moisture content on the oil absorption behavior of potato chips during frying process. Food Bioprocess Technol 9(2):331–340
Zhou J, Faubion JM, Walker CE (2011) Evaluation of different types of fats for use in high-ratio layer cakes. Food Sci Technol 44:1802–1808

Food Additives

7

Manisha Medhi, Arun Kumar Gupta, Subhamoy Dhua, and Poonam Mishra

7.1 Introduction

Since ancient times the use of food additives has been practiced by the people to increase the taste and other attributes of food. One of the most important criteria of using food additives is to enhance the shelf life of foods. Different regulations and food-related laws define food additives in different ways. According to the Codex Alimentarius Commission, a food additive is defined as "any substance which is not normally consumed as a food and not used as a characteristic ingredient of the food, whether or not it has nutritive value, the intentional addition of the same to food for a technological (including organoleptic) purpose while manufacture, processing, preparation, treatment, packaging, and transport in becoming a component or otherwise affecting the characteristics of such foods." As per FSSAI (Food Safety and Standards Authority of India) regulations, food additives are used for maintaining or improving the nutritive value of food, should not include contaminants or substances or not to be present beyond the upper limit of acceptance.

In general, food additives can be defined as any substance which is of food grade and can be added to food intentionally within the permissible limit to increase the shelf life, inhibits the microbial growth, delay various chemical reactions, enhances sensory properties like color, flavor, texture, appearance, taste and food safety. Food additives are categorized under two following heads (Fig. 7.1).

The major law that governs food additives is the Food Additives Amendment to the Federal Food, Drug and Cosmetic Act of 1958. In India, FSSR-3 describes about food additives and their standards. Toxicity and Hazard are the two terms that require special attention while dealing with food additives. Toxicity describes the capacity of any substance to cause injury whereas hazard describes the probability of causing

M. Medhi · A. K. Gupta · S. Dhua · P. Mishra (✉)
Department of Food Engineering and Technology, Tezpur University, Tezpur, Assam, India
e-mail: poonam@tezu.ernet.in

O. P. Chauhan (ed.), *Advances in Food Chemistry*,
https://doi.org/10.1007/978-981-19-4796-4_7

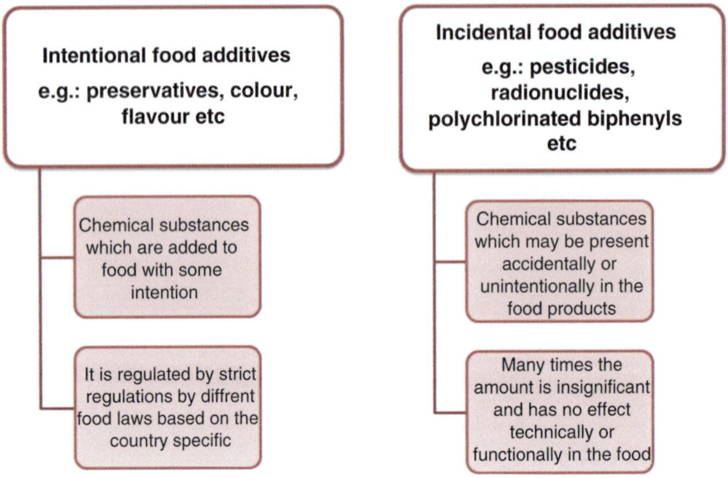

Fig. 7.1 Major classification of food additives. (Adapted from de Man 1999)

injury to its intended use. Therefore, the concept of GRAS (Generally Recognized as Safe) and permissible limit plays an important role here. Most of the additives are chemicals and uses of such additives rely on both GRAS and permissible limit. The discussion of this chapter is focused on different food additives, their uses, mechanism, E numbers, safety, testing methods, intake assessment, and their risk.

7.2 E Numbers (International Numbering System) of Food Additives

Food additives are associated with specific E no., given by the European food safety authority. Codex committee on food additives and contaminants (CCFAC) has prepared the international numbering system (INS) for different food additives and has been prepared to identify food additives in the ingredients list. Codex general standard for the labeling of prepared foods (CODEX STAN 1-1985) shows the need for the identification of food additives on food labels. The general list of E numbers of food additives is given in Table 7.1 and permitted levels of additives are given in Table 7.2.

7.3 Objectives and Principles of Food Additives

The most important and potential objectives of the addition of food additives in edible products are as follows:

Table 7.1 E numbers of food additives

Block of numbers	Food additives
E100–E199	Colors
E200 E299	Preservatives
E300–E399	Antioxidants and acidity regulators
E400–E499	Thickeners, stabilizers, and emulsifiers
E500–E599	Anti-caking agent
E600 E699	Flavor enhancers
E700–E799	Antibiotics
E900–E999	Glazing agents and sweeteners
E1000–E1599	Additional chemicals

Source: https://www.food.gov.uk/science/additives/enumberlist#toc-1

1. **To maintain and improve the nutritive content**
   Nutritional aspect of food is considered as the most important factor. The loss of vitamins and other heat-sensitive compounds need to be addressed. Incorporation of food additives helps in overcoming concern of addition of antioxidants in fats and oils to prevent the loss of vitamin A and beta-carotene.
2. **To enhance the shelf life of food**
   A wide range of food additives are available to improve the shelf life of food products. Food additives used to enhance the shelf life include antioxidants, anti-caking agents, humectants, curing agents, and many others.
3. **To increase consumer acceptability**
   Nonnutritive substances such as emulsifying agents, stabilizers, colorants, and bleaching agents, when added to food products, improve the palatability and appeal of the food products.

7.4 Classification of Additives

Based on the functional properties and their uses, additives can be classified as shown in Fig. 7.2.

7.4.1 Preservatives

Preservatives are substances which are added to food to enhance their shelf life. Class II preservatives are synthetic one which are generally recognized as safe and extends the shelf life of food. The increasing population is related to the increasing demand for food supply. Food additives play an important role in increasing the shelf life of the product. These may be natural and artificial or synthetic. Natural and artificial preservatives are called as Class I and Class II preservatives. Class I preservatives are the natural food substances which are generally used in our day-to-day cooking and preservation activities like salt, sugar, oil, spices, lemon when

Table 7.2 Maximum permissible limit for different food additives in different food

S. No.	Additive name	E No.	Type food in which used	Maximum permissible limit (mg/kg)	Reference
1	Acesulfame potassium	E950	Canned or bottled (pasteurized) fruit	350	Codex Alimentarius Commission (2019)
2	Acetic acid, glacial	E260	Complementary foods for infants and young children	5000	
3	Adipic acid	E355	Fermented milk	1500	
4	Agar	E406	Fermented milk	GMP	
5	Alginic acid	E400	Fresh pasta and noodles	GMP	
6	Allura red AC	E129	Dairy-based desserts (pudding, yogurt, etc.)	300	
7	Amaranth	E123	Flavored milk	50	
8	Ammonium alginate	E403	Frozen fish, fish fillets	GMP	
9	Ammonium carbonate	E503 (i)	Complementary foods for infants and young children	GMP	
10	Ammonium hydroxide	E527	Fermented milk	GMP	
11	Ammonium salts of phosphatidic acid	E442	Dairy-based desserts (pudding, yogurt, etc.)	5000	
12	Ascorbic acid, L-	E300	Complementary foods for infants and young children	500	
13	Beeswax	E901	Fine bakery wares (sweet, salty, savory)	GMP	
14	Brilliant black	E151	Confectionary (hard and soft candy)	100	
15	Brilliant blue FCF	E133	Soups and broths	50	
16	Butylated hydroxyanisole	E320	Butter oil, anhydrous milk fat, ghee	175	
17	Calcium acetate	E263	Complementary foods for infants and young children	GMP	

(continued)

Table 7.2 (continued)

S. No.	Additive name	E No.	Type food in which used	Maximum permissible limit (mg/kg)	Reference
18	Calcium alginate	E404	Fermented milk	GMP	
19	Calcium ascorbate	E302	Complementary foods for infants and young children	200	
20	Calcium carbonate	E170 (i)	Fresh meat, poultry	GMP	
21	Calcium chloride	E509	Salt substitutes	GMP	
22	Calcium hydroxide	E526	Butter	GMP	
23	Calcium lactate	E327	Fermented vegetable	10,000	
24	Calcium oxide	E529	Fermented milk	GMP	
25	Calcium silicate	E552	Dried whey and whey products	10,000	
26	Carrageenan	E407	Infant formulae	300	
27	Chlorine	E925	Flours	2500	
28	Citric acid	E330	Fruit juice	3000	
29	Diacetyltartaric and fatty acid esters of glycerol	E472e	Candied fruit	1000	
30	Dimethyl dicarbonate	E242	Grape wines	200	
31	Dioctyl sodium sulfosuccinate	E480	Fruit-based desserts	15	Codex Alimentarius Commission (2019)
32	Fast green FCF	E143	Water-based flavored drinks	100	
33	Ferrous gluconate	E579	Vegetables	150	
34	Ferrous lactate	E585	Vegetables	150	
35	Fumaric acid	E297	Fresh pasta and noodles	700	
36	Guar gum	E412	Complementary foods for infants and young children	2000	
37	Gum arabic	E414	Complementary foods for infants and young children	10,000	
38	Hexamethylenetetramine	E239	Ripened cheese	25	
39	Hydroxypropyl cellulose	E463	Pasteurized cream	GMP	
40	Hydroxypropyl methyl cellulose	E464	Smoked, dried, fermented fish	GMP	

(continued)

Table 7.2 (continued)

S. No.	Additive name	E No.	Type food in which used	Maximum permissible limit (mg/kg)	Reference
41	Lactic acid, L-, D-, DL-	E270	Complementary foods for infants and young children	2000	
42	Lecithin	E322 (i)	Cooked fish and fish products	GMP	
43	Lysozyme	E1105	Cider and perry	500	
44	Malic acid, DL-	E296	Fruit juice	GMP	
45	Maltol	E636	Dairy-based desserts (flavored yogurt, pudding)	200	
46	Mannitol	E421	Frozen fish, fish fillets	GMP	
47	Methyl cellulose	E461	Fresh meat, poultry	GMP	
48	Monosodium L-glutamate	E621	Fermented vegetables	GMP	
49	Nisin	E234	Cereal and starch-based desserts (tapioca)	3	
50	Pectins	E440	Concentrates for fruit juice	GMP	
51	Potassium alginate	E402	Dried pasta and noodles	GMP	
52	Quinoline yellow	E104	Flavored fluid milk drinks	10	
53	Riboflavins	–	Fermented soybean paste	30	
54	Sodium aluminum silicate	E554	Chewing gum	100	
55	Sucralose	E955	Fermented fruit products	150	
56	Sunset yellow FCF	E110	Fruit-based desserts	50	
57	Tartrazine	E102	Fully preserved, including canned or fermented fish	30	
58	Tertiary butylhydroquinone (TBHQ)	E319	Beverage whiteners	100	

(continued)

Table 7.2 (continued)

S. No.	Additive name	E No.	Type food in which used	Maximum permissible limit (mg/kg)	Reference
59	Tragacanth gum	E413	Dried pasta and noodles	GMP	
60	Xanthan gum	E415	Complementary foods for infants and young children	10,000	
61	Zeaxanthin, synthetic	E161h (i)	Flavored fluid milk drinks	100	

Abbreviation: *GMP* good manufacturing practices

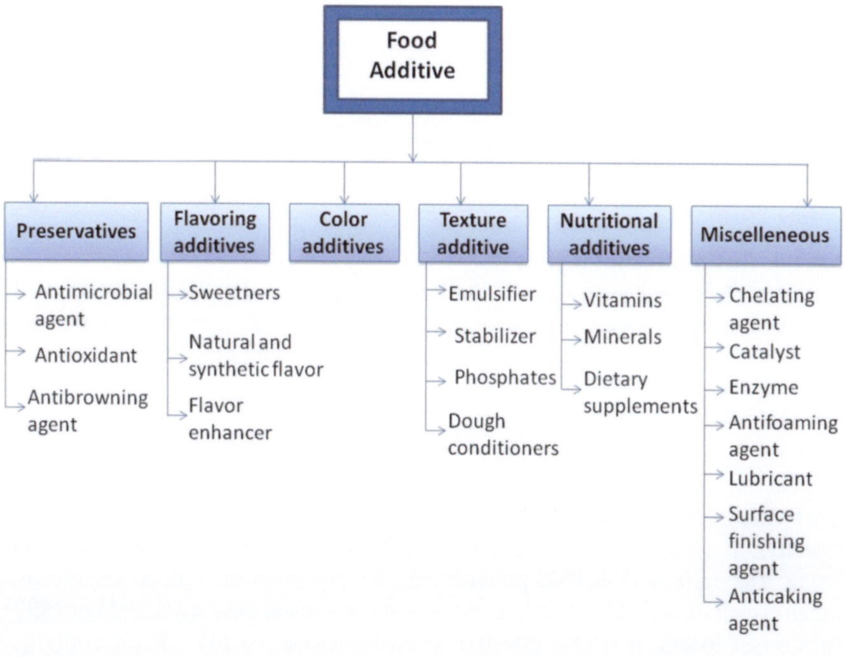

Fig. 7.2 Classification of Food Additive based on their functional properties and uses. (Source: Branen et al. 2001)

used in intended way and generally do not cause any harm to our health and class II are generally artificial chemical substances which are used within prescribed limit like sorbic acid, potassium sorbate, nitrites, etc. These preservatives work on the principles of (1) retardation of microbial growth, (2) slowing down chemical reactions and the changes, and (3) making unavailability of free water (Fig. 7.3).

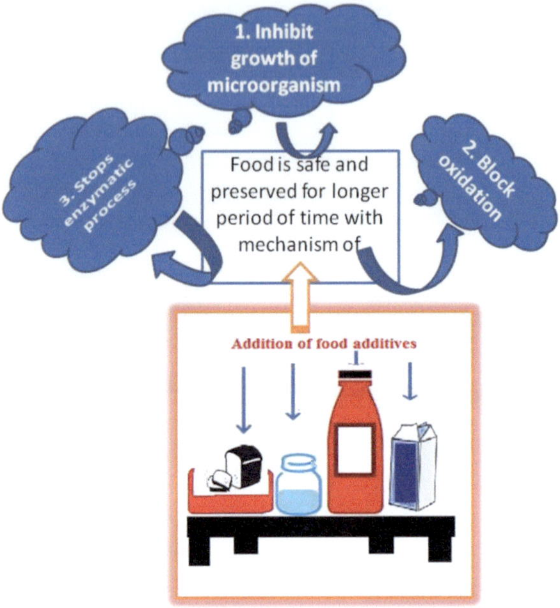

Fig. 7.3 Mechanism of action of food preservatives

7.4.1.1 Antimicrobial Agent

Antimicrobial agents are used in food for retarding the growth of microorganisms in food, and, hence, prevent food spoilage. This antimicrobial component falls under the E and INS numbers ranging from 200 to 290. This subgroup generally contains benzoate and sorbate-like compounds.

7.4.1.1.1 Benzoic Acid/Benzoates

In general, it is used as either benzoic acid ($C_7H_6O_2$) or sodium benzoate ($C_7H_5NaO_2$). The molecular weight is 122.12 and 144.11, respectively. Main objective of using this preservative is to inhibit the growth of yeast and mold (Jay 1995). They are used in food products like bakery products, ciders, mayonnaise, tomato ketchup, soft drinks, jam, pickles, and carbonated beverages (deMan 1999). To a lesser extent, it is also effective against bacteria. At pH 4.5, the antifungal property is likely to be most effective, as the property directly depends upon the undissociated molecule. Benzoic acid is a white crystal with slight odor, soluble in water and ethanol while sodium benzoate is a white odourless powder, soluble in water (Jay 1995). It inhibits the enzyme and works by disrupting the functioning of the cell membrane.

7.4.1.1.2 Propionic Acid/Propionates

Propionic acid ($C_3H_6O_2$), sodium propionate ($C_3H_5NaO_2$), and calcium propionate ($C_6H_{10}CaO_4$) are forms of preservatives in this category with molecular weights 74.08, 96.06, and 186.22, respectively. Their function is more prominent against mold by disrupting their cell membrane and commonly used as a mold inhibitor in bakery products like cake, bread, and in cheese items. The most effective pH range is 5.5 (Doores 1983). Activities of Calcium propionate as a preservative in terms of inhibiting fungal growth in fermented acidic rye bread (sourdough) and sponge cake (intermediated moisture product and alkaline in pH scale) were studied by Suhr and Nielsen (2004). The factors considered were preservatives concentration (0%, 0.003%, 0.03%, 0.3%), a_w (0.80–0.95), and pH (4.4–4.8) and reported that calcium propionates in concentration 0.3% at all conditions except at high a_w of 0.97 and high pH 4.8 worked at its best in inhibiting fungal growth for a period of 14 days except for *Eurotium rubrum*, *Penicillium commune*, and *Penicillium roqueforti*.

7.4.1.1.3 Sorbic Acid/Sorbates

Another mold inhibitor used in food is sorbic acid ($C_6H_8O_2$) and potassium sorbate ($C_6H_7KO_2$) which has molecular weight of 112.12 and 150.22, respectively. It is effective till pH 6.5, after which it becomes inefficient. It is more effective against mold compared to propionate and benzoate (Jay 1995). Compared to the acid form the salt form is more soluble in water and hence used more profoundly in bakery products like cake. Different food categories in which sorbic acid and their salt-like potassium sorbates are used like syrups, jellies, salad dressing, fruit juice, dried fruits, cake, cake mixes, baked food, pies, cheese, fish products, butter, wine, etc. (Saranraj and Geetha 2012). Marín et al. (2002) conducted a study on the intermediate moisture-containing bakery products against the spoilage causing microorganisms like *Eurotium* sp., *Aspergillus* sp., *Penicillium* sp., with respect to certain factors like different levels of weak acid preservatives, pH, and water activity (a_w). The weak acid preservatives were potassium sorbate, calcium propionate, and sodium benzoate at 0–0.3%, pH 4.5–5.5, a_w 0.8–0.9. The study reported that considering all the factors, potassium sorbate showed the best result in controlling the fungal spoilage at the mentioned a_w range at a concentration of 0.3%, but slightly reduced when the pH range reached 5.5. Although it is effective against bacteria however potassium sorbate is most commonly used as fungistatic against a group of fungi from the species *Penicillium* and *Aspergillus niger*, resulting in increasing the shelf-life period of nearly 14 days with uses of 0.5% potassium sorbate as reported by many studies (Sauer and Burroughs 1993; Ray and Bullerman 2001). Sorbic acid acts by disrupting the cell membrane of microorganisms, disrupting bacterial spore germination, and inhibiting enzyme.

7.4.1.1.4 Parabens/Esters of *p*-Hydroxybenzoate

Alkyl esters of *p*-hydroxybenzoic acid (PABA) are known as parabens. Esters may be of methyl, ethyl, propyl, butyl, or heptyl. The alkyl group present in the ester is associated with its solubility. Length of alkyl group is indirectly proportional to its solubility and directly proportional to the antimicrobial activity. Its effect is more pronounced in mold and yeast compared to bacteria. Their inhibitory effects are generally due to their action on membrane transport and mitochondrial processes (Seetaramaiah et al. 2011). It finds application in various products including fruit cakes, fillings, pastries, frozen dairy products, tomato puree, pulp, jelly, gelatine, fruit juice, grain products, soft drinks, seasoning, puddings, candy, etc. (Seetaramaiah et al. 2011). It is effective even at a high pH value of 8 and above. In recent days, parabens are considered as largely used preservatives as they have a wide range of antimicrobial activity, low toxicity, stable in terms of a large range of pH, nonirritating, non-sensitizing, generally soluble in water, and as a whole safer for use. The chain length of paraben's ester groups is directly proportional to the antimicrobial activity.

7.4.1.1.5 Sulfur Dioxide/Sulfites

Sulfur dioxide (SO_2) is a colourless gas and used for many functions such as anti-browning agent, antimicrobial agent, and bleaching agent. In gaseous form, the SO_2 is used in bottled wines and for other products such as dried fruits, molasses, lemon juice, desiccated coconut, and sausages where it is used as a salt in the form of sodium sulfite (Na_2SO_3), potassium sulfite (K_2SO_3), or potassium metabisulfite ($K_2S_2O_5$). It works against bacteria and wild yeast produced during fermentation. But, the use of SO_2 is restricted in foods that contain vitamin B_1, thiamine, etc., as it destroys these vitamins. Sulfide possesses its antimicrobial activity only in free form. Generally, this kind of preservative inhibits enzymes and forms additional compounds and act on bacteria, yeast, and molds.

7.4.1.1.6 Nitrites/Nitrates

Both nitrites and nitrates have antimicrobial property. Since ancient times it is used for curing meat and meat products. The curing salt nitrite and nitrates, e.g., salt of sodium or potassium nitrite/nitrate is commonly used in meat to retain its color. Sodium nitrite ($NaNO_2$) is a powder (white to slightly yellow), soluble in water and ethanol with a molecular weight of 69 g/mol. These can be used directly in food or can be used in solution. Mechanism of action against bacteria especially *Clostridium botulinum* is disruption of cell membrane functioning along with inhibition of the enzyme by reacting with enzymes in vegetative cells, germinating spores, obstructing membrane permeability.

7.4.1.1.7 Nisin

Nisin ($C_{143}H_{230}N_{42}O_{37}S_7$) is the most widely used polypeptide antibiotic with a molecular weight of 3354.25 g/mol created by some *Lactococcus lactis* strains originally present in dairy products. It has properties like resistance to heat, solubility in dilute acids, and majorly effective against strain like *Clostridium botulinum*, a gram-positive bacterium along with the lactic acid producing bacteria by disrupting their cell membrane function. It is applied to canned fruits and vegetables and hard cheeses.

7.4.1.1.8 Hydrogen Peroxide

Hydrogen peroxide is a colorless liquid which makes a homogenous mixture when mixed with water and has a molecular weight of 34.01 g/mol. It has a preserving power over raw milk which is generally used for cheese making and liquid egg white. In aseptic packaging, hot hydrogen peroxide is used for sterilization purpose.

The mode of action as preservatives, generally lactoperoxidase enzyme catalyzes the oxidation reaction where oxidation of thiocyanate ion by hydrogen peroxide facilitated with the production of hypothiocyanite kind weak acid which is bacteriostatic in nature.

7.4.1.1.9 Acetic Acid and Lactic Acid

Acetic acid ($C_2H_4O_2$) has a molecular weight of 60.05 g/mol and has a pungent odor with solubility in water and ethanol to make a homogeneous solution. It is a constituent of mayonnaise and produced during fermentation of pickles. For meat carcasses, acetic acid containing scald water sprays is used. Lactic acid ($C_3H_6O_3$) is hygroscopic in nature which is easily soluble with water and ethanol. It consists of a mixture of lactic acid and lactic anhydride and is a colorless or yellowish liquid. During fermentation process of making yogurt and sauerkraut, lactic acid is produced by the fermentative organism.

7.4.2 Antioxidant

Antioxidants, which cover the E and INS number from 300 to 326, are used to prevent the auto-oxidation reaction. Addition of antioxidants to foods helps in retarding the lipid or vitamin oxidation which leads to the development of rancidity and off flavor. Here another term arises, which is the onset of rancidity which is known as reversion. Generally, the antioxidant helps in dealing with the oxidation process by scavenging the free radical formed during the oxidation process. Antioxidants prevent auto-oxidation which leads to off-flavor formation. It can be classified under two heads—natural and synthetic. Vitamin C and E are the natural antioxidants and examples of some synthetic antioxidants are butylated hydroxyl anisole (BHA), butylated hydroxyl toluene (BHT), ethanol, formaldehyde, etc. The mechanism of action of antioxidant in stopping auto-oxidation is graphically represented in Fig. 7.4.

7.4.3 Anti-browning Agent

Browning is a chemical reaction that is responsible for turning food into brown color. Mostly it is undesirable, e.g., browning of sliced fruits and vegetables. But, sometimes browning in food products is desirable, e.g., brown crust of baked goods. Browning is categorized into two types, i.e., enzymatic and non-enzymatic browning (Fig. 7.5).

Anti-browning agents are the chemicals which are used to overcome both enzymatic and nonenzymatic browning reactions during food processing and preservation. Some examples of anti-browning agents are vitamin C (E300), citric acid (E330), sodium sulfite (E221), etc. In food processing industry, many canned products are dipped in different solutions (like sugar, brine, etc.) to avoid the browning reactions in foods

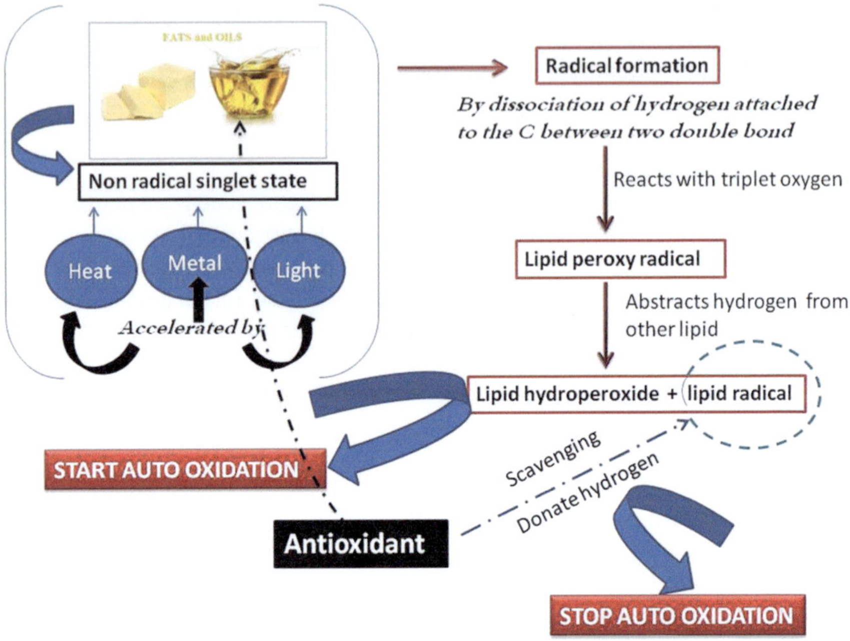

Fig. 7.4 Mechanism of interaction of antioxidant as food additives in scavenging oxidation

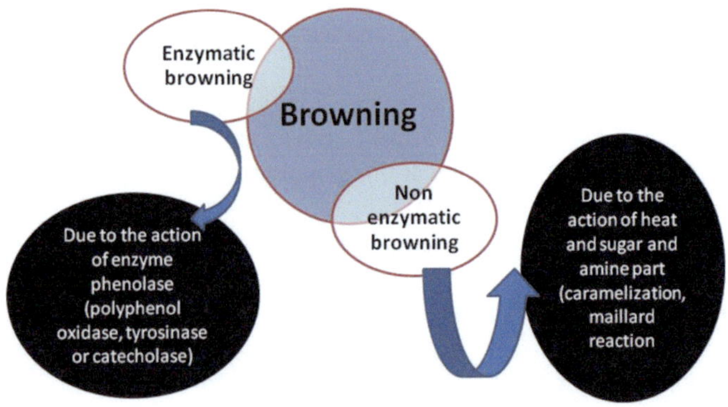

Fig. 7.5 Types of browning

7.5 Flavouring Additives

Flavoring additive is a substance which has predominantly odor-producing properties and which possibly affects the taste (Council of Europe). Generally, it is a chemical composition of a single or mixed blend, either natural or synthetic in origin. It can be summarized as follows:

7.5.1 Sweeteners

Sweetness is measured in relation to sucrose (more specifically reference sugar) and is the most important aspect of sweeteners. Based upon origin or intrinsic properties, sweeteners are classified. However, governing bodies like EFSA (European Food Safety Authority) of the EU (European Union) follow the classification based on intrinsic properties (nutritive value, sweetening power, etc.). Nutritive (mannitol (E421), sorbitol (E420), etc.) and intensive (saccharin (E954), glycyrrhizin, etc.) are the two-classification based on intrinsic properties. Based on origin, sweeteners are classified as natural (monellin, neohesperidine dihydrochalcone-E959, etc.) and synthetic (thaumatin-E957, dulcin, etc.). Sugar is the most commonly used sweetening agent in the world. Type II diabetes, breast cancer, colon cancer, obesity, etc., are some examples of diseases caused by sweeteners (Carocho et al. 2017). According to the United States Food and Drug Administration (FDA), aspartame (E951), acesulfame K (E950), cyclamates (E952), sucralose (E955), tagatose, etc., are considered "generally recommended as safe" (GRAS). Alencar et al. (2017) reported that the addition of sweeteners (E950, E955, and stevia) is highly acceptable to the consumers due to improved sensory attributes (taste, aroma, texture).

7.5.2 Natural and Synthetic Flavor

To substitute the flavor of food, mixture of several chemicals, i.e., natural and synthetic flavors are used. Generally, these chemical mixtures mimic the natural flavor. Nevertheless, besides performing as a flavor additive, the group of chemicals also functions as antimicrobial, gelling agent, etc. More than 1700 flavor compounds (natural and synthetic) are available to impart flavor to foods. Acidulants (more specifically, organic acids) perform important functions by lowering the overall pH of food. Fumaric acid (E297), acetic acid (E260), lactic acid (E270), malic acid (E296), etc., are some common examples of acidulants used in food.

7.5.3 Flavor Enhancer

The combined sensation of taste and the olfactory perception of foods are known as flavor. Flavor enhancers are used to magnify and enhance the flavor of food. However, flavor enhancers do not contribute their own flavor. Monosodium glutamate (MSG: E621), glutamic acid (E620), disodium inosinate (E631), etc., are the most commonly used flavor enhancer. MSG is a salt of glutamic acid (naturally occurring amino acid) responsible for producing the umami flavor (savory flavor). It enhances the natural flavor of food solely or in combination with disodium guanylate or disodium inosinate. However, glutamic acid behaves like an excite toxin which can destroy nerve cells by excessive stimulation. Sanabria et al. (2017) reported that MSG leads to lipid peroxidation of food stuffs and overweight development in Chinese adults.

7.5.4 Color Additives

Color additives are the compounds which are used in food to confer or alter colors to food. The main objective behind the color additive application is to improve the appearance and to increase the attractiveness of food towards the customers. Different natural and synthetic color additives are used in the food industry since ancient times which are categorized in E system from E100 to E180 and in the INS system from 100 to 182. Calcium carbonate (E170) is the only dye with *quantum satis* status and confers white color to food. Color additives are grouped into five categories: azo compounds, xanthenes group, indigo colorants, chinophthalon derivatives, and triarylmethane group.

7.5.4.1 Azo Compounds

Azo compounds are the compounds which contain azo group and it exhibit different colors. The functional group R–N=N–R′ (R and R′ can be either alkyl or aryl group) is common for all azo compounds. The most commonly used azo compounds are tartrazine (E102), sunset yellow (E110), allura red (E129), etc. However, the azo compounds containing N=N and aromatic rings form aromatic amines, some of these are mutagenic, toxic, and carcinogenic in nature. Moreover, some studies have revealed that tartrazine is toxic to human lymphocytes (4 mM) and binds to DNA (Carocho et al. 2014). However, Tanaka et al. (2008) reported that tartrazine is safe to consume within acceptable limit of daily intake.

7.5.4.2 Xanthenes Group

Xanthene group comprises of Erythrosine, eosines, rhodamines, and fluoresceine. Erythrosine is a polyiodinated compound. It alters childhood behavior and thyroid function as it contains a higher amount of iodine. Moreover, a study has revealed that it is toxic to human lymphocytes, inhibits protein-protein binding, and binds DNA (Carocho et al. 2014).

7.5.4.3 Indigo Colorants

Indigo dye which is known as FD&C Blue No. 2 in the United States was extracted from the shrub *Indigofera tinctoria*. From the indigo dye the indigo colorants are derived, but, now-a-days indigo colorants are chemically produced.

7.5.4.4 Chinophthalon Derivatives

Sodium monosulfates, disulfates, and trisulfates are mixed to prepare quinopthalone synthetic dye (more specifically quinolone yellow: E104) chemically. However, asthma, rashes, urticaria, hyperactivity, etc., can be caused by this compound. It has been reported that this compound alters the conformation of bovine serum albumin and cause skin eruption (Carocho et al. 2014).

7.5.4.5 Triarylmethane Group

Triphenylmethane is the backbone material of triarylmethane group and it produces different compounds like brilliant black (E151), brilliant blue (E133), fast green (E143) etc. These compounds are slowly absorbed by the body. Therefore, 95% of these compounds are present in faeces. These compounds are used in lollipops (brilliant blue, patent blue etc.). However, these are dangerous to children health (Carocho et al. 2014).

7.5.5 Texturizing Agents

Texturizing agents are the substances which are added to food commodities to alter the texture and consistency of the food. An optimum amount of these substances is added to food commodities so that, they can possess proper texture and consistency to get easily accepted by the consumers. Generally, texturizing additives are used at a greater amount than other food additives. Therefore, texture additives impart creaminess, viscosity, thickness, transparency, puffiness, etc. As a consequence, texture, consistency, and other properties like mouth feeling are improved. Texture additives include emulsifiers, stabilizers, phosphates, dough conditioners, etc., which plays an important role in modifying the texture of foods.

7.5.5.1 Emulsifiers

The worldwide production of emulsifiers is about 0.5 million metric tons approximately. Emulsifiers are substances which help to form a fine mixture of two immiscible liquids in which one liquid is dispersed over the other liquid. Emulsifiers are used to hold the mixture strongly so that the constituent materials do not get separated from each other. They show amphiphilic characteristics, i.e., both lipophilic and hydrophilic characteristics. Due to amphiphilic nature, emulsifiers can hold both oil phase and aqueous phase at the same time. Triglycerides (natural fats and oils) are used to form food-grade emulsifiers. Moreover, sucrose, glycerol, and sorbitol are considered as the other important materials for emulsifier formation. Etherification, esterification, and interesterification are the most commonly used methods for food-grade emulsifier formation. Food emulsifiers are classified as nonionic, anionic, cationic, and zwitter ionic based on their fundamental nature. Non-ionic is the most prevailing type of emulsifier used in food application (Norn 2015). Emulsifiers are also classified as natural (lecithin, mono, and diglycerides) and synthetic (glyceryl monostearate, carboxymethyl cellulose). Emulsifiers perform numerous functions like foam stabilization in cakes (propylene glycol esters) by the rapid movement of surface-active compound towards air-water interface and forming a fluid-like film (α-Gels: hexagonally packed lamellar crystalline mesophases consisting of lipid bilayers) which stabilizes the gas cells. It also plays a role as an anti-sticking agent in candies (lecithin) by wetting the contact surface of emulsion, viscosity modifier in chocolate (lecithin) due to change in concentration subject to emulsion concentration, dispersion stabilizer in peanut butter (mono/diglycerides) and it also controls fat agglomeration in ice cream (polysorbate 80) due to prevention of droplet coalescence by imparting static charge to the droplets which keep them away from each other due to repulsion or by creating a physical barrier (Hasenhuettl 2008). Therefore, as a consequence of those functions emulsifier results in the modified texture of food which give proper mouthfeel.

7.5.5.2 Stabilizers

The application of stabilizer in foods has started more than 60 years ago. Stabilizers are substances which increase stability and thickness of food, and, therefore, help in maintaining the emulsion and retention of its physical characteristics. They stabilize the emulsion by either adsorbing to the outer surface of oil droplets or by increasing the viscosity of water phase. Their stabilizing effect increases the viscosity of the emulsion system which reduces the kinetic motion of the droplets responsible for flocculation and coalescence. Therefore, it helps to improve the structure and texture of the food when added. Generally, small amount of stabilizers are added to food as it exasperates the activity of emulsifier. Stabilizer allows the flavoring compounds to disperse homogeneously to give food having uniform nature. The main functions of stabilizer are (1) maintaining optimum consistency of food and (2) holding ingredients bounded by emulsifiers. Xanthan gum (E415), agar-agar, pectin, gelatin, alginic acid, etc., are some examples of food stabilizers. In dairy industry stabilizers (carrageenans, alginates, gelatins, and gums) are used to give kinetic stability to the food emulsion to exhibit good textural properties during the shelf life of the products. High water binding capacity of stabilizers gives thickening and gelling effect to dairy products (Lal et al. 2006). Hydrocolloids are another example of a stabilizer which is used to improve the texture as well as mouthfeel of food (organoleptic properties). It retards droplet creaming and particle sedimentation caused by bulk viscosity effects, more specifically by enhanced viscosity and gel characteristics (Kumar and Mishra 2004). Hydrocolloids also inhibit aggregation by steric or electrosteric forces and may get absorbed onto the surface of the droplets.

7.5.5.3 Phosphates

Phosphates are polyvalent ions which can form structures containing ring and chain structures (Lampila and Godber 2002). Ring phosphates and chain phosphates are the two basic forms of phosphates available where the former is commonly used in food. However, ring phosphates are generally used in food. Phosphates are generally used to adjust pH, increase water holding capacity, enhance shelf life, improve texture and other sensory attributes (tenderness, juiciness, etc.). Therefore, it helps to improve the overall quality of food. Alteration in ionic charge distribution inside the system is responsible for altering quality attributes. Phosphates are used to modify the texture of protein or starch-containing foods. They are also used to stabilize emulsion in meat and dairy products. Sodium monophosphate, disodium phosphate, sodium diphosphate, and sodium tripolyphosphate are some examples of phosphates generally used in meat products. Moreover, diphosphates are mostly the functional phosphates among all the phosphates. They find application in improving the quality of seafood. Nevertheless, phosphates help to fulfill the requirement of phosphorus in the human body. Inorganic phosphates are effectively used in meats, cheese, canned fish, and baked products and are generally regarded as safe (GRAS) by the United States Food and Drug Administration (FDA).

7.5.5.4 Dough Conditioners

Dough conditioners are also called as dough improvers. Dough conditioners are defined as substances which improve the processing of dough and the quality of baked products. Dough conditioners perform functions such as improvement of crumb structure and texture, enhancement of volume of baked goods, improvement of baked products' sliceability, and improvement in shelf life.

Dough conditioners function as:

1. Emulsifiers by consolidating the gluten and add to its tolerance while being handled. Because of their water- and fat-soluble regions, they are able to make a more balanced environment in the dough which leads to a more uniform dough and desirable crumb consistency.
2. Enzymes as it breaks down certain molecules in the dough and feed the yeast to expedite the fermentation process and gas production.
3. pH regulators by increasing the growth of yeast to expedite the rising process.
4. Reducing agent by breaking down the protein network in dough and restructure the gluten which leads to less mixing and proofing time requirement.
5. Oxidants encourage gas retention by strengthening the dough through disulphide bonding. Steroyl-2-lactylate and sodium silico-aluminate are examples of dough conditioners which modifies texture in terms of enhanced crumb softening and dough stability. Carboxymethyl cellulose and sodium caseinate are used as an additive which helps to increase the extensibility of chapatti (Gujral and Pathak 2002).

7.5.6 Nutritional Additives

The nutritional additives are the compounds which enrich the food stuff and increase its nutritive value. Nutritional additives have gained more importance with increasing concern of the public towards a healthier life due to their functional activities and nutraceutical behavior leading to improvement of health. Vitamins, polyphenols, fibers, fatty acids, amino acids, etc., are some examples of nutritional additives. Nutritional additives can be classified as natural and synthetic depending upon the origin.

7.5.6.1 Vitamins

The vitamins present in nature play a distinct role in terms of physical, chemical, and biological functions in the human body. However, all vitamins are highly reactive in nature. Moreover, all vitamins are not suitable to be used as food additives. Only five vitamins have found application in the food industry. These five vitamins are as follows: ascorbic acid (vitamin C), carotenoids (provitamins A), DL-α-tocopherol (vitamin E), nicotinic acid and its amide, and riboflavin and its phosphate sodium salt (vitamin B_2). The principal application of vitamin C is in soft drinks, bread manufacturing (dough improver), and meat processing to improve the nutritional quality. Fucoxanthin is the most abundant carotenoid found in marine algae. Carotenoids are used in the food industry as a colorant, especially in fruits, candy, jelly, milk products, etc. Niacin is used in meat processing in conjunction with ascorbic acid and nicotinic acid is used as an acidifying agent in soft drinks. Vitamin E is basically odorless and yellow in color, viscous and oily in nature. It is a natural food antioxidant. Vitamin B_2 is less soluble in water and it is used as coloring agent for sweets, sugar-coated products, salad cream, etc. (Counsell 1993).

7.5.6.2 Minerals

Naturally, food is an ample source of minerals. However, some minerals (iron, calcium, copper, magnesium, and zinc) are not readily available in all types of food. Therefore, there is always a need of fortifying those minerals in food. Salt iodization is very important to overcome iodine deficiency in iodine-deficient areas (Belitz et al. 2009). Calcium salt may function like texturizing agent. Moreover, addition of minerals into food provides significant health benefits to the human being. Kwak et al. (2003) reported the microencapsulation of iron which reduces the astringent and metallic taste of iron-fortified milk. Romita et al. (2011) reported that coating is needed while fortifying two minerals (iron as coated ferrous fumarate salt and iodine as iodized salt) together.

7.5.6.3 Dietary Supplements

Dietary supplements are also known as food or nutritional supplements and belong to a special category of nutritional and health products and include OTC (over-the-counter) formulations. The available research publications on dietary supplements in food show their important role in prevention of some diseases like issues related to iron and vitamin deficiency, bowel syndrome, stress, hypertension, mental retardation, etc. Consumption of a can fulfill the nutrient (vitamins, minerals) requirement as a supplement of normal diet. These dietary supplements are sometimes referred as nutraceuticals or functional food and are available in the form of capsule, powder, tablet, or liquid. These are generally used as complementary or alternative medical treatments. Dietary supplements are categorized as follows: vitamins (vitamin C, vitamin E, B vitamins, vitamin A and beta-carotene, niacin, folic acid, etc.), minerals (calcium, magnesium, zinc, iron, etc.), herbs and botanicals (echinacea, garlic, ginseng, ginkgo biloba, etc.), sports (creatine, amino acids, protein formulas, fat-burners, ribose, etc.), and specialty (melatonin, chondroitin, pro- and pre-biotics, colostrum, enzymes, etc.) (Augustin and Sanguansri 2012). The primary function of these kinds of additives is therapeutic or physical condition maintaining performance (Branen et al. 2001). The addition of flavors like sweeteners aspartame, saccharine, etc., especially the food acids like citric acid, malic acid have an important role in masking the bitterness, increasing the intensity of flavor and palatability. Improving palatability with the addition of the food acids is due to their ability to chelate metal ions and act as a shield in preventing the metal ion from reaching the flavor receptors (Motekaitis and Martell 1984; Branen et al. 2001). Amino acid methionine, glutamic acids, and cysteine have odor, bad taste, and are acidic in nature which is not acceptable from consumer point of view and improvement for the same is done by derivatization of the side chains of the component which blocks the undesirable flavor and taste producing component (Fürst 1994). To slow down or to change the diffusion properties or to control the release in targeted part, microencapsulation is done. The dietary additives are sometimes coated with this technology to prevent the release of unpleasant taste and flavor at the starting point. Same way to inhibit the bitter nature, inhibitors are used which inhibit the respective properties by entrapping the nutritive compound inside the inhibitors. Mostly used inhibitors are cyclodextrin, maltol (E635–E636), etc., which inhibit the unwanted property of amino acid and B vitamin complex, respectively (Pazur 1991; Murray et al. 1995). The ecosystem of the bowel plays an important role in a healthy lifestyle. For many years prebiotics and probiotics are gaining momentum. There are many bacterial strains which are probiotic and promote bowel health, e.g., *Lactobacilllus acidophilus, Lactobacillus planterum, Bifidobacterium lactis*, etc. These organisms are widely used as dietary food additives which have general therapeutic value when added to functional foods as one of the compositions.

7.5.7 Miscellaneous Additives

There are numerous other chemical substances, which although do not belong to any specific category, are being added to food for their unique functions. Therefore, compared to other additives their use is limited. Chelating agents, catalysts, enzymes, antifoaming agents, surface finishing agents, lubricants, and anti-caking agents are some examples of processing aids of additives.

7.5.7.1 Chelating Agents

Chelating agents are the chemical substances which performs chelation, i.e., formation of stable water-soluble complexes (chelates) with free metal ions (calcium, magnesium, etc.). Chelating agents release metal ions in a controlled manner for nutritional purposes or for gelation of thickeners. Therefore, it eliminates some undesirable activities like insolubility of free metal ions in food system, color compound formation, and food components degradation which results in nutrition loss, rancidity, discoloration, and precipitation (Nauta 1991). Chelating agents bind with metal ions (calcium, magnesium, etc.) which influence the activity of enzymes as cofactors. Therefore, chelating agents protect food products during processing and storage from deterioration due to enzymatic reactions. Stability constant (K), which is defined as the ratio of chelated to unchelated metal ions, is a performance measure of chelating agents. Some examples of chelating agents are glycine, EDTA, citric acid, tartaric acid, etc. Moreover, the United States Food and Drug Administration has permitted maximum levels (25–800 ppm) of EDTA doses.

7.5.7.2 Catalyst

Catalysts are substances which increase the rate of chemical reactions in biological systems by lowering the energy needed to carry out the overall process. Enzyme is a biological catalyst which is used as a food additive. A little amount of enzyme can increase the rate of reaction million times subjected to some specific conditions (temperature and pH). It improves baking properties, brewing attributes, milk fermentation, food digestibility, nutritional value of food, etc. $AlK(SO_4)_2$ is a catalyst which is commonly known as alum and is used as a food additive.

7.5.7.3 Enzyme

Enzymes are the biological substances which are protein in nature, solely related to food and considered as an important element in food processing. Activity of enzymes depends upon system properties such as temperature, pH, concentration of substrate, inhibitors, and residence time. Enzymes are grouped into six major categories such as (1) oxidoreductases, (2) transferases, (3) hydrolases, (4) lyases, (5) isomerases, and (6) ligases depending upon their catalyze reaction.

Enzymes have the following applications in food systems:

1. To improve the baking properties (accelerated fermentation, increased loaf volume, improve crumb structure and crust color, increased yield, reduced dough mixing time, improved gluten properties).
2. Glucose production, brewing (sugar from starch conversion, removal of polyphenolics).
3. Nutritional quality improvement by degrading the protein followed by assimilation of amino acids.
4. Amylases convert starch to dextrins, oligosaccharides, and sugar maltose. Maltose produces fermentable sugar for the yeast. Hence, fermentation is accelerated. Darker crust color, better flavor, improved loaf volume and symmetry, and extended shelf life are other benefits.
5. Proteases break down the protein like gluten, and thus increase dough extensibility and flow, improve gas retention, reduce mix times, and produce a darker crust as well as a better crumb grain and texture, etc.

7.5.7.4 Antifoaming Agents

Antifoaming agents are the chemical substances which are used to control undesirable foaming. Antifoaming agents possess high surface activity responsible for lowering the surface tension of local film when its micro droplet contacts liquid film. The local area of low surface tension will lose elasticity, self-restoring ability, and extends all around; thus, causing rupturing of the liquid film. Antifoaming agents also play an important role, they can also reduce the surface viscosity of film, enhancing the drainage velocity, and raising the diffusion velocity of gas, finally leading to the rupturing of the liquid film. At industrial level, undesired foam formation reduces the capacity and efficiency of the equipment. Moreover, it also hampers the quality of the final product. An antifoaming agent must possess attributes like, low solubility, low surface tension, quick dispersibility, inert in nature, left no residue, etc. However, the selection of antifoaming agent depends on the system properties (chemical nature of foam-forming agent, viscosity, pH, temperature, concentration, etc.). Antifoaming agents are categorized as aqueous (emulsions) and non-aqueous (fluids and compounds). AF70, AF72, AF75, SF18-350 are some examples of antifoaming agents. Silicone antifoams are mostly used to overcome process and product-related problems (Zotto 1991).

7.5.7.5 Lubricants

Generally, food-grade oils or lipids are used as lubricants in food processing operations. A very common example is oil which is sprayed inside the bread pans so that baked bread loaves can be easily removed from the bread pans without any structural damage (Maga 1995). The greasing of bread pans done at home scale also performs the same function.

7.5.7.6 Surface Finishing Agents

These are the substances which are added to food to make the surface glossy, retain the original color and appearance, inhibit discoloration of other foods, etc. Therefore, these help to maintain the palatability of foods. Waxes, polishes, surface coatings, and food-grade glazes are used as surface finishing agents. These are commonly applied in fruits and vegetables like apple and cucumber (Maga 1995). Surface finishing agents retards moisture loss of the food during transportation and storage by providing a physical barrier to the surface coating materials. As a consequence of this, it improves the shelf life of food and reduces the chances of staling within a short period of time.

7.5.7.7 Anti-caking Agent

Anti-caking agents are the substances which are added to powders and granulated materials to keep them free-flowing and prevent lump formation during storage and use. Anti-caking agents absorb water from the granulated materials or form a water repellent coating over the granulated materials therefore the granulated materials retain their free-flowing nature, inhibit lump formation and agglomeration (Maga 1995). Anti-caking agents can be of two types, natural and synthetic. Silicon dioxide, magnesium carbonate, calcium silicate, etc., are some examples of synthetic anti-caking agents; whereas, corn starch and magnesium silicate are natural anti-caking agents. Calcium silicate is a common anti-caking agent which is usually added to table salt.

7.6 Food Additive Intake Assessment

Evaluation of the safety of food additives is done by both international and local organizations. Listed below are three major goals which are accomplished by intake assessment of food additives:

1. Chemical intakes are controlled so that they can be linked to the ADI values.
2. Identification of the consumers who may be at risk by the consumption of food additives nearer or higher than ADI values.
3. Regulatory bodies get the information provided by intake assessment which helps them in reassessing the food additive regulation when there is a high intake of additives in a few or all groups.

The main aim of food additive intake assessment is to safeguard the health of consumers and provide assistance in the development of food additive regulations (Branen et al. 2001). Figure 7.6 depicts the methods adopted for dietary intake assessment where one phase method, two phase method, and food consumption data are considered important to assess the same. Exposure assessment of the food additives is of great importance as the concern regarding excessive use of chemicals in the production and processing of food is increasing in recent times. To verify whether these food additives can cause health hazards after consumption, their assessment is needed and their intake levels need to be checked. Identification of the food additive characteristic, relevant food consumption data survey, determining and attaining the concentration data of food additive, and picking up the most appropriate exposure analysis methods are crucial steps that must be given attention in designing a study of food additive intake assessment. Exposure to food additives is analyzed by various methods such as per capita approach and studies of total diet. These methods were basically used at the population level. Food consumption data are collected individually using four dietary survey tools like 24 h dietary recall,

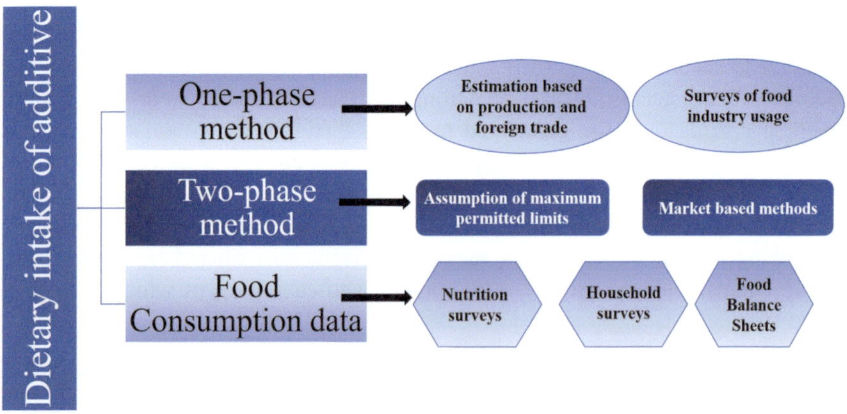

Fig. 7.6 Methods adopted for assessment of dietary intake of additive

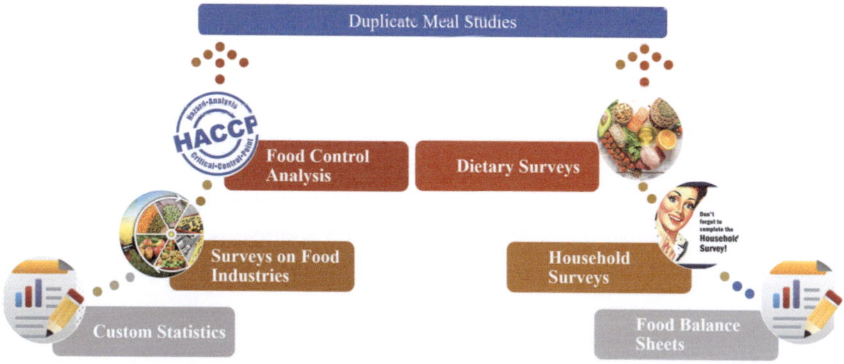

Fig. 7.7 The methods used for assessing the intakes of food components and food additives. (The combination of the methods from top-down indicate increasing cost and accuracy in intake assessment, Source: Salminen and Tahvonen 2001)

dietary history, food frequency questionnaire (FFQ), and Food records. Among all these methods, dietary records are considered to be the gold standard for the survey of diet generally. There is no method which may be called as best for the dietary assessment.

The decision of selecting a method mainly depends on the purpose of investigation, the food of prime concern, the requirement for group versus individual data, the need for absolute intake versus relative intake, population characteristics (age, sex, education, motivation, culture, etc.), the time frame of interest, the level of specificity needed for food description, and the availability of resources. Dietary data on food consumption have been used mostly. Meanwhile, for the studies on food additives, 24 h recall, and semi-quantitative FFQ have been used in most of the studies. The variability of daily decides the number of days for which a recall/record should be done. In case of high variability, the days of the survey are increased to determine the food products and its variety consumption.

In many countries for carrying out large-scale surveys, total diet study for single food items and market-based approaches are followed. Retained fidelity card scheme is one of the modern methods which is easy to use but has a drawback of low precision. They are helpful in population-based studies and for examining the health risks associated with the consumption of additives above ADI. Figure 7.7 shows the scheme of tools used for the assessment of ADI.

Possible study methods that can be used for estimation of the intake of food additives are the following:

7.6.1 Per Capita Estimates

It requires an estimate of the total amount of food additives entering in food supply in a given country or a region. This amount divided by the number of consumers gives a figure for the average intake. The calculation is simple but it gives no information about the distribution of intake within the population. The method is used for monitoring trends in additive usage. For example, it is possible to determine whether the use of an additive is increasing or decreasing by repeating the usage survey at regular intervals. By using sales data from *Gfl* (*Gesellschaft fur Konsument markt und absatzforschung*) it is possible to estimate the total consumption of a certain food product. With these available data, potential high consumption can be identified and it indicates whether the ADI of a specific food additive is exceeded or in right amount (Ilbäck and Busk 2000).

7.6.2 Duplicate Diet Studies

A selected group of individuals are asked to retain duplicate portions of all the food they consume over the study period. These samples are then aggregated and analyzed for the substance(s) of interest. This method requires a high degree of cooperation and considerable effort from the participants. Practically this study can only be undertaken over short periods. It is especially useful for studying specific population groups assumed to have a high intake of the substance(s) of interest (Ilbäck and Busk 2000).

7.6.3 Calculations Using Data on Food Consumption and Additive Occurrence

The most common approach to estimate the intake of food additives is to combine the information on the occurrence of the additive and the consumption of the specific food that contains the additive. Additive occurrence can be obtained from several sources, such as maximum permitted levels, technological levels, manufactures data, and analytical data. Food consumption, information can be obtained from trade data (including market research), national diet studies at the level of the household, national diet studies at the individual level, whether by 24-h recall, diary record, dietary history or food frequency questionnaires, as well as from other more limited dietary surveys concentrating on population subgroups, geographic regions and/or specific foods (Ilbäck and Busk 2000).

7.6.4 Biological Marker of Exposure

Saccharin, for example, is not metabolized in the body and is excreted unchanged in the urine. It is therefore possible to estimate the intake by measuring the amount present in urine over a 24-h period since it will approximately be equal to the amount present in that individual's diet. The parent compound, metabolites, or other biological markers of exposure can be used in these studies (Ilbäck and Busk 2000).

7.6.5 Budget Method

Originally developed in Denmark, it is a simple, inexpensive way of assessing proposed maximum use levels of food additives to ensure that the ADIs are not exceeded. The method is designed to cover the worst-case scenario, and, therefore, exaggerates potential additive intake. The method is applicable for assessing potential maximum intake in countries with different eating habits. The budget method is however unsuitable for predicting what consumers are actually ingesting. Where more precise information on intake is considered necessary, more precise calculations should be performed. This method is valid internationally as it is based on the fact that there is a physiological upper limit to the amount of food and drink, and thus of food additives, that can be consumed each day. A further assumption is that only a certain proportion of the diet is likely to contain food additives. Using appropriate estimates for the upper limits of food and drink, which can be consumed each day, the budget method allows regulators to allocate the ADI of a food additive between food and beverages and also to set maximum levels of use (Ilbäck and Busk 2000).

7.6.6 New Methods Under Development

There is a future need for new and refined methods. Stochastic modeling is a method in which there is a growing interest by agencies (personal communication J. Lambe). The US Environmental Protection Agency (EPA) has stated that probabilistic analysis techniques can be viable statistical tools for analyzing variability and uncertainty in risk assessments (Ilbäck and Busk 2000).

7.7 Testing Methods

Compounds of chemical origin can cause risk to the consumer's health. So, in order to use food additives, we need a rigorous food safety policy. The growing utilization of food additives can cause allergies, diabetes, obesity, and metabolic disorders. Analytical techniques are used for the identification and quantification of different types of food additives in food products. Hence, it is important to use those techniques for evaluation which provide reliability, selectivity, fastness, sensitivity, and are less expensive (Wrolstad et al. 2005). For an efficient chemical analysis of the food additives, there is a need to select an appropriate instrumental technique and evaluation of sampling and pretreatment methods (Mitra 2004).

The sampling protocol is developed through statistical techniques to ensure the selection of sample and also for the reliability of analytical results. The sampling protocol depends on the size, variability, and cost of analysis of each sample (Nielsen 2014). Pretreatment like mixing, homogenization, dilution, centrifugation, distillation, simple solvent extraction, supercritical fluid extraction, pressurized-fluid extraction, microwave-assisted extraction, and Soxhlet extraction of food samples should be performed in order to remove the interfering compounds, such as fats, oils, lipids, proteins, carbohydrates, salts, pigments, emulsions, and turbidity, as it can interrupt in evaluation (Wrolstad et al. 2005). The validation of analytical data in food samples is done by using analytical parameters such as linearity, accuracy, precision, sensitivity, range concentration, limit of detection, limit of quantitation, and robustness (Wrolstad et al. 2005). The analytical technique used for analysis depends on the type of sample. The following are the main analytical methods used in the analysis of food additives:

7.7.1 Spectroscopy Techniques

The interaction between electromagnetic radiation and matter is known as spectroscopy technique. These techniques are used to measure the amount of radiated energy either absorbed or emitted by the analyte. The main methods used by the food industries are ultraviolet/visible radiation spectroscopy and infrared spectroscopy (Wrolstad et al. 2005). The UV-VIS is a technique that measures (from 190 to 800 nm wavelength) fraction of radiation transmitted through the sample. The instrumentation of UV/VIS spectroscopy contains a stable source of radiant energy, a wavelength selector that isolates a limited region of the spectrum for measurement, one or more sample containers, and a radiation detector which converts radiant energy to a measurable electrical signal. According to Beer's law, the transmittance of radiation is related to the absorbance and concentration of absorbing species (Christian et al. 2014).

UV/VIS spectroscopy is used in the determination of different food additives and this method is considered suitable for online measurement because of its efficiency, low cost, environment-friendly approach, and operational simplicity. During the analysis of food additives, some other chemical compounds which are present in the sample absorb the electromagnetic radiation and promote undesirable selectivity. That is why pretreatment of the sample is required for selectivity and efficiency (Skoog et al. 2013).

IR spectroscopy is the measurement of the wavelength and intensity of the absorption of infrared light by a sample (Putzig et al. 1994). It is widely used to determine the functional groups present in solid, liquid, and gaseous samples. For qualitative analysis, the near-IR is used where the absorption of radiation measures from 0.8 to 2.5 μm wavelength while for the quantitative analysis, mid-IR spectroscopy is used and the radiation measures from 2.5 to 15 μm wavelength.

NIR spectroscopy is a spectroscopic method that uses the near-infrared region of the electromagnetic spectrum (from 780 to 2500 nm). For quantitative analysis, the near-IR spectral region (0.7–2.5 μm, equal to 700–2500 nm) are used. NIR spectroscopy has the ability to measure the composition of solid food products directly by the use of diffuse reflection techniques. NIR spectroscopy is a rapid and nondestructive method used for the evaluation of the quality of food. The instrumentation of NIR spectroscopy requires statistical techniques and specific calibration for each product to measure the quantity of different types of food additives. High initial cost of the instrument is a disadvantage.

Fourier transform infrared (FTIR) spectroscopy is used to identify organic materials. This is a nondestructive process and it gives a rapid and accurate data. With the help of different kinds of software algorithms, this technique allows data analysis which increases its use in food industries (Babushkin et al. 2016). It has the same disadvantage as NIR spectroscopy, that is, initial cost (Amir et al. 2013; Van de Voort 1992).

Raman spectroscopy, a spectroscopic technique used for the detection of rotational, vibrational, and other states in a molecular system, is capable of examining the chemical composition of the material. This technique is based on the scattering of light. For the qualitative and quantitative estimation, Raman spectra can be used and this can be formed by inelastic collision between the incident monochromatic radiation and sample molecules. This spectroscopic technique releases radiation in Near-IR wavelength, and the incident light measures the scattered light intensity at different frequencies. Because of its ease in sample preparation, this analysis is fast and cheap (Bumbrah and Sharma 2016; Nielsen 2014). Fluorescence spectroscopy is a device which gives us information at a nanoscopic level with exceptional sensitivity and is used for examining and receiving data on the structure and properties of materials with high precision. This is a technique based on the emission of electromagnetic radiation by chemical compounds. The fluorescence spectra are made by the excitation and emission wavelength and will depend on the chemical structure of the target compound. The chemical compounds which have rings and are made up of rigid chemical structure do not permit the release of energy absorbed by the paths and emit electromagnetic radiation with substantial intensity (Christian et al. 2014). The instrumentation of fluorescence spectroscopy is the same as UV/Vis spectroscopy but the sensitivity of fluorometry will be 10–100 time better than UV/Vis spectrometry. The high sensitivity of fluorescence makes it one of the best methods available for trace analysis (Itagaki 2000). The use of this technique is decreasing in the food industry because it is used only for the analysis of rigid chemical structure compounds (Albani 2006).

7.7.2 Chromatographic Techniques

Chromatography is a separation technique used to separate the molecules from a mixture of components. This is used for analysis, isolation, and purification. This is done by the distribution of sample between two phases; one phase is present in form of a porous bed, layer, or film which is generally immobile (stationary phase) and the second one is a liquid (mobile phase) that permeates through the stationary phase. Almost all chromatographic system contains the transportation of solute zones in the mobile phase (Poole 2000).

The classifications of chromatographic techniques are based on the physicochemical principles which are involved in the separation of components. In adsorption chromatography, the components of the sample are adsorbed on the solid stationary phase. On the other hand, in partition chromatography, the stationary phase is liquid and supported on solid. In ion-exchange chromatography the stationary phase is an ion exchange support. The exclusion chromatography separates solvated molecules by penetrating porous pockets and passages in the stationary phase. These separation mechanisms operate in especially two types of main chromatography which is used in food additive analysis, gas chromatography, and liquid chromatography (Jonsson 1987).

Gas chromatography is a versatile technique which uses inert gas as a mobile phase and either a solid or immobilized liquid packed in a closed tube, as a stationary phase (Nielsen 2014). The instrumentation of GC comprises of pressure and flow control regulators, injection port of the samples, an inert gas supply, detector, column, and a data recording and processing system. The column oven is capable of maintaining a constant temperature. The simplest method of injecting sample into GC inlet is through micro syringe (Poole 2000). For separation of molecules according to different types of properties like boiling point, molecular size, and polarity, first of all the sample is changed in the form of vapor state and with the help of a controlled temperature gradient the sample is injected into a column. The sample reaches the detector with a constant flow rate and the specific response depends upon the analyte. There are various types of detectors available such as flame ionization, thermal conductivity, electron capture, flame photometric, photoionization, electrolytic conductivity, and mass spectrometry. The selection of detector used for analysis totally depends on nature of sample to be analyzed. The use of detector provides sensitivity and selectivity during sample analysis. The area under the peak is proportional to the concentration and the amount of substance can be detected by construction of calibration curve (Wrolstad et al. 2005). Recently, a vacuum UV spectrophotometer is developed to be used as a GC detector, which is used for measuring the absorption of gas-phase chemical species in a range of 120–240 nm (Santos and Schug 2011). The sample should be volatile and stable at the operational temperature, mainly at 50–300°C for the use of GC. Instrumentation of GC requires a high cost in initial acquisition, maintenance, and in utilization of equipment. In spite of the advantages of GC, almost 80% of known components are not sufficiently volatile so they cannot be separated by GC. In place of GC, the use of liquid chromatography (LC), especially high-pressure liquid chromatography (HPLC) is suitable. In HPLC, the mobile phase is a liquid whereas the stationary phase is solid, which can be used for improvement in the separation and identification of various chemical classes of organic and inorganic compounds, based on the polarity of target compounds. The liquid chromatography analysis is dependent on adequate selection of mobile phase, composed of organic solvents and buffer solutions, and it promotes suitable separation of target components. The classical LC work at near atmospheric pressure by using a peristaltic pump or gravity flow to keep a constant flow of mobile phase liquid. By using pumps and specific types of detectors, which promote advancement in separation, identification, and quantification of organic components, HPLC originated from classic LC (Wrolstad et al. 2005). The basic instrumentation of HPLC consists of an injector, pump, column, detector, and data system. The mobile phase used in the HPLC, called eluent, utilizes highly purified organic solvents and buffer solutions, and by using column these solutions are pumped at a constant flux, which promotes the separation of components by differentiation in affinities between the mobile and stationary phase. The pumps permit isocratic elution arrangements or gradient elution (Christian et al. 2014). The pumping system of HPLC is able to run a gradient which contains almost four components and is used highly in quality control. To introduce the sample into column with mobile phase, injectors are used. For injection of a large number of samples in food control analysis, highly used injector is valve-type injector and auto-samplers. The column of HPLC is made up of high purity silica particles with a diameter less than 2 μm. The highly pure silica particles cause high back pressure and create minor loss of efficiency at high flow rates, known for rapid separation. For the analysis of food through HPLC, a wide range of detectors are used, that change the flux which consists of mobile phase and sample components to an electric signal proportional to its concentration. Among all of these detectors, the UV/Vis absorption detector measures the absorption of radiation through chromophore-containing compounds. The emitted lights are measured by fluorescence detectors. The change in the refractive index of the mobile phase due to the dissolved analyte is measured by a refractive index detector. Electrochemical detectors are based on either change in conductivity of eluent or on electrochemical oxidation-reduction of the analyte. The most sensitive and specific detector used in HPLC and GC is the MS detector which identify and separate ions present in the gas phase (Christian et al. 2014). The simultaneous detection of various types of food additives in food products is performed by coupling HPLC and GC with MS (Logue et al. 2017). The most important advantage of HPLC is its efficiency, various types of stationary phases, because of different types of detectors the sensitivity is very high and also the recovery of sample with suitable resolution. On the other hand, the instrumentation cost is very high, requirement of a technical operator and release of toxic residue from organic solvents are the disadvantage of HPLC. The different types of columns used in chromatography also define the type of chromatography. Normal phase chromatography employs a polar stationary phase and a nonpolar mobile phase. Reverse-phase chromatography utilizes polar organic solvents, like acetonitrile and methanol joint buffer solvents. Ion exchange chromatography employs ion exchange resins which are made up of particles that carry fixed negative or positive charges. In anion exchange, stationary phase consists a large number of positively charged groups and resin contains negatively charged groups which travel more slowly than column than positively charged species. Positively charged species will get attracted towards cation exchange resin's negatively charged groups. In ion-exchange chromatography micro-particulate ion exchangers are used and identification and quantification are done by conductometric detection. The separation of molecules on the basis of size is performed by size exclusion chromatography (Christian et al. 2014).

Capillary electrophoresis has been used as a versatile and high-performance tool that made the separation fast and efficient. It uses fewer amounts of sample, solvent, and reagent. The coupling of capillary electrophoresis separation with HPLC or MS gives suitable sensitivity and selectivity in analysis of food (Le et al. 2017).

7.7.3 Electroanalytical Techniques

Due to the operational simplicity, minimum cost, and simple preparation of sample, the demand of electroanalytical technique in food quality control has been increased to a large extent in the last 20 years. To identify and quantify organic compounds, electrical properties like current, potential, charge, resistance, conductance, impedance, and conductivity are measured by this technique. For identification and quantification of food additives in complex samples like feedstock and foodstuffs, voltammetric techniques are used. During food analysis, the potential electroanalytical techniques improve the reaction of electron transfer followed by current measurement (Wang 2001).

In voltammetric techniques, a voltammogram is used for the identification and quantification of target compounds but only when the compound is electroactive. Voltammogram is produced by using potential change which promotes an electron transfer reaction and the current generated by a potential change is recorded as a function of applied potential. The classification of voltammetric technique is based on the way potential is imposed on the electrode. Differential pulse voltammetry and square wave voltammetry are mainly used for the analytical process because responses are based on superior elimination of capacitive current, which boosts the sensitivity in comparison to chromatographic techniques (Scholz 2010). In differential pulse voltammetry, small pulses of fixed amplitude superimposed on a linear potential ramp are put in the working electrode where the reaction of interest is found. Measurement of current is done before and after the application of pulse, which produces a signal that employs against the applied potential, and contains peak current, with height directly proportional to the concentration of target compound and peak potential which will be used for its identification (Dahmen 1986). In square wave voltammetry, a waveform made up of symmetrical square wave, superimposed on a base staircase potential, is used in the working electrode. The current which has been measured produces excellent sensitivity because the peak currents are measured one time at the end of forward peak and one time at the end of reverse peak, generating a net signal bigger than either forward or reverse signals, producing sensitivity higher than the differential pulse voltammetry (Wang 2001). The analytical benefit of differential pulse voltammetry (DPV) and square wave voltammetry (SWV) incorporate excellent sensitivity along with a large effective linear concentration range for organic as well as inorganic compounds, a number of effective solvents and electrolytes, a broad range of temperatures, fast estimation time, concurrent detection of various analytes, and the potential to estimate kinetic and mechanistic parameters (Mirceski et al. 2007). The DPV and SWV need to employ an electrochemical cell made up of two or three electrodes, known as electric conductor, also an inert electrolytic solution, known as ionic conductor or supporting electrolyte. To overcome the electro-migration effects, to manage the ionic strengths of medium, and for lowering the resistance of solution, inert electrolytic solution is required.

During analysis pure supporting electrolytes such as inorganic salts, buffer solutions, and mineral acids are used and this electrolyte should not be frequently reduced or oxidized in the potential range of analysis. The instrument set up for this technique includes two economical integrated circuits, one consists of polarizing circuit which employs the potential among working and reference electrodes, another is measuring circuit which specifies the current that is generated between auxiliary and working electrodes. Hence, the estimation and instrumentation show comparably modest costs and a significant operational facility, which are the principal benefits of the electroanalytical techniques (Wang 2001). The change in working electrode encourage changes in redox reaction, so changing the current values is measured, because target compounds alter their charges at the surface of the electrode, through the exchange of one or more electrons with conductor. In an electrochemical reaction, the reduced and oxidized compounds both are present in the solution, on the other hand, conductor is chemically inert and used for electrons. The selection of working electrode based on redox behavior of target compound and the surrounding current over the potential zone is applied in the analysis. The electrical conductivity, surface reproducibility, potential window, mechanical properties, cost, availability, and toxicity are also the major concern in the execution of this technique (Scholz 2010).

Mercury, as dropping mercury electrode, hanging dropping mercury electrode, and mercury film electrode, was the most employed material as it increases the negative potential window, presents high reproducibility, is frequently renewable, and have plain surface. However, in the last 20 years, it is exchanged by other surfaces, because of the limitation of only cathodic potential zones and high toxicity (Scholz 2010). Solid amalgam electrodes, made from the mixture of a metal powder and mercury liquid followed by amalgamation process, show very close proprieties to the mercury electrode but lacking of mercury residues (De Souza et al. 2011). These electrodes are a good substitute in the growth of sensing tool for food estimation applications. Solid electrodes made up of carbon and noble metals have been used as a working electrode, to observe oxidizable compounds. However, the analytical reactions are based on the condition of the surface electrode that needs pretreatment and polishing steps to get suitable responses. Carbon-based electrodes, like graphite powder with liquid or solid binders, carbon fibers, highly oriented pyrolytic graphite, carbon nanotubes, and boron-doped diamond, have been estimated as adequate surfaces in electroanalysis. Platinum and gold electrodes show high analytical responses but a limited range of applications, because of surrounding current attached with oxide production on their surfaces (Wang 2001). These electrodes show planned surface changes, to modify the analytical production and solve different electroanalytical problems, like sensitivity and selectivity. However, the reproducibility linked with these changes is the main disadvantage in the application of food control analysis (Wang 2001; Sierra-Rosales et al. 2017).

In the last few years, the use of electrochemical aptasensors, while DNA or RNA are aptamers with sufficient secondary structures which work as ligands in various transduction systems, are capable of binding to a target molecule with high affinity and specificity (Amaya-González et al. 2013). Development in nanotechnology has generated aptasensors with sufficient stability and surface coverage by the aptamers, whereas maintaining a high binding affinity in solution, gives permission for their use in food detection (Vasilescu et al. 2018). The use of porous paper as a substrate for DNA or RNA aptamers immobilization and construction of aptasensors for food estimation has also been known (Vasilescu et al. 2018).

7.8 Advantages of Food Additive

Advantages of food additives have been recognized for ages. Since ancient times, food additives like salt and sugar were used for preservation purpose. There are several benefits of food additives which makes it a safe and sound compound for use in edible and drinkable products for human consumption. The most important benefits are listed below.

7.8.1 Enhancement in the Safety and Nutritional Content of the Food

Additives like antioxidant, antimicrobial agents, and several more additives affect the nutritive value when added to the food product. Antioxidants when added to some food product not only protect it but also aids in recovering the losses made during processing.

7.8.2 Availability of Low-Cost and Affordable Products

Recent studies have shown that the addition of food additives affects the overall cost of food products. Addition of additives increases the shelf life of the products thus affecting their price. Some researchers have reported that on removal of additives from margarine, purchasing of butter and other spreads as an alternative increased. Another case was reported that if the food additives were removed from products like wieners, breads, and processed cheese it will lead to the enhancement of the cost of refrigeration, food processing, and many other required factors which is important for the availability of the product. Alternatively, there are techniques and technologies like processing methods and packaging materials that can be used to extend the shelf life of the product and provide a safe product but the most important drawback of these alternatives is their high cost. Food additives are cheaper in context. Addition of food additives helps in the availability of cheaper products with extended shelf life.

7.8.3 Availability of Variety of Foods

With the application of food additives, a wide range of food products with appealing flavor, color, texture, and many other benefits are available. Convenience food which is said to be the most easily available and low-cost product has an extended shelf life and exceptional taste due to added food additives. Snacking items have gained popularity because of their mouth-watering flavor and aroma. These items go through high-temperature treatments and also contain food additives in the form of preservatives for extending their shelf life. Food additives like cyclamate and saccharine are widely used in the present time because consumer preference is shifting from high-calorie junk food to low-calorie diet food. High-calorie sweeteners are not fully but partially replaced by zero or low-calorie sweeteners like aspartame. Application of artificial color and aroma has attracted customers worldwide. The future use of the food additives is most likely to be in nutraceuticals and functional food industry.

7.9 Safety of Additives

The amount of food additive addition in food is generally low. Addition of food additive may enhance the stability and shelf life; it also upgrades the organoleptic properties of the food. Due to the demand of consumers and its different functions in food technology, the use of Food additives in processing operations has been increasing day by day. This growing consumption of food additives may cause severe health hazards to the consumer. Hence, it is mandatory that it should be safe and acceptable for their intended use. The approval of any additive should rely on several types of research focusing on the detrimental effects on human health including carcinogenicity, reproductive toxicity, and metabolism of the particular additive (Fig. 7.8). The approved additive by research thus undergoes to the various organization in order to achieve the safety where governmental bodies and regulations are important pillars (Griffiths and Borzelleca 2014). Various growing bodies have rigorous command on them. Additionally, various food safety governing bodies have set the limits of food additives for their use in all food industries in order to lower the health risk.

In the European Union, the commanding bodies are the European Food Safety Authority (EFSA) and the European Commission, Parliament and council. All of these bodies are responsible for the safety evaluation, which involve toxicological studies and dietary exposure estimation, authorization which involve preserving and publishing data of food additives that are used in the European Union, and Control which is associated with legislation and labeling of Food additives. The most important governing body of food additives in the USA is US food and drug administration (USFDA) and almost all the countries have their own food governing bodies.

Food and Agriculture Organisation (FAO) and the World Health Organisation (WHO) both governing bodies work in the international arena through a Joint Expert on Food additives (JECFA). Joint expert committee on Food additives work on acceptable daily intake level (Karunaratne and Pamunuwa 2019).

All approved food additives on the Swedish market have been evaluated for safety either by Joint FAO and WHO Expert Committee on Food Additives

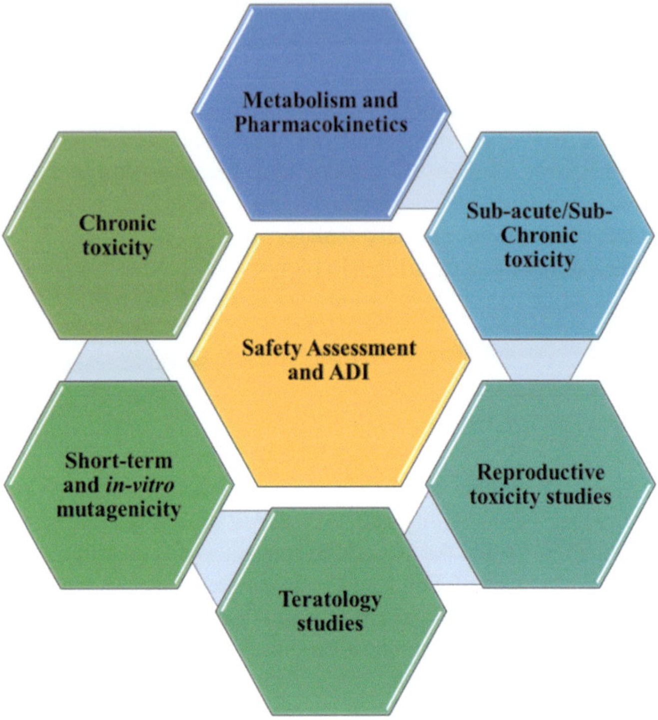

Fig. 7.8 Safety assessment and ADI examination of Food additive

(JECFA) or the Scientific Committee for Food (SCF), both proposing ADI values (The Acceptable Daily Intake). In the US, the Food and Drug Administration (FDA) does the corresponding work. The JECFA evaluations are continuously published as monographs, whereas the earlier work of SCF was not officially published in detail. However, SCF reports (opinions) from recent years are available on the Internet. Thus, the most important work regarding the evaluation of use, intake, and safety of food additives in Sweden is today performed by the SCF and their Working Group. This working Group consists of about 15 independent scientists, half of them are members of the SCF and the other half are ***ad*** *hoc* experts. The author (NGI) participates as ad *hoc* expert (Ilbäck and Busk 2000).

7.10 Risks Associated with Food Additives

Intentional or unintentional addition of additives has demonstrated their usefulness to a food processor or consumer, but in some cases, they may lead to harmful toxicological effects when consumed by sensitive population groups or when taken in excess amount. Australia has reported some adverse effects of certain types of food additives. Both the short-term and lifelong cumulative exposures to food additives may result in acute or chronic effects which is detrimental to health.

Potential health risks include digestive disorders (including colic pain and diarrhea), respiratory disorders (including sinusitis and asthma), neural disorders (including insomnia and hyperactivity), and skin problems (including rashes and itching). The nitrates present in the curing agents get converted into nitrites which bind with hemoglobin and form met-hemoglobin which is a substance responsible for causing loss of consciousness and even death in infants. The use of artificial colors especially allura red and tartrazine in food items along with benzoates as preservatives are responsible for hyperactivity in infants. In pregnant women, nitrates and benzoates have shown harmful effects as it lowers hematocrit value and hemoglobin thus leading to the considerable lowering of length and weight of the infant. Monosodium glutamate (MSG) which is mainly used in all most all frozen and canned foods causes heart palpitations, weakness, and numbness (Amit 2017).

Considerations for evaluating the overall risk of using a food additive must include whether it has specific hazardous properties, a prediction of the likelihood of adverse effects based on exposure, and an estimation of the amount of exposure. To protect individuals from the possible adverse effects of these substances, studies to assess the risk of exposure to chemical residues should be performed. Other adverse health effects may be observed after observation of short-term or high-level exposure to a particular additive. In other words, the use of some additives may be hazardous when consumed in specific quantities.

Toxicology studies may be designed to identify the dose just above the threshold level where effects are observed [lowest observed effect level (LOEL)] and the dose just below the threshold at which no effects are seen [no observed effect level (NOEL) or no observed adverse effect level (NOAEL)]. Often, an uncertainty factor has been applied to the NOEL to give a value known as the acceptable daily intake (ADI). This term may be expressed as the acceptable chemical exposure per amount of body weight per day. The ADI is usually calculated as either the NOAEL divided by 100 when the NOAEL is derived from animal studies, or as the NOAEL divided by 10 when the NOAEL relates to human data (Renwick 1996; WHO 1987). The decision to incorporate a specific quantity of an additive should consider not only the ADI level but also the minimum amount that is deemed necessary to achieve a desired technical effect.

For carcinogens, estimated cancer risks are obtained by multiplying exposure estimates by cancer potency factors. This practice often results in numerical cancer risks that may describe the frequency of people (e.g., one in one million) who would be expected to develop new cancers after long-term exposure to the food additive. For noncarcinogens, risk characterization typically relates the estimated exposure to the acceptable daily intake. Exposures at the level of the ADI represent a very low risk. Increasing chemical exposures above the ADI would result in an increased risk or increased probability of an adverse health consequence. Due to the considerable uncertainties and wide ranges of data used to estimate risks, a risk characterization should include qualitative evaluations of risk (Winters and Francis 1997).

7.11 Conclusion

Now-a-days people are more concerned about their health and at the same time most of the additives are of chemical nature and require tight regulations over their use. Hence, there lies a controversy and need for a higher range of research and findings together with strong interpretation from the governments for their proper use. Till date, more than 2500 additives fall under intentional additive categories. For utilization of additives in food processing, it should be non-toxic, food grade, and non-reactive. For the better use of the additives, there is a need for more research to know the exact mechanism of their inhibiting property along with the migration property. Then only the food additives can be used in food for the betterment of the society by giving healthier and safe food to its consumer.

References

Albani JR (2006) Fluorescence spectroscopy in food analysis. In: Encyclopedia of analytical chemistry: applications, theory and instrumentation. Wiley, Hoboken, NJ

Alencar NM, de Morais EC, Steel CJ et al (2017) Sensory characterisation of gluten-free bread with addition of quinoa, amaranth flour and sweeteners as an alternative for coeliac patients. Int J Food Sci Technol 52(4):872–879

Amaya-González S, De-los-Santos-Álvarez N, Miranda-Ordieres AJ et al (2013) Aptamer-based analysis: a promising alternative for food safety control. Sensors 13(12):16292–16311

Amir RM, Anjum FM, Khan MI et al (2013) Application of Fourier transform infrared (FTIR) spectroscopy for the identification of wheat varieties. J Food Sci Technol 50(5):1018–1023

Amit SK, Uddin MM, Rahman R et al (2017) A review on mechanisms and commercial aspects of food preservation and processing. Agric Food Secur 6(1):1–22

Augustin MA, Sanguansri L (2012) Challenges in developing delivery systems for food additives, nutraceuticals and dietary supplements. In: Garti N, McClements DJ (eds) Encapsulation technologies and delivery systems for food ingredients and nutraceuticals. Woodhead, Cambridge, UK, pp 19–48

Babushkin V, Spiridonov A, Kozhukhar A (2016) Application of NIR and FTIR in food analysis. J Phys Sci Appl 6(2):47–50

Belitz HD, Grosch W, Schieberle P (2009) Food additives. Food Chem 2009:429–466

Branen AL, Davidson PM, Salminen S, Thorngate J (eds) (2001) Food additives. CRC Press, Boca Raton, FL

Bumbrah GS, Sharma RM (2016) Raman spectroscopy–basic principle, instrumentation and selected applications for the characterization of drugs of abuse. Egypt J Forensic Sci 6(3): 209–215

Carocho M, Barreiro MF, Morales P, Ferreira IC (2014) Adding molecules to food, pros and cons: a review on synthetic and natural food additives. Compr Rev Food Sci Food Saf 13(4):377–399

Carocho M, Morales P, Ferreira IC (2017) Sweeteners as food additives in the XXI century: a review of what is known, and what is to come. Food Chem Toxicol 107:302–317

Christian GD, Dasgupta PK, Schug KA (2014) Analytical chemistry, 7th edn. Wiley, Hoboken, NJ

Codex Alimentarius Commission (2019). General Standard for Food Additives (GSFA) Online. https://www.fao.org/gsfaonline/index.html. Accessed 25 May 2021

Counsell JN (1993) Vitamins as food additives. In: The technology of vitamins in food. Springer, Boston, MA, pp 143–171

Dahmen EA (1986) Electroanalysis: theory and applications in aqueous and non-aqueous media and in automated chemical control. Elsevier, Amsterdam deMan JM (1999) Additives and contaminants. In Principles of food chemistry. Springer, Boston, MA, pp 429–473

De Souza D, Mascaro LH, Fatibello-Filho O (2011) The effect of composition of solid silver amalgam electrodes on their electrochemical response. J Solid State Electrochem 15(9): 2023–2029

Doores S (1983) Organic acids. In: Branen AL, Davidson PM (eds) Antimicrobials in Foods. Marcel Dekker, New York, pp 75–109

Fürst P (1994) New parenteral substrates in clinical nutrition. Part I. Introduction. New substrates in protein nutrition. Eur J Clin Nutr 48(9):607–616

Griffiths JC, Borzelleca JF (2014) Food additives. In: Encyclopaedia of toxicology, 3rd edn. Academic, San Diego, CA, pp 622–627

Gujral HS, Pathak A (2002) Effect of composite flours and additives on the texture of chapati. J Food Eng 55(2):173–179

Hasenhuettl GL (2008) Overview of food emulsifiers. In: Food emulsifiers and their applications. Springer, New York, pp 1–9

Ilbäck NG, Busk L (2000) Food additives. Use, intake and safety. Näringsforskning 44(1):141–149

Itagaki H (2000) Fluorescence spectroscopy. In: Experimental methods in polymer science. Academic, San Diego, CA, pp 155–260

Jay JM (1995) Antimicrobial food preservatives. In: Handbook of biocide and preservative use. Springer, Dordrecht, pp 334–348

Jonsson JA (1987) Chromatographic theory and basic principles. CRC Press, Boca Raton, FL

Karunaratne DN, Pamunuwa GK (2019) Introductory Chapter: Introduction to food additives. In: Food additives. Intech Open, London

Kumar P, Mishra HN (2004) Mango soy fortified set yoghurt: effect of stabilizer addition on physicochemical, sensory and textural properties. Food Chem 87(4):501–507

Kwak HS, Yang KM, Ahn J (2003) Microencapsulated iron for milk fortification. J Agric Food Chem 51(26):7770–7774

Lal SN, O'Connor CJ, Eyres L (2006) Application of emulsifiers/stabilizers in dairy products of high rheology. Adv Colloid Interface Sci 123:433–437

Lampila LE, Godber JP (2002) Food phosphates. In: Branen AL, Davidson PM, Salminen S, Thorngate JH III (eds) Food additives, 2nd edn. Marcel Dekker Inc, New York

Le TH, Nguyen TQ, Tran CS et al (2017) Screening determination of food additives using capillary electrophoresis coupled with contactless conductivity detection: a case study in Vietnam. Food Cont 77:281–289

Logue C, Dowey LR, Strain JJ et al (2017) Application of liquid chromatography–tandem mass spectrometry to determine urinary concentrations of five commonly used low-calorie sweeteners: a novel biomarker approach for assessing recent intakes. J Agric Food Chem 65(22):4516–4525

Maga JA (1995) Types of food additives. Marcel Dekker Inc, New York

Marín S, Guynot ME, Sanchis V et al (2002) Aspergillus flavus, Aspergillus niger, and Penicillium corylophilum spoilage prevention of bakery products by means of weak-acid preservatives. J Food Sci 67(6):2271–2277

Mirceski V, Komorsky-Lovric S, Lovric M (2007) Square-wave voltammetry: theory and application. Springer Science & Business Media, New York

Mitra S (ed) (2004) Sample preparation techniques in analytical chemistry. Wiley, Hoboken, NJ

Motekaitis RJ, Martell AE (1984) Complexes of aluminum (III) with hydroxy carboxylic acids. Inorg Chem 23(1):18–23

Murray PR, Webb MG, Stagnitti G (1995) Advances in maltol and ethyl maltol applications. Food Technol Int Eur 1995:53–55

Nauta T (1991) Chelating agents. In: Food additive user's handbook. Springer, Boston, MA, pp 273–279

Nielsen SS (2014) Food analysis, 4th edn. Springer, Berlin

Norn V (2015) Emulsifiers in food technology. Wiley Blackwell, Chichester

Pazur J (1991) Enzymatic synthesis and use of cyclic dextrins and linear oligosaccharides of the amylodextrin type. In: Friedman R (ed) Biotechnology of amylodextrin oligosaccharides, ACS Symp. series, vol 458. ACS, Washington, DC, p 51

Poole CF (2000) Chromatography. In: Encyclopedia of separation science. Academic, San Diego, CA, pp 40–64

Putzig CL, Leugers MA, McKelvy ML et al (1994) Infrared spectroscopy. Anal Chem 66(12): 26–66

Ray L, Bullerman LB (2001) Preventing growth of potentially toxic molds using antifungal agents. J Food Prot 45:953–963

Renwick AG (1996) Needs and methods for priority setting for estimating the intake of food additives. Food Addit Contam 13(4):467–475

Romita D, Cheng YL, Diosady LL (2011) Microencapsulation of ferrous fumarate for the production of salt double fortified with iron and iodine. Int J Food Eng 7(3)

Salminen, S Tahvonen R (2001) Food Additive intake assessment. pp 11–27

Sanabria ER, Pereira MF, Dolnikoff MS et al (2017) Deficit in hippocampal long-term potentiation in monosodium glutamate-treated rats. Brain Res Bull 59(1):47–51

Santos IC, Schug KA (2011) Recent advances and applications of gas chromatography vacuum ultraviolet spectroscopy. J Sep Sci 40(1):138–151

Saranraj P, Geetha M (2012) Microbial spoilage of bakery products and its control by preservatives. Int J Pharm Biol Arch 3(1):38–48

Sauer DB, Burroughs R (1993) Efficacy of various chemicals as grain mold inhibitors. Tram Am Soc Agric 17:357–559

Scholz F (2010) Thermodynamics of electrochemical reactions. In: Electroanalytical methods. Springer, Berlin, pp 11–31

Seetaramaiah K, Smith AA, Murali R et al (2011) Preservatives in food products-review. Int J Pharm Biol Arch 2(2):583–599

Sierra-Rosales P, Toledo-Neira C, Squella JA (2017) Electrochemical determination of food colorants in soft drinks using MWCNT-modified GCEs. Sensors Actuators B Chem 240: 1257–1264

Skoog DA, West DM, Holler FJ et al (2013) Fundamentals of analytical chemistry. Nelson Education, Toronto

Suhr KI, Nielsen PV (2004) Effect of weak acid preservatives on growth of bakery product spoilage fungi at different water activities and pH values. Int J Food Microbiol 95(1):67–78

Tanaka T, Takahashi O, Oishi S et al (2008) Effects of tartrazine on exploratory behavior in a three-generation toxicity study in mice. Reprod Toxicol 26(2):156–163

Van de Voort FR (1992) Fourier transform infrared spectroscopy applied to food analysis. Food Res Int 25(5):397–403

Vasilescu A, Hayat A, Gáspár S et al (2018) Advantages of carbon nanomaterials in electrochemical aptasensors for food analysis. Electroanalysis 30(1):2–19

Wang J (2001) Analytical electrochemistry, 2nd edn. Wiley, New York

Winter CK, Francis FJ (1997) Assessing, managing, and communicating chemical food risks. Food Technol 51(5):85–92

World Health Organization (1987) Principles for the safety assessment of food additives and contaminants in food-environmental health criteria. World Health Organization, Geneva, p 70

Wrolstad RE, Acree TE, Decker EA et al (eds) (2005) Handbook of food analytical chemistry, Volume 1: Water, proteins, enzymes, lipids, and carbohydrates. Wiley, Hoboken, NJ

Zotto AA (1991) Antifoams and release agents. In: Food additive user's handbook. Springer, Boston, MA, pp 236–241

Pigments and Colors

8

Rudragoud Policegoudra, Smitha M., O. P. Chauhan, and A. D. Semwal

8.1 Introduction

Color is considered as one of the most delightful as well as impressive qualities of food products, which influences the eating desires of the consumers along with their preference and choice (Delgado-Vargas and Paredes-Lopez 2003). Food color is any dye, pigment, or substance which on addition to the food gives it color. Adding color elevates the look of the food and makes it more attractive and also influences our perception. Sometimes, the loss of color during processing is made up by adding food colors. Natural colors have been an important part of our diet. They are added to foods for good appeal as well as identity (Griffiths 2005). The trend is increasing in the markets to ensure expectations of consumers with more diverse food products with different textures, colors, and tastes are available (Ayala-Zavala et al. 2011).

Since the pre-historical era, human beings have left their identity in the nature that can be observed in painted images, whether it may be either simple handprints or fine artworks (Barnett et al. 2006). A pigment produces the colors which we observe in the leaves, flowers, fruits, and roots of the plants and is also present in the skin, hair, eyes, and other parts of the animals (Delgado-Vargas et al. 2000).

For decades, we have been associated with different colors of food products. The natural pigments have a very limited scope which is being permitted for use in foods, and they are under strict regulations (Clydesdale 1993).

R. Policegoudra (✉) · S. M. · A. D. Semwal
Fruit and Vegetable Technology Division, Defence Food Research Laboratory (DFRL), Mysore, India
e-mail: policegoudra.dfrl@gov.in

O. P. Chauhan
Defence Food Research Laboratory, DRDO, Mysore, India

O. P. Chauhan (ed.), *Advances in Food Chemistry*,
https://doi.org/10.1007/978-981-19-4796-4_8

Structurally diversified natural food colorants which are widely distributed, are grouped into tetra-pyrrols, tetra-terpenoids, and flavonoids. Chlorophyll is the most important pigment, which is commonly found in leaves. Carotenoids are tetra-terpenoids that are responsible for the yellow, orange, and red color of many fruits and vegetables, whereas the red/purple shade of many fruits is due to the presence of anthocyanins. Other important classes of pigments are the anthraquinones found in carmine and madder, and the betalains found in beetroot.

8.2 Pigments

Pigments are the compounds which are generally responsible for the color of many products including food products. We observe various pigments in our day-to-day life since these compounds are widespread in every living organism. Among them, plants are the major principal producers (Delgado-Vargas et al. 2000). Hundreds of different structures can be found only in anthocyanin group. These compounds display various hues of colors like yellow, brown, blue, black, green, red, orange, pink, etc. Currently, pigments are used in various fields like food, medicine, cloths, cosmetics, furniture, and in other areas (Hari et al. 1994). However, along with imparting beauty to the products, pigments have also involved in most important functions of the living organism (Mortensen 2006). Quinones play a very significant role in the conversion of light energy into chemical energy. Since time immemorial, people used to distinguish the quality of the product by their color, especially in the case of meat (Koes et al. 1994; Mol et al. 1996).

Naturally occurring biological pigments are grouped as isoprenoids, tetrapyrroles, benzopyrans, quinines, metalloproteins, and N-heterocyclic compounds. Among them, carotenoids and chlorophylls are the most abundant pigments in nature. Plants, protozoa, and photosynthetic bacteria provide organic material, which is required for the growth of other animals (Hari et al. 1994). The pigments found in animals are presented in Table 8.1.

8.2.1 Classification of Pigments

1. **Natural Pigments**
   These are pigments produced mainly by plants, animals, and microorganisms and they have low stability and are found in low concentration (Hari et al. 1994).
2. **Synthetic Pigments**
   Synthetic pigments are produced by chemical reactions. They are highly stable and occur in high concentrations. Synthetic pigments can be separated into water-soluble and water-insoluble based on their solubility and they may be either organic or inorganic in nature (Amchova et al. 2015).

Table 8.1 Pigments in animals

Pigments	Distribution
Heme proteins 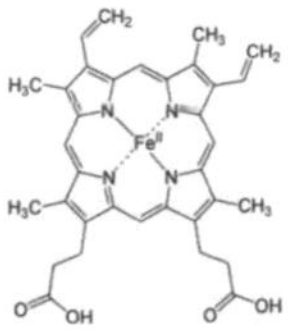	Crustacea, Mollusks, Arachnida malacostraca, annelids
Melanins 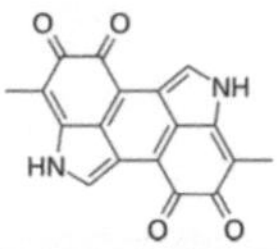	Wide distribution including crustacea, Echinoderms, insects, malacostraca
Carotenoids 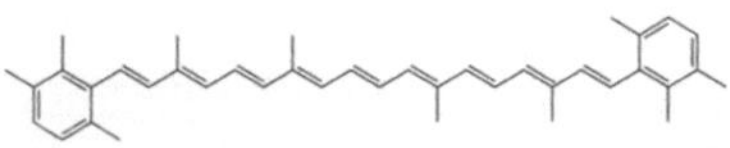	Mammals, birds, reptiles, fishes, amphibians, Echinoderms, insects, etc.
Riboflavin 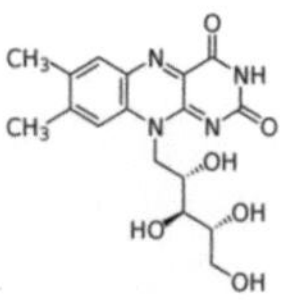	Reptiles, amphibians, fish
Flavonoids 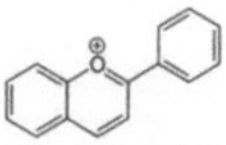	Crustacea, insects
Quinones 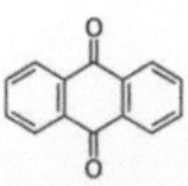	Echinoderms, insects, and arachnida

8.2.2 Extraction of Pigments

The most common and easy method of coloring any food is to add strongly colored food to the product which needs to be colored. This approach is generally used in cooking, where spices may impart color along with flavor. However, for industrial food application, this approach shows problems like low concentration of pigments, provides unwanted flavors, and insoluble matter like peel and seed, which is unacceptable in beverages (Delgado-Vargas et al. 2000). Due to their low concentration, these pigments are extracted with organic solvents for fat-soluble pigments, and anthocyanins are extracted with water or alcohol (Mortensen 2006).

8.2.3 Major Classes of Natural Pigments

8.2.3.1 Chlorophylls

These are fat-soluble green pigments present in the plastids of plants, algae, and few bacteria. Chlorophylls are involved in the process of photosynthesis for the biosynthesis of complex substances from simpler ones like H_2O and CO_2. These pigments are widely distributed in all green leafy vegetables and green algae (Simpson et al. 2012). Among several chlorophylls, chlorophyll *a* and *b* are playing a very important role in food coloration as they are very common in green plants (Fig. 8.1).

Chlorophyll contains tetrapyrrole or porphyrin ring system which is very much similar to myoglobin (Mb) and hemoglobin (Hb) structures. The four pyrrole rings are connected together with the central Mg^{2+} ions to make a porphyrin ring. This porphyrin ring along with phytol makes the complete chlorophyll structure.

When chlorophyll-containing foods like vegetables and green leaves are cooked, chlorophyll can undergo changes with respect to the color and solubility. The color may be turned to dull green, brownish, or bright green. This color change is mainly due to changes or reactions of chlorophyll molecule, which may be due to the loss of the phytol side chain by the action of the enzymes called chlorophyllase and/or acidic conditions, or the removal of the central Mg^{2+} atom when exposed to heat treatment and acids (Patek 1936).

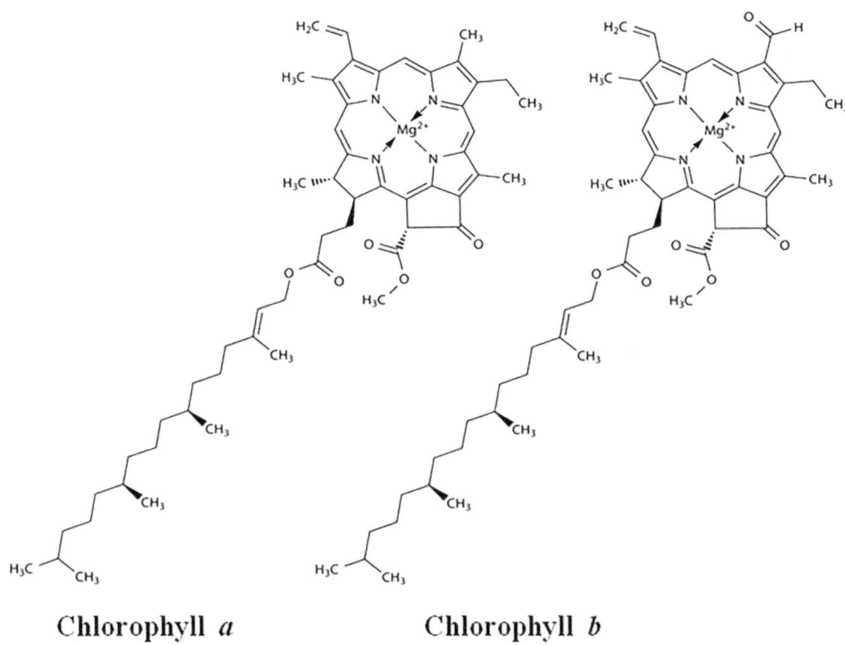

Fig. 8.1 The structure of chlorophyll *a* and Chlorophyll *b*

Chlorophyll exists in nature in various forms such as chlorophyll *a*, *b*, *c1*, *c2*, and *d*. Chlorophyll is generally extracted from plants like spinach, nettles or grass, and alfalfa using acetone and hexane in darkness to avoid degradation. Exposure to light, heat, extreme pH, and air cause adverse effect on its stability. The stability of chlorophyll increases when complexes with copper ion as well as by de-esterification (Colio and Babb 1948).

8.2.3.1.1 Chlorophyll Degradation

The degradation of chlorophyll is an important biochemical process which occurs during fruit ripening and also leaf senescence. Figure 8.2 shows the chlorophyll degradation pathway. During this process, the enzyme called chlorophyllase catalyzes the chlorophyll by removing the phytol group which results in the formation of the chlorophyllide. Jaques et al. (2001) reviewed the degradation of chlorophyll when green cellular tissues are subjected to various processing conditions. During heat treatment, freezing, and storage the chlorophyll degrades into pheophytin, which is having a color of olive brown. During processing, the duration of treatment, pH of the medium, temperature, and the release of acids speed up the chlorophyll degradation.

Maximum chlorophyll degradation takes place under longer and hotter processing conditions. For example, when a kiwifruit was exposed to the processing temperature of 100 °C for 5 min, chlorophylls were completely damaged (Robertson 1985). Chlorophyll is more stable at higher pH conditions. Hence, sodium bicarbonate is added to increase the pH of water during the cooking of peas and beans (Schwartz and Lorenzo 1990). Chlorophyll *a* is known to break down quickly than chlorophyll *b* during heat processing. But, reverse action happens during the storage of vegetables in flexible containers.

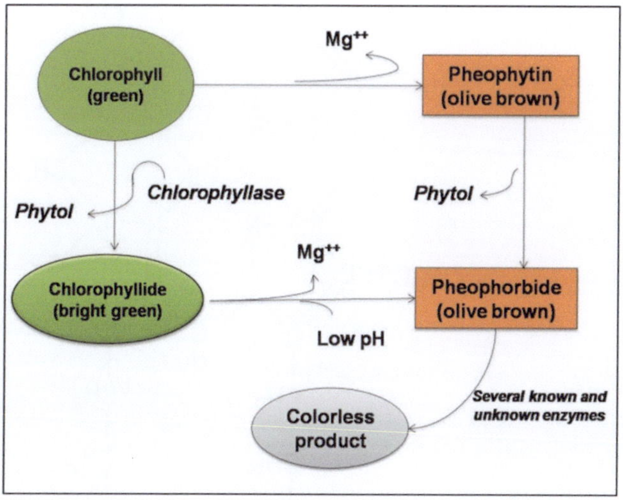

Fig. 8.2 Chlorophyll degradation pathway

8.2.3.1.2 Health Benefits

Chlorophyll has numerous health benefits for human beings. Firstly, they have been found capable of rebuilding the bloodstream without side effects when administrated in high doses through oral, intravenous, and intramuscular (Patek 1936). Chlorophyll also encourages the fertility rate by regulating sex hormones. They can also be used as antibacterial agents (Bowers 1947). They are capable to clean the deposition of drugs, deactivation of toxins in the body, reduce problem associated with blood sugar, and cleanses liver (Colio and Babb 1948). Chlorophyll also plays a major role in inhibiting oral bacterial infections in deodorizers, which promotes healing of rectal sores and reduces typhoid fever (Offenkrantz 1950).

8.2.3.2 Heme (blood) Pigments

Heme is a basic chemical (Fig. 8.3) responsible for the red coloration of hemoglobin and myoglobin. The color of the red meat is due to the presence of myoglobin. The other color compounds of muscles like vitamin B_2, cytochromes, and flavoproteins do not contribute much to red meat (Simpson et al. 2012).

The major function of heme pigments is the transportation of oxygen for generation of energy. The central Fe atom attaches to four nitrogen atoms in the porphyrin ring and the fifth coordinates to join with the nitrogen atom of histidine residue of globin (Weber et al. 1974). In terms of concentration, the major pigment of meat muscle is Mb, approximately 80%, and remaining 20% is Hb, which is common in the blood vessels for the transportation of oxygen (Kim et al. 2003).

Fig. 8.3 Structure of heme

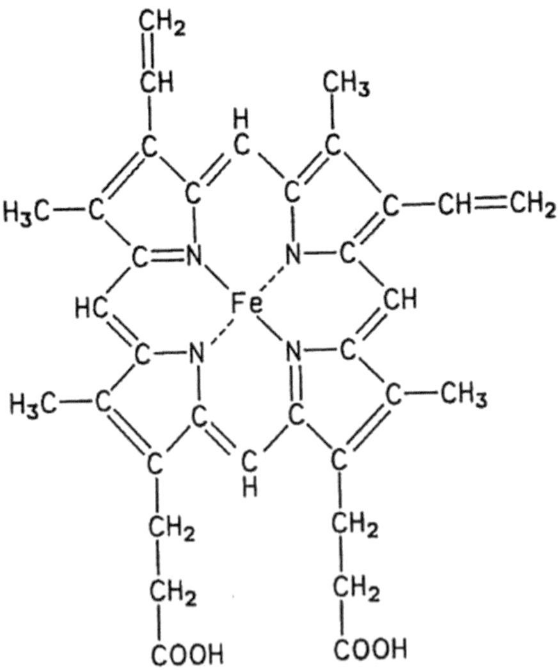

8.2.3.2.1 Health Benefits

Red meats are abundant in iron, which is very much essential for the human body to produce red blood cells as well as to regulate the temperature of the body. If a person is having Iron deficiency, it may lead to a decline in cognitive abilities, predominantly in children.

8.2.3.3 Anthocyanins

These pigments belong to the flavonoid family. These groups of reddish water-soluble pigments are present in flowers, fruits, and vegetables. Anthocyanin color mainly depends on pH condition. Anthocyanins are in red color under acidic condition; whereas, they appear blue and purple under basic and neutral pH, respectively (Solymosi et al. 2015).

Anthocyanins commonly occur in fruits, especially berries, nuts, vegetables, roots, grains, and flowers (De Brito et al. 2007; Einbond et al. 2004). Major sources are purple, raspberry, strawberry, cherry, blueberry, plum, red cabbage, etc. Based on the chemical structures, they are classified into two types, i.e., anthocyanidin aglycones and true anthocyanins, presented in Fig. 8.4.

Various factors such as cultivar or variety (Lee and Finn 2007), maturity (Ahmadiani et al. 2014), growing area, season, cultivation practices (Kovacevic et al. 2015), and storage conditions may affect the composition of anthocyanins. The effect of processing and storage on the stability of anthocyanins has been studied and it was found that effect of processing conditions and storage on the polyphenol content is negligible.

Anthocyanins are commonly used to color food products like jams, jellies, drinks, pastries, and confectionaries since they impart blue or red colors. However, they are vulnerable to pH changes. Molecules in the anthocyanin may get some protection against degradation due to the presence of sugar under ambient conditions. The harmful effects can be reduced by storing at low temperature (Simpson et al. 2012).

8.2.3.3.1 Health Benefits

The toxicity of anthocyanins has not been reported but it is believed that anthocyanins are nontoxic at high temperature. They are potential antioxidants and play various health benefits like improved visual perception, coronary heart disease, antiviral activity, etc. The anthocyanins like delphinidin, petunidin, and malvidin enhance the activity of glutamate decarboxylase, the enzyme that acts as a catalyst in

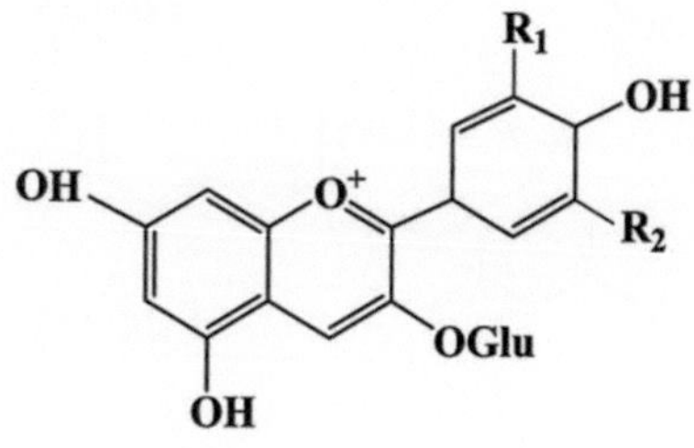

Fig. 8.4 Structure of anthocyanin

the decarboxylation of glutamic acid to gamma amino benzoic acid (Simpson et al. 2012).

8.2.3.4 Carotenoids

Carotenoids are commonly present in fruits and vegetables, yellow-colored flowers, animal species like crustaceans, birds, fishes, insects, and seaweeds (Britton 1996). Ripening in many fruits like citrus fruits, apricots, and tomatoes is mainly associated with the accumulation of carotenoids and the disappearance of chlorophyll. The oxygenated carotenoid counterparts like lutein, astaxanthin, cryptoxanthin, canthoxanthin, and zeaxanthin are formed by hydroxylation (Tanaka et al. 1976) (Fig. 8.5).

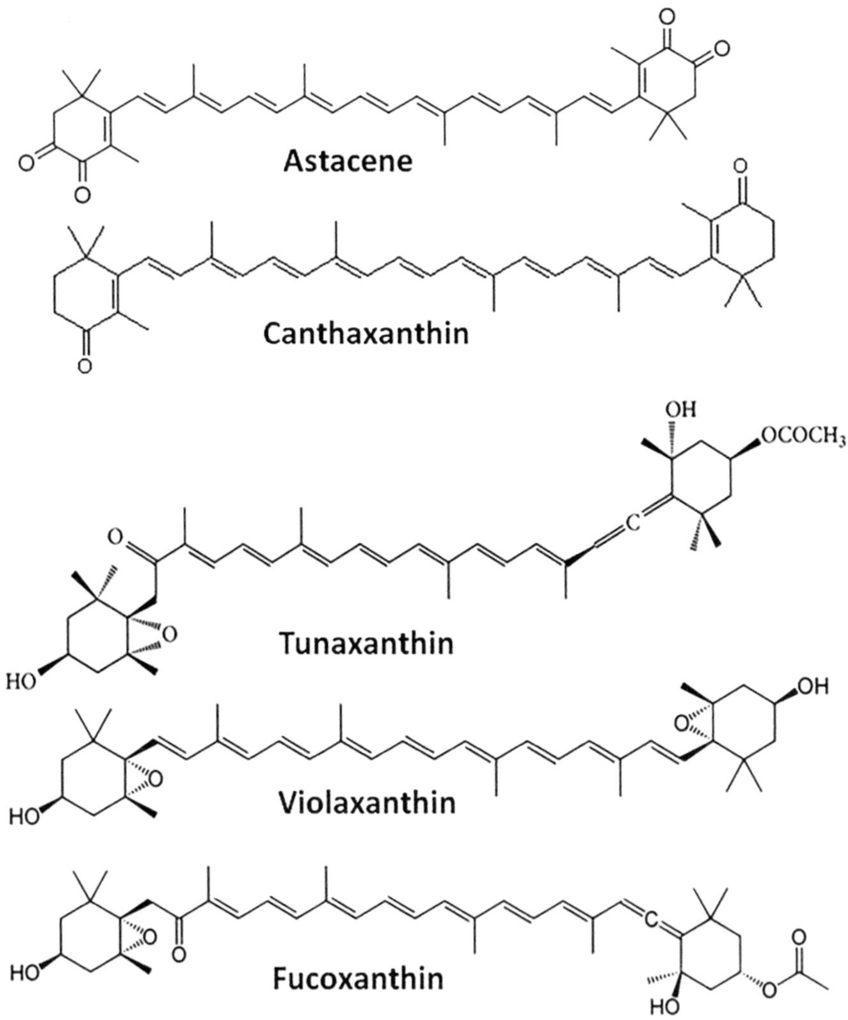

Fig. 8.5 Chemical structure of some common carotenoids

8.2.3.4.1 Health Benefits

Carotenoids are very important and most essential in the retina of the eye for vision and are eye disorders (Krinsky and Johnson 2005; Dembinska-Kiec 2005). Carotenoids also inhibit oxidation of low-density lipoprotein and cardiovascular diseases (Hadley et al. 2003). Carotenoids boost the immune system and also reduce the adverse side effects of cyclooxygenase inhibitor drugs (Kearney et al. 2006).

8.2.3.5 Flavonoids

Flavonoids are water-soluble compounds that exhibit shades of yellow to colorless appearance. They are removed very fast from the body which leads to inadequate absorption and low bioavailability. Isoflavones are the most bioavailable pigment while the flavanols are the least bioavailable among different flavonoids (Manach et al. 2005). Few flavonoids are esterified and occur in plants and foodstuffs. The major flavanols are catechin, epicatechin, epigallocatechin, epicatechin gallate, epigallocatechin gallate, and teaflavins. Flavanones are butin, hesperidin, hesperetin, naringenin, and naringin (Fig. 8.6) (Simpson et al. 2012).

8.2.3.6 Betalains

Betalains are water-soluble pigments and they are classified into two classes, betacyanin and betaxanthin. Betacyanins exhibit reddish to violet color, which includes amaranthine, isoamaranthine, betalain, isobetalain, phyllocactin, and isophyllocactin. Betaxanthin exhibit yellow to orange color, which comprises dopaxanthin, miraxanthin, indicaxanthin, portulaxanthin, portulacaxanthin, and vulgaxanthin (Fig. 8.7) (Simpson et al. 2012).

8.2.3.6.1 Health Benefits of Betalains

Betalain extracts are used as food colorant in wines and juices. The red beetroot is the most common source of betalain-based food colorant. Betalains are very less resistant to light and temperature. Hence, color may change from red to yellow color, based on the surrounding environment.

8.2.3.7 Astaxanthin

It is an orange-red colored keto-carotenoid produced and accumulated in few red algae, green algae, and bacteria. The most common astaxanthin accumulated in animal parts are flamingo feathers and crustacean shells. Astaxanthin is most commonly used in animal feed, in marine aquaculture including ornamental fish. It is also ideal in dietary supplements such as tablets, capsules, syrups, and soft gels (Solymosi et al. 2015).

8.2.3.8 Lycopene

Lycopene is an expensive pigment and is highly prone to oxidative degradation and is found in plants containing β carotene, in low concentration since lycopene act as a

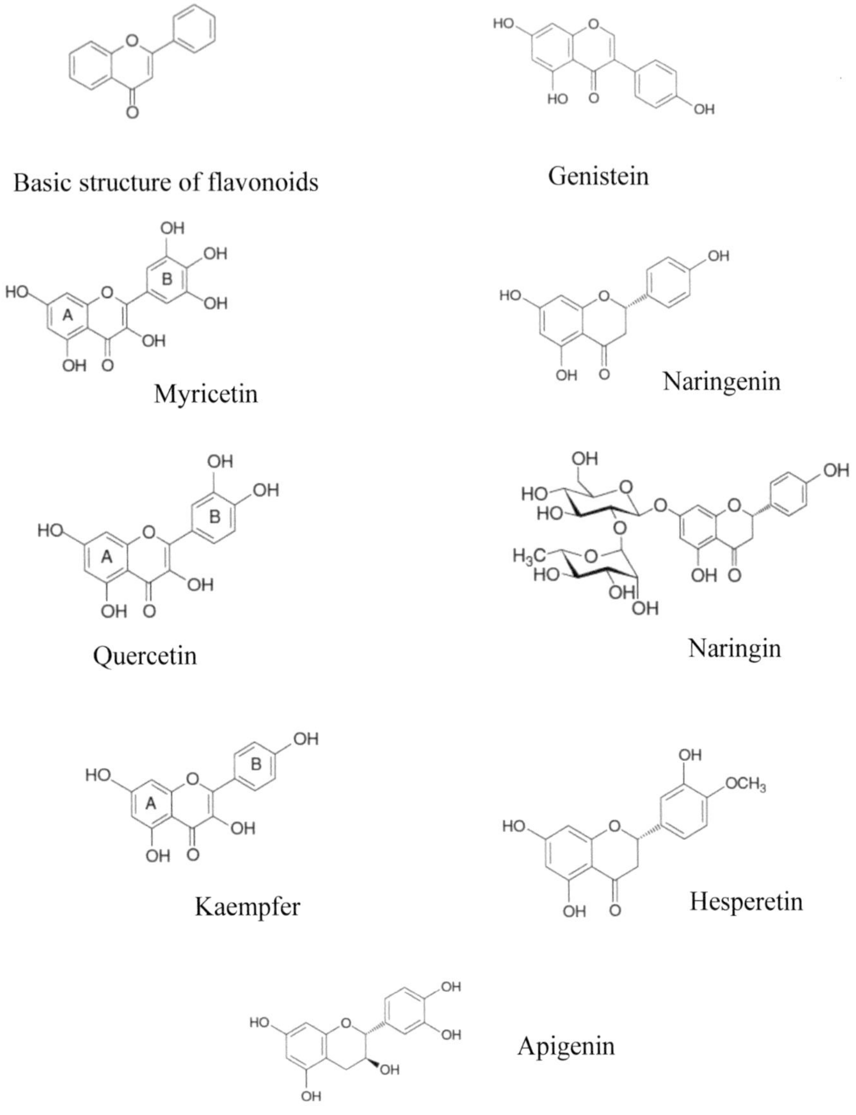

Fig. 8.6 Chemical structure of common flavonoids

precursor in the synthesis of β carotene. Lycopene is very common in tomatoes, watermelon, grapes, etc. It is a powerful antioxidant which reduces the risk of prostate cancer and ischaemic heart disease (Solymosi et al. 2015).

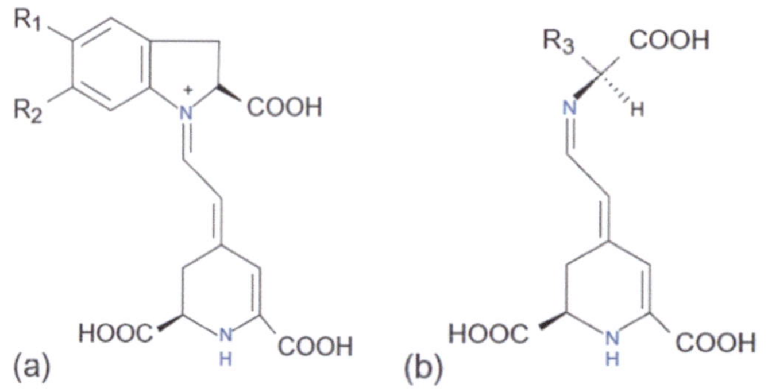

Fig. 8.7 Structure of (**a**) Betalain (**b**) Betacyanins

8.3 Food Colors

There are varieties of foods available in nature and each of them is recognized by its own taste, texture, smell, and color. Among these, food color is one of the major factor which influences consumers' attention. Color makes food items very attractive and appealing to taste. The reality is that most of the color pigments of any food product are unstable. Hence, whenever the food is exposed to harsh conditions during processing, color pigments can be destroyed (Kumari and Meghwal 2016).

Color is one of the important selection criteria for food preference. Recent studies show their importance and role in selection of colors might modify among the population over a period of time (Clydesdale 1993). Color is the first noticeable characteristic of the food and helps to predetermine our expectation regarding that particular food, either taste or flavor. For example, the consumer perceives that yellow indicates the lemon flavor and pink goes with the grapefruit (Griffiths 2005). In our daily life, consumers get the opportunity to inspect any kind of food visually before tasting and buying (Spence 2015). Studies have found that by changing the intensity or hue of the color, food items can make use of a drastic impact on consumers' expectation and sensory attributes (Clydesdale and Walford 1984).

Food coloring is defined as any dye, pigment, or substance which is added to food items, either solid or liquid, to impart the required color. A food coloring agent exists in various forms like solid powders, liquids, gel or pastes, etc. Other than food industry, food colorants are used in cosmetics, pharmaceuticals, medical devices, etc. The global turnover of food colorants is about 8000 tons per year and in that India accounts for only 2% of output (Solymosi et al. 2015). There are set of laws and regulations given by the FDA and other regulatory organizations regarding the use of food color.

The natural color of plant-origin foods is mainly due to four groups of pigments such as green-colored chlorophylls, yellow-red-orange of carotenoids, red-blue-purple of anthocyanins, and red-colored betacyanin. Due to the health benefits of natural colorants, consumers prefer them over artificial food dyes. However, natural colorants are less stable and high cost when compared to synthetic dyes, besides having limited range of hues (Rodriguez-Amaya 2018). The usage of synthetic colors in the food industry has been started in the 1800s for decorative purposes as well as to mask low-quality food products (Sulz 1888).

8.4 Major Classification of Colors

Colorants are divided into two groups based on their source as natural and synthetic (Demirag and Uysal 2006).

8.4.1 Natural Color

Natural food color is any pigment extracted from plants, animals, or any other natural sources capable of coloring food. There are reports that in Europe food colorants were used during the Bronze age (Lakshmi 2014). Natural food colorants are safer than synthetic colorants and they are accepted worldwide due to their health benefits and biological potential along with their reliability as well as functionality (Martins et al. 2016). The natural colorants from beetroot, carrot, grape, paprika, and cabbage are very popular and safe. Apart from chinoides, flavonoids, betalains, isoprenoids, and porphyrins, there are few natural pigments such as curcumin and caramel, which are equally significant and commonly used in many food products (Solymosi et al. 2015).

8.4.1.1 Organic Natural Colors

Anatto: It is yellow-orange color food additive which is widely used in cosmetics and food industries like beverages, bakery, and dairy products. Recently, usage of nitrile in sausage is being replaced by Annato powder (Wrolstad and Culver 2012).

Carotenes: The main coloring chemical of carotene is β-carotene. These are highly preferred in various food products with more fatty acid content (Solymosi et al. 2015).

Caramel: Caramel accounts for more than 80% of all food colorants. Here, Class I has the lightest shade, whereas Class IV has the darkest shade (Sengar and Sharma 2014).

Carmine (Natural Red 4): It is a red dye extracted from insects like *Kermes vermilii*, *Porphyrophora polonica*, *Dactylopius coccus*, etc. (Mortensen 2006). At low pH, carmine is orange color at low pH and violet at alkaline pH.

Curcumin (Turmeric): In Turmeric, curcumin basically exists in 1,3-di-keto and enol forms, but also in the form of demethoxy curcumin, bis-demethoxycurcumin and cyclo-curcumin. The potential of curcumin is very high, because of its antioxidative and anti-inflammatory properties (Nielsen and Holst 2002).

Lycopene: The color of lycopene depends upon its concentration. It is red when accumulated in high concentration and orange in a dilute form. It is a very powerful antioxidant that reduces the risk of prostate cancer and ischaemic heart disease (Solymosi et al. 2015).

Marennine: It is a very popular natural color with a blue pigment and exhibits an anti-proliferative effect and presents antiviral and anticoagulant properties.

Melanins: Melanins are extensively distributed in animals, fungi, bacteria and play an important role in the protection against environmental stresses (Solymosi et al. 2015).

Riboflavin: It is a most effective natural yellow water-soluble colorant for powdered and solid food applications. It is being extracted from the fungi *Eremothecium ashbyii*. Its major drawback is that it is sensitive to light and vulnerable to oxidation thus resulting in limited applications (Solymosi et al. 2015).

8.4.1.2 Inorganic Natural Color

Silver gray colored Aluminium dust, red and brown colored iron oxide, gold, titanium dioxide and calcium carbonate are the main inorganic natural colorants. These inorganic colorants are commonly used in chocolate, bread, confectionery, etc. (Emerton and Choi 2008).

8.4.2 Nature-Identical Color

Nature-identical colors are synthetic chemicals and certification is not required from FDA. They are identical in chemical and functional properties when compared with natural colors.

8.4.3 Synthetic Colors

These are produced by chemical synthesis and do not occur in nature. Synthetic food coloring was originally manufactured from coal tar. People have started using synthetic food colors, though there are many natural colors in nature. Cost is a very big reason to go with synthetic color. Artificial food dyes are highly stable than natural ones of same color and there is a limitation for the application of natural colors as food dye (Andrade et al. 2014). In India, at present, eight synthetic colors, i.e., Sunset Yellow FCF, Tartrazine, Ponceau 4R, Carmosine, Erythrosine, Brilliant blue FCF, Fast Green FCF, and Indigo carmine are permitted to add in food items (FSSAI). The maximum permissible limit of all food colors is 100 ppm, which can be either individual or in combination.

Synthetic food colors are extensively used in many food products and in many instances they exceed the permissible level (Andrade et al. 2014; Kiseleva et al. 2003). Synthetic dyes such as Sunset Yellow, Tartrazine, Amaranth, and Brilliant Blue are vastly used in beverages (Al-Degs 2009). Many synthetic food colors are injurious to human health because they have azo group, which is known for genotoxicity (Lopez-de-Alba et al. 2001; Combes and Haveland-Smith 1982).

8.4.3.1 Water-Soluble Synthetic Colors

Allura Red AC: It is generally obtained from insects and used in the manufacture of various food products, viz., wine, soups, sauces, gums, snacks, carbonated drinks, etc. European Union permitted it, while many countries like Australia, Denmark, France, Switzerland, Norway, Belgium, and Sweden have banned it (Pandey and Upadhyay 2012).

Amaranth: This synthetic color is reddish-brown in color with water-soluble properties (Demirag and Uysal 2006).

Sunset Yellow: This synthetic chemical is orange-red in color and commonly used in drinks, beverages, sweet powders, ice cream, cereals, snacks, etc. (Branen and Haggaerty 2001).

Brilliant Blue FCF: This granular form synthetic color is water soluble with blue and black colors. It is used in different products like beverages, cheese, wine, sauce, etc. (Martins et al. 2016).

Tartrazine: It is a yellow colored synthetic chemical used to color cream, bread, cereal, beverages, confectionery, ice cream, peanuts, and canned food (FDA and US 2010).

Erythrosine: It is a xanthene-class colorant and added to flavored milk and pudding, ice cream, jelly, etc. (Karaali and Ozcelik 1993).

8.4.3.2 Fat-Soluble Synthetic Colors

These are soluble in oil or organic solvents and they are banned for food coloring as they have toxic properties. The Penso SX is an oil-soluble chemical used to color butter, whereas Yellow AB and Oil red XO are used in the coloring of orange peel (Demirag and Uysal 2006).

8.4.3.3 Lake Colors

These are water-insoluble precipitation of aluminum hydrate substrate and are produced in the form of fine powders. The color tone of the powder is decided based on dye content and particle size (Downham and Collins 2000). Since they are not soluble in water, oil, and other solvents, they are dispersed directly into food matrix to impart color. Lake colors are being used in cakes, biscuit fillings, confectionery, powder drinks, sweets, soups, and spice mixtures.

Today, most of the synthetic food colors are obtained from crude oil. The list of permitted and non-permitted synthetic colors is provided in Tables 8.2 and 8.3.

Table 8.2 Artificial food colors permitted in India as per FSSAI (2011)

S. No.	Color	Common name	Chemical class
1	Red	Panacea 4R	Azo
		Carmoisine	Azo
		Erythrosine	Xanthene
2	Yellow	Tartrazine	Pyrazolone
		Sunset yellow	Azo
3	Blue	Indigo caramine	Indigoid
		Brilliant blue	Triaryimethane
4	Green	Fast green FCF	Triaryimethane

Table 8.3 List of non-permitted synthetic food colors in India

S. No.	Color	Color index name	Application
1	Amaranth	C. I. Food Red 9	Crushed ice, cut and glazed fruits, jam and jellies, milk sweets
2	Fast Red E	C. I. Food Red 4	Hard-boiled sugar confectionery, crushed ice, sweetened supari
3	Metanil Yellow	C. I. Acid Yellow 36	Bakery products, savories, biryani, non-milk sweets, turmeric powder, dhokla
4	Orange II	C. I. Acid Orange 7	Milk sweets, non-milk sweets, bakery products, sugar toys, turmeric powder
5	Auramine	C. I. Basic Yellow 2	Non-milk sweets, sugar toys, savories
6	Blue VRS	C. I. Food Blue 3	Ice cream cones, glazed fruits, biryani, savories
7	Green S	C. I. Food Green 4	Milk sweets, non-milk sweets
8	Malachite green	C. I. Basic Green 4	Fresh vegetables, fresh coriander leaves, green peas, milk, and non milk sweets
9	Rhodamine B	C. I. Basic Violet 10	Milk-based and non milk-based sweets, confectionery, sugar toys, colored tamarind

8.5 Regulatory Status and Labeling

As per the inventory of colored eatables, more colored food products are common in the urban areas when compared to the rural areas (Tripathi et al. 2007). About 69% of colored food products are manufactured using permitted color while 31% of samples revealed the presence of non-permitted colors. The use of non-permitted colors was more in rural areas (38%) as compared to urban areas (25%). However, the permitted colors in food products were within the approved limit of 100 ppm in urban markets (73%) than in rural areas (50%) (Tripathi et al. 2007).

Nowadays, many synthetic colors which are meant for paper and textiles are being used in food products. In spite of regulatory supervision, the non-permitted synthetic colors are being used in some of the local food vendors or non-branded food products. In other instances, the permitted food colors are being used beyond the permissible limit. The disproportionate application of non-food grade colors needs to be regulated (Tripathi et al. 2007).

In spite of strict regulation for synthetic food colors, the uses, status, doses, acceptable daily intake (ADI), labeling requirements, and applications are continuously being reevaluated. Many recognized authorities have made regulations list of approved food color additives and limitations for use in food products. Many countries follow the specifications provided by the Codex Alimentarius or FAO for the usage of color additives and to set allowable doses in foods (Corradini 2018).

In the EU, the European Food Safety Authority (EFSA) established its ADI for the evaluation and safety of synthetic colors. According to EU regulations, stating either the name of synthetic color or its corresponding additive E number on the label is required. Additionally, if any of the six synthetic colors (Allura Red AC, Azorubin, Ponceau 4R, Quinoline yellow, Sunset Yellow FCF, and Tartrazine) is being added to any product, its label should include a warning. Within the EU, the RASFF (Rapid Alert System for Food and Feed) has been developed to assist the exchange of information on adulterated foods (Corradini 2018).

Permission for food colorants is stringently regulated by detailed laws at both national and international levels. The natural colorants permitted by FSSAI for use in India are carotene and carotenoids, chlorophyll, riboflavin, caramel, annatto, saffron, turmeric, or curcumin. The list of natural colors permitted in the US are listed in Table 8.4.

Table 8.4 Application limit for natural color as per US regulations

S. No.	Food product	Colorant	Maximum limit (ppm)
01	Breakfast cereals	Carminic acid Caramel color-III, Carotenes, Bixin, Norbixin, Capsanthin, Betanin, Anthocyanins	25–200
02	Butter	Carrot extract	NL
03	Malt bread	Caramel color (E150a–d)	NL
04	Chips (potato)	Curcumin, Turmeric	NL
05	Cheese	Carotenes, Bixin, Norbixin	1.5–50 for E160b
06	Margarine, emulsions of fat in water	Curcumin, Turmeric	10 for E160b
07	Vinegars	Caramel colors (E150a–d), Anthocyanins	NL
08	Bitter beverages	Curcumin, Turmeric, Riboflavin, Carminic acid	100 mg/L
09	Fruit and vegetable juices	Carotenes, Lycopene, β-apo-8-carotenal	NL
10	Beer and Cider	Caramel colors (E150a–d)	NL
11	Preserved vegetables	Riboflavin, Chlorophyll, Caramel colors (E150a–d), Chlorophyllins, carotenes, Anthocyanins	NL
12	Jams and Marmalade	Curcumin, Turmeric, Chlorophyll, Chlorophyllins, Capsanthin, Paprika, Lutein, Betanin, Anthocyanins, Carminic acid	100 only for synthetic colorants
13	Sausages, meat pastes, fish products	Curcumin, Turmeric, Riboflavin, Carminic acid, Caramel colors (Ea–d), Carotenes, Capsanthin, Paprika, Betanin	20–100
14	Hamburger	Carminic acid, Caramel colors (Ea–d)	100 only for E120

NL no limit specified

8.6 Health Benefits of Pigments

We consume a wide variety of foods which includes carotenoids, chlorophylls, and anthocyanins which are present in many colorful tubers, seeds, fruits, and vegetables. However, compared to synthetic food colors, consumers predominantly favor natural colors due to their health benefits and nutrition (Downham and Collins 2000). The findings of Southampton University confirmed a quantifiable change in diverse behavioral issues like hyperactivity and attention period when consumed food products spiked with food colors and preservatives (McCann et al. 2007; Bonan et al. 2013). Anthocyanin and their aglycones like cyanidin, delphinidin, malvin, and peonidin demonstrated anti-proliferative and proapoptotic activities in gastric adenocarcinoma (Duluc et al. 2014).

There are many reports to link the adverse side effects like carcinogenicity, behavior problems like hyperactivity in children, allergy, asthma, etc., when food products with synthetic food colors are consumed (Hashem et al. 2010; Kobylewski and Jacobson 2010; Tripathi et al. 2007). Lycopene and lutein are associated with a reduced risk of prostate cancer because of their high antioxidant properties (Zhao et al. 2017; Jia et al. 2017; Akhtar and Bryan 2008; Aghajanpour et al. 2017; Li et al. 2018). The potential health risks are increasing due to the adulteration of wines and juices with synthetic food colors instead of using natural sources (Komissarchik and Nyanikova 2014; Kobylewski and Jacobson 2010). Consumption of synthetic food colors causes some potential adverse health effects like allergenicity, behavioral problems like hyperactivity syndrome in children, neurotoxicity, genotoxicity, and carcinogenicity. Recent studies show that adverse effects like allergy, asthma, hyperactivity, and even cancer can be linked to the intake of synthetic food dyes (Hashem et al. 2010; Kobylewski and Jacobson 2010; Tripathi et al. 2007; McCann et al. 2007).

8.7 Conclusion

In recent years, food additives which may be either natural or synthetic are used to preserve food products to improve their appearance, taste, texture, and color. Natural or synthetic food colors are used in many food products either to restore or substitute the color to reduce batch variation and to attract consumers. Natural food colors are in great demand because of their health benefits without any side effects when compared to synthetic food colors, which are identified for many side effects. There is a lot of scope and opportunity to explore new sources of natural colors and their application specific to the products as well as the processes involved. New technologies may overcome the short shelf life of natural food colors and to impart color stability along with their health-promoting activities of food products.

References

Aghajanpour M, Nazer MR, Obeidavi Z et al (2017) Functional foods and their role in cancer prevention and health promotion: a comprehensive review. Am J Cancer Res 7(4):740–769

Ahmadiani N, Robbins RJ, Collins TM et al (2014) Anthocyanins contents, profiles, and color characteristics of red cabbage extracts from different cultivars and maturity stages. J Agric Food Chem 62:7524–7531

Akhtar MH, Bryan M (2008) Extraction and quantification of major carotenoids in processed foods and supplements by liquid chromatography. Food Chem 111(1):255–261

Al-Degs YS (2009) Determination of three dyes in commercial soft drinks using HLA/GO and liquid chromatography. Food Chem 117(3):485–490

Amchova P, Kotolova H, Ruda-Kucerova J (2015) Health safety issues of synthetic food colorants. Regul Toxicol Pharmcol 73(3):914–922

Andrade FID, Guedes MIF, Vieira IGP et al (2014) Determination of synthetic food dyes in commercial soft drinks by TLC and ion-pair HPLC. Food Chem 157:193–198

Ayala-Zavala JFN, Vega-Vega V, Rosas-Domínguez C et al (2011) Agro-industrial potential of exotic fruit byproducts as a source of food additives. Int Food Res J 44(7):1866–1874

Barnett JR, Miller S, Pearce E (2006) Colour and art: a brief history of pigments. Opt Laser Technol 38(5):445–453

Bonan S, Fedrizzi G, Menotta S, Elisabetta C (2013) Simultaneous determination of synthetic dyes in foodstuffs and beverages by high-performance liquid chromatography coupled with diode-array detector. Dyes Pigments 99:36–40

Bowers WF (1947) Chlorophyll in wound healing and suppurative disease. Am J Surg 71:37–50

Branen AL, Haggaerty RJ (2001) Introduction to food additives. In: Branen AL, Davidson PM, Salminen S, Thorngate J (eds) Food additives. CRC Press, Boca Raton, FL, pp 1–11

Britton G (1996) Carotenoids. In: Hendry GAF, Houghton JD (eds) Natural food colorants, 2nd edn. Blackie Academic & Professional, London, pp 197–243

Clydesdale FM (1993) Color as a factor in food choice. Crit Rev Food Sci Nutr 33(1):83–101

Clydesdale FM, Walford J (eds) (1984) Developments in food colors–2. Elsevier Applied Sciences, London, UK, pp 75–112

Colio LG, Babb V (1948) Study of a new stimulatory growth factor. J Biol Chem 174:405–409

Combes RD, Haveland-Smith RB (1982) A review of the genotoxicity of food, drug and cosmetic colors and other azo, triphenylmethane and xanthene dyes. Mutat Res Rev Genet 98(2):101–243

Corradini MG (2018) Synthetic food colors. Reference module in food science, University of Massachusetts, Amherst, pp 291–296

De Brito ES, de Araujo MCP, Alves RE et al (2007) Anthocyanins present in selected tropical fruits: acerola, jambolao, jussara and guajiru. J Agric Food Chem 55:9389–9394

Delgado-Vargas F, Jimenez AR, Paredes-Lopez O (2000) Natural pigments: carotenoids, anthocyanins, and betalains - characteristics, biosynthesis, processing and stability. Crit Rev Food Sci Nutr 40(3):173–289

Delgado-Vargas, Paredes-Lopez (2003) Natural colorants for food and nutraceutical uses. CRC Press LLC

Dembinska-Kiec A (2005) Carotenoids: risk or benefit for health. Biochim Biophys Acta Mol basis Dis 1740(2):93–94

Demirag K, Uysal V (2006) Renklendiriciler. In: Altug T (ed) Gida katkımaddeleri. Meta Basim, Izmir, pp 169–191

Downham A, Collins P (2000) Coloring our foods in the last and next millennium. Int J Food Sci Technol 35:5–22

Duluc L, Jacques C, Soleti R et al (2014) Delphinidin inhibits VEGF induced-mitochondrial biogenesis and Akt activation in endothelial cells. Int J Biochem 53:9–14

Einbond LS, Reynertson KA, Luo XD et al (2004) Anthocyanin antioxidants from edible fruits. Food Chem 84:23–28

Emerton V, Choi E (2008) Essential guide to food additives, 4th edn. Leatherhead Publishing, Cambridge, UK
FDA (2010) Food and Drug Administration & US Department of Health and Human Services. Food ingredients and colors. International Food Information Council (IFIC) and US Food and Drug Administration, Washington, DC
FSSAI (Food Safety Standards Authority of India) (2011) Food Safety and Standards (contaminants, toxins and residues) regulations, 2011, available at: http://www.fssai.gov.in
Griffiths JC (2005) Coloring foods and beverages. Food Technol 59:38–44
Hadley CW, Clinton SK, Schwartz SJ (2003) The consumption of processed tomato products enhances plasma lycopene concentrations in association with reduced lipoprotein sensitivity to oxidative damage. J Nutr 133:727–732
Hari RK, Patel TR, Martin AM (1994) An overview of pigment production in biological systems: functions, biosynthesis, and applications in food industry. Food Rev Int 10(1):49–70
Hashem MM, Atta AH, Arbid MS et al (2010) Immunological studies on Amaranth, Sunset Yellow and Curcumin as food coloring agents in albino rats. Food Chem Toxicol 48(6):1581–1586
Jaques AG, Bortik K, Hau J et al (2001) Improved method to track chlorophyll degradation. J Agric Food Chem 49:1117–1122
Jia YP, Sun L, Yu HS et al (2017) The pharmacological effects of lutein and zeaxanthin on visual disorders and cognition diseases. Molecules 22(4):610
Karaali A, Ozcelik B (1993) Gıda Katkısı Olarak Dogal ve Sentetik Boyalar. GIDA 18(6):389–396
Kearney PM, Baigent C, Godwin J et al (2006) Do selective cyclo-oxygenase-2 inhibitors and traditional non-steroidal anti-inflammatory drugs increase the risk of atherothrombosis? Meta-analysis of randomized trials. BMJ 332(7553):1302–1308
Kim YS, Yoon SK, Song YH et al (2003) Effect of season on color of Hanwoo (Korean native cattle) beef. Meat Sci 63(4):509–513
Kiseleva MG, Pimenova VV, Eller KI (2003) Optimization of conditions for the HPLC determination of synthetic dyes in food. J Anal Chem 58(7):685–690
Kobylewski S, Jacobson MF (2010) Food dyes: a rainbow of risks. Centre for Science in the Public Interest, Washington, DC
Koes REK, Quattrocchio F, Mol JNM (1994) The flavonoid biosynthetic pathway in plants: function and evolution. BioEssays 16(2):123–132
Komissarchik S, Nyanikova G (2014) Test systems and a method for express detection of synthetic food dyes in drinks. LWT-Food Sci Techol 58(2):315–320
Kovacevic DB, Putnik P, Dragovic-Uzelac V et al (2015) Influences of organically and conventionally grown strawberry cultivars on anthocyanin content and color in purees and low-sugar jams. Food Chem 181:94–100
Krinsky NI, Johnson EJ (2005) Carotenoid actions and their relation to health and disease. Mol Asp Med 26(6):459–516
Kumari R, Meghwal M (2016) Benefits of food colors and safety in usage. Food 6:22–24
Lakshmi GC (2014) Food coloring: the natural way. Res J Chem Sci 4(2):87–96
Lee J, Finn CE (2007) Anthocyanin and other polyphenolics in American elderberry (Sambucus canadensis) and European elderberry (S. nigra) cultivars. J Sci Food Agric 87:2665–2675
Li X, Wang Z, Zhang G et al (2018) Improving lycopene production in Saccharomyces cerevisiae through optimizing pathway and chassis metabolism. Chem Eng Sci 193:364–369
Lopez-de-Alba PL, Lopez-Martinez L, Cerda V et al (2001) Simultaneous determination of tartrazine, sunset yellow and allura red in commercial soft drinks by multivariate spectral analysis. Quimica Analitica Bellaterra 20(2):63–72
Manach C, Williamson G, Morand C et al (2005) Bioavailability and bio-efficacy of polyphenols in humans. I. Review of 97 bioavailability studies. Am J Clin Nutr 81(1):230S–242S
Martins N, Roriz CL, Morales P et al (2016) Food colorants: challenges, opportunities and current desires of agro-industries to ensure consumer expectations and regulatory practices. Trend Food Sci 52:1–15

McCann D, Barrett A, Cooper A, Crumpler D et al (2007) Food additives and hyperactive behaviour in 3-year-old and 8/9-year-old children in the community: a randomised, double-blinded, placebo-controlled trial. Lancet 370(9598):1560–1567

Mol J, Jenkins G, Schäfer E et al (1996) Signal perception, transduction and gene expression involved in anthocyanin biosynthesis. Crit Rev Plant Sci 155(6):525–557

Mortensen A (2006) Carotenoids and other pigments as natural colorants. Pure Appl Chem 78(8): 1477–1491

Nielsen SR, Holst S (2002) Developments in natural colourings. In: MacDougall DB (ed) Color in food improving quality. Woodhead, Cambridge, UK, pp 331–352

Offenkrantz W (1950) Water-soluble chlorophyll in ulcers of long duration. Rev Gastroenterol 17: 359–367

Pandey RM, Upadhyay SK (2012) In: El-Samragy Y (ed) Food additive. Intech Open, London, pp 1–30

Patek A (1936) Chlorophyll and regeneration of blood. Arch Int Med 57:73–84

Robertson GL (1985) Changes in the chlorophyll and pheophytin concentrations of kiwifruit during processing and storage. Food Chem 17(1):25–32

Rodriguez-Amaya DB (2018) Natural food pigments and colorants. In: Bioactive molecules in food. Springer, Cham, pp 1–35

Schwartz SJ, Lorenzo TV (1990) Chlorophylls in foods. Crit Rev Food Sci Nutr 29:1–17

Sengar G, Sharma HK (2014) Food caramels: a review. J Food Sci Technol 51:1686–1696

Simpson BK, Benjakul S, Klomklao S (2012) Natural food pigments. In: Simpson BK, Nollet MLN, Toldra F et al (eds) Food biochemistry and food processing, 2nd edn. Wiley, Hoboken, NJ, pp 705–722

Solymosi K, Latruffe N, Morant-Manceau A et al (2015) Food color additives of natural origin. In: Color additives for foods and beverages. Woodhead, Cambridge, UK, pp 1–34

Spence C (2015) On the psychological impact of food color. Spence Flavour 4:1–16

Sulz CH (1888) A treatise on beverages, or, the complete practical bottler: Full Instructions for Laboratory Work with Original Practical Recipes for All Kinds of Carbonated Drinks, Mineral Waters, Flavorings, Extracts, Syrups. EtcDick & Fitzgerald Sulz & Co., New York

Tanaka Y, Matsuguchi H, Katayama T et al (1976) The biosynthesis of astaxanthin-XVI. The carotenoids in crustacea. Comp Biochem Physiol 54B:391–393

Tripathi M, Subhash K, Das KM (2007) Surveillance on use of synthetic colors in eatables vis a vis Prevention of Food Adulteration Act of India. Food Control 18:211–219

Weber RE, Hemmingsen EA, Johansen K (1974) Functional and biochemical studies of penguin myoglobin. Comp Biochem Physiol B Biochem Mol Biol 49(2):197–204

Wrolstad RE, Culver CA (2012) Alternatives to those artificial FDC food colorants. Annu Rev Food Sci Technol 3:59–77

Zhao L, Temelli F, Curtis JM et al (2017) Encapsulation of lutein in liposomes using supercritical carbon dioxide. Food Res Int 100:168–179

Flavor Chemistry

9

Siddharth Priyadarshi, Kanchan Goyal,
and Madeneni Madhava Naidu

9.1 Introduction

The root of utilizing aromatic substances for delight or medicinal purpose is as old as humanity. Individuals have consistently been keen on the odor. The herbs and spices are rich in essential oil and the utilization of volatile oils are accredited to many flavor components they contain along with the antiseptic and bacteriostatic properties. The utilization of volatile oil is linked to human history. The origin of the fragrance and flavor industry began over 150 years before. This duration is marked by a large technological breakthrough in the field of chemistry where various flavor compounds were developed. The flavor and fragrance industry gained a profitable market with time. It supports beverage, cosmetics, fragrance, food, household products markets, and toiletry industries. The private labeling and food service companies are the most profitable sector in this business. The total market of cosmetics, fragrances, and flavor sectors is estimated to be €15 billion whereas alone flavor and fragrance share around € 6.5 billion each. The largest markets for the flavors industry are Europe, Africa, the Middle East region (36%), and NorthAmerica (32%), followed by Asia-Pacific (26%) and South America (6%) while emerging markets of this industry are in Central America, China, India, and Russia. About 60% of the total world market is shared only by eight major multinational companies (Guentert 2007). Flavorists and perfumers study and exploit materials which potentially affect the sense of chemesthesis, smell, and taste in human. Flavorists work with the materials that are either chemically synthesized or obtained directly or indirectly from animal or plant origin. The products developed by flavorists are intended for the food and beverages industries. While the perfumers

S. Priyadarshi · K. Goyal · M. M. Naidu (✉)
Plantation Products, Spices and Flavour Technology, CSIR-Central Food Technological Research Institute, Mysuru, India
e-mail: mmnaidu@cftri.res.in

O. P. Chauhan (ed.), *Advances in Food Chemistry*,
https://doi.org/10.1007/978-981-19-4796-4_9

work with the substances of animal, plant, or petrochemical origin intended to develop scented personal care products, perfumes, and fragranced household goods. The instrumental analysis in the fragrance and flavor industry is very significant and sophisticated. The sophisticated instruments used in the aroma industries are Fourier transform IR spectroscopy (FTIR), gas chromatography (GC), GC-FTIR, GC-mass spectrometry (MS), high-performance liquid chromatography (HPLC), HPLC-MS, GC-MS-MS, nuclear magnetic resonance (NMR), and HPLC-NMR. These instruments empower companies to isolate or separate complex mixes and elucidate the chemical structure of unknown compounds (Guentert 2007).

9.2 Flavor

Flavor is defined as the sum of those characteristics of any material taken in the mouth, perceived principally by the senses of taste and smell and interpreted by the brain. Flavor sensation is also perceived by the general temperature, tactile, and pain receptors in the mouth. Flavor also denotes the combination of the characteristics of the material which produces that sensation. Flavor is one of three significant sensory properties that helps to select, accept, and ingest any food material. Flavor is a sensation derived from odor. The odor is created by aromatic substances which are present in every living matter grown in nature. Taste is a perception which depends on odor. Thus, the taste is the odor component of flavor. The specific molecules of food stimulate the chemical senses of smell and taste, which gives flavor to the food. The specialized cells in the taste buds carry out taste reception. The five basic taste sensation is bitter, salty, sour, sweet, and umami, and they are detected in the mouth, throat, and tongue. Taste cells are specific for certain flavor molecules (e.g., sweeteners). The flavoring molecules apart from providing basic taste also stimulate specific olfactory (smell) cells in the nasal cavity. The olfactory cells have the potential to detect more than 10,000 different stimuli, fine-tuning the flavor sensation of food (Rogers 1969).

9.3 Flavorants or Flavoring Substances

Flavorant is a substance that gives flavor, altering the solute's characteristics and causing it to become sweet, sour, tangy, etc. Flavorants are volatile organic compounds, not necessarily present in foods but useful in the formulation of food flavorings. Flavorants are the edible extracts and chemicals that modify any food stuff's flavor through the sense of smell. Its smell mainly determines the flavor of any food. The taste of food is limited to bitter, salty, sour, sweet, and umami while the smell is unlimited. So, a food flavor can be easily modified by altering its smell while keeping its taste similar. Flavorings are used to enhance or change the flavors of natural food products (such as meats and vegetables) or create a flavor for food products that do not have the desired flavor (such as candies and other snacks). Flavorings mainly are focused on the taste and smell of a food product. Flavorings are concentrated ingredients intentionally added in small quantity to impart characteristic flavor to the food and feeding stuff. They should not be consumed as such without blending with other food matrices. Flavorings may contain flavoring adjuvants, flavoring preparations, flavoring substances, process flavorings, and smoke flavorings. A flavor additive may contain a single compound or combinations of chemicals of synthetic or natural origin which offers all or part of the flavor impact of food. Flavor additives are used to develop new products and substitute the flavor lost during processing. Flavorings with more than 1200 chemicals form the largest group of food additives exploited commercially. Flavorings are categorized as natural or synthetic based on their origin. Natural flavorings generally originate from animals, microbial fermentation, herbs, plants, or spices while artificial flavorings are blends of synthetic molecules which are chemically identical to natural ones. The scarcity, insufficient potency, and high cost of natural flavorings provide a barrier for its commercial exploitation; hence, artificial flavorings are preferred over natural flavorings (Rogers 1969). Flavor enhancers are compounds used to enhance or supplement the natural flavor of food. The use of flavor enhancers originated from Asia, where seaweed was mixed with soup stocks to give a better flavor to foods. The flavor-enhancement property of seaweed is due to the presence of monosodium glutamate (MSG) and L-glutamate. These were the first flavor enhancers to be used at the industrial scale. The flavor linked with L-glutamate was called umami. Other commercially used flavor enhancers include guanosine monophosphate (GMP), hydrolyzed vegetable protein, inosine monophosphate (IMP), 5′-ribonucleotides and yeast extract. Flavor enhancers are generally used in broths, canned and frozen vegetables, flavoring and spice blends, gravies, meat, sauces and soups (Rogers 1969). Flavoring substances are chemically defined ingredients with a characteristic aroma and primary flavorings used in foods are categorized as natural, nature-identical, and artificial flavoring substances (EU Flavour Directive 1988; Muller 2007). Flavors are classified based on source, the extent of usage, and chemical structure (Fig.9.1).

9.3.1 Natural Flavoring Substances

Natural flavoring substances are the material obtained from animal or plant origin, by enzymatic, microbiological, or physical processes (Berger 2015). These substances have been detected/identified in a natural material of plant or animal source and not further chemically modified. They can be processed or utilized in their natural form for human consumption, but cannot contain any nature-identical or artificial flavoring substances. During the production of natural flavors, the flavorant is first extracted from source material by distillation, solvent extraction, or other physical methods; then purified and subsequently added to food products. Natural flavors are generally obtained from aromatic seeds (aniseed, cumin), fruits (orange, lemon), herbs (basil, mint), spices (cardamom, clove, turmeric), vegetables (peas, onions, garlic), etc. Natural flavors are complex mixtures of various compounds. Natural flavoring agents are predominantly aromatic organic compounds present as

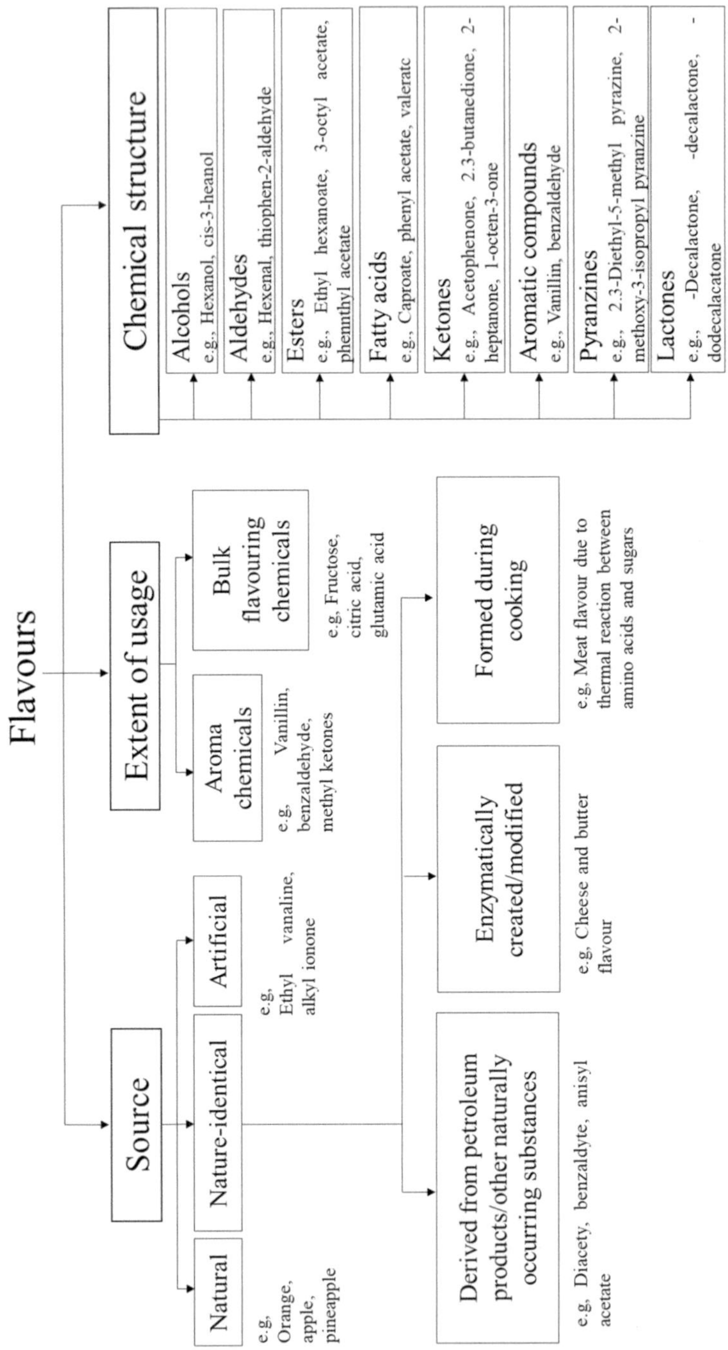

Fig. 9.1 Classification of flavors

volatile/essential oils or nonvolatile constituents such as resins and oleoresins formed in the plant during normal plant metabolism. The essential oils of fruits and vegetables contain volatile acids, alcohols, aldehydes, esters, ether, and ketones, which provide characteristic aroma and flavor.

9.3.2 Nature-Identical Flavoring Substances

Nature-identical flavoring substances are produced by chemical synthesis, which is chemically identical to natural flavoring substances. They are chemically similar to natural substances but obtained by a chemical process or chemical modification from natural substances. They lack any artificial flavoring substances. Due to the high cost or unavailability of natural flavor extracts, most commercial flavorants are nature-identical, which means that they are synthesized chemically rather than extracted from plants or animal sources.

9.3.3 Artificial Flavoring Substances

Flavoring substances which are formed by chemical synthesis are defined as Artificial/Synthetic flavoring substances. These are the substances obtained by chemical synthesis or chemical modification of natural substances but are not present in natural products. They are not identified in a natural product intended for human consumption. Flavor manufacturers must find out the naturally occurring aroma chemicals, mix them appropriately to make the desired flavor or create a novel nontoxic synthetic compound that gives a specific flavor for the production of artificial flavor.

The increased concern among the consumers has raised the demand for natural and minimally processed food. To meet these ever-increasing demands food industries, implement many practices, including the use of natural, plant-based ingredients, etc., and they also provide additional functional/nutraceutical benefits. Natural ingredients are derived or extracted from plants, spices, herbs, animals, or microbial fermentation. Spices are grown principally in East and West Indies, the Malay Archipelago, India, China, Indo-China, Japan, Europe, Africa, and North America. Spices are endowed with flavor substances which are utilized to add flavor and aroma to the foods and overhaul the enjoyment of eating. The aromatic and pungent principles which render spices valuable are found in their volatile oils and resins. The active principles are separated in the form of essential oil or volatile oils, oleoresins, and extracts. They are often used as flavor instead of spice. The aromatic and flavor properties of various condiment spices are present in the form of essential oils or volatile oils and distributed in the different organs such as in bulbs (onion and garlic), flowers (turmeric, orange, lavender, and clove), fruits (tamarind, star, peppers, and anise), leaves (rosemary, parsley, oregano, mint, marjoram, and basil), resin (myrrh), roots (turmeric and ginger), seeds and grains (sesame, nutmeg, fennel, cumin, and coriander), and twigs or bark of trees (cinnamon). Essential oils

Table 9.1 Important flavor compounds identified from the various food sources

Source	Important flavor compounds
Almonds, cherries, plums	Benzaldehyde
Allspice (pimento)	Eugenol, β-caryophyllene
Anise	(*E*)-Anethole, methyl chavicol
Beetroot	Geosmin
Butter	Acetoin, diacetyl, γ-decalactone
Capsicum peppers	Capsaicin, dihydrocapsaicin
Caraway	*d*-Carvone, carvone derivatives
Cardamom	*α*-Terpinylacetate, 1,8-cineole, linalool
Cheese	Fatty acids and others, e.g., methyl ketones
Cinnamon, cassia	Eugenol, cinnamaldehyde
Clove	Eugenol, eugenyl acetate
Coffee	2-Furfurylthiol
Cottage cheese	Acetoin, acetaldehyde
Cucumber	(E,Z)-2,6-Nonadienal
Fennel	(*E*)-Anethole, fenchone
Grapefruit	(R)-1-p-Menthene
Ginger	Gingerol, shogaol, neral, geranial
Hazelnuts	Trans-5-Methyl-2-hepten-4-one (filbertone)
Lemon	Neral/geranial
Mace	1-terpenin-4-ol, *α*-pinene, sabinene
Milk	Dimethylsulfide
Milk fat and meat defect	δ-Decalactone, γ-decalactone
Mustard	Allyl isothiocyanate
Nutmeg	*α*-Pinene, myristicin, sabinene
Orange	(R)-limonene
Parsley	Apiol
Pepper	δ-3-carene, β-caryophyllene, piperine
Raspberry	1-(p-Hydroxy- phenyl)-3-butanone (raspberry ketone)
Saffron	Safranal
Turmeric	1,8-cineole, turmerone, zingiberene
Vanilla	*p*-OH-benzyl methyl ether, vanillin
Yogurt and other sour milk products	Acetaldehyde/lactic acid/ethanol

Source: Robert (2017), Cheetham (1997), Nijssen et al. (1999)

are aromatic complex mixtures of different volatile components (Diniz do Nascimento et al. 2020). Essential oils and their components, commonly used as a flavoring in the food industry, also provide antibacterial, antifungal, and antioxidant properties. Essential oils contain 85–99% volatile and remaining being nonvolatile components. The molecular weight of volatile constituents is generally low and consists of a blend of terpenes, terpenoids, aromatic, and aliphatic constituents. The pungent and flavor principles in spices stimulate the flow of the digestive juices and enhance the enjoyment of food in addition to having a pleasing effect upon the organs of taste and smell. Table 9.1 contains important flavor compounds distributed in the different food sources (Robert 2017; Cheetham 1997; Nijssen et al. 1999) and the structure of some important flavor compounds from plants are shown in Fig.9.2.

Volatile oils are natural products with generally recognized as safe (GRAS) status and used for their unique flavor in perfumery, pharmaceutical, food and beverages industries (Kabara 1991). They are a complex mixture of volatile components synthesized by living organisms and also known as aetheroleum, essence, essential oil, or etheric oil which act as the raw material for flavor and fragrances (Sangwan et al. 2001). Volatile oils are mainly found in aromatic plants such as herbs and spices and occur in glandular hairs, oil cells, secretory cavities or ducts; mainly composed of low molecular weight lipophilic volatile compounds and usually responsible for the flavor in various organs of the plants (Lawrence 2004; Edris 2007; Turek and Stintzing 2013; Smeriglio et al. 2018). They are also bound with carbohydrates and present in the form of glycosides. The glycosidic linkage is hydrolyzed to liberate such volatile compounds which can be achieved by allowing enzymatic reactions during wilting before distillation (Dewick 2002; Evans 2002; Baser and Demirci 2007). The chemical composition of volatile oil differs according to the distillation method, environmental and geographic conditions, harvesting time, age, and organ of the plant (Burt 2004; Boukhatem et al. 2014). The exact reason for their occurrence in the natural system is still not completely known which paves a way for further research. However, some reports suggest that they are produced by the organism as a secondary metabolite for defense mechanism or signaling (Belitz et al. 2004; Breitmaier 2005; Baser and Demirci 2007). The European Council defines "essential oil" as a product obtained from "vegetable raw material" (Anonymous 2000). The well-known essential oil-bearing families are *Zygophyllaceae*, *Zingiberaceae*, *Santalaceae*, *Rutaceae*, *Piperaceae*, *Pinaceae*, *Myrtaceae*, *Lauraceae*, *Lamiaceae*, *Hypericaceae*, *Cupressaceae*, *Asteraceae,* and *Apiaceae* (Baser and Demirci 2007).

The volatile oils are secondary metabolites produced by the plant system in response to stress conditions or fighting parasitic and infectious agents (Rauha et al. 2000). They are obtained by distillation of buds, flowers, fruits, leaves, and seeds and employed in food and beverages industries (as flavoring and preservatives), perfumes (fragrances and aftershaves), and pharmaceuticals for their therapeutic potential (Zygadlo and Juliani 2000). Volatile oils from spices and culinary herbs possess antimicrobial and antifungal activities; hence, used in traditional medicine (Hammer et al. 1998, Kamble and Patil 2008). The broad-spectrum antimicrobial activity is linked with the phenolic compounds and purified volatile components such as thymol, linalool, eugenol, and carvacrol (Bagamboula et al. 2004). Ten major aromatic crops contribute to 80% of the global market of volatile oil while the rest 20% is provided by 150 crops. The major essential oil-producing countries are Mexico, Indonesia, India, Guatemala, Egypt, China, and Brazil while the main consumers include Japan, the US, and Western Europe (Anonymous 2003). The demand for essential oil in the global market was estimated to be 247.08 kilotons in the year 2020 and the compound annual growth rate (CAGR) from 2020 to 2027 is expected to be 7.5% (EOMSA Report 2020). India contributes around US$ 500 million from US$ 24.10 billion to the global fragrance

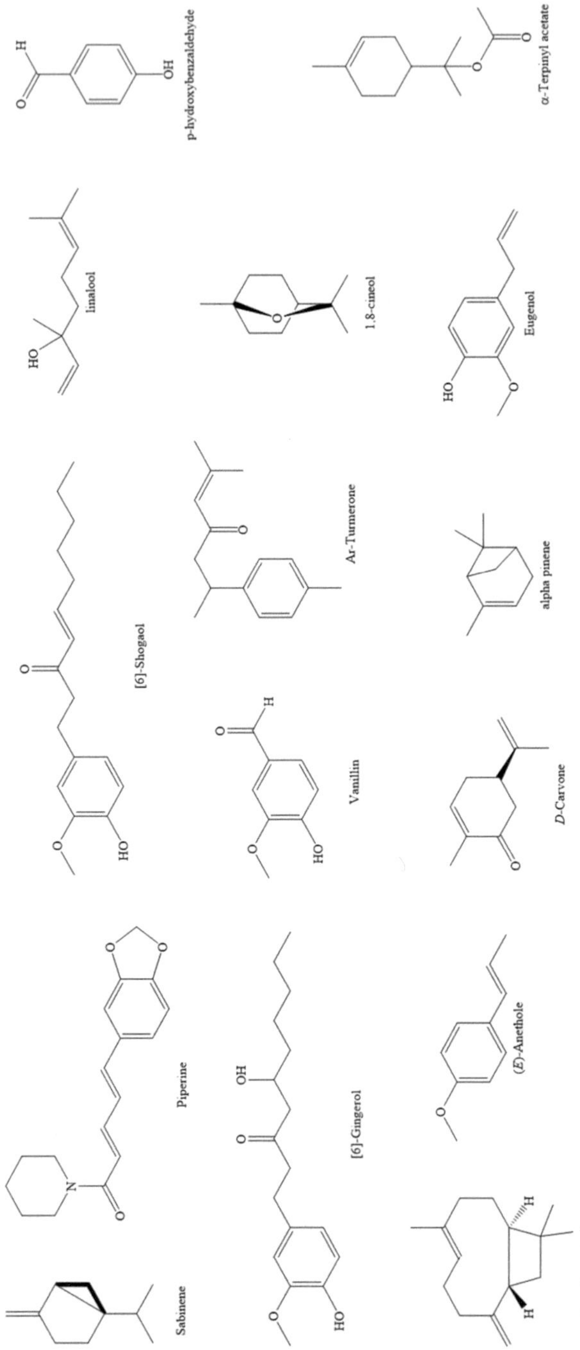

Fig. 9.2 Structure of some important flavor compounds

and flavor industry. However, India has had a growth rate of 11% in the last few years and is expected to grow exponentially in the upcoming years (MSME 2020). The global production of essential oils for exotic fragrances, processing, and flavors was estimated to be 4344, 80,410, and 21,670 tons, respectively, and India contributed around 4%, 21%, and 14% of the global production, respectively (Sanganeria 2010).

9.4 Origin of Flavoring Compounds

Volatile oils are mainly used in aromatherapy, cosmetics and toiletries, fine chemicals, flavor and fragrance, food and beverages, perfumery, and pharmaceutical industries. Essential oils mainly contain volatile components of terpenoid or non-terpenoid origin. The volatile compounds are hydrocarbons or their oxygenated derivatives and some may contain sulfur or nitrogen derivatives. The essential oil is composed of mono, di, and sesquiterpenes and exists as acids, alcohols, aldehydes, amines, epoxides, esters, ketones, sulfides, etc. Some fatty acid, phenylpropanoids, fatty acid esters, or their products are also regarded as volatiles (Mann et al. 1994; Torssell 1997; Pybus and Sell 1999; Steward 2005; Baser and Demirci 2007). Volatile oils are classified as terpenoids and non-terpenoid hydrocarbons.

9.4.1 Non-terpenoid Hydrocarbons

Non-terpenoid hydrocarbons include short-chain aldehydes and alcohols which are formed by metabolic degradation or conversion of fatty acids and phospholipids. They consist of carbon, hydrogen, and sometimes also contain nitrogen, oxygen, or sulfur (Croteau and Karp 1991).

9.4.2 Terpenoids

Terpenes, also known as isoprenoids, are formed by rearrangements of two or more isoprene units and form an essential component of volatile oil. They are one of the largest classes of natural compounds with over 30,000 terpenoids isolated from animals, microorganisms, and plant sources (Dev 1989; Barton et al. 1999; Dewick 2002; Theis and Lerdau 2003; Anonymous 2006). They are classified as hemiterpenes, monoterpenes, sesquiterpenes, diterpenes, etc. Table 9.2 includes types, molecular formula, and examples of various terpenes (Sticher 1977).

Hemiterpenes contain one isoprene unit hence isoprene itself is the only compound in this category; however, compounds such as isovaleric acid and prenol containing oxygen molecules are also considered as hemiterpenoids (Dev 1989; Barton et al. 1999; Dewick 2002; Torssell 1997; Dey et al. 1991; Baser and Demirci 2007). Generally, volatile oil does not contain heavier terpenes like diterpenes. Ruzicka (1953) proposed the "biogenetic isoprene rule" which states that the

Table 9.2 Types and examples of various terpenoids

Terpenoid types	Number of isoprene units	Molecular formula	Example
Hemiterpenes	1	C_5H_8	In combination with other compounds such as coumarins, quinines
Monoterpenes	2	$C_{10}H_{16}$	Volatile oil, geraniol, iridoids, limonene, linalool, myrcene, pinene, terpineol
Sesquiterpenes	3	$C_{15}H_{24}$	Volatile oil, bitter principles, farnesenes, farnesol, humulene
Diterpenes	4	$C_{20}H_{32}$	Resin aids, cafestol, cembrene, gibberellins, kahweol, phytol, retinal, retinol, taxadiene, vitamin A
Triterpenes	6	$C_{30}H_{48}$	Sterols, steroids, saponins, squalene
Tetraterpenes	8	$C_{40}H_{64}$	Carotenoids, lycopene
Polyterpenes	n	$(C_5H_8)_n$	Rubber, gutta

Source: Sticher (1977)

terpenes are formed by the head-to-tail rearrangement of two or more isoprene units. This rule also specifies that terpenes are formed from aliphatic precursors such as monoterpenes from geraniol, sesquiterpenes from farnesol, diterpenes from geranylgeraniol, and triterpenes from squalene.

9.5 Biosynthesis of Terpenes

The biosynthesis of almost all terpenoids takes place from isopentenyl pyrophosphate (IPP) and dimethylallyl pyrophosphate (DMAPP). These precursors contain simple five-carbon molecules. They are synthesized by the mevalonate pathway (MVA) from two acetyl-CoA molecules through mevalonic acid. A second pathway for the biosynthesis of IPP and DMAPP was discovered which does not involve the formation of mevalonic acid; hence, named mevalonate-independent or non-mevalonate (or deoxyxylulose phosphate) pathway. The 1-deoxy-D-xylulose-5-phosphate (DXP) and 2 C-methyl-D-erythritol-4-phosphate (MEP) act as the precursor for the biosynthesis of IPP and DMAPP in the non-mevalonate pathway. This pathway starts with the condensation of glyceraldehyde 3-phosphate and pyruvic acid to form DXP and later IPP and DMAPP are formed following a series of reactions (Fig. 9.3) (Theis and Lerdau 2003; Rohdich et al. 2005; Baser and Demirci 2007).

IPP and DMAPP lead to the formation of an immediate precursor of monoterpenoids geranyl pyrophosphate (GPP) which on following further reactions leads to the formation of nerylpyrophosphate (NPP). NPP acts as a precursor for various acyclic, cyclic, bicyclic, or tricyclic compounds. The various reactions such as hydration, oxidation, rearrangement, and reduction in terpene cyclases lead to the formation of several terpene derivatives. GPP and IPP condensed to form an

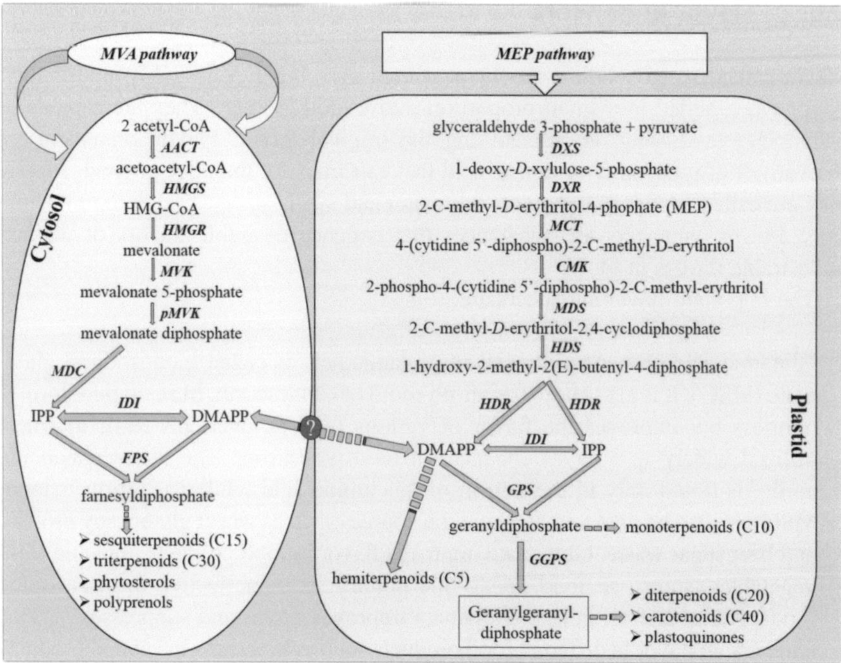

Fig. 9.3 The two pathways for terpenoid synthesis pathways, MVA in the cytosol and the MEP in the plastid. The single step is denoted by solid arrows while multiple steps are represented by dashed arrows. *AACT* Acetoacetyl-coenzyme A thiolase, *CMK* 4-Diphosphocytidyl-2-Cmethyl-D-erythritol kinase, *DMAPP* Dimethylallyl diphosphate, *DXR* 1-Deoxy-D-xylulose-5-phosphate reductoisomerase, *DXS* 1-Deoxyxylulose-5-phosphate synthase, *FPS* Farnesyl diphosphate synthase, *GGPS* Geranylgeranyl diphosphate synthase, *GPS* Geranyl diphosphate synthase, *HDR* 4-Hydroxy-3-methylbut-2-enyl diphosphate reductase, *HDS* 4-Hydroxy-3-methylbut-2-enyl diphosphate synthase, *HMGR* HMG-CoA reductase, *HMGS* 3-Hydroxy-3-methylglutaryl (HMG)-CoA synthase, *IDI* IPP/DMAPP isomerase, *IPP* Isopentenyl diphosphate, *MCT* 2-C-methyl-Derythritol 4-phosphate cytidylyltransferase, *MDC* Mevalonate diphosphate decarboxylase, *MDS* 2C-methyl-D-erythritol 2,4-cyclodiphosphate synthase, *MVK* Mevalonate kinase, *pMVK* Phosphomevalonate kinase. *Source:* Champagne and Boutry (2016)

immediate precursor of sesquiterpene farnesyl pyrophosphate (FPP). Likewise, FPP and IPP are condensed to form diterpenoids (Baser and Demirci 2007). The mevalonate pathway is situated in the cytoplasm of the higher plants where the synthesis of the most sesquiterpenoids occurs while non-mevalonate pathway is located in the chloroplast of all phototropic organisms as well as most bacteria and accountable for the biosynthesis of plastid-related isoprenoids (mono and diterpenes). However, both mevalonate and mevalonate-independent run independently in most algae and higher plants and their inter-relationship is shown in Fig.9.3 (Rohmer 2003; Steinbacher et al. 2003; Rohdich et al. 2005).

9.6 Flavor Enhancers

Flavor enhancers have little or no flavor of their own, but they desirably modify their flavor when added in a small proportion into a food product. They are capable of enhancing, modifying, or intensifying the original flavor. Flavor enhancers are derivatives of nucleotide or amino acid that can improve the odor of food. Mostly they are called as savory flavorants or umami and synthesized as calcium or sodium salts. Flavor enhancers are substances that enhance desirable flavors or depress undesirable flavors in foods.

The important flavor enhancers are:

1. Glutamic acid salts: Sodium salt of glutamic acid is called monosodium glutamate (MSG). It is also known as aji-no-moto or Chinese salt. MSG in pure form is odorless but improves the flavor of various food products hence of the most utilized widely as a flavor enhancer in food processing. The flavor enhancing ability is found only in the L-form of this amino acid while its D-form is inert. MSG can also be prepared from natural sources such as wheat gluten, soy protein, and beet sugar waste. Glutamate improves flavor in food products obtained from vegetables, soups, seafood, meat, and poultry. It is ineffective in sweet-spicy foods, fruits, or fruit juices. Glutamate improves flavor and suppresses various undesirable flavor in different food products such as bitterness in canned products of fish, meat, stews, soups, earthiness of potatoes and rawness of many vegetables, the sharpness of onion. Glutamates stimulate our taste receptor, tactile sense, and increase salivation. The excessive consumption of MSG causes "Chinese restaurant syndrome" with symptoms like stiffening joints, stomach ache, headache, and drowsiness.
2. 5′-nucleotides: Generally, 5′-guanylate and 5′-inosinate enhance flavor similar to MSG, but they do not affect sweet and sour flavor. They also improve the viscosity of liquid foods. They also act synergistically with glutamate. It is generally used in processed foods like canned ketchup, sauces, soups, chips, dry, peanuts, potato, etc.
3. Glycine salts: A simple amino acid that is generally used along with glutamic acid as a flavorant.
4. Guanylic acid salts: Nucleotide salts that are generally used along with glutamic acid as a flavorant.
5. Organic acids: They are generally not considered and regulated as flavorants by law but can impart different taste that alters the flavor of food:
   (a) Acetic acid: It imparts vinegar's sour taste and distinctive smell.
   (b) Citric acid: It is found in citrus fruits and responsible for their sour taste.
   (c) Lactic acid: It is found in various milk products and offers a rich tartness to milk products.
   (d) Malic acid: It is located in apples and responsible for their tart or sour taste.
   (e) Tartaric acid: It is present in grapes and wines and impart their tart taste.

9.7 Encapsulation of Flavors

The quality and acceptability of any food are determined mainly by its flavor. The addition of even a small quantity of flavor improves smell, taste, consumer satisfaction, and the final product's market price. However, most flavors are highly volatile, sensitive to light and heat, and susceptible to oxidation and other chemical reactions. To overcome this problem, flavors are coated with a protective carrier material that provides resistance to evaporation and undesirable changes during processing and food storage. Encapsulation helps to retain flavor in food, protects against interaction with other food ingredients, improves resistance against light-induced reactions and oxidation, and allows sustained release (Sun et al. 2013). Encapsulation is a method in which a material or a mixture of materials is entrapped within another material. The coated material is known as core material, while the coating material is called wall material or shell or carrier (Madene et al. 2006). Encapsulation protects flavor degradation from acidity, chemical changes, heat, ingredients, moisture, etc. A schematic representation of various encapsulation processes is shown in Fig. 9.4 (Madene et al. 2006).

The polysaccharides such as cyclodextrins, gum arabic, maltodextrins, and modified starches are used as coating material during encapsulation. The encapsulation proceeds via extrusion, spray drying, or formation of inclusion complexes. The application of various encapsulation methods in the food industry is given in Table 9.3. During spray drying, the flavoring substances are emulsified in a solution or suspension of the polysaccharide, which contains solutizer in addition to the emulsifying agent. In preparation for extrusion, a melt of wall material, flavoring

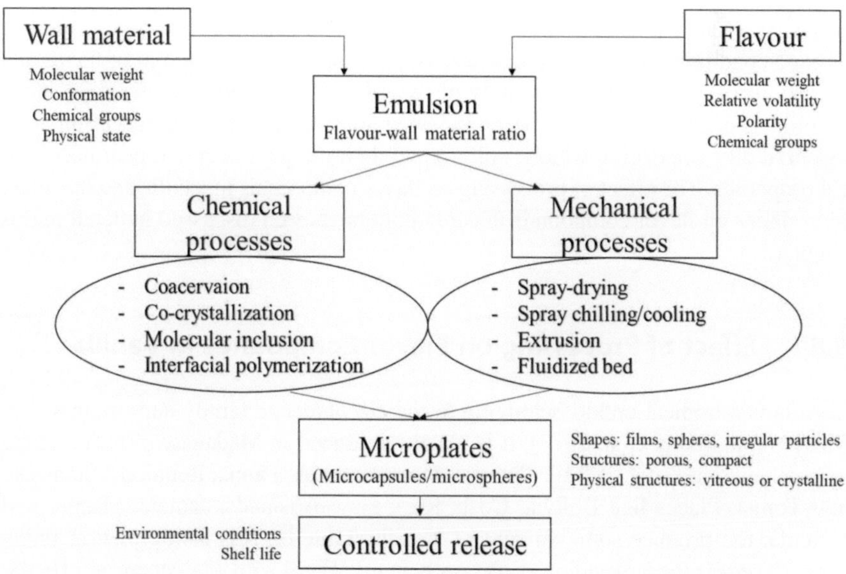

Fig. 9.4 A schematic representation of various encapsulation process (Madene et al. 2006)

Table 9.3 Application of various encapsulation methods in the food industry

Encapsulation techniques	Applications of different encapsulation method in food industry Encapsulation form	Application area
Coacervation	Paste/powder/capsule	Bakery items, chewing gum, toothpaste,
Extrusion	Powder/granule	Confectionery, instant beverages, tea
Fluid bed drying	Powder/granule	Confectionery, prepared dishes
Molecular inclusion	Powder	Confectionery, extruded snacks, instant drinks
Spray cooling/ chilling	Powder	Ices, prepared dishes
Spray drying	Powder	Confectionery, instant beverages, instant desserts, food flavors, milk powder

Source: Madene et al. (2006)

substances, and emulsifiers are produced. The extrusion is conducted in a cooling bath, e. g., isopropanol while β-Cyclodextrins, generally used to develop inclusion complexes. Together with the flavoring substances, they are dissolved in a water/ ethanol mixture by heating. The complexes precipitate out of the cooling solution and are removed by filtration and dried. Criteria for evaluating encapsulated flavors include the amount of flavoring substance adhering to the surface of the capsule, average diameter of the capsules, the concentration of flavoring substance, and stability of the flavor (Madene et al. 2006).

The volatile components are responsible for characteristic flavor in a natural plant material which is affected by environmental conditions, variety, genetic factors, maturity at harvest, production practices, postharvest handling, and processing and storage conditions (Goncalves et al. 2018). For a better understanding of flavor and its chemistry, the detailed case study of two globally utilized aromatic spice crops (vanilla and cardamom) has been reviewed to interpret the differences in their essential oil composition which contribute to their unique flavor and pharmacological properties. The effect of processing on flavor compounds in vanilla and the effect of varieties on flavor compounds in cardamom have been discussed in detail in this chapter.

9.8 Effect of Processing on Flavor Compounds in Vanilla

Vanilla is a tropical orchid belonging to the *Orchidaceae* family and originated in Mexico (Anuradha et al. 2013). It is widely produced in Madagascar (80%), along with other places such as Bali, Comoro, Jawa, Lombok, Papua, Reunions, Sulawesi, and Tonga. Places like Bolivia, Costa Rica, Hawaii islands, Jamaica, Tahiti, and Uganda, too produce some amount of Vanilla. Latterly, it is also grown in India. *Orchidaceae* is the largest flowering family in the World with 788 genera and 18,500 species. The three most important species out of 110 identified vanilla used for business and agriculture are *Vanilla tahitensis*, *Vanilla pompona,* and *Vanilla fragrans* also known as *Vanilla planifolia* (Ranadive 1994; Reineccius 1994; Webster 1995). But among them, *V. planifolia* is rampantly cultivated for its vital role as flavoring and food additives (Purseglove et al. 1988). The various aromatic compounds formed in different stages of processing/curing are responsible for its characteristic flavor/aroma, among which vanillin, the principal component comprises one-third of total flavor. However, the rest is contributed by other volatile and nonvolatile components (Ranadive 1994). Vanilla and its extracts play a major role in confectionery, food and beverage, flavor and fragrance as well as pharmaceutical industries (Funk and Brodelius 1994).

Curing is a vital process for the development of the aroma in odorless green vanilla beans which consists of drying of vanilla beans and making it undergo chemical and enzymatic reactions (Dignum et al. 2001). Every country which cultivates vanilla has invented its own curing process, but the most common steps in all processes are scalding, sunning/sweating, drying, and conditioning which generally takes around 6–8 months (Rao and Ravishankar 2000). The characteristic vanilla flavor is developed by an enzymatic reaction during curing of beans (Dignum et al. 2001). The *Vanilla planifolia* extract has a range of related phenylpropanoid compounds along with vanillin which gives it a unique vanilla flavor (Clark 1990). A small amount of free vanillin is present in green vanilla beans (pods) but during the curing, most of the aroma precursors (β-D-glucosides) combine with β-D-glucosidases (Walton et al. 2003).

Globally, people are moving towards natural products; hence, the demand for vanilla as a major food flavoring agent will surely increase. The farmers too are enthusiastic to cultivate it more because of high price offers from western countries. But, the lack of proper scientific knowledge of cultivation and processing among the farmers leads to low-quality beans, which abstains them from the expected profits (Anuradha et al. 2013).

9.8.1 Curing and Processing Technology of Vanilla

Green vanilla beans (pods) are generally odorless and flavorless even though they contain a small proportion of vanillin. During harvest after maturity, they develop a faint phenolic odor, unlike cured vanilla. At commercial usages, the curing of mature pods brings out the typical aroma, color, and flavor and it is dried before storage to prevent its spoilage. The curing process enables biochemical, chemical, and physical changes which contribute to the desired characteristics. Curing is a type of ripening which involves alternative sweating and drying of green beans in a controlled environmental condition until they lose 80% of their moisture as in Mexican Beans (Correll 1953). Curing is a crucial stage of vanilla production where pods undergo an enzymatic reaction which aids in the development of its typical flavor and aroma (Ranadive 1994). Every region with vanilla cultivation has developed its own appropriate method of curing based on the available resources and expertise gained during its processing. The significant quality to be achieved through curing is primarily the characteristic aroma/flavor, along with its appearance, flexibility, size of the bean, and vanillin content. Despite different curing methods, the key steps involved in proper curing are killing, sweating, drying, and conditioning (Anuradha et al. 2013).

9.8.1.1 Killing

This stage involves blocking the further vegetative development of fresh beans. The enzymatic reaction is initiated for the development of the typical aroma and flavor. The cell structure may be disrupted by the methods such as ethylene gas treatment, freezing, hot water scalding, scarification, sun and oven wilting but the most practiced method is hot water, oven, and sun killing. The process is known as killing because the physiological death induced by disrupting the respiratory function by destroying cell walls and cell membranes ultimately kills the bean tissue (Ranadive 1994).

9.8.1.2 Sweating

This method succeeds the killing process and should be carried out with great precision to prevent the production of inferior quality beans. In this step, initially, moisture is allowed to escape rapidly to reduce the risk of spoilage, but sufficient moisture is retained for further enzymatic reactions (Ranadive 1994). The process is carried out in sweat boxes or enclosed rooms for over 7–10 days. This is the main step where the development of the typical aroma, color, and flavor takes place in vanilla beans.

9.8.1.3 Drying

After the sweating process, the cured beans are brown with a characteristic aroma, still contains 60–70% of moisture, which needs further drying to achieve moisture content between 25 and 32% to protect beans from microbial spoilage and allow further beneficial chemical reactions. The lower moisture content also halts undesirable enzymatic reactions and biochemical changes in the beans (Anuradha et al. 2013).

9.8.1.4 Conditioning

The final step in the curing process is conditioning, which involves storing beans in closed boxes from 1 month to several months. This environment aids in the plethora of biochemical and chemical reactions like etherification, esterification, oxidative degradation, etc., which results in the production of varied aromatic components and enhances the overall flavor quality of cured beans (Anuradha et al. 2013).

9.8.2 Various Vanilla Curing Process

Different curing methods are named after its place of origin, like Bourbon process, Guyana process, Mexican process, and Peruvian process. Bourbon or Mexican process is the most suitable curing method for geographic and environmental conditions in India (Anuradha et al. 2013). A slight modification in the curing process may be incorporated for the production of the best quality cured beans. The beans must be protected from external odors during each stage of processing; hence, the bins, blankets, boxes, drying hall, and planks should be thoroughly cleaned for the best quality cured beans.

9.8.2.1 Mexican Curing Process

The specialist curing firms producing the majority of the vanilla crop in Mexico mostly employ the traditional form of curing such as oven and sun wilting in Mexican curing process (Merory 1968; Theodose 1973). The Mexican curing process is shown in Flow chart 9.1.

In this method, the pods are spread in shade for 7 days which enables them to lose some moisture and shrink. Then sweating is allowed by spreading the shrunken pods on a thick dry woolen blanket under the open, hot sun but the woolen blanket is folded to cover the beans perfectly during the mid-day after beans have absorbed sufficient heat and kept as such open till evening. But later in the evening, small bundles of beans are made and stored overnight in sweating boxes which are lined by the woolen blanket. This is the stage when the fermentation process is initiated and pods undergo fermentation and sweating in the sweating boxes. This whole process starting with spreading of pods under the hot sun till storing the bundles overnight in sweating boxes has to be repeated for 2–3 days so that pods become more flexible and softer with chocolate brown color. The drying process follows after completion of the sweating steps, where the beans are dried by spreading on grass mats under the hot sun for 6–7 days but in the evening the beans should be packed and kept in sweating boxes. Later they are completely dried in shade and arranged in groups of 2 or 3 sizes, tied into bundles of 50 as per size for further conditioning, and finally stored in closed boxes for 3 months or more for the development of their characteristic aroma and flavor. The overall quality of the beans increases with the longer aging period (more than 3 months) and aged beans are preferred over fresh cured beans. The vanillin content in the Mexican cured beans varies between 4.15 and 4.40% (Anuradha et al. 2013).

9.8.2.2 Bourbon Curing Process

The Bourbon curing method is named after the Island of Reunion, which was formerly known as Bourbon. Madagascar, which is the largest producer of vanilla in the globe follows Bourbon method for curing vanilla beans. This curing method differs from the Mexican process wherein scalding in hot water is employed for the killing of beans followed by sweating steps. The higher moisture in the Bourbon as compared to the corresponding Mexican grade propels it for frequent frosting. A slight modification of this procedure is employed in various parts of the world (Lionnett 1959). In Bourbon process, either hot water or hot water vapor pre-curing is followed. The Bourbon curing process is shown in Flow chart 9.2.

Some variations are incorporated in the traditional methods help to achieve better yield irrespective of weather condition. The first crucial step in vanilla curing is harvesting which should be done when the apical end of the bean turns yellow

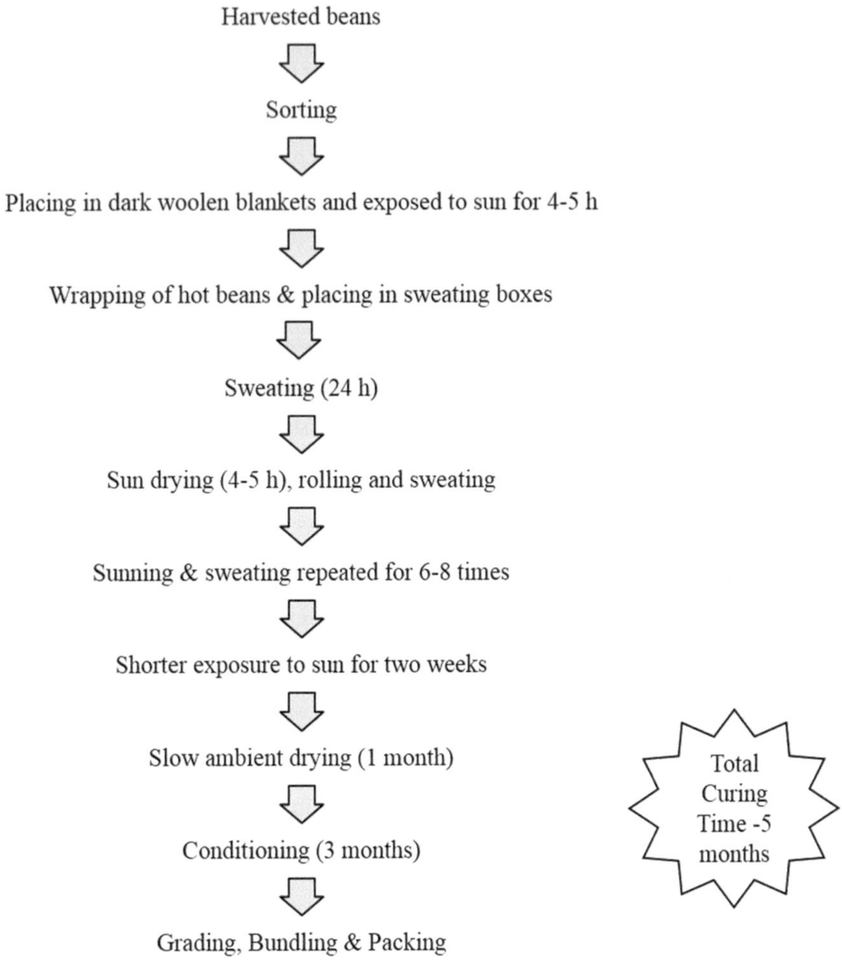

Flow Chart 9.1 Mexican curing process. (*Source:* Anuradha et al. 2013)

because the glucovanillin content was reported to be highest at full maturity (Ranadive 1994). Secondarily, cutting beans into small pieces or scratching their surface increases the surface area which results in higher contacts between substrates and enzymes (Anuradha et al. 2013). Furthermore, hot water scalding (at 65 °C for 2–3 min) of green bean followed by cutting (0.5–1 cm) and wrapping of cut pieces in wax paper and their subsequent sweating in closed containers (38 °C for 48 h) is desirable to obtain cured beans of comparable flavor to traditionally processed beans. Consequently, the beans are dried for 48 h at 38 °C on open trays and finally conditioned in a closed container for 2–3 months at 38 °C (Broderick 1956).

Madhava Naidu et al. (2009) proposed a newer, easy, and faster method for vanilla extract preparation from vanilla pods wherein after the size reduction, vanilla

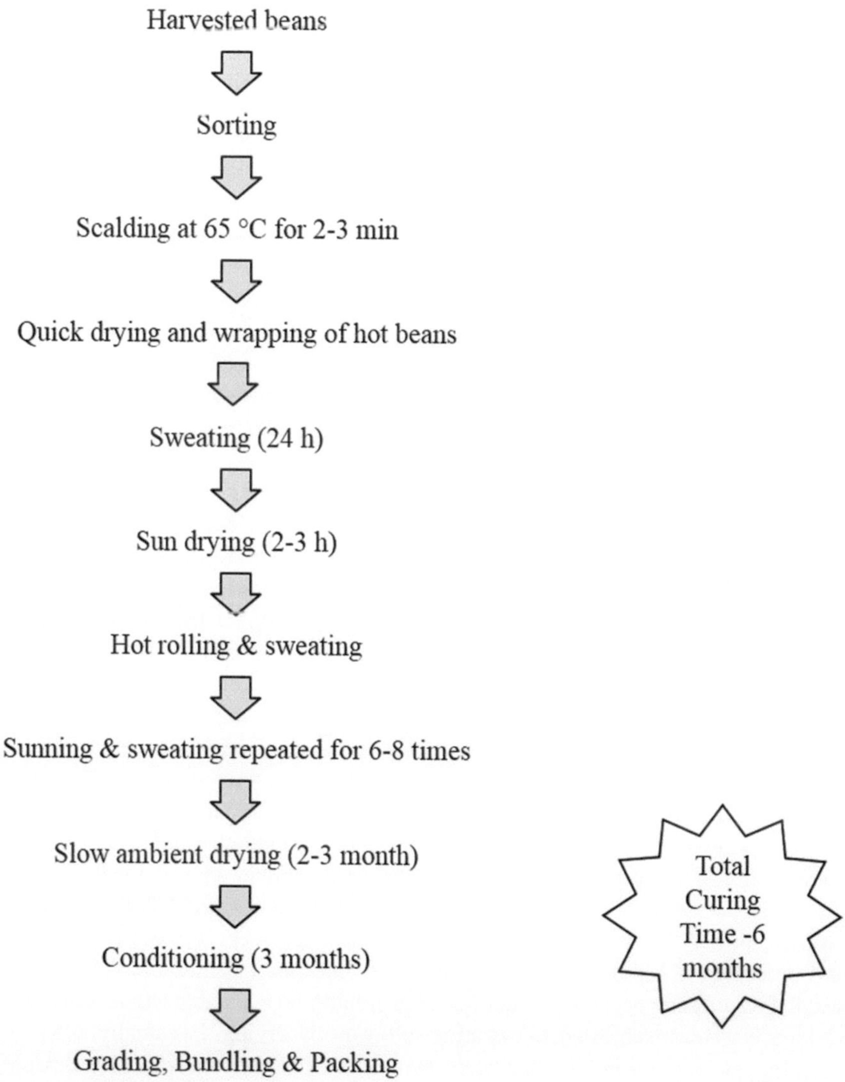

Flow Chart 9.2 Bourbon curing process. (*Source:* Anuradha et al. 2013)

beans were mixed with tea leaf enzyme extract (TLEE) in a suitable ratio and incubated for enzymatic actions on vanilla flavor precursors. Afterward, the mix was squeezed and filtered, and the filtrate was treated with ethanol to obtain vanilla flavor. The extracts obtained after TLEE treatment were reported to have a higher vanilla flavor intensity, vanillin content (4.2%), as well as floral and sweet notes in comparison with viscozyme treated extract (2.4%). The electronic nose analysis further differentiated both the extracts. It was established that TLEE is an easier method which avoids time-consuming and labor-intensive traditional curing method and helps in obtaining a higher vanilla extract and superior quality vanilla flavor, which is much easier than the traditional methods.

9.8.2.3 Traditional Curing Process and Improvements

Curing is a laborious process of drying followed by chemical and enzymatic reactions which result in the development of flavor in the vanilla pods. Despite various curing methods developed by different vanilla-growing countries, the four major steps involved in curing are scalding or killing, sweating, drying, and conditioning. The scalding involves stopping the vegetative phase in the fresh beans by submerging them in the hot water for 2–3 min and cell disruption causes contact of the enzyme with the substrate which results in the development of the characteristic aroma and flavor. In the sweating/sunning stage, the activity of oxidase and glucosidase are high which offer characteristic chocolate brown color to the beans. Sunning means spreading of beans under the hot sun until they absorb sufficient heat, then they are covered with sheets and stored overnight in air-tight boxes for sweating. The process of sunning is repeated for 1–2 weeks (Ranadive 1994). During the drying process, the beans are dried to achieve lower moisture content which prevents beans from microbial spoilage and allow other beneficial reactions for flavor development. Indoor drying which usually avoids direct sunlight is carried out indoors for 2–4 weeks and air circulation is maintained using fans. Conditioning is the most lengthy step where the beans are preserved in a conditioned room for about 6 months. This environment helps in etherification, esterification, and oxidation reactions which result in enhanced flavor quality in the cured beans (Anuradha et al. 2013).

Recently, Central Food Technological Institute, Mysore, India has devised a newer, easy, and faster process know-how of curing vanilla beans for commercialization as shown in Flow chart 3 (Sreedhar et al. 2007; Madhava Naidu et al. 2008). It incorporates hot air drying along with the conventional curing method in an organized curing facility that ensures the same quality at a lower labor cost. In this process, the harvested vanilla pods are manually arranged in groups of different sizes and processed abruptly. The graded beans are filled in a bamboo basket and are scalded by submerging in hot water for a few minutes at a suitable temperature. The scalded beans are subjected to sweating followed by drying. Eventually, after the sweating and drying process, the beans become half of their weight, very flexible, turn dark brown with wrinkles, and contain enhanced vanilla flavor. Then the beans are allowed for slow drying in a properly ventilated room for 6–8 days. Finally, the conditioning takes place for about 10–15 days where beans are stored in polythene covers or air-tight chests or wax paper-lined metal containers to develop the characteristic aroma in the vanilla (Flow chart 9.3). The exact nature of reactions happening in this complex phenomenon during vanilla curing is still a mystery (Odoux 2010).

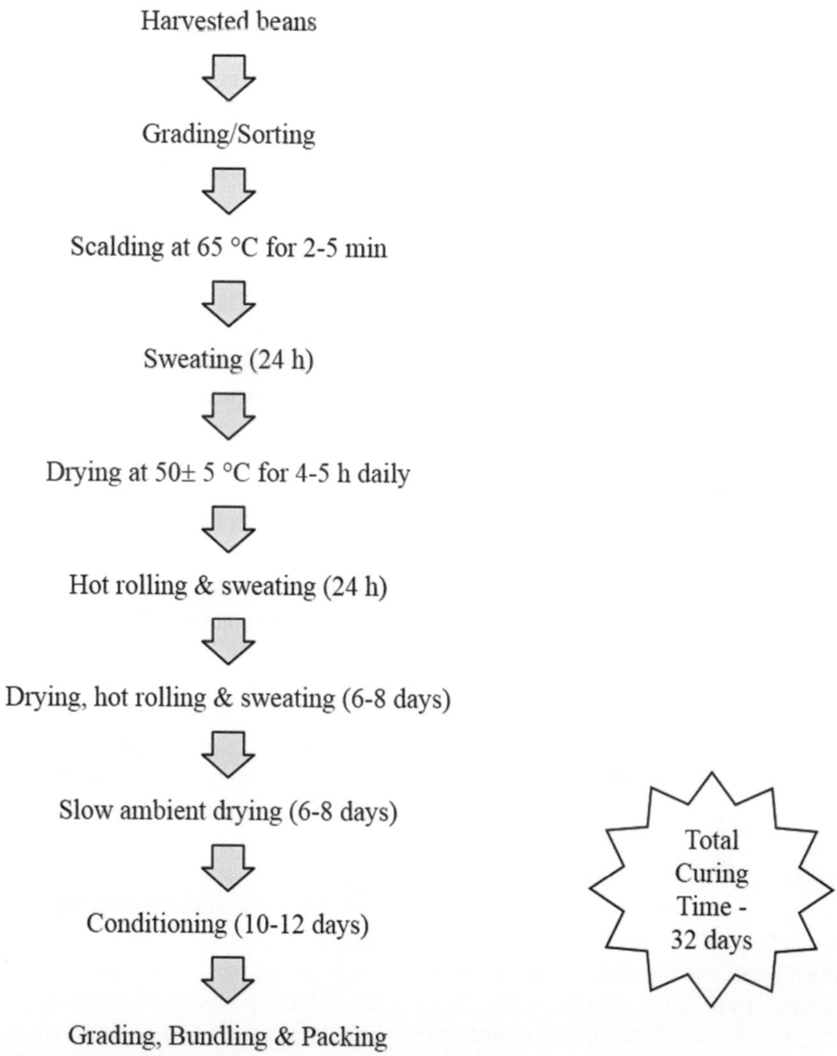

Flow Chart 9.3 Faster curing process. (Source: Anuradha et al. 2013)

9.8.3 Chemistry of Vanilla

9.8.3.1 Biosynthesis of Flavor Compounds in Vanilla

The glucovanillin concentration increases continuously from the base of the beans to its apical end. The maximal concentration of glucovanillin is present in the center of the beans where seeds and placental tissue are present; however, β-glucosidase is mostly concentrated in the outer part of the beans. Therefore, during the initial process of curing, the glucovanillin present in the central part has to be diffused to the outer region of the bean to come in contact with the hydrolyzing enzyme for their complete hydrolysis (Anuradha et al. 2013).

After the third month of pollination, beans become fully matured where the development of glucovanillin begins. Kanisawa (1993) discovered glucovanillin during a growth experiment of beans on the vine. This was further confirmed by another researcher during the hydrolyzing action of β-glucosidase on glycosides (Brodelius 1994). After the third month of pollination, the enzymatic activities of glucosidase, peroxidase, and polyphenol oxidase also increase while the enzymatic activity of proteinase is lost during this period; hence, does not show any effect on the curing process (Wild-Altamirano 1969). An acceleration in the oxidation process is observed in the injured tissues after the scalding step. This results in the production of a higher amount of carbon dioxide by beans. Thus, peroxidases have a pivotal role in curing. Researchers found that a crucial role is also played by oxidases which produce condensed stable pigments and quinines during the curing process (Arana 1944; Jones and Vincente 1949). Although all enzymatic activities are ceased during heating at 120 °C, however during the later stage of curing some peroxidase activity can be seen and the beans remain green without any vanilla flavor. Similarly, after 2 days of sweating when beans are heated no formation of vanilla flavor is noticed, but after a few weeks peroxidase activity is regained. This proves that crucial enzymatic changes occur during the first week of curing, even though it was believed that nonenzymatic actions during the lengthy conditioning period are the major cause of aroma formation (Jones and Vincente 1949). There is another presumption that the peroxidases carry out further oxidation of these aroma compounds and produce other compounds having different flavor. So, this explains that the quality of vanilla does not depend solely on vanillin content (Kanisawa et al. 1994). During the first week of curing, the enzymatic activity of glucosidase is high and then plummets rapidly however the oxidase activity drops more slowly (Anuradha et al. 2013).

The investigation on the effect of scalding on the enzymatic activities revealed that geographic origin does not influence the maturation and vanillin biosynthesis in green beans and the observed variations between the cured ones occur during curing process itself. It has also been reported that the soluble protein content decreases after scalding at 65 °C. After 7 months of harvest, the concentration of glucovanillin and β-glucosidase is found to be very high. Moreover, glucosidases are susceptible to heat and most of its enzymatic activity is lost after scalding and decreases drastically after 1 day of scalding (Ranadive 1994). If the scalding step inhibits β-glucosidase activity then it can be concluded that nonenzymatic reactions play a major role in the production of characteristic vanilla flavor.

The most noticeable feature observed during curing is the formation of free vanillin and other related compounds such as 4-hydroxy-benzaldehyde when β-D-glucosidase reacts with the vanillin β-D-glucoside and other related β-D-glucoside (Ramachandra and Ravishankar 2000; Dignum et al. 2001). Cured beans are reported to contain 2–2.5% of vanillin and over 200 other minor compounds.

The biosynthesis of β-D-glucoside is very complicated unlike the development of vanillin during the curing process and many biosynthetic pathways have been proposed for its production. But, the formation of *β*-D-glucoside from hydroxycinnamic acid precursor involving chain-shortening and other reactions is still not very clear. An earlier study proposed the route for the formation of vanillin and vanillic acid from ferulic acid utilizing radioactively labeled ferulic and vanillic acids (Anuradha et al. 2013). The formation of vanilloyl-CoA was reported to occur through the CoA-dependent β-oxidative cleavage of feruloyl-CoA which on reduction directly forms vanillin or it would undergo deacylation reaction to form vanillic acid (Anuradha et al. 2013).

The biosynthesis of glucovanillin or vanillin in conifer occurs either through glucovanillin or coniferyl alcohol. The results of thin-layer chromatography (TLC) and Infrared (IR) spectroscopy confirmed that the precursor of vanillin is ferulic acid and it is not the precursor of coniferin. A previous investigation utilizing (O-$^{14}CH_3$)-labeled ferulic acid confirmed the formation of 100% vanillin after administering labeled ferulic acid to the bean (Anuradha et al. 2013). The vanillic acid which was not derived from vanillin was found to contain labeled ferulic acid. The glucosylation of the aroma compounds is still unknown.

Various researchers isolated and identified around 30 glycosides from green beans either directly by nuclear magnetic resonance (NMR) and high-performance liquid chromatography (HPLC) or after β-glucosidase treatment (Kanisawa 1993). A pathway was proposed for the biosynthesis of vanillin and other phenolic compounds in *Vanilla planifolia* beans (Kanisawa et al. 1994). This pathway leads to the formation of glucovanillin from 4-coumaric acid through *p*-hydroxybenzaldehyde glucoside (Fig.9.5). This biosynthetic pathway agrees with the previous outcomes using labeled ferulic acid but contradicts the findings of conifer. The vanillin biosynthetic pathways using metabolic engineering tools were reported in *Vanilla planifolia* (Havkin-Frenkel and Belanger 2007).

9.8.3.2 Identification of Flavor Compounds in Green Vanilla Beans

Mostly glycosides are the major flavor component (including vanillin) in green vanilla beans and were isolated using methanol (Kanisawa 1993). The green vanilla bean extract was selectively eluted using an Amberlite XAD-2 column, HPLC and silica gel chromatography and mass spectrometry (MS), as well as NMR, were used for identification of the major glucosides (Table 9.4). The most abundant glucosides identified in green vanilla beans were glucoside A (bis {4-(β-glucopyranosyloxy)-benzyl}-2-isopropyltartarate), glucoside B (bis {4-(β-glucopyranosyloxy)-benzyl}-2-(2-butyl)-tartarate), and glucovanillin. Various other plants of *Orchidaceae* family have been reported to contain Glucosides A and B which belong to the loroglossins (Kanisawa et al. 1994). The green vanilla bean extract was treated with β-glucosidase and the minor glucosides (Table 9.4) were identified using gas chromatography-mass spectrometry (GC-MS).

9.8.3.3 Identification of Flavor Compounds in Cured Vanilla Beans

Vanillin is the most important flavor component in vanilla. A plethora of other volatile compounds in vanilla is also accountable for its characteristic flavor with other notes such as balsamic, creamy, fruity, herbaceous, phenolic, spicy, sweet,

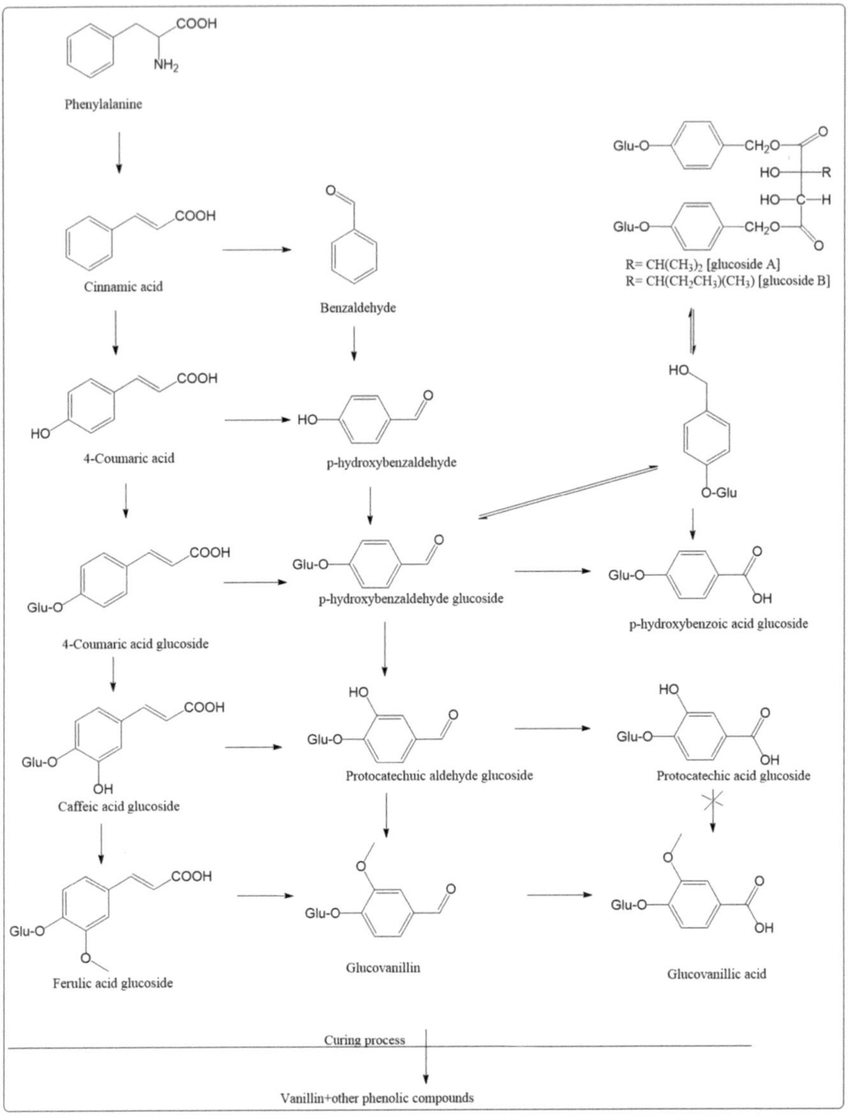

Fig. 9.5 The proposed pathway for the biosynthesis of vanillin and other phenolic compounds in the *Vanilla planifolia* beans. (Source: Dignum et al. 2001)

woody, and cinnamon-like. Previous studies have reported a varied amount of volatile constituents (Maarse et al. 1994; Ranadive 1994) and identified components in cured *V. planifolia* beans are presented in Table 9.5 (Dignum et al. 2001). The major flavor compounds identified in cured vanilla beans were vanillin (~2%), acetic acid (0.02%), and *p*-hydroxy benzyl methyl ether (0.02%) while other minor compounds identified were present below 10 ppm level (Klimes and Lamparsky

Table 9.4 Glucosides identified in vanilla

Glucosides identified directly in green beans extract	
4-formylphenyl-K-D-glucopyranoside, 4-hydroxymethylphenyl-*β* D-glucopyranoside, 4 methylphenyl-*β*-D-glucopyranoside, β-phenylehtyl-*β*-D-glucopyranoside, bis{4-(*β*-D-glucopyranosyloxy)-benzyl}-2-(2-butyl)tartarate (glucoside B), bis{4-(*β*-D-glucopyranosyloxy)-benzyl}-2-isopropyltartarate (glucoside A), glucovanillin, methyl-2-*β*-D-glucopyranosyloxybenzoate, methyl-4-*β*-D-glucopyranosyloxybenzoate, methyl-4-*β*-glucopyranosyloxyferulate	
Glycosides identified after β-glucosidase treatment of the extract	
Mono substitutes	3-Phenylpropanol, phenethyl alcohol, cinnamic alcohol, cinnamic acid, benzylalcohol
Di substitutes	4-Vinylphenol, methyl salicylate, methyl-*p*-hydroxy cinnamate, methyl-*p*-hydroxybenzoate, *p*-cresol, *p*-hydroxy benzylalcohol, *p*-hydroxybenzoicacid, *p*-hydroxy benzylmethylether, *p*-hydroxybenzaldehyde, *p*-hydroxycinnamic acid
Tri substitutes	3,4-Dihydroxybenzoic acid, vanillyl alcohol, 4-vinylguaiacol, vanillin caffeic acid, vanillic acid, methyl-3,4-dihydroxy cinnamate, methyl vanillate, methyl ferulate, homovanillyl alcohol, ferulic acid, acetovanillone, ethyl-4-hydroxy-3-methoxyphenylacetate, 2-methoxy-4-cresol

Source: Kanisawa et al. (1994)

1976). Adedeji et al. (1993) compared the flavor components in the extract of cured beans obtained from 10 different origins (including Tahiti beans). Several components were identified and some components specific to beans obtained from a particular origin were characterized. The Bali beans were reported to contain 2.0% vanillin whereas Java contained only 0.34%. The cured beans were also found to contain vanillin and glucovanillin (Voisine et al. 1995).

9.8.4 Nutraceutical Properties of Vanilla

The advancement in pharmacology and chemistry has paved the way to identify the functional properties of vanilla. The advancement in basic science research is concentrated on the medicinal properties of vanillin. However, currently, a spectrum of health benefits of vanillin are identified such as it inhibits tumor formation (anticarcinogenic), prevents chromosome breakage (anticlastogen), and inhibits sickling of red blood cells in sickle cell anemia patients. Vanilla has varied medicinal effects on ailments such as virus and water retention, tumor, sickle cell anemia, rhinosis, polyp, pain, nervousness, inflammation, immune depression, hysteria, heptoses, fever, dysmenorrheal, cramp, and caries. Vanillic acid is reported as antibacterial, anti-fatigue, anti-inflammatory, antioxidant, antitumor, ascaricide, choleretic, immunosuppressive, laxative, vermifuge, and also suppresses sickle cell production in sickle cell anemia while vanillin as antipolio, antiviral, choleretic, immunosuppressive, and irritant. O-Vanillin forms adduct amino groups, and hence inhibits the gelation of hemoglobin (Ravindran 2006). The various pharmacological properties of bioactive compounds from vanilla is shown in Table 9.6. Vanilla enhances brain activities, increases muscular energy, prevents sleep, and enhances

Table 9.5 Compounds identified in cured vanilla beans of *Vanilla planifolia*

Compound	Reference
Acetic acid, *p*-hydroxybenzaldehyde, vanillic acid, vanillin	Morison (1964)
β-Bisabolol, *β*-cyclocitral, *α*-terpineol, *α*-muurolene toluene, α-curcumene, vanillin 2,3-butyleneglycolacetal, valeric acid, trimethyl benzene, terpinen-4-ol, styrene, salicylicaldehyde, salicylic acid, *p*-vinylphenol, *p*-vinylguaiacol, protocatechuic acid, protocatechualdehyde, propylbenzene, propyl valerate, propionic acid, prenol, *p*-methoxybenzylalcohol, *p*-methoxybenzylalcohol, *p*-methoxybenzaldehyde, *p*-methoxybenzaldehyde, *p*-hydroxybenzyl methyl ether, *p*-hydroxybenzylethylether, *p*-hydroxybenzylalcohol, phenyethylalcohol, phenol, *p*-ethyltoluene, *p*-ethylguaiacol, pentanol, pentanal, pentan-2-ol, *p*-cymene, *p*-cresyl isopropyl ether, *p*-cresol, octanol, octanoicacid, octan-2-one, octa-4,7-dien-3-one, nonane, nonan-2-one, nonan-2-ol, nerol, naphthalate, myrtenol, myristic acid, myrcene, methyl pentadecanoate, methyl nonanoate, methyl heptanoate, methyl cyclohexane carboxylate, methyl *cis*-cinnamate, methyl caproate, methyl butyrolactone, methoxy acetic acid, linalool, lauric acid, lactic acid, isovaleric acid, isopropyl valerate, isobutyric acid, isobutyl valerate, hexanol, hexan-2-one, heptanol, heptan-2-one, heliotropin, guaiacol, glycolic acid, geraniol, furfurylalcohol, furfural, furfural hydroxyl-methylketone, formic acid, ethyl decanoate, ethyl benzene, ethyl 2-methyl butyrate, iphenylether, di-*n*-propyl phthalate, di-*n*-butyl phtalate, dimethyl benzene, dihydroactinidiolid, diacetyl, decnol, decenoicacid, decanoicacid, decan-2-one, cresol, caproic acid, butyric acid, butane-2,3-diol, benzylethyl ether, benzyl butyrate, benzyl benzoate, benzylalcohol, benzyl acetate, benzoic acid, benzene, benzaldehyde, anisol, anisic acid, acrolein, acetophenone, acetaldehyde, 7,10,14-trimethylpentadecan-2-one, 5-piperidone, 5-methylfurfural, 5-hydroxy heptan-2-one, 4-hydroxy-3-methoxy, 4-hydroxy-3-methoxy benzyl methyl ether, 3-penten-2-one, 3-octen-2-one, 3-methylpentanol, 3-methylbutanol, 3-hydroxybutan-2-one, 2-methylbutanol, 2-hydroxyethyl-5-methylfuran, 2-acetylpyrrole, 2-acetylfuran, 1-hydrxy hexan-2-one, 1-hydroxy pentan-2-one, 1,2-dimethoxy benzene	Klimes and Lamparsky (1976)
p-hydroxy benzylmethylether, vanillyl ethyl ether, vanillylmethylether	Galetto and Hoffman (1978)
Anisyl formate, bornyl acetate, butyl hexanoate, citronellylisobutyrate, ethyl salicylate, fenchyl acetate, hexylbutanoate, hexyl salicylateisobornylacetate, linalylacetatementhylacetate, phenethyl formate, pentylsalicylate, *α*-terpinylacetate, trans-*α*-ionone	Werkoff and Guntert (1997)
n-1-alkenes, Δ^5-avenasterol, brassicasterol, 5-ethylalkanes, β-sitosterol, vanillicalcohol	Ramaroson-Raonizafinimanana et al. (1997)
Δ^7-avenasterol, stigmasta-5,22,25-trien-3*β*-ol, ergosta-7,24(28)-dien-3*β*-ol	Ramaroson-Raonizafinimanana et al. (1998)
Dihydromethylpyranones, *cis*/*trans*-vitispirane	Ramaroson-Raonizafinimanana et al. (1999)

Table 9.6 Pharmacological property of bioactive compounds from vanilla

Pharmacological property	Bioactive compounds
Ornithine decarboxylase inhibitor	Ferulic acid
Hepatoprotective	Ferulic acid, catechin
Immunostimulant	Ferulic acid, catechin, benzaldehyde
Antitumor	Vanillin, vanillic acid, ferulic acid, benzaldehyde
Antiperoxidant	Protocatechuic acid
Antioxidant	Vanillin, vanillic acid, syringaldehyde, salicylic acid, protocatechuic acid, ferulic acid, catechol, catechin
Antinitrosaminic	Ferulic acid
Antineoplastic	Ferulic acid
Antimutagenic	Vanillin, ferulic acid, anisaldehyde
Antilipoperoxidant	Catechin
Anti-inflammatory	Vanillic acid, umbelliferone, syringaldehyde, n-hentriacontane, ferulic acid, catechin
Antihepatotoxic	Protoatechic acid, ferulic acid, catechin
Antiestrogenic	Ferulic acid
Antiedemic	Syringaldehyde, coumarin, catechin
Anticancer	Vanillin, vanillic acid, syringaldehyde, salicylic acid, ferulic acid, ethyl ester, *p*-cresol, cinnamic acid, catechol, catechin, benzaldehyde
Antiaggregant	Ferulic acid, catechin, annisyl alcohol
Anesthetic	Benzylalcohol, benzaldehyde

Source: Ravindran (2006)

sexual properties. It is beneficial for hysteria, infusion, rheumatism, and mild fevers. It also acts as an aphrodisiac and excites reproductive systems. Recently, 4-hydroxybenzyl alcohol and 4-hydroxy-3- methoxybenzyl alcohol have been reported to exhibit higher oxidative property than vanillin while vanilla extract showed a commendable antioxidative property (Shyamala et al. 2007). Hence, vanilla extracts possessing high antioxidant activity can be utilized for food preservative and in nutraceuticals. Vanilla extract has vanillin and other compounds which has nutraceutical and antioxidant properties. Advanced clinical researches on vanilla on its basic constituent and their mechanism of action can enhance its medicinal future.

9.9 Effect of Varieties on Flavor Compounds in Cardamom

Cardamom is a rhizomatous monocot of *Elettaria* genus from *Zingiberaceae* family. Cardamom belongs to the species *cardamomum* (Maton). It is a highly valued crop from ancient times. It has a very pleasant flavor and taste hence also known as the "Queen of Spices." The genus *Elettaria* is derived from the Tamil language meaning cardamom seeds which contain six species and *E. cardamomum* (Maton) is the only commercially important species in India (Mabberley 1987). The Sri Lankan (Ceylon) wild cardamom consist of *E. ensal* (Gaertn.) Abeywick. *(E. major* Thaiw.). The Sri Lankan cardamom plants are comparatively larger and stronger and bear an inferior aroma and taste to the true cardamom. Cardamom is differentiated into *Malabar*, *Mysore*, and *Vazhukka* varieties based on the nature of its panicle. The panicle in *Malabar*, *Mysore*, and *Vazhukka* varieties are prostrate, erect, and semi-erect or flexuous, respectively.

Cardamom is the third most costly spice after saffron and vanilla in the world. It is used as medicinal herb since fourth century BC and originated from the Western Ghats of southern India. India had a monopoly of cardamom but now, it is cultivated in Costa Rica, El Salvador, Guatemala, Laos, Mexico, Nepal, Sri Lanka, Tanzania, Thailand, and Vietnam (Mehra 2001; Korikanthimath et al. 2002). Cardamom is mainly exported from Brazil, Costa Rica, Guatemala, India, Indonesia, Nicaragua, Nigeria, South Africa, and Thailand. The major cardamom importing countries are the US, United Arab Emirates, Singapore, Saudi Arabia, the Netherlands, Kuwait, Japan, Hong Kong, and China. However, the single largest importer of cardamom is Saudi Arabia followed by Kuwait (Chempakam and Sindhu 2008).

E. cardamomum also known as "small cardamom" is valued worldwide due to the presence of its characteristic sweet aromatic flavor. Cardamom is an export-oriented crop due to its unique aroma composition. Cardamom is utilized in the form of decorticated seeds, ground, and whole. The whole cardamom has a high demand in the International market while ground spice has very negligible trade and the decorticated seed has a small market. The essential oil from spices is obtained by distillation while oleoresin by solvent extraction. Recently harvested spices provide more oil yield as compared to stored ones (Ravindran 2002). The essential oil is located as one layer in the cells just below the epidermis in cardamom seeds. The flavor components can be recovered as essential oil which can be used as beverage and food additive. The cardamom and eucalyptus essential oil has a similar spicy odor. Cardamom contains 3–8% of volatile oil which is influenced by various factors such as commercial-grade, distillation efficiency, drying procedure, duration and conditions of storage, environmental conditions, genetic variations, the freshness of the sample, green or bleached, maturity at harvest, region, and varieties (Zachariah 2002).

Cardamom oil is a wonderful gift of nature with a perfect blend of ethers, esters, terpenes, terpene alcohol, and other compounds, which confront precise analytical techniques. The dried fruit of cardamom contains calcium oxalate, cellulose, fixed (fatty) oil, minerals, pentosans, pigments, proteins, silica, starch, steam-volatile oil, and sugars. The major component of cardamom seed is starch (up to 50%) while fruit husk is rich in crude fiber (up to 31%). The quality of cardamom is determined by the content and composition of essential oil. The flavor and odor in cardamom are donated by its volatile oil and many studies have been conducted to determine the composition of essential oil. The relative concentration of the compounds of essential oil contributes to the characteristic flavor and aroma of cardamom. The intrinsic organoleptic properties of cardamom are independent of the color of the fruit. However, cardamom stored for a longer period loses its color as well as essential oil which in turn is responsible for the deterioration in its organoleptic properties (Purseglove et al. 1981).

Cardamom seed has a warm, slightly pungent aromatic flavor and used as a stimulant, aromatic and carminative agent. It is utilized largely as a flavoring agent in food and tea preparations. Cardamom oil is a valuable component in beverages, foods, health foods, medicines, and perfumery industries. Indians generally use cardamom for various conditions such as vomiting, urinary tract disorders, pulmonary disease, nausea, kidney stones, indigestion, general debility, digestive upsets, bronchitis, asthma, and anorexia. It is also used for sweetening the breath, soothing a spastic colon, relieving flatulence, preventing stomach pain and griping, detoxifying caffeine and constipation (Chempakam and Sindhu 2008).

9.9.1 Biosynthesis of Flavor Compounds in Cardamom

9.9.1.1 Sites of Synthesis

The secretory structures such as glandular epidermis or oil or resin ducts, glandular trichomes, and oil cells generally secrete or accumulate monoterpenes and sesquiterpenes in plants. Many studies have indicated that these secretary structures are the primary sites for the biosynthesis of mono and sesquiterpene (Francis and O'Connell 1969). Essential oil plants are blessed with the special extracytoplasmic cavity in these secretory structures where the toxic terpenoid cells and resins are sequestered while it is absent in other plants which result in rapid metabolization or volatilization of terpenes (Chempakam and Sindhu 2008).

9.9.1.2 Early Biosynthetic Steps and Acyclic Precursors

Generally, the isoprenoid pathway is employed in the biosynthesis of essential substances in the plant system. The mono and sesquiterpenes are regarded as diverging at the C10 and C15 stages respectively in biosynthetic pathways. The first step of the isoprenoid pathway is the condensation of three acetyl-CoA to acetoacetyl-CoA which on a subsequent condensation forms 3-hydroxy-3-methyl-glutaryl-CoA (HMG-CoA). The HMG-CoA is reduced to mevalonic acid which is regarded as the precursor of all isoprenoids. Mevalonate phosphorylates twice and is finally decarboxylated to form isopentenyl pyrophosphate (IPP) (McCaskill and Croteau 1995). IPP is isomerized to dimethylallyl pyrophosphate (DMAPP) which further leads to the formation of an immediate precursor of monoterpenoids geranyl pyrophosphate (GPP). GPP and IPP condensed to form an immediate precursor of sesquiterpene farnesyl pyrophosphate (FPP). Likewise, FPP and IPP are condensed to form diterpenoids. Almost all the processes during biosynthesis of terpenoid are associated with cell organelles while magnesium and calcium play an important role during the biosynthesis of sesquiterpenes (Preisig and Moreau 1994). McCaskill and Croteau (1995) indicated that the mevalonic acid pathway in the cytoplasm is blocked at HMG-CoA reductase and both monoterpene and sesquiterpene are synthesized from IPP in the plastids only.

Several studies on monoterpene cyclases responsible for the production of 1,8-cineole, α-terpinene, and γ-terpinene have been reported. The synthesis of other monoterpenoids such as limonene and sabinene from cyclization of geranyl pyrophosphate are also well studied (Croteau and Sood 1985). The process of cyclization of farnesyl pyrophosphate and geranyl phosphate to the corresponding sesquiterpenoids and monoterpenoids are different. However, sesquiterpene and monoterpene cyclases are unable to synthesize smaller and larger analogs (Ravindran 2002).

An extensive study has been reported on the biosynthesis of pinene where three monoterpene cyclases (synthases) are responsible for catalyzing the conversion of GPP. The monocyclic and acyclic olefins, (+)-β-pinene and bicyclic (+)-α-pinene are synthesized from FPP with the help of pinene cyclase I (Bramley 1997). GPP is utilized for the synthesis of carvone, monoterpenes, and limonene. Monoterpene synthase cyclizes GPP into (+)-limonene via formation of an intermediate product. The intermediate thus formed is either stored in the essential oil ducts without any metabolism or converted to (+)-trans carveol by limonene-6-hydroxylase which on further oxidation by a dehydrogenase forms (+)-carvone (Bouwmeester et al. 1998). The formation of 1,8-cineole takes place from linalyl pyrophosphate. 1,8-cineole (or eucalyptol) is the end product in many systems which results in accumulation of this compound in large quantities in many plants (Clark et al. 2000).

9.9.2 Chemistry of Cardamom Volatile Oil

The potential flavor compounds in cardamom essential oil are predominantly oxygenated compounds and little mono or sesquiterpenoids. While volatiles from various foods and spice oils are found abundant with alcohols, aldehydes, and esters. The preeminence of esters (linalyl acetates and α-terpinyl acetate) and ether (1,8-cineole) in essential oil makes cardamom a unique combination (Lewis et al. 1966; Salzer 1975; Korikanthimath et al. 1997). The essential oil in cardamom seeds are described as citrusy, lightly camphorated, spicy, sweet, and warm and oil yield varies between 2 and 5% which is mainly dependent on duration and conditions of storage (Boiswert and Hubert 1998). The major and minor components in cardamom oil are given in Table 9.7. The essential oil mainly contains 1,8-cineole (36.3%), α-terpinyl acetate (31.3%), limonene (11.6%), linalool (3.0%), sabinene (2.8%), trans-nerolidol (2.7%), α-terpineol (2.6%), linalyl acetate (2.5%), myrcene (1.6%), α-pinene (1.5%), terpinen-4-ol (0.9%), γ-terpinene (0.7%), nerol (0.5%), geraniol (0.5%), terpinolene (0.5%), citronellol (0.3%), β-pinene (0.2%), α-phellandrene (0.2%), methyl eugenol (0.2%), *p*-cymene (0.1%) (Lawrence 1978; Govindarajan et al. 1982; Korikanthimath et al. 1999). Some of the trace compounds present in cardamom volatile oil are hydrocarbons (α-ylangene, α-terpinene, α-thujene, and γ-cadinene), acids, alcohols, and phenols (acetic acid, cuminyl alcohol, thymol, and *p*-cresol) and carbonyl compounds (α-terpinyl propionate, carvone, and pinole). The combination of 1,8-cineole and α-terpinyl acetate in volatile oil provides a unique

Table 9.7 Major and minor components of cardamom volatile oil

Major components		
1,8-cineole (36.3%), α-terpinyl acetate (31.3%), limonene (11.6%), linalool (3.0%), sabinene (2.8%), trans-nerolidol (2.7%), α-terpineol (2.6%), linalyl acetate (2.5%), myrcene (1.6%), α-pinene (1.5%), terpinen-4-ol (0.9%), γ-terpinene (0.7%), nerol (0.5%), geraniol (0.5%), terpinolene (0.5%), citronellol (0.3%), β-pinene (0.2%), α-phellandrene (0.2%), methyl eugenol (0.2%), *p*-cymene (0.1%)		
Minor components		
Hydrocarbons	Acids	Carbonyls
Camphene	2-Methyl butyric acid	*α*-Terpinyl propionate
Cis-ocimene	3-Methyl butanol	2-Methyl butanal
Cyclosativene	3-Methyl butyric acid	3-Methyl butanal
Toluene	Acetic acid	8-Acetoxy carvotanacetone
Trans-ocimene	Alcohols and phenols	Carvone
α-Copaene	Butyric acid	Cuminaldehyde
α-Terpinene	Cuminyl alcohol	Dihydro-α-terpinyl acetate
α-Thujene	Perillyl alcohol	Furfural
α-Ylangene	Propionic acid	Pentanal
γ-Cadinene	Thymol	Pinole
Δ-cadinene	p-Cresol	Terpinene-4-yl-acetate
p-dimethylstyrene	p-Methyl-3-en-l-ol	–

Source: Lawrence (1978), Govindarajan et al. (1982)

aroma to cardamom (Wijesekera and Jayawardena 1973; Korikanthimath et al. 1999).

There is no significant difference in the volatile oil from freshly separated seeds and whole capsules (husk and seeds) (Govindarajan et al. 1982). The highest content of essential oil is obtained from the cardamom harvested 20–25 days before full maturity. The essential oil from spices is obtained by distillation for about 4 h. During distillation, the early fractions are abundant with 1,8-cineole and lower boiling terpenes while later fractions are lavished with esters. The ratio of α-terpinyl acetate and 1,8-cineole governs the critical aroma of cardamom volatile oil. The α-terpinyl acetate is the key aroma component of cardamom belonging to the class of isoprenoids. α-terpinyl acetate constitutes the largest and most widely distributed class of secondary metabolites.

The composition of essential oil is highly influenced by the variety and difference in the composition was reported in var. *Malabar* and *Mysore* (Table 9.8). The detailed analysis of cardamom essential oil revealed a large difference in 1,8-cineole content and showed var. *Malabar* contains higher (41%) while var. *Mysore* contains a lower (26.5%) content of 1,8-cineole (Govindarajan et al. 1982; Zachariah 2002). The *Malabar* variety is known as “Coorg greens” and contains a relatively higher content of 1,8-cineole which subsidize to “more camphory” flavor and makes it ideal for soft drinks. The early fractions during oil extraction are dominant with 1,8-cineole and low-boiling monoterpenes (Ravindran 2002). On the other hand, var. *Mysore*, also known as “Alleppey green” is “mild spicy” in

Table 9.8 Composition of volatile oils (%) of cardamom from different sources

Volatile component	Coorg (var. Malabar)	Alleppey green (var. Mysore)	Imported	Ceylon (Oleoresin)
α-Pinene	1.4	1.6	0.8	1.1
Camphene	–	–	0.1	–
Sabinene	3.2	4.9	2.5	1.9
Myrcene	0.2	1.9	2.6	0.7
1,4-Cineole	–	–	0.2	–
D-limonene	2.4	2.5	2.7	1.7
1,8-Cineole	41.0	34.2	30.4	22.2
p-cymene	0.5	0.5	0.6	–
Methyl heptenone	1.2	0.1	0.4	–
γ-Terpinene	–	–	–	0.6
Linalool	0.4	6.1	4.1	2.3
Linalyl acetate	1.6	3.1	2.8	5.3
β-Terpineoi	0.8	1.7	0.2	–
α-Terpineol	0.8	1.7	4.3	2.0
α-Terpinyl acetate	30.0	34.5	39.7	53.2
Neryl acetate	1.1	1.2	0.6	–
Geranyl acetate	–	–	1.1	–
Geranoil	0.7	0.7	1.2	1.1
Nerol	1.4	0.6	0.3	–
Nerolidol	0.3	0.7	1.6	–

Source: Govindarajan et al. (1982)

aroma due to the presence of higher content of α- terpinyl acetate, linalool and linalyl acetate, and a lower amount of 1,8-cineole. This unique oil composition is responsible for the relatively pleasant mellow flavor of *Mysore* variety which makes it highest selling cardamom in India (Chempakam and Sindhu 2008). The better flavor perception in essential oil from Alleppey green does not depend on the relative concentration of only one compound. This flavor perception provides superior sensory qualities to *Mysore* variety thus raising its demand in markets.

The flavor compounds in essential oil determine the importance of cardamom in beverage and foods. The difference in volatile components is influenced by extraction techniques, maturity at harvest, processing conditions, varieties, and duration and conditions of storage. The major oxygenated compounds which constitute almost 90% of flavor components in cardamom volatile oil in order of their importance are 1,8-cineole, α-terpinyl acetate, linalool, linalyl acetate, α-terpineol, and terpin-4-ol (Chempakam and Sindhu 2008). Govindarajan et al. (1982) have elaborated the range of concentration of major flavor constituents, their flavor description, and effect on flavor use and the aroma characteristics of some important volatile components of cardamom are given in Table 9.9. The interaction of chemical components in the food matrix with taste perception of the user provides a unique

Table 9.9 Flavor characteristics of volatile components in cardamom

Volatile components	Flavor description	Effect in flavor use	Range of concentration in cardamom oil (%)
Alcohols			
Borneol	Dry camphoraceous woody, peppery	Background herbaceous camphory	–
Citronellol	Fresh rosy odor and floral rosy bitter taste	Extensive use in perfumes and fruit flavors	<0.01–0.04
Geraniol	Floral, rosy with warm Dry tones	Berry and sweet-spicy flavors	0.29–0.64
Linalool	Floral, woody with a citrusy note: Creamy floral taste at low levels	Peculiar pleasant taste effect at low levels	1.4–4.5
Nerol	Sweet rosy, fruity	Berry and sweet-spicy flavors	0.29–0.64
Nerolidol	Woody, floral slightly green	Excellent tenacity	0.28–1.6
α-Terpineol	Delicately floral, sweet, lilac-like	Citrus and spice compositions	1.4–3.3
4-Terpineol	Warm peppery wood with earthy, musty notes pleasantly green	Mainly used in citrus and spice compositions; warm, herbaceous effects	0.14–0.87
Carbonyls			
Camphor	Warm minty odor, bitter warm and then cool mouth feel	Modifier diffusive	0.02–0.04
Citral	Powerful lemon fruity odor	Widely used at high dilutions in citrus and spice compositions	0.19–0.26
Citronellal	Powerful fresh green, citrusy, slightly woody	Citrus and spice composition	–
Esters			
Geranyl acetate	Sweet, floral-fruity with green; note stronger than geraniol	Sweetener, modifier in citrus fragrances and fruit, citrus and spice flavors	0.17–0.23
Linalyl acetate	Sweet, floral, fruity odor and taste. Poor tenacity but stronger than terpinyl acetate	Fresh, sweet modifier in perfume and berry flavors	0.7–6.3
Neryl acetate	Very sweet; fruity floral	Effective in berry flavors, higher levels in perfumery	0.17–0.23
α-Terpinyl acetate	Mildly herbaceous sweet-spicy, piney variation in odor warm, mild spicy taste	To stretch cardamom herbal spice, imitation citrus, and cherry peach flavors	34–52
Ethers			
1,8-cineole	Fresh, camphoraeous cool and odor and taste very diffusive, poor tenacity	Refreshing effect extensively used in perfume and flavors	23–51

Source: Zachariah (2002)

aroma to the food. The characteristic aroma of cardamom essential oil is described as aromatic, camphory and sweet, and rich in oxygenated compounds (Zachariah 2002).

9.9.3 Medicinal and Pharmacological Properties of Cardamom

The volatile oil from cardamom has been used traditionally to aid digestion, relieve spam, colitis, irritable bowel syndrome, cramps, and indigestion. It also can relieve morning sickness during pregnancy and nausea. It is reported as an analgesic, anti-inflammatory, carminative, neuromuscular antispasmodic, stimulant, stomachic, and strong tonic and effective against postoperative vomiting and nausea. The important pharmacological properties of cardamom volatile oil include tonic and antispasmodic, stomachic, stimulant, diuretic, digestive, carminative, antiseptic, antimicrobial, and anti-inflammatory (Al-Zuhair et al. 1996; de Pradier 2006; Chempakam and Sindhu 2008).

9.10 Summary

Flavor is the sensory impression of food and is determined by the chemical senses of taste and smell. The entire market flavor sectors is enormous (around € 6.5 billion). Flavor is one of the primary sensory property that helps in the selection, acceptance, and ingestion of any food material. Flavorants are the edible extracts and chemicals that modify the flavor of any food material. In contrast, flavor enhancers have little or no taste of their own, but they desirably change their flavor when added in a small proportion. The primary flavorings used in foods are classified as natural, nature-identical, and artificial flavoring substances, and they generally originate from the plant, animal, and microbial sources. Volatile oils are natural products used for their unique flavor in perfumery, pharmaceutical, food, and beverage industries. Volatile oils are classified as terpenoids and non-terpenoid hydrocarbons and terpenoids are generally synthesized from the mevalonate pathway in the cytosol and 2C-methyl-d-erythritol-4-phosphate in the plastid. The content and composition of volatile oils are influenced by the distillation method, environmental and geographic conditions, harvesting time, age, the organ and variety of the plant, genetic factors, production practices, postharvest handling, and processing and storage conditions. The effect of processing (like curing) plays a vital process in developing the aroma of odorless green vanilla beans.

There are various curing methods such as Bourbon process, Guyana process, Mexican process, and Peruvian process. Different curing methods follow different steps during processing and require different time, ultimately affecting the flavor profile and gives a characteristic aroma to the vanilla beans. Cured vanilla beans (vanillin, acetic acid, and *p*-hydroxy benzyl methyl ether) have different flavor profiles than green pods (glucoside A, glucoside B, and glucovanillin). The flavor components also vary mainly due to varietal difference (e.g., cardamom). Cardamom is differentiated into *Malabar, Mysore,* and *Vazhukka* varieties. The preeminence of linalyl acetates, α-terpinyl acetate, and 8-cineole in essential oil makes cardamom a unique combination. The *Malabar* variety contains a higher content of 1,8-cineole (41%), which subsidizes more camphory flavor. In contrast, *Mysore* variety is mild spicy in aroma due to higher content of α-terpinyl acetate, linalool and linalyl acetate, and a lower amount of 1,8-cineole (26.5%).

References

Adedeji J, Hartman TG, Ho CT (1993) Flavour characterization of different varieties of vanilla beans. Perf Flav 18:25–29

Al-Zuhair H, El-Sayeh B, Ameen HA et al (1996) Pharmacological studies of cardamom in animals. Pharmacol Res 34:79–82

Anonymous (2000) Natural sources of flavourings. Report no 1. Council of Europe Publishing, Strasbourg

Anonymous (2003) Alberta agriculture, food and rural development, Edmonton

Anonymous (2006) Dictionary of natural products on CD-ROM, version. 14.1 1982–2006. Chapman & Hall, CRC, New York

Anuradha K, Shyamala BN, Naidu MM (2013) Vanilla-its science of cultivation, curing, chemistry, and nutraceutical properties. Crit Rev Food Sci Nutr 53(12):1250–1276

Arana FE (1944) Vanilla curing and its chemistry. Federal experiment station of the USDA, Mayaquez, Puerto Rico., Bulletin No. 42, Washington DC

Bagamboula CF, Uyttendaele M, Debevere J (2004) Inhibitory effect of thyme and basil essential oils, carvacrol, thymol, estragol, linalool and p-cymene towards Shigella sonnei and Shigella flexneri. Food Microbiol 21(1):33–42

Barton DHR, Meth-Cohn O, Nakanishi K (eds) (1999) Comprehensive natural products chemistry. Isoprenoids including carotenoids and steroids, vol 2. Pergamon, Amsterdam

Baser KHC, Demirci F (2007) Chemistry of essential oils. In: Berger RG (ed) Flavours and fragrances: chemistry, bioprocessing and sustainability. Springer, Germany, pp 43–86

Belitz HD, Grosch W, Schieberle P (2004) Food chemistry, 3rd edn. Springer, Berlin Heidelberg New York

Berger RG (2015) Biotechnology as a source of natural volatile flavours. Curr Opin Food Sci 1:38–43

Boiswert C, Hubert A (eds) (1998) L'ABCdaire des épices. Flammarion, Paris

Boukhatem MN, Ferhat MA, Kameli A et al (2014) Liquid and vapour phase antibacterial activity of eucalyptus globulus essential oil ¼ susceptibility of selected respiratory tract pathogens. Am J Infect Dis 10(3):105–117

Bouwmeester HJ, Gershenzon J, Konings et al (1998) Biosynthesis of the monoterpenes limonene and carvone in the fruit of caraway. I. Demonstration of enzyme activities and their changes with development. Plant Physiol 117:901–912

Bramley PM (1997) Isoprenoid metabolism. In: Dey PM, Harborne JB (eds) Plant biochemistry. Academic Press, California, pp 417–434

Breitmaier E (2005) Terpene: Aromen, Dufte, Pharmaka, Pheromone. Wiley-VCH, Weinheim

Brodelius PE (1994) The phenylalanine ammonia-lyase gene family in raspberry. Phytochem Anal 5:27–29

Broderick JJ (1956) The science of curing vanilla. Food Tech 10:184–187

Burt S (2004) Essential oils: their antibacterial properties and potential applications in foods-a review. Int J Food Microbiol 94(3):223–253

Champagne A, Boutry M (2016) Proteomics of terpenoid biosynthesis and secretion in trichomes of higher plant species. Biochem Biophys Acta 1864(8):1039–1049

Cheetham PSJ (1997) Combining the technical push and the business pull for natural flavours. In: Berger RG et al (eds) Biotechnology of aroma compounds. Advances in biochemical engineering/biotechnology, vol 55. Springer, Berlin, Heidelberg

Chempakam B, Sindhu S (2008) Small Cardamom. In: Parthasarathy VA, Chempakam B, Zachariah TJ (eds) Chemistry of spices. CABI, Nosworthy Way, pp 41–58

Clark GS (1990) Vanillin. Perf Flav 15:45–54

Clark G, Stuart C, Easton MD (2000) Eucalyptol. Perf Flav 25:6–16

Correll DS (1953) Vanilla-its botany, history, cultivation and economic import. Eco Bot 7:291–358

Croteau R, Karp F (1991) Origin of natural odorants. In: Muller PM, Lamparsky D (eds) Perfumes-art, science and technology. Elsevier, London, pp 101–126

Croteau R, Sood VK (1985) Metabolism of monoterpenes: evidence for the function of monoterpene catabolism in peppermint (*Mentha piperita*) rhizomes. Plant Physiol 77:801–806

Dev S (1989) Natural products of Woody plants. In: Rowe JW (ed) Chemicals extraneous to the lignocellulosic cell wall, vol 2. Springer, Berlin, Heidelberg New York, pp 691–807

Dewick PM (2002) Medicinal natural products: a biosynthetic approach, 2nd edn. Wiley, Chichester

Dey P, Harborne J, Banthorpe D (1991) Methods in plant biochemistry terpenoids, vol 7. Academic, London

Dignum MJW, Kerler J, Verpoorte R (2001) Vanilla production: technological, chemical, and biosynthetic aspects. Food Rev Int 17:199–219

Diniz do Nascimento L, Moraes AAB, Costa KSD et al (2020) Bioactive natural compounds and antioxidant activity of essential oils from spice plants: new findings and potential applications. Biomol Ther 10(7):988–1022

Edris AE (2007) Pharmaceutical and therapeutic potentials of essential oils and their individual volatile constituents: a review. Phytother Res 21(4):308–323

EOMSA Report (2020) Essential oils market size, share & trends analysis report by application (food & beverages, spa & relaxation), by product (orange, peppermint), by sales channel, and segment forecasts, 2020–2027, pp. 1–153

EU Flavour Directive (1988) Council directive of 22 June 1988 on the approximation of the laws of the member states relating to flavourings for use in foodstuffs and to source materials for their production (88/388/EEC). Official Journal of the European Communities No L 184, 15 July 1988

Evans WC (2002) Trease and Evans' pharmacognosy, 15th edn. Saunders, London

Francis MJO, O'Connell M (1969) The incorporation of mevalonic acid into rose petal monoterpenes. Phytochemistry 8:1705–1708

Funk C, Brodelius P (1994) *Vanilla planifolia* Andrews: in vitro biosynthesis of vanillin and other phenylpropanoid derivatives. In: Bajaj YPS (ed) Biotechnology in agriculture and forestry, medicinal and aromatic plants VI. Springer, Berlin, pp 377–402

Galetto WG, Hoffman PG (1978) Some benzyl ethers present in the extract of vanilla (*Vanilla planifolia*). J Agric Food Chem 26:195–197

Goncalves B, Oliveira I, Bacelar E et al (2018) Aromas and Flavours of fruits. In: Vilela A (ed) Generation of aromas and flavours. IntechOpen, London

Govindarajan VS, Shanti N, Raghuveer KG et al (1982) Cardamom-production, technology chemistry and quality. CRC Crit Rev Food Sci Nutr 16(3):229–326

Guentert M (2007) The flavor and fragrance industry- past, present and future. In: Berger RG (ed) Flavours and fragrances: chemistry, bioprocessing and sustainability. Springer, Germany, pp 1–14

Hammer KA, Carson CF, Riley TV (1998) In vitro activity of essential oils in particular Melaleuca alternifolia (tea tree) oil and tea tree oil products, against Candida spp. J Antimicrob Chemother 42(5):591–595

Havkin-Frenkel D, Belanger FC (2007) Application of metabolic engineering to vanillin biosynthetic pathways in *vanilla Planifolia*. In: Verpoorte R, Alfermann AW, Johnson TS (eds) Applications of plant metabolic engineering. Springer, The Netherlands, pp 175–196

Jones MA, Vincente GC (1949) Vanilla production: technological, chemical and biosynthetic aspects. J Agric Res 78:435–441
Kabara JJ (1991) Phenols and chelators. In: Russell NJ, Gould GW (eds) Food preservatives. Blackie, London, pp 200–214
Kamble VA, Patil SD (2008) Spice-derived essential oils: effective antifungal and possible therapeutic agents. Int J Geogr Inf Syst 14(3–4):129–143
Kanisawa T (1993) Flavour development in vanilla beans. Kouryou 180:113–123
Kanisawa T, Tokoro K, Kawahara S (1994) Flavour development in the beans of *vanilla Planifolia*. In: Kurihara K, Suzuki N, Ogawa H (eds) Olfaction taste IX, proceeding of the international symposium. Springer, Tokyo
Klimes I, Lamparsky D (1976) Vanilla volatiles a comprehensive analysis. Int Flavours:272–278
Korikanthimath VS, Mulge R, Zachariah TJ (1997) Variation in yield and quality characters of cardamom clones. J Med Aroma Plant Sci 19(4):1024–1027
Korikanthimath VS, Mulge R, Zachariah TJ (1999) Variations in essential oil constituents in high yielding selections of cardamom. J Plant Crops 27:230–232
Korikanthimath VS, Govardhan R, Hiremath GM (2002) Cultivation of cardamom (*Eletteria cardamomum*) in valley bottoms under evergreen forest shade. J Med Aroma Plant Sci 24: 53–59
Lawrence BM (1978) Major tropical spices-cardamom (*Elettaria cardamomum*). In: Franz von B (ed) HagersHandbuch der pharmazeutischen Praxis. Allured Publishers, Wheaton, IL, pp 105–155
Lawrence BM (2004) Aromatic plants for the flavor and fragrance industries. In: Goodman RM (ed) Encyclopedia of plant and crop science. CRC Press, Boca Raton, FL, pp 58–64
Lewis YS, Nambudiri ES, Phillip T (1966) Composition of cardamom oils perfume. J Essent Oil Res 57:623–626
Lionnett JFG (1959) Seychelles vanilla-II, curing and marketing of the crop. World Crops 11:15–17
Maarse H, Visscher CA, Willlemse LC et al (1994) Volatile compounds in food. In: Qualitative and quantitative data, supplement 4. TNO, Zeist, Netherlands, pp 219–227
Mabberley DJ (1987) Plant book. Cambridge University Press, Cambridge
Madene A, Jacquot M, Scher JI et al (2006) Flavour encapsulation and controlled release—a review. Int J Food Sci Technol 41:1–21
Madhava Naidu M, Sampath SR, Raghavan B et al (2008) A process for faster curing of vanilla beans and products thereby. 0811/DEL/2008
Madhava Naidu M, Sujith Kumar PV, Shymala BN et al (2009) Enzyme-assisted process for production of superior quality vanilla extracts from Green vanilla pods using tea leaf enzymes. Food Bioprocess Technol 5:527–532
Mann J, Davidson RS, Hobbs JB et al (1994) Natural products: their chemistry and biological significance. Longman, London
McCaskill D, Croteau R (1995) Monoterpene and sesquiterpene biosynthesis in glandular trichomes of peppermint (*Mentha x piperita*) rely exclusively on plastid-derived isopentenyl diphosphate. Planta 197:49–56
Mehra K (2001) Indian system of innovation in biotechnology-a case study of cardamom. Technovation 21:15–23
Merory J (1968) Food Flavourings, consumption, manufacture and use. Avi Pub. Co., Westfort, CT
Morison SD (1964) Flavors and non-alcoholic beverages: determination of compounds related to vanillin in vanilla extract. J Assoc Off Anal Chem 47:808–815
MSME (2020) Ministry of micro, small and medium enterprises, fragrance and flavour industry in India, Government of India
Muller DA (2007) Flavours: the legal framework. In: Berger RG (ed) Flavours and fragrances: chemistry, bioprocessing and sustainability. Springer, Germany, pp 15–24
Nijssen LM, Visscher CA, Maarse H et al (1999) Volatile compounds in food. Qualitative and quantitative data, 7th Edition and supplements 1 and 2. TNO Nutrition and Food Research Institute, Zeist, The Netherlands

Odoux E (2010) Vanilla curing. In: Odoux E, Grisoni M (eds) Vanilla. CRC Press, Taylor and Francis Group, Boca Raton, FL

de Pradier E (2006) A trial of a mixture of three essential oils in the treatment of postoperative nausea and vomiting. Int J Aromather 16(1):15–20

Preisig CL, Moreau RA (1994) Effects of potential signal transduction antagonists on phytoalexin accumulation in tobacco. Phytochemistry 36:857–863

Purseglove JW, Brown EG, Green CL et al (1981) Spices, vol I. Longman, New York, pp 174–228

Purseglove JW, Brown EG, Green CL et al (1988) Spices, vol 2. Scientific and Technical, Longman, New York

Pybus D, Sell C (1999) The chemistry of fragrances. Royal Society of Chemistry, London

Ramachandra R, Ravishankar GA (2000) Vanilla flavor: production by conventional and biotechnological routes. J Sci Food Agric 80:289–304

Ramaroson-Raonizafinimanana B, Gaydou E et al (1997) Hydrocarbons from three vanilla beans species: *V. fragrans*, *V. madagascariensis*, and *V. tahitensis*. J Agric Food Chem 45:2542–2545

Ramaroson-Raonizafinimanana B, Gaydou EM et al (1998) 4-Demethylsterols and triterpene alcohols from two vanilla bean species: *V. fragrans* and *V. tahitensis*. Am Oil Chem Soc 75: 51–55

Ramaroson-Raonizafinimanana B, Gaydou EM, Bombarda I (1999) Long-chain γ-pyrons in epicuticular wax of two vanilla bean species: *V. fragrans* and *V. tahitensis*. J Agric Food Chem 47: 3202–3205

Ranadive AS (1994) Vanilla-cultivation, curing, chemistry, technology and commercial products. In: Charalambous G (ed) Spices, herbs and edible fungi. Elsevier Science BV, Amsterdam, pp 517–577

Rao RS, Ravishankar GA (2000) Vanilla flavor: production by conventional and biotechnological routes. J Sci Food Agric 80:289–304

Rauha JP, Remes S, Heinonen M et al (2000) Antimicrobial effects of Finnish plant extracts containing flavonoids and other phenolic compounds. Int J Food Microbiol 56(1):3–12

Ravindran PN (2002) Introduction. In: Ravindran PN, Madhusoodanan KJ (eds) Cardamom: the genus *Elettaria*. Taylor & Francis, New York, pp 1–10

Ravindran PN (2006) Medicinal properties of vanilla. Spice India 19:8–11

Reineccius G (1994) Source book of flavors, 2nd edn. Chapman and Hall, New York, pp 351–361

Robert CL (2017) Flavors: flavors of herbs and spices. In: Srinivasan D, Kirk LP (eds) Fennema's food chemistry. CRC Press, Taylor & Francis Group, pp 784–793

Rogers JA (1969) Advances in spice flavor and oleoresin chemistry. In: Gould RF (ed) Flavor chemistry. American Chemical Society, Washington DC, pp 203–224

Rohdich F, Bacher A, Eisenreich W (2005) Isoprenoid biosynthetic pathways as anti-infective drug targets. Biochem Soc Trans 33(4):785–791

Rohmer M (2003) Mevalonate-independent methylerythritol phosphate pathway for isoprenoid biosynthesis. Elucidation and distribution. Pure Appl Chem 75(2–3):375–387

Ruzicka L (1953) The isoprene rule and the biogenesis of terpenic compounds. Experientia 9:357–367

Salzer UJ (1975) Analytical evaluation of seasoning extracts (oleoresins) and essential oil from seasoning. Int Flavours Food Addit 6(3):151–157

Sanganeria S (2010) Market overview: vibrant India, opportunities for the flavour and fragrance industry. Ultra International Limited

Sangwan NS, Farooqi AHA, Shabih F et al (2001) Regulation of essential oil production in plants. Plant Growth Regul:34, 3–21

Shyamala BN, Madhava Naidu M, Sulochanamma G et al (2007) Studies on the antioxidant activities of natural vanilla extract and its constituents compounds through in vitro models. J Agric Food Chem 55:7738–7743

Smeriglio A, Alloisio S, Raimondo FM et al (2018) Essential oil of citrus lumniarisso: phytochemical profile, antioxidant properties and activity on the central nervous system. Food Chem Toxicol 119:407–416

Sreedhar RV, Roohie K, Venkatachalam L et al (2007) Specific pretreatments reduce curing period of vanilla beans. J Agric Food Chem 55:2947–2955

Steinbacher S, Kaiser J, Eisenreich W et al (2003) Structural basis of Fosmidomycin action revealed by the complex with 2-*C*-methyl-D-erythritol 4-phosphate synthase (IspC): implications for the catalytic mechanism and anti-malaria drug development. J Biol Chem 278:18401–18407

Steward D (2005) The chemistry of essential oils made simple. Care Publications, Marble Hill

Sticher O (1977) Plant mono-, di- and sesquiterpenoids with pharmacological or therapeutical activity. In: Wagner H, Wolff P (eds) New natural products and plant drugs with pharmacological, biological or therapeutical activity. Proceedings in life sciences. Springer, Berlin, Heidelberg

Sun P, Zeng M, He Z, Qin F, Chen J (2013) Controlled release of fluidized bed-coated menthol powder with a gelatin coating. Drying Technol 31:1619–1626

Theis N, Lerdau M (2003) The evolution of function in plant secondary metabolites. Int J Plant Sci 164(S3):S93–S102

Theodose R (1973) Traditional methods of vanilla preparation and their improvement. Trop Sci 15: 45–47

Torssell K (1997) Natural products chemistry: a mechanistic, biosynthetic and ecological approach. Swedish Pharmaceutical Press, Stockholm

Turek C, Stintzing F (2013) Stability of essential oils: a review. Compr Rev Food Sci 12(1):40–53

Voisine R, Carmichel L, Chalier P et al (1995) Determination of glucovanillin and vanillin in cured vanilla pods. J Agric Food Chem 43:2658–2661

Walton J, Mayuer MJ, Narbad A (2003) Molecules of interest: vanillin. Phytochem 63:505–515

Webster TM (1995) New perspectives on vanilla. Cereal Foods World 40:198–200

Werkoff P, Guntert M (1997) Identification of some ester compounds in Bourbon vanilla beans. LWT-Food Sci Technol 30:429–431

Wijesekera ROB, Jayawardena AL (1973) Recent developments in the production of spices and their essential oils in Ceylon. In: Proceedings of the conference on spices. Tropical Product Institute, London, pp 159–167

Wild-Altamirano C (1969) Enzymic activity during growth of vanilla fruit 1. Proteinase, glucosidase, peroxidase and polyphenoloxidase. J Food Sci 34(3):235–238

Zachariah TJ (2002) Chemistry of cardamom. In: Ravindran PN, Madhusoodanan KJ (eds) Cardamom: the genus *Elettaria*. Taylor & Francis, New York, pp 69–90

Zygadlo JA, Juliani HR (2000) Bioactivity of essential oil components. Curr Top Phytochem 3: 203–214

Antioxidants

10

V. Chauhan, A. Chandel, and O. P. Chauhan

10.1 Introduction

Originally, the term "antioxidant" was used to refer to a molecule that prevented oxygen consumption. The use of antioxidants in significant industrial processes, such as the vulcanization of rubber, the prevention of metal corrosion, and the polymerization of fuels in the fouling of internal combustion engines, received a lot of attention in the late nineteenth and early twentieth centuries (Botterweck et al. 2000). The use of antioxidants to prevent the oxidation of unsaturated fats, which causes rancidity, was the focus of early studies (Randhawa and Bahna 2009). Simply placing the fat in a sealed container with oxygen and monitoring the rate of oxygen consumption might be used to determine antioxidant activity. On the other hand, the discovery of vitamins A, C, and E as antioxidants, transformed the discipline and led to the recognition of antioxidants' role in the biochemistry of living beings (Bleve et al. 2008; Kornienko et al. 2019). When it was realized that a compound with antioxidative activity is expected to be one that is easily oxidized, the possible mechanisms of action of antioxidants were initially investigated (Wang et al. 2015). Antioxidants were discovered as reducing agents that prevent oxidative processes, often by scavenging reactive oxygen species (ROS) before they could even damage cells, thanks to research into how vitamin E slows the process of lipid peroxidation (Wang and Kannan 2018).

V. Chauhan (✉)
Department of Home Science, Air Force Golden Jubilee Institute, New Delhi, India

A. Chandel · O. P. Chauhan
Defence Food Research Laboratory, Mysore, India

O. P. Chauhan (ed.), *Advances in Food Chemistry*,
https://doi.org/10.1007/978-981-19-4796-4_10

10.1.1 Definition

An antioxidant is a compound that is stable enough to give an electron to a rogue free radical, neutralizing it and limiting the free radical's ability to cause damage. These antioxidants work by scavenging free radicals and thereby delaying or preventing cellular damage (Halliwell 1995). These low molecular weight antioxidants can safely react with free radicals and stop the chain reaction from causing harm to critical components. Some antioxidants, such as uric acid, ubiquinol, and glutathione, are produced by the body's natural metabolism (Shi et al. 1999). Although the body has multiple enzyme systems that scavenge free radicals, vitamin E (-α-tocopherol), vitamin C (ascorbic acid), and β-carotene are the most important micronutrient (vitamin) antioxidants (Levine et al. 1991). Because the body cannot produce certain micronutrients, they must be obtained through the diet.

According to the US Food and Drug Administration (FDA), antioxidants are compounds used to preserve food by preventing rancidity, degradation, or discoloration caused by oxidation. Antioxidants are essential to biologists and clinicians because they may help protect the human body from diseases caused by reactive oxygen species (ROS) by regulating ROS-related enzymes. Reactive oxygen species (ROS) and reactive nitrogen species (RNS) are produced as a result of normal metabolic interactions, increasing exposure to the environment, and the maximum level of xenobiotics (RNS). ROS and RNS are responsible for oxidative stress in several pathophysiological conditions. In oxidative stress situations, our body's cellular elements are altered, resulting in a variety of illness states. Increased cellular defenses in the form of antioxidants can effectively mitigate this oxidative damage (Nimse and Pal 2015).

10.1.2 Types of Antioxidants

Antioxidants are classified in several ways:

1. They are classified as enzymatic or nonenzymatic antioxidants based on their activity.
   Free radicals are broken down and removed by enzyme antioxidants. In the presence of cofactors, such as copper, iron, manganese, and zinc, antioxidant enzymes convert harmful oxidative products to hydrogen peroxide and then to water in a multistage process.
   Carotenoids, vitamin E, vitamin C, glutathione, and plant polyphenols are examples of nonenzymatic antioxidants that work by disrupting free radical chain reactions.
2. They can be classified as water-soluble antioxidants or lipid-soluble antioxidants based on their solubility.
   Water-soluble antioxidants like vitamin C are present in cellular fluids like cytosol and cytoplasmic matrix.

Vitamin E, carotenoids, and lipoic acid are lipid-soluble antioxidants that are mostly found in cell membranes.

3. They are divided into two categories based on their size: small-molecule antioxidants and large-molecule antioxidants.
   Small molecule antioxidants, including glutathione (GSH), carotenoids, vitamin E, and vitamin C, neutralize ROS and take them away in a process known as radical scavenging.
   Large molecule antioxidants, such as SOD (superoxide dismutases), CAT (catalase), and GPx (glutathione peroxidase), are enzymes and sacrificial proteins that absorb ROS and prevent them from damaging other important proteins (glutathione peroxidase) (Shahidi and Zhong 2010).
4. Antioxidants can be characterized kinetically as follows (Flora 2009):
   (a) Antioxidants, such as phenols, naphthol, hydroxyquinone, aromatic amines, and aminophenols, can break chains by reacting with peroxyl radicals having weak O-H or N-H bonds.
   (b) Quinines, nitrones, and iminoquinones are antioxidants that interact with alkyl radicals to break chains.
   (c) Aromatic amines, nitroxyl radicals, and variable valence metal compounds are examples of antioxidants that end cyclic chains.
   (d) Hydroperoxide-decomposing antioxidants such as sulfide, phosphide, and thiophosphate.
   (e) Diamines, hydroxyl acids, and bifunctional chemicals are examples of metal-deactivating antioxidants.
5. Depending on their occurrence, they are classified as natural antioxidants or synthetic antioxidants (Sofia et al. 2019).
   Chain-breaking antioxidants react with radicals and transform them into more stable molecules. This group of antioxidants has a phenolic structure and includes:
   (a) Antioxidant minerals include selenium, iron, manganese, zinc, and copper, which are cofactors for antioxidant enzymes. The absence of cofactors will undoubtedly improve the metabolism of several macromolecules, such as carbohydrates.
   (b) Antioxidant vitamins, such as Vitamin C, E, and B, are crucial and required for most human metabolic functions.
   (c) Phytochemicals: These are phenolic compound derivatives that are neither minerals nor vitamins, such as garlic, cloves, black pepper, ginger, curcumin, and derivatives, as well as lycopene, catechins, carotenoids, flavonoids, and carotenes.

Synthetic antioxidants are phenolic substances that capture free radicals and halt the chain reaction from continuing. Nordihydroguaiaretic acid (NDGA), butylated hydroxyl anisole (BHA), propyl gallate (PG), butylated hydroxytoluene (BHT), tertiary butyl hydroquinone (TBHQ), and metal chelating agent (EDTA) are some of these chemicals.

10.1.3 Natural Antioxidants in Treatment of Diseases

Free radical plays a deteriorating role when produced in large amount in the human body during the metabolic process and due to contact with the external environment. These free radicals attack macromolecules like nucleic acid, fatty acids, and proteins leading to oxidative destruction to cells or tissues and even cause gene mutation. "Oxidative stress" can be caused due to high concentration of free radicals in the human body thus destroying internal redox balance leading to a variety of chronic diseases and even premature senility. Hence, to get rid of oxidative stress and various chronic and cardiovascular diseases, the consumption of antioxidants is essential. Antioxidants protect cells in many ways, such as converting reactive oxygen species (ROS) to nonradical species (depending on the antioxidant), stopping the auto-oxidative chain reaction begun by ROS, and lowering localized oxygen concentrations (Oroian and Escriche 2015; Dorman et al. 2003). Commonly consumed beverages, cereals, fruits and vegetables, and other food products are rich in exogenous antioxidants, such as α-tocopherol (vitamin E), ascorbic acid (vitamin C), polyphenols, and carotenoids, which may help to promote the antioxidative defense (Valavanidis et al. 2013; Zunino et al. 2007; Möller and Loft 2006; Sikora et al. 2008).

10.2 Natural Food Antioxidants

Vitamins: The main function of vitamins is to maintain the physiological function of the human body. These are essential trace elements and can be taken from foods only. The two well-known antioxidants are vitamin E and vitamin C.

Vitamin C (ascorbic acid) is the most powerful antioxidant, promoting the eradication of free radicals produced by both the human body and external sources. Ascorbic acid has a critical function in regulating free radical activity, which affects skin health and can cause skin patches, wrinkles, dryness, and even the development of skin cancer. One of the most important causes of skin aging is the activity of free radicals on skin cells. Because it is water-soluble, ascorbic acid can fight free radical damage both inside and outside the cells by donating electrons to free radicals and neutralizing their reactivity (Victor et al. 2001).

Ascorbic acid also protects the central nervous system, lungs, and DNA from free radical and mutagen attack, preventing damaging genetic alterations within the cells and protecting lymphocytes from chromosome mutations (Sasazuki et al. 2008).

Vitamin B6 (folic acid) and Ascorbic acid interact to prevent free radical damage to lipids, lowering blood pressure, cholesterol levels, and lipid peroxidation, as well as thinning and protecting blood from oxidation (Sarita et al. 2007). When ascorbate is combined with α-tocopherol (vitamin E), carotene and selenium are produced, which aids in stroke prevention (Konturek et al. 2007) and in the treatment of pancreatitis (pancreatic inflammation) (Christine et al. 2005). Ascorbic acid (1 g/day) ingestion protects against low-density lipoprotein (LDL) cholesterol (Böhm et al. 2007). Vitamin C also helps to prevent atherosclerosis by preventing the oxidation of LDLs and the formation of collagen in the arterial walls (Traber 2007).

The reducing capacity of vitamin C reflects its antioxidant properties, as it can be dehydrogenated directly and quickly with superoxide ion O_2^- and singlet oxygen such as HOO^- or OH^- to produce dehydroascorbate. It can also act as an indirect antioxidant by reducing oxidized vitamin E and thiols as hydrogen donor. The chemical equation of reaction can be expressed as:

$$RO + C_6H_7O_6^- \rightarrow RO^- + C_6H_7O_6 \rightarrow ROH + C_6H_6O_6$$

Vitamin C is a water-soluble vitamin that acts as an antioxidant by circulating in the blood, body fluids, and cells, and therefore protecting cells and tissues from free radicals (Johnson 1979). The level of low-density lipoprotein (LDL) in the blood is reduced when cardiovascular patients are given continuous vitamin C at a dose of 500 mg for 10 weeks. Because LDL is the primary cause of oxidative damage to blood vessels, vitamin C may have a protective effect against cardiovascular disease (Shidfar et al. 2003).

Vitamin E: Vitamin E refers to a collection of fat-soluble substances that have different antioxidant properties (Sen et al. 2006). Vitamin E is found in eight chemical forms in nature (α-, β-, γ-, and δ-tocotrienol and α-, β-, γ-, and δ-tocopherol) with varying levels of biological activity, with alpha-tocopherol being the only recognized form of vitamin E with sufficient blood concentrations compared to other forms (Isaac et al. 2008). Vitamin E reduces the severity of damage to tissues and macromolecules by inhibiting the production of reactive oxygen species (ROS) during fat oxidation by neutralizing free radicals. Vitamin E can prevent or delay chronic diseases like retinal degeneration, Alzheimer's, prostate enlargement, and osteoarthritis that are caused by free radicals. Tocopherol, in combination with high levels of selenium and tocopherol, serves as a strong anti-inflammatory agent and protects the skin from sun exposure, lowering the risk of prostate cancer (Maiani et al. 2008). The key mechanism of tocopherol and other tocopherols' suppressing free radicals is currently being researched. Vitamin E has a wide range of functions, including defending biological membranes in the human body and nucleic acids in cells against free radical damage (Böhm et al. 2007). Together with glutathione peroxidase (GSH-Px), vitamin E can directly quench singlet oxygen, superoxide dismutase (SOD), and can remove O_2^- and build an antioxidant system in the human body. Vitamin E acts as an antioxidant by delivering protons to halt the lipid peroxidation chain reaction when it reacts with lipid oxygen radicals and lipid peroxyl free radicals (Fig.10.1). Fruits and vegetables like kiwi, almonds, walnuts, and vegetable oil are good sources of vitamin E. Vitamin E has been discovered to have a tumor-suppressing effect.

Fig. 10.1 Principle of scavenging lipid oxygen free radicals by α-tocopherol. R represents lipid oxygen free radicals

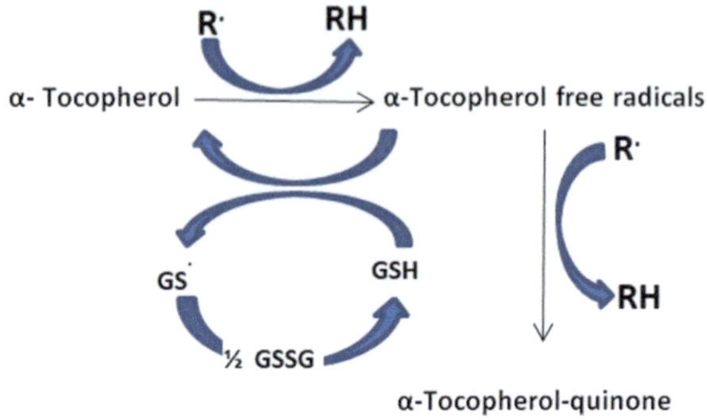

10.2.1 Mechanism of Vitamin E as Antioxidants

Vitamin E's antioxidant action is dependent on the prevention of free radical chain formation. Polyunsaturated fatty acid autoxidation consists of three reactions: initiation, chain propagation, and chain termination (Kamal-Eldin and Appelqvist 1996). The initiation reaction is slow and rate-limiting, and it is prompted by light, heat, or traces of metals. This reaction is most likely the slow ("induction") reaction reported by Cummings and Mattill (1931):

$$I + LH \rightarrow L^{\bullet} + IH\ (\text{slow})$$

where I is the initiator, LH is the fatty acid, and $L^{\bullet}$ is the alkyl radical formed from the fatty acid. The propagation through a chain reaction is as follows:

$$L^{\bullet} + O_2 \rightarrow LOO^{\bullet}\ (\text{fast})$$

$$LOO^{\bullet} + LH \rightarrow LOOH + L^{\bullet}\ (\text{slower})$$

where $LOO^{\bullet}$ is the peroxyl free radical and LOOH is the fatty acid's stable hydroperoxide. The chain is then broken and terminated by tocopherol in the following way:

$$LOO^{\bullet} + TOH \rightarrow LOOH + TO^{\bullet}$$

where TOH is tocopherol and $TO^{\bullet}$ is the relatively stable tocopheroxyl radical, which breaks the chain reaction. Tocopherylquinone is formed when the tocopheroxyl radical reacts with another peroxyl radical.

10.2.2 Carotenoids (β-Carotene, Lycopene, and Astaxanthin)

Carotenoid is a fat-soluble natural pigment found in red, yellow, and dark green fruits and vegetables. It is also a general term for polyunsaturated hydrocarbons with 40 carbon atoms. Carotenoids have a lot of double bonds. They can physically quench singlet oxygen and react with oxygen free radicals in three ways: electron transfer, hydrogen atom transfer, and radical coupling. In plants, the most common carotenoids are β-carotene, lycopene, and γ-carotene, and in animals, astaxanthin.

β-carotene: It is also known as pro-vitamin A, and is a powerful antioxidant that efficiently protects cells against a variety of cancers, particularly lung cancer. Apart from that, carotene protects the skin from an inherited skin illness called erythropoietic protoporphyria, in which the skin is susceptible to sunlight due to free radical damage, and improves immunological response by boosting T helper cells (Sen et al. 2006). Beta-carotene alone neutralizes free radicals in the body, particularly singlet oxygen. β-Carotene totally removes oxygen species before they may turn into skin or lung cancer with the help of Vitamin E. The richest sources of β-carotene are turnip greens, peaches, butternut squash, red peppers, beet greens, collard greens, kale, melons, prunes, tomatoes, apricots, and sweet potato.

Lycopene and Lutein: They are a member of the carotenoid family. Carotenoids are potent antioxidants that neutralize free radicals, particularly singlet oxygen, which is formed when the skin is exposed to UV light and is the principal cause of skin aging (Berneburg et al. 1999). The molecular processes underlying lycopene's protective effects such as function regulation and antioxidant characteristics are only partially understood. According to recent research, lutein consumption is linked to a lower risk of coronary heart disease and certain types of cancer (Berendschot et al. 2000). Red fruits contain lycopene. Lycopene prevents cancer in the lungs, stomach, and prostate by activating cancer-fighting enzymes including phase II detoxification enzymes, which eliminate dangerous carcinogens (Giovannucci et al. 2002). Lycopene protects lymphocytes against NO_2 damage when it is present in lung tissues (Rao and Rao 2007). Studies show that lycopene can effectively protect against the damage of neurons; administration of lycopene in a rotenone-induced mouse model with Parkinson's disease could result in a significant increase in the number of dopaminergic neurons and reduced activity of oxidative stress indicators, such as MDA, SOD, GSH-Px, and CAT, indicating that the damage level of oxidative stress is mitigated due to the application of lycopene (Liu et al. 2013). Alzheimer's disease is a neurodegenerative disease that is closely linked to oxidative stress and can have serious consequences for patients' health and quality of life. A study based on a nutrition and health survey and corresponding analysis of 6958 elderly people over the age of 50 found a significantly negative correlation between serum levels of lycopene and lutein and the risk of Alzheimer's disease, indicating that increasing dietary intake of foods rich in lycopene and lutein can lower the risk of Alzheimer's disease. (Min and Min 2014).

Astaxanthin: Many studies have found that it has anti-aging and anti-inflammatory properties. In the aging model mice induced by D-galactose, astaxanthin treatment can restore GSH-Px and SOD activities, increase GSH content and reduce oxidative stress, improve pathological injury of the hippocampus, and increase the expression level of BNDF, achieving an anti-aging role (Wu et al. 2014).

10.2.3 Polyphenols (Chocolate Polyphenols, Tea Polyphenols, and Red Wine Polyphenols)

Polyphenols are organic compounds found naturally in fruits and vegetables, tea, red wine, honey, and cocoa beans. To carry out healthcare functions, polyphenols with multiple hydroxyl groups can effectively remove free radicals such as O_2 and singlet oxygen. Similarly, previous research has shown that cocoa polyphenols can significantly reduce the level of oxidative stress in the alcoholic fatty liver (Suzuki et al. 2013). Furthermore, cocoa polyphenols inhibit the oxidation of low-density lipoprotein thereby preventing arteriosclerosis, coronary heart disease, and myocardial infarction. The content of LDL in the blood was significantly reduced after a month of administration of cocoa polyphenols to model rabbits with high cholesterol. The damaged area of arteriosclerosis was significantly smaller in mice treated with cocoa polyphenols than in the control group. The levels of cholesterol and TBARS, as well as oxidative stress in tissues, were significantly lower in the experimental group than in the control group (Kurosawa et al. 2005).

10.2.4 Flavonoids (Flavonoids, Isoflavones, Xanthones, and Anthocyanins)

Plants contain a wide variety of flavonoids, and flavonoids play a critical role in plant growth and disease prevention. Flavones, isoflavones, anthocyanins, and xanthonoids are examples of common flavonoid compounds. After supplying hydrogen to lipid compound radicals, flavonoids clear free radicals by converting them into phenolic radicals (inert). In Europe, prostate cancer is the first malignant disease that has a significant impact on men's health. Many studies have shown that soybeans have high isoflavone content, and Asians consume more soybeans in their diets than Europeans and Americans. Furthermore, the content of isoflavones in Japanese serum was found to be 10–100 times higher than in Europeans. As a result, the incidence of prostate cancer in Asia is much lower than in Europe and the United States, implying that isoflavone has a prostate cancer preventive effect (Aufderklamm et al. 2014).

Isoflavones are flavonoids found in soybeans that have anticancer properties. Isoflavones have been shown to help prevent ovarian, cervical, and breast cancer. A previous pathological study found that patients with ovarian cancer consumed 75.3 ± 53.6 g of soybean foods per day, which was significantly lower than the control group (110.7 + 88.8 g/d). According to logistic regression analysis, consuming soybean diets can significantly reduce the incidence of ovarian cancer. (Wu et al. 1998; Lee et al. 2014).

Anthocyanins are natural water-soluble pigments found in plants that have high antioxidant activity. Anthocyanins are abundant in dark-colored foods such as purple sweet potatoes, black rice, blueberries, grapes, mulberries, and so on. According to recent research, anthocyanins play an important role in the prevention and treatment of cardiovascular diseases, neurodegenerative diseases, and cancer (Zafra-Stone et al. 2007; de Pascual-Teresa 2014).

Mangosteen contains a high concentration of xanthone compounds, which are natural antioxidants. Neuroprotection, anti-inflammation, antioxidative stress, and anti-DNA damage are all effects of xanthones. It has been reported that mangosteen extract can protect against DNA damage by scavenging OH and DPPH free radicals (Lin et al. 2014; Phyu and Tangpong 2014). Currently, several different flavonoids extracted from plants are being used in the development of pharmaceutical and healthcare products.

Superoxide dismutase (SOD): It is a crucial enzyme that serves as a cellular antioxidant. It is found as isoenzymes in various organelles, such as copper-zinc SOD in the cytoplasm and manganese SOD in the mitochondria, in order to maintain a low superoxide anion concentration (Chaitanya et al. 2010). In addition to the intracellular forms of superoxide dismutase, there is an external version of the enzyme found in plasma, lymph, and synovial fluid. It is possible that the extracellular enzyme works on cell surfaces. SOD is an enzyme that catalyzes the dismutation of superoxide anion, and its absence is fatal. Specific redox-sensitive genes in cells regulate the amount of superoxide dismutase produced (Felicity and Cecilia 2005).

Catalase: It is a heme-containing protein that catalyzes the detoxification of hydrogen peroxide. Catalase is a cytoplasmic enzyme that is normally present in cells' peroxisomes and is expressed in all types of cells except erythrocytes, which lack peroxisomes. Catalase plays a similar protective role to glutathione peroxidase in that both are key hydrogen peroxide removal enzymes (Ho et al. 2004).

Hydrogen peroxide detoxification requires both catalase and glutathione peroxidase.

Glutathione peroxidase is a cytoplasmic and mitochondrial enzyme that is involved in the detoxification of H_2O_2 in nearly all cells. It is a seleno protein, meaning it has a seleno-cysteine amino acid in place of a regular cysteine at the active site (Stefan et al. 2003). The selenium replaces the typical sulfur in amino acid and improved nucleophilic characteristics and ionizes more quickly to release a proton. In the reaction catalyzed by this enzyme, it is a significantly more effective catalyst.

The flavoprotein glutathione reductase utilizes the reducing capacity of the pentose phosphate pathway (NADPH) to maintain a substantially reduced glutathione pool in the cell. This enzyme is particularly effective at reducing the cellular glutathione pool even when there is a lot of hydrogen peroxide present. The net effect of this cycle is that NADPH is used to convert hydrogen peroxide to water involving two electrons (Kamerbeek et al. 2007).

10.2.5 Bioactive Components in Chinese Herbs

Chinese herbs are a plentiful legacy from our forefathers. Recent research has also discovered that many bioactive components in Chinese herbs have strong antioxidant properties. Taiwanese researchers screened and identified 195 different types of Chinese herbs with high antioxidant activity, accounting for half of the Chinese herbs currently in use. Panaxnotoginseng flavonoids have potent antioxidant and antibacterial properties (Hong et al. 2014). Salvia miltiorrhiza also contains numerous bioactive components that have antioxidant and anti-inflammatory properties.

10.2.6 Beneficial Effects of Antioxidants in Food Product

Oxidation damage can occur not only in the human body but also in many food products when exposed to air (oxygen) and/or heat or light. The deterioration of food products is closely related to oxidation reactions and the decomposition of oxidation products. As a result, antioxidants play an important role in the overall quality of the products. Lipid peroxidation (for example, in margarine, mayonnaise, and frying oils) is a common deterioration process (Jayaprakasha et al. 2001; Carocho et al. 2014). It results in the production of undesirable chemical compounds such as aldehydes, ketones, and organic acids, which reduces the storage life and nutritional value of lipid-containing food products. Rancidity is the sensory impact of lipid oxidation, and it is responsible for changes in flavor properties. Furthermore, oxidative lipid deterioration can cause bleaching in foods due to reactions with pigments, particularly carotenoids (Li et al. 2015; Böttcher et al. 2015). Another oxidative phenomenon that occurs during the maturation, processing, and storage of food products is enzymatic browning. The enzymatic oxidation of phenolic compounds results in the formation of dark pigments. In the presence of oxygen, phenolic compounds act as substrates for oxidoreductases, specifically polyphenol oxidases (Galanakis et al. 2018a, b) and peroxidize (Oroian and Escriche 2015; López-Serrano and Ros Barceló 2002; Tripodo et al. 2018; Jiang et al. 2004). These enzymes are major contributors to color changes and, ultimately, the final quality of many fruits and vegetables (Soto-Vaca et al. 2012; Caleja et al. 2015).

Because food products are not consumed immediately after manufacture, necessitating storage and transportation, suppliers must ensure that they are supplied to consumers in a safe and sanitary manner, with nutritional content similar to or greater than when they were produced (Peschel et al. 2006). The introduction of antioxidants is one way to delay the oxidation of biomolecules; there is a continuing quest for strategies to improve food items' overall quality and shelf life, in many cases by minimizing or blocking oxidative damage. Natural antioxidants, which can be easily derived from natural sources, have a lot of potential for replacing synthetic preservatives (Xiu-Qin et al. 2009).

10.3 Natural Antioxidants and Synthetic Antioxidants

Butylated hydroxytoluene (BHT), tert-butylhydroquinone (TBHQ), butylated hydroxyanisole (BHA), and propyl gallate (PG) are the most commonly used synthetic antioxidants in the food industry. In fruits and vegetables, other often used antioxidants are 2,4-dichlorophenoxyacetic acid (2,4-DA), 4-phenylphenol (OPP), and 2-naphthol (2NL)(Xiu-Qin et al. 2009).

Synthetic antioxidants in high amounts may damage DNA and accelerate ageing (Kornienko et al. 2019). In animal studies, BHA and BHT have already been linked to negative effects on the liver and carcinogenesis (Saad et al. 2007; Botterweck et al. 2000). Furthermore, nothing is known about the prevalence and fate of these chemicals in the environment (Wang et al. 2015; Wang and Kannan 2018). Natural antioxidants are increasingly being used to replace synthetic antioxidants (Tajkarimi et al. 2010). Consumer perceptions of the hazards linked with the use of synthetic substances for coloring and preserving food goods have been investigated. Customers are worried about being exposed to synthetic substances in their daily diet, with a significant inclination toward natural compounds, according to the findings (Dickson-Spillmann et al. 2011; Sebranek et al. 2005; Kobus-Cisowska et al. 2014). Furthermore, the use of natural antioxidants allows manufacturers to meet customer desires for cleaner-label products containing only natural components. It should be noted however that the fact that they are of natural origin does not automatically make them safe. Toxicology research for these substances is still needed to determine the conditions of their usage in food products.

Plant antioxidants can be divided into three categories: phenolic chemicals, vitamins, and carotenoids (Dorman et al. 2003; Sikora et al. 2008; Hong et al. 2014). In addition to being the most abundant plant components with antioxidant activity, several phenolic compounds also have antibacterial and antifungal properties, as well as significant impacts on the flavors and textures of food (Soto-Vaca et al. 2012). From simple molecules (e.g., ferulic acid, vanillin, gallic acid, and caffeic acid) to polyphenols like tannins and flavonoids, phenolic compounds have a wide range of configurations (Abbas et al. 2017). Vitamins E and C are the most important in terms of vitamins. Safety issues of antioxidants for human consumption:

1. Regulatory organizations (GRAS level) must approve food-grade antioxidants.
2. They must not have an adverse effect on color, odour, or flavor.
3. They should work in low concentrations (0.001–0.01%).
4. They must be food-compatible and easy to use.
5. They must be stable during processing and storage, as well as cost-effective.
6. Antioxidants should also have LD50 values of less than 1000 mg/kg body weight and should not have any significant effects on the growth of an experimental animal in long-term experiments at a level 100 times higher than that recommended for human consumption, among other features (Taghvaei and Jafari 2015).

7. The organoleptic qualities of the food product are taken into account when selecting natural extracts from plants, to minimize consumer rejection owing to their distinctive colors or flavors (Mansour and Khalil 2000).

The majority of natural antioxidants come from plant sources such as fruits, vegetables, herbs, and spices (Bansal et al. 2013; Dimitrios 2006; Jiang and Xiong 2016). According to Halvorsen et al. (2002), the *Empetraceae, Rosaceae, Ericaceae, Juglandaceae, Asteraceae, Grossulariaceae, Punicaceae, and Zingiberaceae* families of plants contain antioxidant-rich compounds, which include fruits like strawberries, blueberries, blackberries, black currants, pomegranates, walnuts, and others. Essential oils from spices and herbs including dittany, oregano, thyme, rosemary, marjoram, and lavender have also been shown to be great sources of natural antioxidant molecules, although their applications are limited due to their strong flavor qualities (Tajkarimi et al. 2010; Embuscado 2015). Because they contain many components such as catechins, tannins, and other flavonoids, aqueous tea extracts have also been employed as natural antioxidant sources, with the advantage of not having a strong flavor like essential oils (Yin et al. 2012).

By-products of fruit and legume processing can be a source of functional compounds, such as apple pomace, which is a good source of polyphenols, particularly the peel (Fernandes et al. 2019); and grape pomace, which has a high concentration of anthocyanins, catechins, flavanols, phenolic acids, and stilbenes (Fernandes et al. 2019; Brezoiu et al. 2019). Pomegranates are an example of a fruit with a high concentration of antioxidants, primarily polyphenols, in both edible and nonedible components. According to the literature, nine tonnes of by-products are produced for every tonne of pomegranate juice produced, which can be exploited as a natural source of bioactive chemicals. (Diamanti et al. 2017).

The valorization of by-products through the recovery of antioxidant-rich extracts is all the more intriguing when you consider that the nonedible parts of fruits frequently have higher bioactive content than the edible parts. The peels of several citrus fruits, such as lemons, oranges, and grapefruit, contained 15% more phenolic compounds than the peeled fruits, according to Gorinstein et al. (2001). When comparing the total phenolic content of pineapple by-products (13.79 mg of gallic acid equivalents per 100 g) to that of fresh pulp (2.71 mg of gallic acid equivalents per 100 g), pineapple by-products had a higher total phenolic content (13.79 mg of gallic acid equivalents per 100 g) (da Silva et al. 2013). In pineapple rinds and cores, Freitas and Moldo-Martins (Freitas et al. 2015) found high levels of beneficial chemicals, such as carotenoids (carotene) and vitamin C, which have antioxidant potential.

10.4 Antioxidants and Free Radicals

Free radicals are produced both outside and inside the body and are often described as electron-hungry molecules and are produced when oxygen is burnt in the body. Free radicals bring about various changes in the body like aging, several chronic, and cardiovascular disease. They are highly reactive and unstable compounds. They have both useful and harmful effects on the body, causing diseases and death. A suitable balance between the amount of free radicals produced and the amount of antioxidants is required for health benefits. Due to this balance, the reduction and oxidation of free radical occurs which is called redox. If this balance is disrupted, it results in oxidative stress which can cause several diseases. Free radicals are produced in our body due to various processes that occur inside the body and it also enters our body through environmental factors like pesticides, cigarette smoking, pollutants, radiations, etc. One can avoid free radicals by increasing the amount of antioxidants in the body through diet rich in antioxidants and also by reducing environmental exposure to avoid ultraviolet rays and air pollution that may increase the proportion of free radicals in the body (Chinwe Elochukwu 2015).

10.5 Generation of Free Radicals

Harman introduced the "free radical theory" in the 1950s, claiming that free radical damage to cellular macromolecules is a primary predictor of life span in aerobic species (Kregel and Zhang 2007). Both external and endogenous sources of free radicals exist. Endogenous free radicals are produced in the body by a variety of mechanisms:

1. As a result of regular oxygen-dependent food metabolism: Mitochondria is an intracellular powerhouse that generates universal energy molecules. Adenosine triphosphate (ATP) generally consumes oxygen and converts it to water in the process. However, due to incomplete reduction of the oxygen molecule, undesirable by-products such as hydrogen peroxide and hydroxyl radical are invariably formed. During normal metabolism, each cell is expected to produce more than 20 billion molecules of oxidants per day. It's strange that oxygen, which is necessary for life, may have various negative and devastating impacts on the human body in some circumstances. Regardless, humans must breathe oxygen-rich air in order to live. This is similar to burning wood in a fireplace when smoke is produced as a by-product. Similarly, when food is converted to energy, oxygen oxidizes (or burns) the food to produce energy. This method does not produce smoke like burning wood in a fireplace, but it can produce potentially harmful by-products called free radicals (Colbert 2000).
2. White blood cells use oxidants (free radicals) including nitric oxide, superoxide, and hydrogen peroxide to kill parasites, germs, and viruses. As part of the body's defense system against diseases, free radicals are released to fight invading pathogenic bacteria, but with electrons unhinged, free radicals roam the body, wreaking havoc. In order to acquire stability, the free radical attacks surrounding molecules in order to obtain another electron, and the attacked molecules are destroyed as a result. When a molecule is attacked and loses an electron, it becomes a free radical, which starts a chain reaction. Once the process has begun, it has the potential to cascade, culminating in the disruption of live

cells, which then rip through the tissues, causing tissue damage (Dhalla et al. 2000).

3. Cytochrome P450, a cellular enzyme, is one of the body's principal defenses against hazardous substances consumed through food. The generation of oxidant by-products is caused by the induction of these enzymes to protect against harm from toxic foreign substances such as medicines and pesticides. These free radicals, behaving like biological terrorists ripping through our bodies, are constantly attacking virtually all organs and tissues in the body, and they must be stopped as soon as possible. Antioxidants are a type of defense system that the body uses to avoid free radical damage. Antioxidants are chemicals that may safely interact with free radicals and stop chain reactions from causing damage to important molecules (Colbert 2000).

10.5.1 Generation of ROS

The generation of ROS requires rapid uptake of oxygen, activation of NADPH oxidase, and the production of superoxide anion radical ($O_2^{\bullet-}$) (Table 10.1),

$$2O_2 + \text{NADPH} \overset{(\text{oxidase})}{\rightarrow} 2O_2^- + \text{NADP}^+ + H^+$$

The $O_2^{\bullet-}$ is then rapidly converted to H_2O_2 by SOD.

$$2O_2^- + 2H^+ \rightarrow H_2O_2 + O_2$$

The reactive species can also be produced by the myeloperoxidase–halide–H_2O_2 system. In the neutrophil cytoplasmic granules, the enzyme myeloperoxidase (MPO) is present. H_2O_2 is transformed to hypochlorous (HOCl) in the presence of the ubiquitous chloride ion. HOCl is an oxidant and antibacterial agent (Babior 1999).

$$Cl^- + H_2O_2 + H^+ \overset{(\text{MPO})}{\rightarrow} \text{HOCl} + H_2O$$

Table 10.1 List of the ROS

Name
Alkoxyl radical
Alpha oxygen
Singlet oxygen
Superoxide
Hydroxyl radical
Peroxides
Hydrogen peroxide
Lipid hydroperoxide

Source: Apak et al. (2013)

Fenton and/or Haber–Weiss reactions create ROS from O_2 and H_2O_2 via respiratory burst (Knight 1999).

Reactive nitrogen species (RNS) is produced by the enzyme nitric oxide synthase such as nitric oxide ($NO^{\bullet}$) from arginine.

$$L - Arg + O_2 + NADPH \rightarrow NO^{\bullet} + citrulline$$

An activated nitric oxide synthase (iNOS) is capable of continuously generating a large amount of $NO^{\cdot}$, which acts as an $O_2^{\bullet -}$quencher. When $NO^{\bullet}$ and $O_2^{\bullet -}$ react they generate peroxynitrite ($ONOO^{-}$), a very strong oxidant. Peroxynitrite is a powerful and diverse oxidant that may assault a wide spectrum of biological targets. Both NO and O_2 are weak oxidants (Zhu et al. 1992).

10.5.2 Damaging Reactions of Free Radicals

10.5.2.1 Free Radical and Aging

According to the Free Radical Theory of Aging, aging is the primary mechanism associated with the formation of free radicals in humans (Harman 1956). Atherosclerosis, a sign of aging, is thought to be caused by free radical oxidation. The mitochondrial DNA is the primary site of free radical damage. Damage to mitochondrial DNA is difficult to repair, resulting in the shutting down of mitochondria, resulting in cell death and aging (Speakman et al. 2004). The bombardment of free radicals with metal atoms such as cadmium, mercury, lead, and even pesticides accelerates the formation of free radicals by a factor of a million, resulting in mitochondrial damage. A cascade triggered by Bcl-2 proteins on the surface of mitochondria causes severe mitochondrial damage in cells, which leads to apoptosis. Free radicals' destructive powers will affect not just the aging process but also a wide range of diseases through numerous metabolic activities. Free radical damage to cells, according to research, causes the degenerative changes linked with aging (Ashok and Ali 1999). An increasing number of diseases or disorders, as well as the aging process itself, demonstrate a link either directly or indirectly to these reactive and potentially destructive molecules (Sastre et al. 1996). The major mechanism of aging attributes to DNA or the accumulation of cellular and functional damage (Cantuti-Castelvetri et al. 2000). Reducing or slowing the formation of free radicals may help to slow down the aging process. Some antioxidants found in foods can help to slow down the aging process and protect you from disease. According to these findings, increasing oxidative stress is a regular occurrence during the aging process, and antioxidant status may have a substantial impact on the impacts of oxidative damage associated with growing older. Free radical damage can be minimized with appropriate antioxidant defense, and optimal antioxidant nutrient consumption may contribute to improved quality of life, according to research. According to a recent study, antioxidants may even help people live longer lives.

The word describes the state of oxidative damage that occurs when the crucial balance between free radical formation and antioxidant defenses is disrupted (Rock et al. 1996). Oxidative stress is linked to damage to a wide range of molecular species, including lipids, proteins, and nucleic acids, as a result of an imbalance between free radical generation and antioxidant defenses (Mc Cord 2000). Trauma, illness, heat injury, hypertoxia, toxins, and excessive exercise can all cause short-term oxidative stress in tissues. Increased radical generating enzymes (e.g., xanthine oxidase, lipogenase, cyclooxygenase), phagocyte activation, release of free iron, copper ions, or a disruption of the electron transport chains of oxidative phosphorylation all result in excess ROS in these wounded tissues. The imbalance between ROS and the antioxidant defense system has been related to the initiation, development, and advancement of cancer as well as the negative effects of radiation and chemotherapy. Diabetes mellitus, age-related eye illness, and neurological diseases like Parkinson's disease have all been linked to reactive oxygen species (ROS) (Rao et al. 2006).

10.5.3 Diseases Caused by Free Radical Formation

In the human eye, the accumulation of free radicals causes cataracts. The eye's ability to scavenge free radicals is impaired by age-related insufficient synthesis of antioxidant scavenging systems, resulting in the creation of an opaque patch on the eye lens, which causes blindness (José et al. 1991). In the heart, myocytes are the source of free radical buildup. The sarcoplasmic reticulum's proteins and calcium pumps are damaged by free radicals, leading to calcium buildup. High calcium levels induce myocytes to contract erratically, resulting in arrhythmia. Arrhythmia spreads to other cells, disrupting the heartbeat and producing serious problems (Marczin et al. 2003).

Cancer is caused by free radicals created by external causes, particularly radiation (Dreher and Junoda 1996). The cells receive the majority of the radiation energy, which is then absorbed by the water, splitting one of its oxygen-hydrogen covalent bonds and forming a free radical. This free radical targets and injures the macro molecules of the cell, such as DNA, breaking its strands and creating mutations in its bases, in microseconds of its formation (Curtis et al. 2006). However, free radicals created during combustion may linger in the lungs for a longer time, binding to other contaminants and causing lung cancer.

10.5.4 Ways to Control the Harmful Effects of Free Radicals

As soon as these free radicals are produced, they are regulated by antioxidant enzymes in the body or antioxidant elements in our meals (vitamin A, vitamin C, vitamin E, beta-carotene, and flavonoids are common examples). Antioxidants are the defense systems against free radicals, while oxidants are the free radicals themselves. Antioxidants are compounds that protect tissue from free radical damage. In the dictionary of food and nutrition (or peroxides), antioxidants are defined as food additive or chemical that fights oxidation or inhibits or delays reactions caused by oxygen. According to another definition, antioxidants are chemicals that help defend against free radical damage. They can neutralize free radicals and end the vicious cycle. Antioxidants in biological systems may function in several ways, including catalytic elimination of free radicals, as scavengers of free radicals, or as proteins that reduce the availability of prooxidants like metal ions (Colbert 2000). Some antioxidants, such as glutathione, ubiquinol, and uric acid, are created in the body during normal metabolism. Glutathione may neutralize free radicals multiple times before it is oxidized, and it also returns vitamin C and vitamin E to their reduced forms, allowing them to continue scavenging free radicals. Glutathione also aids in DNA repair and protects DNA from free radical damage. Dietary sources of lighter antioxidants can also be identified. Although around 4000 antioxidants have been identified, vitamin E, vitamin C, and carotenoids are the most well-known (Halliwell et al. 2005). Vitamin C is a scavenger of oxygen. It transforms its hydrogen atoms into oxygen, preventing additional reactions from taking place. Vitamin E and A (including lycopene) are two other dietary vitamin antioxidants that act as bodyguards, protecting the body from free radical damage because the body cannot produce enough antioxidants. When vitamin E is present in food, it works as an antioxidant, protecting other nutrients from oxidation. Vitamin A, for example, is protected by vitamin E. (Onyeka 2005). Coenzyme Q10 is an antioxidant that can rebuild vitamin E from its radical state, as well as scavenge oxygen radicals and protect lipid cell membranes from disruption.

10.6 Oxidative Stress and Human Diseases

Many illnesses, including atherosclerosis, inflammatory disorders, certain malignancies, and the aging process, have been linked to oxidative stress. All inflammatory diseases (lupus erythematosus, vasculitis, adult respiratory diseases syndrome, glomerulonephritis, arthritis), ischemic diseases (intestinal ischemia, heart diseases, stroke), hemochromatosis, acquired immunodeficiency syndrome, gastric ulcers, emphysema, organ transplantation, preeclampsia, and hypertension) and neurological disorders (Alzheimer's disease, Parkinson's disease) are also developed due to oxidative stress (Stefanis et al. 1997). Excessive oxidative stress can cause the oxidation of lipids and proteins, which results in structural and functional alterations.

10.6.1 Cardiovascular Diseases

Heart disease is the leading cause of death, accounting for nearly half of all deaths. Because oxidative events have the capacity to alter cardiovascular disorders, they have the potential to deliver tremendous health and lifespan benefits. Low-density lipoproteins (LDL) contain a large amount of polyunsaturated fatty acids, and the oxidation of these lipid components in LDL plays an important role in atherosclerosis (Esterbauer et al. 1991a, b). Endothelial cells, smooth muscle cells, and macrophages are the three most important cell types in the vascular wall; they can emit free radicals that cause lipid peroxidation (Neuzil et al. 1997). With a high quantity of oxidized lipids in the blood, blood vessel damage to the reaction process continues, which can result in the formation of foam cells and plaque, which are atherosclerosis symptoms. Antiherogenic oxidized LDL is hypothesized to play a role in the production of atherosclerosis plaques. In addition, oxidized LDL is cytotoxic and can harm endothelial cells directly. Antioxidants such as B-carotene and vitamin E are important in preventing cardiovascular disease.

10.6.2 Lipid Peroxidation

Oxidative stress and biomolecular oxidation have a role in a variety of physiological and pathological processes, including aging, atherosclerosis, inflammation, carcinogenesis, and medication toxicity. Lipid peroxidation is a free radical process that involves the production of a secondary free radical, which can then act as a second messenger or directly react with other biomolecules, resulting in the enhancement of biochemical lesions. Following radical chain reaction, lipid peroxidation occurs on polyunsaturated fatty acids found on cell membranes. The hydrogen atom is removed by the hydroxyl radical, which produces a lipid radical, which is then transformed into diene conjugate. It also creates a peroxyl radical when oxygen is added; this extremely reactive radical then attacks another fatty acid, generating lipid hydroperoxide (LOOH) and a new radical. As a result, lipid peroxidation spreads. A variety of chemicals are generated as a result of lipid peroxidation, including alkanes, malonaldehyde, and isoprostanes. These chemicals have been confirmed in a variety of disorders, including neurodegenerative diseases, ischemia reperfusion injury, and diabetes, as indicators in lipid peroxidation assays (Lovell et al. 1995).

10.6.3 Oxidative Damage to DNA

Several investigations have shown that DNA and RNA are vulnerable to oxidative damage. According to reports (Woo et al. 1998), DNA has been identified as a primary target in aging and cancer. During oxidative damage to DNA caused by UV radiation or free radicals, oxidative nucleotides such as glycol, dTG, and 8-hydroxy-2-deoxyguanosine are reported to be elevated. Mitochondrial DNA is more sensitive to oxidative damage, which has been linked to a variety of disorders, including cancer. 8-hydroxy-2-deoxyguanosine has been proposed as a potential biological marker for oxidative stress (Rao and Agarwal 1999).

10.7 Role of Antioxidants in Reducing Oxidative Stress

Antioxidants can reduce oxidative stress-induced carcinogenesis by scavenging reactive oxygen species (ROS) and/or decreasing cell growth due to protein phosphorylation. Because oxidative products can cause genetic harm, antioxidant properties of beta-carotene may protect against cancer. As a result, photoprotective qualities of beta carotene may protect against UV light-induced carcinogenesis. Beta-carotene immuno-enhancement may help to protect against cancer. Beta-carotene may also have an anticarcinogenic impact by changing the effects of carcinogens on liver metabolism (Jenkinson et al. 1999). Vitamin C may aid in the prevention of cancer (Park et al. 2009). Antioxidant effects, inhibiting the development of nitrosamines, enhancing the immunological response, and speeding up the detoxification of liver enzymes are all proposed processes by which vitamin C may affect carcinogenesis. Vitamin E, an important antioxidant, aids immunological competence by boosting humoral antibody protection, bacterial infection resistance, cell-mediated immunity, tumor necrosis factor generation by T-lymphocytes, mutagen inhibition, DNA membrane repair, and microcell line development (Esterbauer et al. 1990). As a result, vitamin E may be useful in cancer prevention and carcinogenesis inhibition by stimulating the immune system. The administration of a combination of the three antioxidants resulted in the greatest reduction in the risk of heart malignancy.

Chronic illnesses such as cancer, coronary heart disease (CHD), and osteoporosis are linked to ROS-induced oxidative stress (Dizdaroglu et al. 2002). Free radicals damage all major types of biomolecules, particularly cell membrane polyunsaturated fatty acids (PUFA). Lipid peroxidation, a type of oxidative damage to PUFA, is particularly harmful because it is a self-perpetuating chain reaction (Halliwell and Gutteridge 2007; Von Sonntag 2006).

Where LH is the target PUFA and R is the initiating, oxidizing radical, the general process of lipid peroxidation is represented below. When PUFAs are oxidized, a fatty acid radical (L) is formed, which quickly adds oxygen to produce a fatty acid peroxyl radical (LOO). The carriers of chain reactions are peroxyl radicals. Peroxyl radicals can further oxidize PUFA molecules and start new chain reactions, resulting in lipid hydroperoxides (LOOH), which can be broken down into even more radical species (Dizdaroglu 1992).

$$\mathrm{LH} + \mathrm{R}^{\bullet} \rightarrow \mathrm{L}^{\bullet} + \mathrm{RH}$$

$$\mathrm{L}^{\bullet} + \mathrm{O_2} \rightarrow \mathrm{LOO}^{\bullet}$$

$$\mathrm{LOO}^{\bullet} + \mathrm{LH} \rightarrow \mathrm{LOOH} + \mathrm{L}^{\bullet}$$

$$\mathrm{LOOH} \rightarrow \mathrm{LO}^{\bullet} + \mathrm{LOO}^{\bullet} + \text{aldehydes}$$

Aldehydes are usually formed when lipid hydroperoxides are broken down. Many of these aldehydes are physiologically active chemicals that can propagate the attack from the initial site of the attack to other areas of the cell (Breen and

Murphy 1995; Duckworth 2001). Lipid peroxidation has been linked to tissue damage and diseases for a long time (Naziroglu and Butterworth 2005).

OH, O_2, and non-radical H_2O_2 are produced during oxygen metabolism. OH is a highly reactive molecule that reacts with biological components, such as DNA, proteins, and lipids, causing chemical changes. Several studies have been published on the oxidative damage of DNA caused by OH (Jialal et al. 2002).

By a variety of processes, the OH reacts with the base pairs of DNA, causing oxidative damage to the heterocyclic and sugar moieties in oligonucleotides. The physiological circumstances of mutagenesis, carcinogenesis, and aging are all linked to this sort of oxidative DNA damage (Lee and Davis 2011; Koo et al. 2005). The addition processes produce OH-adduct radicals of DNA bases, whereas the abstraction reactions produce the allyl radical of thymine and carbon-centered sugar radicals.

10.7.1 Reactions of Free Radicals in the Body

Free radical reactions are believed to cause increasing negative alterations throughout the body as people get older. Fortunately, the body is naturally equipped with antioxidant defense systems to detoxify these harmful substances; nevertheless, as we age, our bodies' defense systems become less effective, resulting in oxidative damage and the development of chronic degenerative diseases. Supplementing with too many antioxidants can tip the oxidant-antioxidant balance in favor of the antioxidants (Joseph et al. 2008). However, there are times when genetic and environmental variances influence this common trend or pattern. When the amount of free radicals produced exceeds the antioxidant enzymes or nutrients' ability to manage them, problems ensue, and damage to the cell membrane occurs.

Free radicals are hazardous and can disrupt important molecules in the body, such as lipids, proteins, and DNA, regardless of how or why they are formed. Oxidative stress is the outcome of these factors. If uncontrolled, it can potentially alter cellular calcium metabolism, resulting in cell damage or death (Bonnefont-Rousselot and Collin 2010). Damage to three primary structures is caused by oxidative stress: DNA, lipids, and proteins. DNA strands can be damaged directly by free radicals in close proximity to the DNA or indirectly by a reduction in the creation of protein required for DNA repair. A crucial element in the development of cancer is a change in DNA (Sandberg et al. 2004). Over time, oxidative damage to DNA can lead to changes in chromosome shape and function, which can lead to cancer and chronic illnesses. Fatty acid side chains of intracellular membranes and lipoproteins could be attacked by free radicals. The result is a chain reaction known as lipid peroxidation. Lipid peroxidation's result can further damage membrane proteins, causing the cell membrane to become "leaky" and finally lose its integrity. Arteriosclerosis is thought to be caused by lipid peroxidation (Collard 2009).

Cellular proteins are the last structures to be harmed by oxidative stress. This contributes to the development of cataracts. Free radicals can disrupt protein function, resulting in a skewed, aberrant metabolism and accelerated aging. In most human diseases, oxidative stress (free radical reactions) occurs. This is not to argue that oxidative stress is to blame for the majority of illnesses. The increase in free radicals could be a side effect of the disease. The disorders listed below are connected with free radical damage in the clinical setting: Arthritis, Alzheimer's disease, hypertension, heart failure, diabetes, amyotrophic lateral sclerosis, arteriosclerosis, coronary artery disease, and a variety of other ailments are just a few of the illnesses that people suffer from. In fact, free radicals are thought to be involved in more than 60 distinct diseases (Elias et al. 2008).

10.8 Application of Antioxidants

10.8.1 Role of Antioxidants in Diabetes

Diabetes is a major global health issue. It is a chronic metabolic condition marked by absolute or relative insulin secretion deficits or non-secretion, resulting in chronic hyperglycemia and carbohydrate, lipid, and protein metabolism abnormalities. Diabetes has a wide range of consequences as a result of metabolic disturbances, including macro- and microvascular dysfunction (Duckworth 2001). Diabetes mellitus has been linked to an increase in the free radical formation and a decrease in antioxidant capability, which leads to a disruption in the balance between radical formation and protection, resulting in oxidative damage to cell components such as proteins, lipids, and nucleic acids. In both insulin-dependent (type 1) and non-insulin-dependent diabetes (type 2), there is an increase in oxidative stress (Naziroglu and Butterworth 2005). Glucose autoxidation is the most important factor in the formation of free radicals among the different variables that cause increased oxidative stress. In addition, cellular oxidation/reduction imbalances and a loss in antioxidant defenses are contributors (including decreased cellular antioxidant levels and a reduction in the activity of enzymes that dispose of free radicals). Increased levels of prooxidants, including ferritin and homocysteine, have also been noted. The interaction of advanced glycation end products (AGEs) with specific cellular receptors known as AGE receptors is another crucial element (RAGE).

Individuals with higher levels of serum antioxidants, particularly serum tocopherol, have a decreased chance of developing type 2 diabetes. Reduced glutathione (GSH) and glutathione peroxidase (GSH-Px) are the cell's major defenses against oxidative stress (Jialal et al. 2002). Low levels of ascorbate, glutathione, and superoxide dismutase are the most prevalent antioxidant deficits found in diabetes. Reduced glutathione concentrations in diabetic neutrophils and monocytes have been found to be lower. Plants, especially those with high quantities of potent antioxidant chemicals, play a vital role in treating oxidative stress-related diseases like diabetes mellitus. There have been several studies on the impact of these plants and their antioxidant contents on diabetes and its complications, with promising results demonstrating the benefits of plants with high antioxidant levels in the management of diabetes mellitus (Rahimi et al. 2005).

10.8.2 Role in Premature Infants

Supplementing newborns with enzymatic and/or nonenzymatic antioxidants may help to reduce the harm caused by excessive ROS generation, especially in conditions like periventricular leukomalacia, retinopathy of prematurity, necrotizing enterocolitis, and bronchopulmonary dysplasia (Lee and Davis 2011).

10.8.3 Enzymatic Antioxidants

Premature babies have lower amounts of enzyme antioxidants than full-term neonates, which is due to gestational regulation. According to numerous models using transformed human alveolar epithelial cells, overexpression of antioxidants may attenuate ROS-induced damage. In lung epithelial cells, increased expression of either MnSOD or CuZnSOD reverses the growth inhibitory effects of hyperoxia (Koo et al. 2005). Overexpression of SOD not only reduced ROS generation but also reduced the activation of the JNK/AP1 pathway, which has been linked to ROS-induced mitochondrial damage and apoptotic cell death (Joseph et al. 2008). Melatonin is a pineal hormone that has both indirect and direct antioxidant effects, such as lipid peroxidation and scavenging oxygen-induced ROS via promoting SOD and glutathione peroxidase activity (Bonnefont-Rousselot and Collin 2010).

10.8.4 Nonenzymatic Antioxidants

Nonenzymatic antioxidants (NAC) are depleted in response to ROS-mediated stress; hence, resistance to oxidative stress relies on nonenzymatic mechanisms. Vitamin A's actions are likely to be mediated through the retinol-binding protein and the retinoic acid receptor. Sandberg et al. (2004) found that NAC is a precursor to glutathione and a major multicenter trial found no difference in survival or the incidence of BPD after 36 weeks of CGA, but it did enhance pulmonary function at term. Ceruloplasmin, transferrin, and ferroxidase all help in iron metabolism, which can be a powerful oxidizer. Reduced function or bioavailability of these proteins may predispose a preterm newborn to produce more reactive oxygen species (ROS) (Lee and Davis 2011; Collard 2009).

10.8.5 Role in Food Systems and Human Body

Antioxidants are essential for reducing oxidative reactions in both food systems and the human body. The application of nutritional antioxidants in food systems can aid in preventing lipid peroxidation and the production of secondary lipid peroxidation products thus preserving the flavor, texture, and color of the food product throughout storage. Antioxidants also aid in the reduction of protein oxidation and the interaction of lipid-derived carbonyls with proteins, which results in a change in protein function (Elias et al. 2008). Natural antioxidants like vitamin C and tocopherols, as well as herbal extracts like rosemary, sage, and tea, have previously been sold as synthetic antioxidant replacements in food systems (Shahidi 2000). In many muscle foods, proteins and protein hydrolysates produced from sources such as soya, milk, fish, and eggs also have antioxidant action (Hagen and Sandnes 2004; Pena-Ramos and Xiong 2003; Sakanaka and Tachibana 2006).

Endogenous antioxidants protect the human body from oxidative damage caused by reactive oxygen and nitrogen species such as hydroxyl radicals (OH^-), peroxyl radicals (OOR^-), superoxide anion (O_2^-), and peroxynitrite ($ONOO^-$). Enzymes like superoxide dismutase, catalase, and glutathione peroxidase, as well as nonenzymatic substances like selenium, α-tocopherol, and vitamin C, make up the endogenous antioxidative systems (Wojcik et al. 2010). Apart from that, amino acids, peptides, and proteins contribute to the general antioxidative capacity of cells as well as the health of biological tissues. For example, blood proteins are thought to scavenge 10–50% of the peroxyl radicals produced in plasma (Frei et al. 1998; Wayner et al. 1987). Endogenous antioxidative action is widely recognized for peptides such as carnosine, anserine, and glutathione (Babizhayev et al. 1994). However, as people become older, the antioxidant-prooxidant balance in their bodies alters, as do other factors including pollution, weariness, excessive alcohol consumption, and high-fat diets. With increasing age, the antioxidant potential of plasma and cells, as well as the absorption of nutrients, including antioxidants, decreases (Elmadfa and Meyer 2008; Rizvi et al. 2006). Researchers have also discovered that the activity of free radicals on proteins causes an increase in protein carbonyls in individuals as they age (Stadtman 2006; Chakravarti and Chakravarti 2007). The use of dietary antioxidants to increase the body's antioxidant load has been identified as a potentially useful way to boost human health.

In today's market, dietary antioxidant supplements and functional foods containing antioxidants such as α-tocopherol, vitamin C, or phytochemicals from plants, including lycopene, lutein, isoflavones, green tea extract, and grape seed extracts, are in high demand (Samaranayaka and Li-Chan 2011).

10.9 Medical Application of Antioxidants

10.9.1 Antioxidants as Anticancerous Agents

Various research studies have established the role of antioxidants such as lanthanides, selenium, flavonoids, lycopene, and glutathione in bio-coordination chemistry as anti-cancerous chemicals. Recent advances in medicinal chemistry have become critical for bettering chemical design, decreasing hazardous side effects, and understanding the mechanisms of action. In the treatment of cancer, several metal-based medicines were used.

Lanthanides are also used as pharmacological agents in radio-immunotherapy and photodynamic therapy, and their therapeutic radioisotopes are of particular interest (Kostova 2005). According to reports (Blot et al. 1993), these lanthanides are coordination compounds with better pharmacological characteristics and a broader range of anti-tumor activities.

Flavonoids, or plant-derived low molecular weight polyphenols, are a class of naturally occurring chemicals. These are abundantly available in the human food supply as fruits and vegetables and are thought to have anticarcinogenic properties (Suzuki et al. 2013). These are thought to be excellent scavengers of free radicals. Around 28 flavonoids have been proposed as potential anti-leukemic drugs, both natural and synthetic. In addition, flavonoids have been proven to have anti-inflammatory, anti-allergic, antiviral, and anticancer activities (Jain et al. 2013).

Lycopene: Dietary changes have long been recognized as an important therapy for cancer prevention and progression inhibition. Lycopene has been shown to reduce the risk of prostate cancer in males significantly. It also helps to prevent malignancies of the pancreas, rectum, oral cavity, cervix, oesophagus, large intestine, ovaries, colon, and mouth. Lycopenes have an important role in preventing heart disease and protecting the skin from sun damage (Sharoni et al. 2000).

Glutathione: A tripeptide thiol molecule is a key intracellular antioxidant in the body. Glutathione has been suggested as a potential therapy for hepatocellular cancer. Another rat study discovered that oral glutathione supplementation resulted in the regression of liver tumors and enhanced the survival of tumor-bearing mice (Novi 1981).

Selenium, a mineral antioxidant, is a component of endogenous enzymes. It is an important trace mineral in the human body. It is a natural antioxidant that protects the body from free radical damage. Selenium has been linked to the prevention of cancer and the management of heart failure, according to studies (Hamid et al. 2010).

10.9.2 Antioxidants as Hepatoprotective Agents

According to previous publications, antioxidants have been used religiously in the treatment of numerous forms of liver illnesses. In several clinical trials (Singal et al. 2011), antioxidants including vitamin C, vitamin E, and others have been shown to be useful in the treatment of hepatocellular carcinoma patients.

10.10 Antioxidant and Nervous System

The cerebellum's function in controlling diverse motor processes in the body is well established, and the developing brain is vulnerable to ROS's harmful effects. Antioxidants have been shown to protect cerebellar development from oxidative damage and play a crucial role in overall health and wellbeing maintenance (Imosemi 2013). In a small number of cases (Cheremisinoff 1989), antioxidants have been used as therapeutic treatments for acute central nervous system injuries.

10.10.1 Antioxidants and Red Blood Cells

During their 120-day lifespan, erythrocytes' primary role is to transfer oxygen and carbon dioxide through the lungs and capillaries. These RBCs are damaged because they are constantly exposed to intracellular ROS resulting from the antioxidation of oxyhaemoglobin. Antioxidant enzymes are found in RBCs to avoid this damage. According to research, Cu, Zn, SOD, and catalase accumulate in the RBC membrane as the first line of defence against oxidative stress, according to research. Glutathione peroxidase and catalase are thought to work together to protect the entire RBC (membrane and cytoplasm) from ROS destruction (Bing 2009)

10.10.2 Antioxidants and Their Therapeutic Usage

Antioxidant consumption from fruits and vegetables, which are good sources of antioxidants, aids in the prevention of cardiovascular illnesses. Antioxidants are also being investigated as potential treatments for neurological illnesses like Alzheimer's, Parkinson's, and amyotrophic lateral sclerosis. Excessive oxidative damage to cells causes rheumatoid arthritis, cardiovascular illnesses, ulcerogenesis, and acquired immunodeficiency diseases, among other pathological conditions. Antioxidants are thought to play a role in the treatment of many diseases and disorders.

10.11 Conclusion

Antioxidants have been extensively studied in the context of oxidative stress in a variety of disorders, including leukemia, thalassemia, ischemic stroke, hemodialysis, rheumatoid arthritis, critically sick patients, postmenopause in women, schizophrenia, and depression. The importance of antioxidants in addressing the problem of male infertility has been established, and the efficacy and safety of antioxidant supplementation in the medical treatment of idiopathic male infertility have been demonstrated. In recent years, various antioxidants have been proven to be useful in preventing hyperoxaluria-mediated nephrolithiasis. Antioxidants have also been discovered to offer a lot of potential in the treatment of nephrolithiasis (Urinary tract stone disease). According to some research, antioxidant supplement therapy as an adjuvant therapy can help people with stress-related mental disorders and generalized anxiety disorders.

References

Abbas M, Saeed F, Anjum FM et al (2017) Natural polyphenols: an overview. Int J Food Prop 20: 1689–1699

Apak R, Gorinstein S, Böhm V et al (2013) Methods of measurement and evaluation of natural antioxidant capacity/activity (IUPAC technical report). Pure Appl Chem 85(5):957–998

Ashok BT, Ali R (1999) The aging paradox: free radical theory of aging. Exp Gerontol 34:293–303
Aufderklamm S, Miller SF, Galasso A et al (2014) Chemoprevention of prostate cancer by isoflavonoids. Recent Results Cancer Res 202:101–108
Babior BM (1999) NADPH oxidase: an update. Blood Am J Hematol 93(5):1464–1476
Babizhayev MA, Seguin MC, Gueyne J et al (1994) L-carnosine (beta-alanyl-l-histidine) and carcinine (beta-alanylhistamine) act as natural antioxidants with hydroxyl-radical-scavenging and lipid-peroxidase activities. Biochem J 304:509–516
Bansal S, Choudhary S, Sharma M et al (2013) Tea: a native source of antimicrobial agents. Food Res Int 53:568–584
Berendschot TT, Goldbohm RA, Klöpping WA et al (2000) Influence of lutein supplementation on macular pigment, assessed with two objective techniques. Invest Ophthalmol Vis Sci 41:3322–3326
Berneburg M, Grether-Beck S, Kurten V et al (1999) Singlet oxygen mediates the UVA-induced generation of the photoaging-associated mitochondrial common deletion. J Biol Chem 274: 15,345–15,349
Bing Li (2009) An antioxidant metabolon at the red blood cell membrane, Thesis Dept of Chemistry and Biochemistry Canada: Concordia University, Montreal, Quebec (July 2009)
Bleve M, Ciurlia L, Erroi E et al (2008) An innovative method for the purification of anthocyanins from grape skin extracts by using liquid and sub-critical carbon dioxide. Sep Purif Technol 64(2):192–197
Blot WJ, Li JY, Taylor PR et al (1993) Nutrition intervention trials in Linxian China: supplementation with specific vitamin/mineral combinations, cancer incidence, and disease-specific mortality in the general population. J Natl Cancer Inst 85(18):1483–1491
Böhm F, Settergren M, Pernow J (2007) Vitamin C blocks vascular dysfunction and release of interleukin-6 induced by endothelin-1 in humans in vivo. Atherosclerosis 190:408–415
Bonnefont-Rousselot D, Collin F (2010) Melatonin: action as antioxidant and potential applications in human disease and aging. Toxicology 278(1):55–67
Böttcher S, Steinhauser U, Drusch S (2015) Off-flavour masking of secondary lipid oxidation products by pea dextrin. Food Chem 169:492–498
Botterweck A, Verhagen H, Goldbohm R et al (2000) Intake of butylated hydroxyanisole and butylated hydroxytoluene and stomach cancer risk: results from analyses in the Netherlands cohort study. Food Chem Toxicol 38:599–605
Breen AP, Murphy JA (1995) Reactions of oxyl radicals with DNA. Free Radical Biol Med 18(6): 1033–1077
Brezoiu AM, Matei C, Deaconu M et al (2019) Polyphenols extract from grape pomace. Characterization and valorisation through encapsulation into mesoporous silica-type matrices. Food ChemToxicol 133:110787
Caleja C, Barros L, Antonio AL et al (2015) Foeniculumvulgare mill as natural conservation enhancer and health promoter by incorporation in cottage cheese. J Funct Foods 12:428–438
Cantuti-Castelvetri I, Shukitt-Hale B, Joseph JA (2000) Neurobehavioral aspects of antioxidants in aging. Int J Dev Neurosci 18:367–381
Carocho M, Barreiro MF, Morales P et al (2014) Adding molecules to food, pros and cons: a review on synthetic and natural food additives. Compr Rev Food Sci Food Safety 13:377–399
Chaitanya KV, Akbar Ali Khan P, Sowmya Spandana M et al (2010) Role of oxidative stress in human health: an overview. J Pharm Res 3:1330–1333
Chakravarti B, Chakravarti DN (2007) Oxidative modification of proteins: age-related changes. J Gerontol 53:128–139
Cheremisinoff NP (1989) Performance properties of plastics and elastomers. Handbook of polymer science and technology, vol II. Marcel Dekker Inc, New York
Chinwe Elochukwu (2015) Generation and reaction of free radicals in the human body: a major cause of aging and chronic degenerative diseases. EC Nutrition 1(3):132–136

Christine F, Thao N-K, Claudine S et al (2005) Effects of oral vitamin C supplementation on oxidative stress and inflammation status in haemodialysis patients. Nephrol Dial Transplant 20: 1874–1879

Colbert MD (2000) What you don't know may be killing you. Siloam Pub

Collard KJ (2009) Iron homeostasis in the neonate. Pediatrics 123(4):1208–1216

Cummings MJ, Mattill HA (1931) The auto-oxidation of fats with reference to their destructive effect on vitamin E. J Nutr 3:421–432

Curtis RE, Freedman DM, Ron E et al (2006) New malignancies among cancer survivors: SEER cancer Registries. Jr NIH Publ 5:5302

Da Silva DI, Nogueira GD, Duzzioni AG et al (2013) Changes of antioxidant constituents in pineapple (Ananascomosus) residue during drying process. Ind Crop Prod 50:557–562

Dhalla NS, Temsah RM, Netticadan T (2000) Role of oxidative stress in cardiovascular diseases. J Hypertens 18(6):655–673

Diamanti AC, Igoumenidis PE, Mourtzinos I et al (2017) Green extraction of polyphenols from whole pomegranate fruit using cyclodextrins. Food Chem 214:61–66

Dickson-Spillmann M, Siegrist M, Keller C (2011) Attitudes toward chemicals are associated with preference for natural food. Food Qual Prefer 22:149–156

Dimitrios B (2006) Sources of natural phenolic antioxidants. Trends Food Sci Technol 17:505–512

Dizdaroglu M (1992) Oxidative damage to DNA in mammalian chromatin. Mutat Res 275(3–6): 331–342

Dizdaroglu M, Jaruga P, Birincioglu M et al (2002) Free radical-induced damage to DNA: mechanisms and measurement. Free Radical Biol Med 32(11):1102–1115

Dorman H, Peltoketo A, Hiltunen R et al (2003) Characterisation of the antioxidant properties of de-odourised aqueous extracts from selected Lamiaceae herbs. Food Chem 83:255–262

Dreher D, Junoda AF (1996) Role of oxygen free radicals in cancer development. Eur J Cancer 32: 30–38

Duckworth WC (2001) Hyperglycaemia and cardiovascular disease. Curr Atheroscle Rep 3:383–391

Elias RJ, Kellerby SS, Decker EA (2008) Antioxidant activity of proteins and peptides. Crit Rev Food Sci Nutr 48:430–441

Elmadfa I, Meyer A (2008) Body composition, changing physiological functions and nutrient requirements of the elderly. Ann Nutr Metab 52(1):2–5

Embuscado ME (2015) Spices and herbs: natural sources of antioxidants—a mini review. J Funct Foods 18:811–819

Esterbauer H, Dieber-Rotheneder M, Waeg G et al (1990) Biochemical structural and functional properties of oxidized low-density lipoprotein. Chem Res Toxicol 3:77–92

Esterbauer H, Pubi H, Dieber-Rothender M (1991a) Effect of antioxidants on oxidative modification of LDL. Ann Med 23:573–581

Esterbauer H, Schanr RJ, Zollen H (1991b) Chemistry and biochemistry of 4-hydroxynonenal, malonaldehyde and related aldehydes. Free Radic Biol Med 11:81–121

Felicity J, Cecilia G (2005) Superoxide dismutases and their impact upon human health. Mol Asp Med 26:340–352

Fernandes PA, Le Bourvellec C, Renard CM et al (2019) Revisiting the chemistry of apple pomace polyphenols. Food Chem 294:9–18

Flora SJS (2009) Structural, chemical and biological aspects of antioxidants for strategies against metal and metalloid exposure. Oxid Med Cell Longev 2(4):191–206

Frei B, Stocker R, Ames BN (1998) Antioxidant defenses and lipid peroxidation in human blood plasma. ProcNatlAcadSci U S A 85:9748–9752

Freitas A, Moldão-Martins M, Costa HS et al (2015) Effect of UV-C radiation on bioactive compounds of pineapple (*Ananascomosus* L. Merr.) by-products. J Sci Food Agric 95:44–52

Galanakis CM, Tsatalas P, Charalambous Z et al (2018a) Control of microbial growth in bakery products fortified with polyphenols recovered from olive mill wastewater. Environ TechnolInnov 10:1–15

Galanakis CM, Tsatalas P, Charalambous Z et al (2018b) Polyphenols recovered from olive mill wastewater as natural preservatives in extra virgin olive oils and refined olive kernel oils. Environ TechnolInnov 10:62–70
Giovannucci E, Willett WC, Stampfer MJ et al (2002) A prospective study of tomato products, lycopene, and prostate cancer risk. J Natl Cancer Inst 94:391–396
Gorinstein S, Martín-Belloso O, Park YS et al (2001) Comparison of some biochemical characteristics of different citrus fruits. Food Chem 74:309–315
Hagen H, Sandnes K (2004) Process for the improvement of meat quality in fish, protein hydrolysate and method of producing a protein hydrolysate. International patent, WO 2004/071202
Halliwell B (1995) How to characterize an antioxidant- an update. BiochemSocSymp 61:73–101
Halliwell B and Gutteridge JMC (2007) Free radicals in biology and medicine, Oxford University Press, Oxford 4th edn
Halliwell B, Rafter J, Jenner A (2005) Health promotion by flavonoids, tocopherols, tocotrienols, and other phenols: direct or indirect effects? Antioxidant or not? Am J Clin Nutr 81(1):268S–276S
Halvorsen BL, Holte K, Myhrstad MCW et al (2002) A systematic screening of Total antioxidants in dietary plants. J Nutr 132:461–471
Hamid AA, Aiyelaagbe OO, Usman LA et al (2010) Antioxidants: its medicinal and pharmacological applications. Afr J Pure Appl Chem 4(8):142–151
Harman D (1956) Aging: a theory based on free radical and radiation chemistry. J Gerontol 11:298–300
Ho YS, Xiong Y, Ma W, Spector A et al (2004) Mice lacking catalase develop normally but show differential sensitivity to oxidant tissue injury. J BiolChem 279:32804–32812
Hong J, Hu JY, Liu JH et al (2014) *In vitro* antioxidant and antimicrobial activities of flavonoids from *Panaxnotoginseng* flower. Nat Prod Res 28:1260–1266
Imosemi O (2013) The role of antioxidants in cerebellar development. A review of literature. Int J Morphol 31(1):203–210
Isaac M, Quinn R, Tabet N (2008) Vitamin E for Alzheimer's disease and mild cognitive impairment. Cochrane Database Syst Rev 16(CD002854)
Jain P, Pareek A, Ratan Y et al (2013) Free radicals and dietary antioxidants: a potential review Int J Pharm Sci Rev Res 18(1):34–48
Jayaprakasha G, Singh R, Sakariah K (2001) Antioxidant activity of grape seed (*Vitis vinifera*) extracts on peroxidation models in vitro. Food Chem 73:285–290
Jenkinson AM, Collins AR, Duthie SJ et al (1999) The effect of increased intakes of polyunsaturated fatty acids and vitamin E on DNA damage in human lymphocytes FASEB J13(15): 2138–2142
Jialal I, Devaraj S, Venugopal SK (2002) Oxidative stress, inflammation, and diabetic vasculopathies: the role of alpha tocopherol therapy. Free Radic Res 36(12):1331–1336
Jiang J, Xiong YL (2016) Natural antioxidants as food and feed additives to promote health benefits and quality of meat products: a review. Meat Sci 120:107–117
Jiang Y, Duan X, Joyce D et al (2004) Advances in understanding of enzymatic browning in harvested litchi fruit. Food Chem 88:443–446
Johnson FC (1979) The antioxidant vitamins CRC. Crit Rev Food SciNutr 11:217–309
José V, Ferrera JS, Federico VP et al (1991) Cataract: a review on free radical related pathogenesis and antioxidant prevention. Arch GerontolGeriatr 13:51–59
Joseph A, Li Y, Koo HC et al (2008) Superoxide dismutase attenuates hyperoxia-induced interleukin-8 induction via AP-1. Free RadicBiol Med 45(8):1143–1149
Kamal-Eldin A, Appelqvist LA (1996) The chemistry and antioxidant properties of tocopherol and tocotrienols. Lipids 31:671–701
Kamerbeek NM, Van Zwieten R et al (2007) Molecular basis of glutathione reductase deficiency in human blood cells. Blood 109:3560–3566
Knight JA (1999) Free radicals, antioxidants, aging and disease. AACC Press, Washington

Kobus-Cisowska J, Flaczyk E, Rudzińska M et al (2014) Antioxidant properties of extracts from Ginkgo biloba leaves in meatballs. Meat Sci 97:174–180

Konturek P, Brzozowski T, Konturek S et al (2007) Role of histamine in ghrelin-induced gastroprotection against acute gastric lesions. J Inflamm Res 56:S25–S26

Koo HC, Davis JM, Li Y et al (2005) Effects of transgene expression of superoxide dismutase and glutathione peroxidase on pulmonary epithelial cell growth in hyperoxia. Am J Physiol Lung Cell MolPhysiol 288(4):718–726

Kornienko JS, Smirnova IS, Pugovkina NA et al (2019) High doses of synthetic antioxidants induce premature senescence in cultivated mesenchymal stem cells. Sci Rep 9:1296

Kostova I (2005) Lanthanides as anticancer agents. Curr Med Chem Anticancer Agents 5(6): 591–602

Kregel KC, Zhang HJ (2007) An integrated view of oxidative stress in aging: basic mechanisms, functional effects and pathological considerations. Am J Phys Regul Integr Comp Phys 292(1): R 18-36

Kurosawa T, Itoh F, Nozaki A et al (2005) Suppressive effects of cacao liquor polyphenols (CLP) on LDL oxidation and the development of atherosclerosis in Kurosawa and Kusanagi-hypercholesterolemic rabbits. Atherosclerosis 179:237–246

Lee JW, Davis JM (2011) Future applications of antioxidants in premature infants. CurrOpinPediatr 23(2):161–166

Lee AH, Su D, Pasalich M et al (2014) Soy and isoflavone intake associated with reduced risk of ovarian cancer in southern Chinese women. Nutr Res 34:302–307

Levine M, Ramsey SC, Daruwara R (1991) Criteria and recommendation for vitamin C intake. JAMA 281:1415–1423

Li J, Solval KM, Alfaro L et al (2015) Effect of blueberry extract from blueberry pomace on the microencapsulated fish oil. J Food Process Preserv 39:199–206

Lin J, Gao Y, Li H et al (2014) DNA protective effect of mangosteen xanthones: an in vitro study on possible mechanisms. AdvPharmBull 4:147–153

Liu C, Wang R, Pan H et al (2013) Effect of lycopene on oxidative stress and behavioral deficits in rotenone induced model of Parkinson's disease. Chin J Appl Physiol 29:380–384

López-Serrano M, Ros Barceló A (2002) Comparative study of the products of the peroxidase-catalyzed and the polyphenoloxidase-catalyzed (+)-catechin oxidation. Their possible implications in strawberry (Fragaria× ananassa) browning reactions. J Agric Food Chem 50: 1218–1224

Lovell MA, Ehmann WD, Buffer BM et al (1995) Elevated thiobarbituric acid reactive substances and antioxidant enzyme activity in the brain in Alzheimer’s disease. Neurology 45:1594–1601

Maiani G, PeriagoCastón MJ, CatastaG et al (2008) Carotenoids: actual knowledge on food sources, intakes, stability and bioavailability and their protective role in humans. MolNutr Food Res 53: S194-S218

Mansour EH, Khalil AH (2000) Evaluation of antioxidant activity of some plant extracts and their application to ground beef patties. Food Chem 69:135–141

Marczin N, El-Habashi N, Royston D (2003) Free radicals and cardiac arrhythmias following coronary surgery: actors of the drama or bystanders of the spectacle?. ActaAnaesthesiologicaScandinavica 47: 639–642

Mc Cord JM (2000) The evolution of free radicals and oxidative stress. Am J Med 108:652–659

Min JY, Min KB (2014) Serum lycopene, lutein and zeaxanthin, and the risk of Alzheimer's disease mortality in older adults dement. GeriatrOgnDisord 37:246–256

Möller P, Loft S (2006) Dietary antioxidants and beneficial effect on oxidatively damaged DNA. Free Radic Boil Med 41:388–415

Naziroglu M, Butterworth P (2005) Protective effects of moderate exercise with dietary vitamin C and E on blood antioxidative defence mechanism in rats with streptozotocin-induced diabetes. Can J Appl Physiol 30(2):172–185

Neuzil J, Thomas SR, Stocker R (1997) Requirement for promotion, or inhibition of α- tocopherol of radical induced initiation of plasma lipoprotein lipid peroxidation. Free RadicBiol Med 22: 57–71

Nimse SB, Pal DK (2015) Free radicals, natural antioxidants and their reaction. RSC Adv 5(35): 27986–28006

Novi AM (1981) Regression of aflatoxin B1-induced hepatocellular carcinomas by reduced glutathione. Science 212:541–542

Onyeka UE (2005) Food and nutrition. Charismatic Forum Pub Owerri, Nigeria

Oroian M, Escriche I (2015) Antioxidants: characterization, natural sources, extraction and analysis. Food Res Int 74:10–36

Park Y, Nam S, Yi HJ et al (2009) Dietary n-3 polyunsaturated fatty acids increase oxidative stress in rats with intracerebral hemorrhagic stroke. Nutr Res 29:812–818

de Pascual-Teresa S (2014) Molecular mechanisms involved in the cardiovascular and neuroprotective effects of anthocyanins. Arch Biochem Biophys 559:68–74

Pena-Ramos EA, Xiong YL (2003) Whey and soy protein hydrolysates inhibit lipid oxidation in cooked pork patties. Meat Sci 64:259–263

Peschel W, Sánchez-Rabaneda F, Diekmann W et al (2006) An industrial approach in the search of natural antioxidants from vegetable and fruit wastes. Food Chem 97:137–150

Phyu MP, Tangpong J (2014) Neuroprotective effects of xanthone derivative of *Garciniamangostana* against lead-induced acetylcholinesterase dysfunction and cognitive impairment. Food Chem Toxicol 70:151–156

Rahimi R, Nikfar S, Larijani B et al (2005) A review on the role of antioxidants in the management of diabetes and its complications. Biomed Pharmacother 59:365–373

Randhawa S, Bahna SL (2009) Hypersensitivity reactions to food additives. Curr Opin Allergy Clin Immunol 9:278–283

Rao AV, Agarwal S (1999) Role of lycopene as antioxidant carotenoid in the prevention of chronic diseases: a review. Nutr Res 19:305–326

Rao AV, Rao LG (2007) Carotenoids and human health. Pharmacol Res 55:207–216

Rao AL, Bharani M, Pallavi V (2006) Role of antioxidants and free radicals in health and disease. Adv Pharmacol Toxicol 7:29–38

Rizvi SI, Jha R, Maurya PK (2006) Erythrocyte plasma membrane redox system in human aging. Rejuvenation Res 9:470–474

Rock CL, Jacob RA, Bowen PE (1996) Update of biological characteristics of the antioxidant micronutrients—vitamin C, vitamin E and the carotenoids. J Am Diet Assoc 96:693–702

Saad B, Sing YY, Nawi MA et al (2007) Determination of synthetic phenolic antioxidants in food items using reversed-phase HPLC. Food Chem 105:389–394

Sakanaka S, Tachibana Y (2006) Active oxygen scavenging activity of egg-yolk protein hydrolysates and their effects on lipid oxidation in beef and tuna homogenates. Food Chem 95:243–249

Samaranayaka A, Li-Chan E (2011) Food-derived peptidic antioxidants: a review of their production, assessment, and potential applications. J Funct Foods 3:229–254

Sandberg K, Fellman V, Stigson L et al (2004) N-acetylcysteine administration during the first week of life does not improve lung function in extremely low birth weight infants. Biol Neonate 86(4): 275–279

Sarita NC, Umesh M, Shruti M et al (2007) Effect of supplementation of vitamin C and oxidative stress on osteoporosis. Indian J Clin Biochem 22:101–105

Sasazuki S, Hayashi T, Nakachi K et al (2008) Protective effect of vitamin C on oxidative stress: a randomized controlled trial. Int J Vitam Nutr Res 78:121–128

Sastre J, Pellardo FV, Vina J (1996) Glutathione, oxidative stress and aging. Age 19:129–139

Sebranek J, Sewalt V, Ketal R (2005) Comparison of a natural rosemary extract and BHA/BHT for relative antioxidant effectiveness in pork sausage. Meat Sci 69:289–296

Sen CK, Khanna S, Roy S (2006) Tocotrienols: vitamin E beyond tocopherols. Life Sci 78:2088–2098

Shahidi F (2000) Antioxidants in food and food antioxidants. Nahrung 44:S158–S163
Shahidi F, Zhong Y (2010) Novel antioxidants in food quality preservation and health promotion. Eur J Lipid Sci Technol 112:930–940
Sharoni Y, Danilenko M, Levy J (2000) Molecular mechanism for the anticancer activity of the carotenoid lycopene. Drug Develop Res 50:448–456
Shi HL, Noguchi N, Niki N (1999) Comparative study on dynamics of antioxidative action of α-tocopheryl hydroquinone, ubiquinol and α- tocopherol, against lipid peroxidation. Free Radic Biol Med 27:334–346
Shidfar F, Keshavarz A, Jallali M et al (2003) Comparison of the effects of simultaneous administration of vitamin C and omega-3 fatty acids on lipoproteins, apo A-I, apo B, and malondialdehyde in hyperlipidemic patients. Int J Vitam Nutr Res 73:163–170
Sikora E, Cieślik E, Topolska K (2008) The sources of natural antioxidants. Acta Sci Pol Technol Aliment 7:5–17
Singal AC, Jampana SC, Weinman SA (2011) Antioxidants as therapeutic agents for liver disease. Liver Int 31(10):1432–1448
Sofia C, Lourenço MM-M et al (2019) Antioxidants of natural plant origins: from sources to food industry applications. Molecules 24(22):4132
Soto-Vaca A, Gutierrez A, Losso JN, Xu Z, Finley JW (2012) Evolution of phenolic compounds from color and flavor problems to health benefits. J Agric Food Chem 60:6658–6677
Speakman JR, Talbot DA, Selman C et al (2004) Uncoupled and surviving: individual mice with high metabolism have greater mitochondrial uncoupling and live longer. Aging Cell 3:87–95
Stadtman ER (2006) Protein oxidation and aging. Free Radic Res 40:1250–1258
Stefan B, Hans J, Rupprecht et al (2003) Glutathione peroxidase 1 activity and cardiovascular events in patients with coronary artery disease. N Engl J Med 349:1605–1613
Stefanis L, Burke RE, Greene LA (1997) Apoptosis in neurodegenerative disorders. Curr Opin Neurol 10:299–305
Suzuki K, Nakagawa K, Miyazawa T et al (2013) Oxidative stress during development of alcoholic fatty liver: therapeutic potential of cacao polyphenol. Biosci Biotechnol Biochem 77:1792–1794
Taghvaei M, Jafari SM (2015) Application and stability of natural antioxidants in edible oils in order to substitute synthetic additives. J Food Sci Technol 52:1272–1282
Tajkarimi M, Ibrahim S, Cliver D (2010) Antimicrobial herb and spice compounds in food. Food Control 21:1199–1218
Traber MG (2007) Vitamin E regulatory mechanisms. Annu Rev Nutr 27:347–362
Tripodo G, Ibáñez E, Cifuentes A et al (2018) Optimization of pressurized liquid extraction by response surface methodology of Goji berry (*Lyciumbarbarum L.)* phenolic bioactive compounds. Electrophoresis 39:1673–1682
Valavanidis A, Vlachogianni T, Fiotakis K et al (2013) Pulmonary oxidative stress, inflammation and cancer: respirable particulate matter, fibrous dusts and ozone as major causes of lung carcinogenesis through reactive oxygen species mechanisms. Int J Environ Res Public Health 10:3886–3907
Victor HG, Juan CV, David WG (2001) Mechanism of vitamin C inhibition of cell death induced by oxidative stress in glutathione-depleted *HL-60*. J Biol Chem 276:40,955–40,961
Von Sonntag C (2006) Free-radical-induced DNA damage and its repair. Springer, Heidelberg
Wang W, Kannan K (2018) Inventory, loading and discharge of synthetic phenolic antioxidants and their metabolites in wastewater treatment plants. Water Res 129:413–418
Wang W, Asimakopoulos AG, Abualnaja KO et al (2015) Synthetic phenolic antioxidants and their metabolites in indoor dust from homes and microenvironments. Environ Sci Technol 50:428–434
Wayner DDM, Burton GW, Ingold KU et al (1987) The relative contributions of vitamin E, urate, ascorbate, and proteins to the total peroxyl radical-trapping antioxidant activity of human blood plasma. Biochem Biophys Acta 924:408–419

Wojcik M, Burzynska-Pedziwiatr I, Wozniak LA (2010) A review of natural and synthetic antioxidants important for health and longevity. Curr Med Chem 17:3262–3288

Woo RA, Melure KG, Lee PW (1998) DNA dependent protein kinase acts upstream of p53 in response to DNA damage. Nature 394:700–704

Wu AH, Ziegler RG, Nomura AM et al (1998) Soy intake and risk of breast cancer in Asians and Asian Americans. Am J Clin Nutr 68:1437S–1443S

Wu W, Wang X, Xiang Q et al (2014) Astaxanthin alleviates brain aging in rats by attenuating oxidative stress and increasing BDNF levels. Food Funct 5:158–166

Xiu-Qin L, Chao J, Yan-Yan S et al (2009) Analysis of synthetic antioxidants and preservatives in edible vegetable oil by HPLC/TOF-MS. Food Chem 113:692–700

Yin J, Becker EM, Andersen ML et al (2012) Green tea extract as food antioxidant. Synergism and antagonism with α-tocopherol in vegetable oils and their colloidal systems. Food Chem 135: 2195–2202

Zafra-Stone S, Yasmin T, Bagchi M et al (2007) Berry anthocyanins as novel antioxidants in human health and disease prevention. Mol Nutr Food Res 51:675–683

Zhu L, Gunn C, Beckman JS (1992) Bactericidal activity of peroxynitrite. Arch Biochem Biophys 298:452–457

Zunino SJ, Storms DH, Stephensen CB (2007) Diets rich in polyphenols and vitamin a inhibit the development of type I autoimmune diabetes in non obese diabetic mice. J Nutr:1–3

Chemistry of Animal Tissues 11

Khudsia Sultana, K. Jayathilakan, and V. A. Sajeevkumar

11.1 Introduction

According to the Codex Alimentarius (2005), "meat" can be defined as the parts of an animal that are categorized as safe and suitable for consumption by humans. In general, the skeletal muscle and its adhering fat is referred to as meat but it also includes some organs like lungs, liver, kidneys, brain, skin, bone marrow, etc. Meat, fish, and egg have a unique role in the human diet as they are a very nutritious and versatile food. They form an indispensable part of the non-vegetarian diet and are liked for their distinctive taste. They are also rich sources of various nutrients, providing fatty acids, B-complex vitamins, good quality animal proteins, essential amino acids, minerals, and trace elements (Singh et al. 2015). In developed nations, meat, fish, and egg products form a significant part of diet, where its consumption culture is influenced by the increase in global trade and innovations in its distribution and advances in preservation techniques (Swatland 2010).

Meat, fish, and egg have a high nutritional value due to its high content of macronutrients, such as highly nutritious and readily digestible proteins and balanced proportion of essential amino acids (EAA's) (Lawrie and Ledward 2006) and micronutrients, such as iron, which makes meat a good product for everyone. Meat, fish, and egg are also good sources of fat-soluble vitamins (A, D, E, and K) and water-soluble B-vitamins but generally low in vitamin C. They are also rich sources of minerals like Zn, K, Cu, Na, Fe, and P. Meat contains iron and zinc bound to heme protein, which is easily absorbed into the body (Neumann et al. 2002).

Meat, fish, and egg products are important food for human beings. Since these products are nutrient-dense, they provide a perfect environment for the development and proliferation of meat spoilage microorganisms and common foodborne

K. Sultana (✉) · K. Jayathilakan · V. A. Sajeevkumar
Freeze Drying and Animal Products Technology Division, Defence Food Research Laboratory, Mysore, India

O. P. Chauhan (ed.), *Advances in Food Chemistry*,
https://doi.org/10.1007/978-981-19-4796-4_11

pathogens. It is therefore necessary that suitable processing techniques have to be optimized with limited additives to preserve their quality and safety (Aymerich et al. 2008) and attain better storage life in terms of physical, chemical, microbiological, and sensory profile. Alternative nonthermal preservation technologies such as cold plasma, high hydrostatic pressure, super chilling, natural bio-preservatives, and active packaging have been proposed and investigated by many food scientists and researchers in response to the increased demand for high quality, fresh appearance, convenient, and extended shelf life in fresh animal produce. The techniques employed in meat and fish preservation are mainly focused on reducing microbial spoilage, although other methods of preservation are required to minimize other deteriorative changes such as color and oxidative changes. All of these alternative technologies aim to be gentle, energy-efficient, and environmentally friendly while removing viruses and spoilage microorganisms (Zhou et al. 2010).

11.2 Composition of Meat

The animal carcass and the fish flesh consists of muscle, fat, connective tissue, bone, and almost 75% water in varying proportions depending on breed, species, seasonal changes, diet, age, temperature of the environment, size, etc. The most variable proportion in the meat and fish flesh is the fat content, which can vary from 2% in some free-living animals to 15–40% in well-reared domesticated animals and can range from 5 to 20% in different species of fishes. The three major components of meat-lean or the myofibrillar components, fat, and connective tissue affect meat and fish flesh quality in different ways (Miller et al. 2001).

A comparative evaluation of the macronutrients from different animal and fish sources has been depicted in Table 11.1.

Table 11.1 Nutritional composition of different meats and fish (per 100 g)

Meat	Moisture	Protein	Fat	Ash	Energy (kcal)
Beef muscle	74.3	22.6	2.6	1.0	114
Buffalo muscle	78.7	19.4	0.9	1.0	86
Fowl	72.2	25.9	0.6	1.3	109
Goat meat	74.2	21.4	3.6	1.1	118
Mutton muscle	71.5	18.5	13.3	1.3	194
Pork muscle	77.4	18.7	4.4	1.0	114
Chicken	75.0	22.8	0.9	1.2	105
Fresh water fishes					
Rohu (*Labeo rohita*)	76.7	16.6	1.4	0.9	97
Catla (*Catla catla*)	73.7	19.5	2.4	1.5	111
Marine water fishes					
Shark (*Stromateus sinesis*)	76.0	21.6	0.4	1.2	93
Sardine (*Sardinella fimbriata*)	78.1	21.0	1.9	1.7	101

Source: Gopalan et al. (1971)

Meat and fish products are an excellent source of essential nutrients such as iron, vitamin-B complex, high biological value proteins, and a variety of other beneficial bioactive components. They deliver a satisfying dining experience, good health, and well-being in addition to the highest quality nutrition. Meat is nutritionally significant because of its high-quality protein, which contains all the required amino acids, polyunsaturated fatty acids, and highly bioavailable minerals and vitamins. Meat contains vitamin B_{12} and iron, both of which are scarce in vegetarian diets. Vitamins A, D, and E are abundant in fatty fish and are more quickly absorbed by the body than plant meals. Adipose tissue, which is used by the animal to store energy and consists of real fat esters of glycerol with a fatty acid, or intramuscular fat, which comprises phospholipids and unsaponifiable components, such as cholesterol, can both be found in meat (Lawrie and Ledward 2006).

Meat consists of about 75% water, 19% protein, 2.5% intramuscular fat, 1.2% carbohydrate, and 2.3% other soluble non-protein substances. These comprise nitrogenous compounds such as amino acids and inorganic substances such as minerals (Lawrie and Ledward 2006). Red meat contains high biological value protein and important vitamins namely B complex vitamins like Vitamin B_1 (thiamin) and Vitamin B_2 (riboflavin) that are essential for our well-being. It also contains various fats, including the essential omega-3 polyunsaturated fats.

11.2.1 Water in Meat

Because adipose tissue contains less moisture, the lower the total water content of an animal's carcass, the fatter it is. Beef muscle from mature, overweight cows can contain as little as 45% moisture, whereas veal muscle from very young, lean animals can contain up to 72% moisture. The moisture content of fish muscle is about 75% of its body mass. But there are certain exceptions like the Bombay duck (*Harpodon nehereus*) which contains upto 90% moisture. Texture, color, and flavor of muscle are influenced by the quantity of water in muscle tissue.

11.2.2 Proteins in Meat

Proteins take part in a particularly important character as they are building blocks of muscle cellular structures and in addition, they constitute parts of enzymes thus they play both active and stable functions. Meat provides high-quality protein having all the essential amino acids, which are far superior to plant proteins due to their very high biological value. Meat protein contains all the required amino acids as well as limiting amino acids (histidine, isoleucine, lysine, threonine, methionine, phenylalanine, tryptophan, leucine, and valine) (Williams 2007). With a maximum possible value of 1.0, the Protein Digestibility Corrected Amino Acid Score (PDCAAS) is a way of measuring protein quality. Animal meats, such as beef, have a score of around 0.9, compared to 0.5–0.7 for most plant diets, and so meat is considered a "complete protein" source (Schaafsma 2000).

Most of the protein is present in the muscle and the connective tissue. Sarcoplasmic and myofibrillar proteins are seen as high-quality proteins. Sarcoplasmic, myofibrillar, and connective tissue proteins constitute around 30–34%, 50–55%, and 10–15%, respectively in total meat protein (Tornberg 2005). Connective tissue proteins have lower levels of tryptophan and sulfur-containing amino acids while collagen is poor in lysine content (Sharma 1999) and muscle food proteins are characterized by high digestibility and bioavailability.

Muscle proteins are considered myofibrillar, sarcoplasmic, and stromal proteins on the basis of their solubility. Sarcoplasmic proteins are soluble in low ionic strength aqueous solution whereas myofibrillar proteins are salt soluble proteins that are extracted by higher ionic strength salt solutions while the stromal proteins consist of proteins of connective tissues which are very fibrous and insoluble (Aberle et al. 2001). Among insoluble proteins, 0.5 proportions of collagen, 0.3 proportions of elastin, and the remaining 0.47 is a combination of a variety of proteins such as reticulin.

Meat proteins can be separated as those involved in the myofibrillar structure and those in the sarcoplasm. The sarcoplasmic proteins are primarily composed of enzymes responsible for muscle function. The major proteins in meat are missing (~50%) and actin (~20%), while troponin and tropomyosin are the most significant proteins in the biochemistry of muscle function, which are associated with actin in the fragile filament.

The proteins of typical mammalian muscle after rigor mortis but before postmortem deteriorative changes contain about:

11.5%—Structural protein (actin and myosin)
5.5%—Soluble sarcoplasmic protein in the muscle juice
2%—Connective tissue (collagen and elastin)
2.5%—Fat deposited among the protein fibers

Muscle food proteins are characterized by high bioavailability, balanced amino acid profile, and higher digestibility in comparison to plant proteins (Berrazaga et al. 2019). Because collagen and elastin lack sulfur-containing amino acids, the nutritional quality of meat rich in connective tissue proteins is low. There are only 0.8 g of each per 100 g of total protein, compared to 2.6 and 1.3 g of collagen and elastin in "excellent beef." When there is a lot of connective tissue in the meat, it tends to be tough.

The meat proteins are characterized by its excellent bioavailability, balanced amino acid profile, and higher digestibility. The damage to a protein caused by cooking is of little practical significance. Meat and meat products are generally more concentrated sources of protein and it is easier to increase the amount of protein in our diet by the consumption of these foods. There is a remarkable similarity in amino acid composition, i.e., protein quality and also between the different tissues. The composition of the protein tissues is under genetic control and diet has no influence.

11.2.3 Fat Content in Muscle Foods

Fatty tissues are a portion of the meat carcass that occurs naturally. Fatty tissues in a living organism serve as energy stores, provide insulation against body temperature loss, and provide protective cushioning to the skeleton and organs, particularly the heart and kidney. Fat cells are usually located under the skin within the endomysium surrounding muscle fibers. When fat is associated with muscle fiber due to its appearance in the meat as wavy lines and resembles the appearance of marble. Hence, it is referred to as marbling of fat.

Animal fat is mainly constituted of neutral fats and phospholipids and has a high amount of FFA. Glycerol esters of straight-chain carboxylic acid of triglycerides, which typically include 16–18 carbon atoms, make up the majority of neutral lipids (Dugan Jr 1971). Animal fats include a modest amount of phospholipids. As structural and functional components of cells and membranes, they perform a critical role. Phospholipids make up a significant portion of the intramuscular lipids found in muscle meals. They normally contain between 0.5 and 1% lean muscle. Phospholipids are detected in higher concentrations in poultry and fish muscle than in red meat (Igene et al. 1985).

Saturated and unsaturated fatty acids are found in equal amounts in meat. Palmitic acid (C16:0) and stearic acid are the two most abundant saturated fatty acids (SFA) in meat (C18:0). Palmitic acid (16:0) accounts for 40% of total fatty acids in the lean component and 48% in the fat component of red meat, with stearic acid accounting for nearly a third of total fatty acids in both the lean and fat components (18:0). The quantities of these two fatty acids are more similar in lamb and mutton. Polyunsaturated fatty acids make up about 11%–29% of total fatty acids. More omega-3 fatty acids are found in lamb than in either chicken or pork (Enser et al. 1996).

Unsaturated fatty acids, MUFAs, and PUFAs are all found in meat. MUFAs are the most common unsaturated fatty acids found in beef, accounting for roughly 40% of the total fat content. Polyunsaturated fatty acids make up about 11%–29% of total fatty acids. Beef and lamb also have more omega-3 fatty acids. They are α-linolenic acid and stearidonic acid. They cannot be synthesized by the body and hence are called essential fatty acids. They are found in valuable quantities in meat. In veal and lamb, the trans-fatty acid (18:1 trans) content in raw muscle meat varies from 22 mg/100 g to 123 mg/100 g respectively but usually has less than 3% of the total fatty acid content. Lamb and mutton have larger levels of both raw and cooked muscular meat than beef and veal (Droulez et al. 2006).

Animal fat is mainly constituted of neutral fats and phospholipids and has a high amount of FFA. Oleic acid is the rich FA in meat. Pork and organ meats are a relatively good source of omega-6 linoleic acid and omega-3 linolenic acid. Fats are a concentrated energy source, and form essential parts of cell membranes and steroid hormones (Warriss 2000a).

11.2.3.1 Saturated Fatty Acids

Meat comprises mixtures of FA as saturated and unsaturated; the levels of SFA are reduced in recent years due to health concern. The most predominant SFA are palmitic and stearic acid; whereas, myristic acid is found to be in lesser quantities which is the utmost atherogenic FA with fourfolds the cholesterol-raising property of palmitic acid (Ulbricht and Southgate 1991).

11.2.3.2 Monounsaturated Fatty Acids

MUFA are the dominant UFA in meat which accounts for approximately 40% of the total fatty acids. When it comes to blood cholesterol levels, these fats are considered neutral. The major MUFA in beef is oleic acid, which may also be found in olive oil and is associated with a healthy Mediterranean diet.

11.2.3.3 Polyunsaturated Fatty Acids

Because PUFAs are prevalent in membrane phospholipids and are involved in eicosanoid production, they have a structural role. PUFA includes n-3 and n-6 FA in meat and meat products.

11.2.3.4 Conjugated Linoleic Acid

The presence of CLA in meat is an emerging dietary benefit from ruminant meat. CLA is a fatty acid that occurs naturally in lamb and beef. CLA has numerous health-promoting properties like tumor reducing (Belury 2002) and atherosclerotic reducing activity (Gavino et al. 2000).

11.2.3.5 Cholesterol

Complex mixtures of triglycerides along with small amounts of cholesterol and phospholipids mainly constitute animal fats. The incorporation of animal fat in food products makes a varying amount of cholesterol. Cholesterol is a steroid lipid molecule normally synthesized by animal cells. It gives membrane structural integrity and fluidity to animal cells. Egg, cheese, poultry, beef, pork, shrimp, and fish are important dietary sources of cholesterol (Patterson et al. 2008). It is susceptible to oxidation forming hydroperoxides, which are degraded to form secondary cholesterol oxidation products (Kubow 1992). Various food processing and storage treatments may cause oxidation of cholesterol yielding cholesterol oxidation products in the presence of light, heat, oxygen, and radiation.

11.2.4 Vitamins and Minerals

Meat and its products are a considerable and good source of B complex group of vitamins including thiamine, riboflavin, niacin, biotin, pyridoxine, cobalamin, pantothenic acid, and folacin. Most of the B-vitamins are lost during cooking as they are heat-labile and the extent of its losses mostly depends on the time and temperature of the cooking method. Vitamin C is more or less absent in meat but a minor amount is present in certain organ meats. Higher levels of niacin and good amounts of thiamine and riboflavin are found in meat. Pork has a good amount of thiamine present than other meats. It contains roughly 5–10 times more thiamine than beef or lamb (Higgs and Mulvihill 2002).

Meat is a principal source of many B-vitamins namely thiamin (B_1), riboflavin (B_2), niacin (B_3), pantothenic acid (B_5), pyridoxine (B_6), and cyanocobalamin (B_{12}). Pork and its products are one of the richest sources of niacin. Half the niacin provided by meat is derived from tryptophan. The liver and kidney are rich in pantothenic acid. The vitamins B_2, B_6, and B_{12} are especially found abundantly in the liver and are also rich in fat-soluble vitamins like vitamin A and supply an appreciable amount of vitamins D, E, and K (Purchas et al. 2003).

Except liver, meat, and meat products were considered poor sources of vitamin D; however, new analytical data for the composition of meat and meat products have revealed that it contains a significant amount of 25-hydroxycholecalciferol (Chan et al. 1995). Meat is now acknowledged as the richest natural dietary source of vitamin D, accounting for around 21% of daily requirements (Gibson and Ashwell 1997). Because 25-OH vitamin D has higher biological activity, 100 g of cooked beef may provide 12% of the daily requirement of 10 g for a 51–70-year-old individual, and cooked lamb may provide more than 25%, making meat an excellent source of this nutrient.

Red meat provides about two-thirds of the daily need for vitamin B12 and around 25% of the recommended dietary intakes for vitamin B2, B3, B6, and B12 per 100 g. (Williams 2007). Chicken breast is a good source of vitamin B3 (about 56% of the daily requirement) and vitamin B6 (roughly 27% of the daily requirement) in poultry, whereas turkey breast provides 31% of the vitamin B3 daily requirement and 29% of the vitamin B6 daily requirement in 100 g. Both provide about 6–8% of our daily needs (USDA 2011).

11.2.5 Micronutrients in Meat

11.2.5.1 Iron

Meat is a rich dietary source of iron for our body. The apprehension about iron deficiency is one nutritional reason for suggesting the consumption of at least some meat in our diet (COMA 2000).

Iron is also present in good quantity that helps in the synthesis of myoglobin and hemoglobin along with certain enzymes. Meat and its products provide about 14% of iron intake. All the meats, in particular beef, are exceptional sources of dietary zinc which are imperative for healing, immune system, reproduction, and growth (Aggett and Comerford 1995). Meat is known to have about 10 mg/100 g of selenium which includes 25% of RDA and acts as an antioxidant.

Heme iron and non-heme iron are two types of iron found in food. Meat is the main source of heme iron, which is generated from hemoglobin and myoglobin, whereas non-heme iron is found in plant sources such as whole grains, legumes, nuts, seeds, fruits, and vegetables. Iron is found in meat in two forms: heme (50–60%) and non-heme. Heme iron (20–30%) is directly absorbed by our systems since it is unaffected by other dietary variables. The quantity of ingested iron that is absorbed and utilized by the body is referred to as iron bioavailability. Vitamin C and meat are two dietary components that increase non-heme iron absorption.

Non-heme iron is absorbed 15–25% better from meat than it is from plants (Hallberg and Hulthén 2000).

11.2.5.2 Zinc

All meats especially beef are excellent sources of dietary zinc. One-third of the overall zinc intake comes from meat and meat products. Zinc absorption is hampered by inhibitors contained in plant diets, such as phytates and oxalates. Meat, on the other hand, aids in the absorption of zinc (Rayman 2000).

11.2.5.3 Selenium

About 10 mg of selenium is found in 100 g of meat which is roughly about 25% of our daily required intake. Beef and pork have higher selenium levels than lamb, which could be attributable to the animal's age, as selenium can accumulate in meat over time. Plant diets were assumed to have higher bioavailability than animal foods, but new research shows that meat, both raw and cooked, is a highly bioavailable source of selenium (USDA 2011).

11.2.5.4 Other Minerals

Meat also contains phosphorus, which is necessary for carbohydrates, fat, and protein metabolism. An average serving meets about 20–25% of an adult's nutritional needs. Copper, magnesium, potassium, iodine, and chlorine are all abundant in meat.

11.3 Nutritional Composition of Fish Muscle

About 50–60% of fish mass comprises of muscle, having 16–21% of proteins as its major constituent, followed by lipids (0.5–2.3%), ash (1.2–1.5%), water (52–82%), and the carbohydrate content is very negligible, around 0.5%.

The essential amino acids in fish proteins are abundant, and they have a high biological value and are easily digestible. The proteins found in the fish muscle are broadly categorized into three types—sarcoplasmic, myofibrillar, and stroma proteins. Sarcoplasmic proteins are water-soluble and most enzymes are made up of sarcoplasmic proteins. They constitute about 20–30% of the fish muscle. They are soluble in low ionic strength salt solutions. The sarcoplasmic protein in fish includes myogen, globulin, etc. Generally, sarcoplasmic proteins are regarded as enzymes of muscle metabolism. They are found to be more in pelagic fishes and lesser in demersal fishes. Myofibrillar proteins or contractile proteins are muscle proteins and constitute about 65–75% of the muscle protein. They are soluble in high ionic strength solutions. They consist mainly of actin, myosin, tropomyosin, actomyosin, alpha-actinin, c-protein, m-protein, beta-actinin, and troponins C and I (Vareltzis 2000). These proteins help in the contraction and relaxation of muscles postmortem and proteolysis causes Myofibrillar breakdown. They also determine the functional characteristics like gelling and rheological properties and hence useful in surimi industry. The stroma or connective tissue proteins constitute about 1–3% of the total muscle proteins. Fish muscles have very less stroma proteins. The negative aspect of stroma proteins are they are hard to digest and are not soluble in water and ionic solutions. As the fish proteins are low in stroma protein content, they are digested easily than other livestock meat.

Fish fat content varies greatly in terms of amount and fatty acid makeup. The significant amount of long chain–3 series unsaturated fatty acids, also known as polyunsaturated fatty acids, distinguishes fish lipids from those of terrestrial species (PUFAs). The fatty acids found in fish lipids are extremely complicated. The fatty acids have 10–22 carbon atoms and its unsaturation varies from 1 to 6 double bonds. Due to the high unsaturation characteristic of these fatty acids, they are prone to lipid oxidation and oxidative degradation. Individual fatty acid ratios in fish vary from species to species and within species, depending on feed intake, season, environment, spawning, migration, and other factors. Fish lipid contains a high concentration of fatty acids. Myristic, palmitic, and stearic acids are key saturated fatty acids in fish lipids, while palmitoleic (16:0) and oleic acid (18:1 n-9) are important monounsaturated fatty acids, and arachidonic (C-20:4 n-6), eicosapentaenoic (EPA, 10:5 n-3), and docosahexaenoic acids (DHA, 22:6 n-3) are major polyunsaturated fatty acids.

Fish meat is a rich source of both water-soluble and fat-soluble vitamins. The fish lipids contain vitamins A, D, and E which are fat-soluble. The liver of cod and shark fish is a rich source of vitamins A and D. The shark liver oil is the richest source of vitamin A and contains upto 400,000 IU/g of oil. Fish and fish products are often regarded as the most important natural vitamin D sources. The vitamin D content varies significantly from species to species. In general, the fishes with higher fat content are found to have more vitamin D content. In addition to the fat-soluble vitamins, the vitamins B_1, B_2, B_6, and B_{12} are also present in good amounts in fish.

Minerals can also be found in fish. The majority of minerals found in seawater can also be found in fish tissue. Individual mineral composition, on the other hand, varies greatly. Calcium, sodium, potassium, phosphorus, magnesium, and other minerals are contained in fish. The vital elements selenium and iodine are the two most significant minerals found mostly in marine fish species. The only major natural source of these minerals is fish.

11.4 Structure of Muscle

The animal carcass is composed of three major constituents, i.e., muscle, bone, and fat. Components of muscle have a vital role in the final meat quality along with it several factors of both antemortem and postmortem affects the tenderness and processing characteristics. Consumers always have some expectations regarding meat quality. These qualities depend on the breed, age, sex, feed of animals, etc. the state of muscle contraction is known to influence the meat texture.

The animal musculature can be categorized into two major types—striated and non-striated muscles. The striated muscles are further classified into cardiac or skeletal muscles. Skeletal muscles have a complex composition because they contain huge amounts of supportive connective tissue, a complete circulatory supply, and a nerve supply that controls each of the billions of muscle fibers. Skeletal muscles also function as lipid storage facilities and contain large amounts of extracellular fluid, the majority of which is water (Hui et al. 2001). All muscles have the same basic structure, which is composed of muscle fibers joined together in bundles. Muscles are made up of numerous tiny spindle-shaped multinucleated cells or fibers covered in a thin membrane of the sarcolemma. A group of muscle fibers are joined by loose connective tissue, endomysium to form a bundle that is further embedded in another connective tissue perimysium and fat deposits (Warriss 2000b). Later epimysium covers the complete muscle. The muscle is supplied with blood vessels and innervated by nerve fibers. These nerve fibers originate from the central nervous system and end at the neuromuscular junction. Muscle fibers contain all the organelles and in young animals, it is found to be less in size compared to aged animals (Lawrie and Ledward 2006).

11.4.1 Microscopic Structure of Muscle Fibres

Meat is postmortem animal skeletal muscle tissue, primarily consisting of water, protein, fat, and minor quantities of polysaccharides. Muscle is converted into meat by a series of biochemical pathways. The tissue comprises of connective tissue, intramuscular fat, and sarcomeres. The diameter of muscle fibers ranges from 10 to 100 μ with either conical or tapering ends with lengths from 1 to 40 mm. Hundreds of cylindrical protein fibers (sarcomeres) ranging in length from 1 to 40 mm are covered by a thick connective tissue casing (epimysium) and are divided into bundles of fibers by a connective tissue network (perimysium). A plasma membrane surrounds these muscle fibers, which is surrounded by connective tissue (endomysium), which is made up of a basement membrane covered by a reticular layer in which a meshwork of fine collage fibrils is embedded in a matrix. (Fig. 11.1

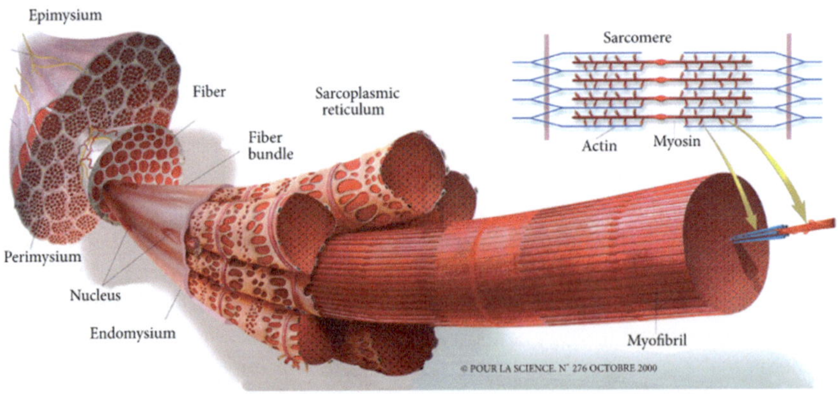

Fig. 11.1 Schematic representations of Muscle (Adopted from Listrat et al. 2016)

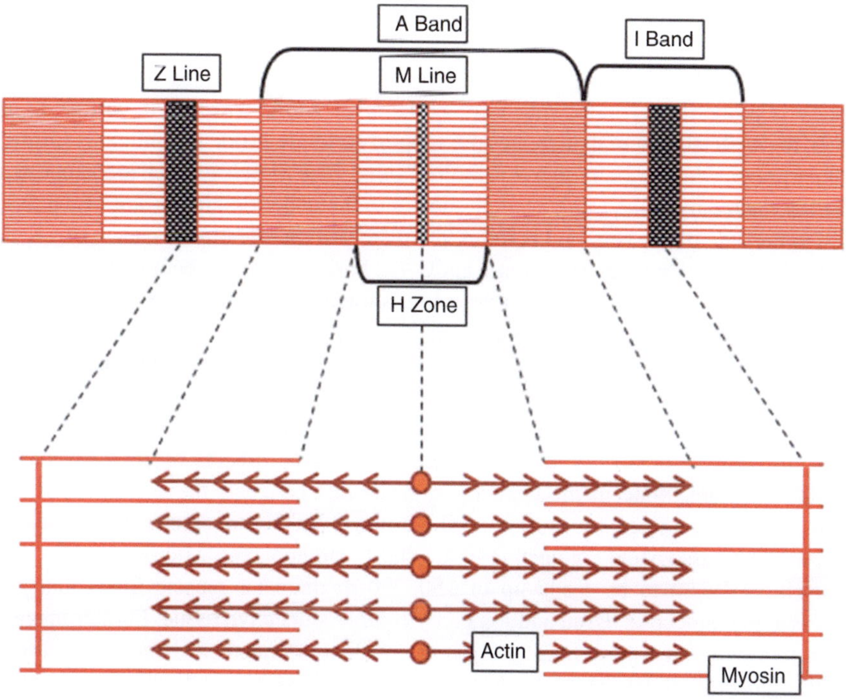

Fig. 11.2 Muscle fibril banding pattern

and 11.2). The sarcomeres have bundles of smaller fibers called myofibrils which are made up of numerous proteins, of which actin and myosin play important functions which are responsible for contraction. Collagen is the major component of connective tissues and it is a strong protein polymer, and found to be 2% of the total muscle proteins. Collagen can be classified into heat soluble and heat insoluble fractions reflecting the degree of cross linkages (Powell et al. 2000). Myofibrils are accountable for the cross striated manifestation of the muscle fiber that remains embedded in the cytoplasm of the muscle fiber. Fibers have all the organelles found in living cells such as nuclei, mitochondria, sarcoplasmic reticulum, and lysosomes (Warriss 2000a). The muscle fibers appear banded under the microscope with a series of bands vertical to the myofibrillar axis. These bands are a result of the overlay of the contractile proteins, with darker regions corresponding to regions of high overlap. Structural proteins that hold the contractile proteins in place appear as dark lines (Z and M lines). The H zone consists mainly of actin whereas the dark regions of the A band comprises primarily of actomyosin complex and the light I band consists mainly of myosin.

11.4.2 Structure of Fish Muscle

Even though the partitions between them are not visible from the outside, a fish's body is primarily separated into three parts: head, trunk, and tail. In cartilaginous fish, the skeletal system that creates the support structure inside the animal is formed of cartilage, while in bony fish it is built of bone. The muscles which are supported by the main part of the trunk account mostly to the major edible portion of the fish. On either side of the vertebral column, there are two muscle bundles, each of which is further divided into an upper mass above the horizontal axial septum and a ventral mass below the septum. In comparison to mammalian muscle, fish muscle has less connective tissue. It is mostly made up of striated muscles. Sarcoplasma, which contains nuclei, glycogen grains, mitochondria, and other components, as well as a number of myofibrils, makes up a muscle cell.

Myotomes, or myomeres, are the longitudinally oriented muscle cells of a fish that are split vertically by sheets of connective tissue. Sheets of collagen called myocomata are found in the connective tissue that connects the skeleton and the skin, while myotomes are muscle mass portions. To the naked eye, the muscular mass and its associated connective tissue sheets are apparent. Microscopically, the muscle is shown to be made up of 150–300 m diameter muscle fibers that are encased in connective tissue and joined by central connective tissue. Further magnification reveals that these muscle fibers are made up of smaller fibers or myofibrils with diameters of 10–20 m. Each myofibril is divided longitudinally into a huge number of sarcomeres, which are identical units. The principal contractile proteins, actin (thin filament) and myosin (thick filament), as well as minor related proteins like troponin and tropomyosin, enzymes like myosin-ATPase, and other components, make up sarcomeres. With age, the length of the cells and the thickness of the myocomata or myoseptum grow. On microscopic observation, these proteins, or filaments, are arranged in a regular alternating pattern, giving the muscle a striated appearance. (Fig. 11.3).

Fig. 11.3 Structure of Salmon trout muscle

11.4.2.1 Contraction of Fish Muscles

A nerve impulse causes Ca++ to be released from the sarcoplasmic reticulum to the myofibrils, which causes muscle contraction. The enzyme ATP-ase is activated when the Ca++ concentration at the active enzyme site on the myosin filament rises. This enzyme releases energy by splitting the ATP present between the actin and myosin filaments. The majority of this energy is used as contractile energy, which causes the actin filament to slip in between the myosin filaments, causing the muscle fiber to contract. The muscle relaxes when the reaction is reversed, i.e., when Ca++ is pumped back into the muscle, the contractile ATP-ase activity stops, and the filaments are free to slip passively past each other. When the muscle is relaxed, ATP acts as both a fuel for contraction and a plasticizer in the presence of Mg++, making it one of the most critical components for muscular contraction. Actin and myosin filaments remain interlocked as actomyosin when myofibrillar ATP is missing. During postmortem rigour mortis, this happens in the rigid muscle.

11.4.2.2 Types of Muscles in Fish

The muscles in fishes are broadly categorized into two types; dark or red muscle and white muscle (Fig. 11.4).

White Muscle: The majority of fish muscle tissue is white, although certain fish have a small percentage of black muscle, depending on the species. White muscle can be found in the majority of freshwater fish. Sluggish, slow-moving, or bottom-feeding fish are common. In the case of marine fish, demersal fish that eat in the middle of the water column or on the bottom and move slowly or irregularly have a higher proportion of white muscles than dark muscles. In comparison to black muscles, white muscles have lower levels of lipids, hemoglobin, glycogen, and vitamins. The white muscle is typically used for quick and rapid movement, which is necessary for avoiding predators or obtaining prey.

Dark Muscle: Many pelagic or mesopelagic rapid swimming species have dark tissue that is brown or reddish in color in addition to white muscles. The chemical reaction of hemoglobin with myofibrillar proteins, known as myoglobin, produces

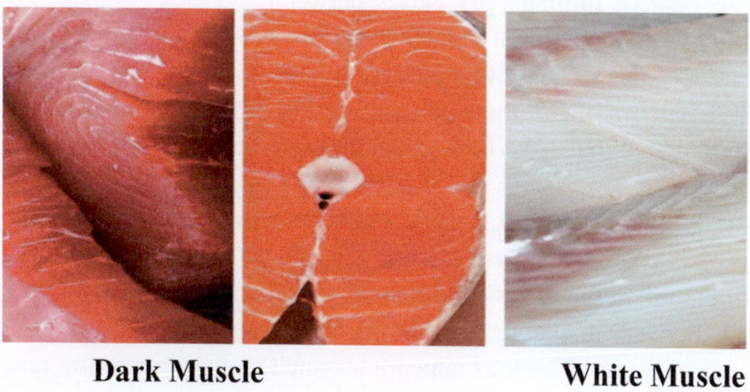

Fig. 11.4 Different types of fish muscle

this blackness. The black muscle runs along the side of the body or lateral line, and in the case of some active species, it also runs in a band towards the spine.

The proportion of dark to white muscles varies depending on the fish's activity. Dark muscle can account for up to 48% of the total weight of pelagic fish, such as herring and mackerel, which swim almost constantly. Lipids, amino acids, hemoglobin, glycogen, trimethylamine oxide, and vitamins are all thought to be abundant in dark muscle. Dark muscle fish have a high fat content, which contributes to their rancidity. Dark muscle also reduces the ability of muscle tissue to produce gels, which is an important feature of fish for heat-treated textured foods. The dark muscle is predominantly a motion muscle, meaning it is responsible for slow, continuous action. The reddish color of the meat found in salmon and sea trout is due to the influence of a red carotenoid known as astaxanthin, not myoglobin. Because fish cannot generate astaxanthin, their red color is determined by the amount of red pigments consumed in their diets.

11.5 Muscle to Meat Conversion

Muscle to meat conversion is a complex process, which is known to affect the meat qualities such as color and flavor that are both dependent on the oxidative changes that occur in meat. After the slaughter, meat muscle is stored at refrigerated conditions for a required length of time to develop organoleptic quality of final product namely "meat" (Ouali et al. 2006).

11.5.1 Postmortem Changes in Meat

11.5.1.1 Acidification of the Muscles After Animal Death

After the death of the animal, oxygen supply ceases thus any further metabolisms undergo through the anaerobic pathway. The glycogen is broken down to lactic acid that is not removed by the blood system thus acidifies the muscle gradually (Fig. 11.5). In an unstressed and well-fed animal, the pH will fall from about 7.2–5.5. The final or ultimate pH varies between muscles (Warriss 2000b). The time required for acidification of muscle varies from animal to animal as reported by Dransfield (1994), the process of acidification in pigs takes 4–8 h, in sheep 12–24 h, and in cattle 15–36 h. The ultimate pH is inversely proportional to the concentration of lactate and the initial glycogen concentration becomes limiting below about 10 mg/g muscle.

As the pH of meat falls, muscle protein gets denatured, and, thus, the water bound to protein is reduced and leads to a lowering in the water-holding capacity (WHC) of muscles. Reduction in the water-holding capacity further leads to an increase in drip loss also an increase in weight loss is observed. Meat pH has an impact on its physical characteristics as well as appearance such as light scattering properties are altered with the change in protein structure.

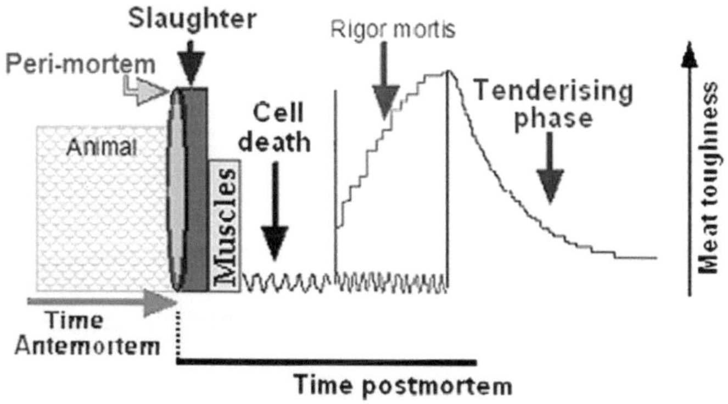

Fig. 11.5 Different phases of conversion of muscle into meat (Reprinted from Ouali et al. 2006 with permission from Elsevier)

11.5.1.2 Rigor Mortis

This is a Latin word meaning "stiffness." As the postmortem glycolysis continues, the muscle becomes inextensible, this stiffening of muscle is termed as rigor mortis. As soon as the animal is dead ATP reserve present in meat gets depleted. Muscle contraction occurs when ATP is hydrolyzed to ADP and rigor mortis eventually starts when the ATP level is low to a certain level that is required to maintain muscle relaxation. As soon as the ATP in muscle is depleted actin and myosin present in muscle form an actomyosin complex. This actomyosin complex is formed irreversibly and extensibility is lost. Firstly, the actomyosin complex formation proceeds slowly in "the delay phase" after that there is a rapid decrease in extensibility in the "Rapid phase." The ATP is slowly lowered with time due to surviving noncontractile muscle ATP-ase activity of myosin. ATP resynthesis in dead animal occurs when there is sufficient glycogen reserve but it cannot be maintained at a level that can prevent actomyosin complex formation.

Before slaughter, the animal is usually starved for a certain period that leads to depletion of glycogen reserve and thus lowers the pH. Excess of oxygen, by stimulating the respiration of animal will delay the onset of rigor mortis. Rigor takes different time to develop in the different species. The onset of rigor mortis is dependent on the ATP and creatinine phosphate reserve, temperature, and initial store of glycogen (Fig. 11.6). The loss of extensibility has a great effect on meat texture (Warriss 2000b).

11.5.1.3 Rigor Mortis in Fish

The fastest onset of rigidity occurs when the fish is stunned and killed by hypothermia (the fish is killed in icy water), while a hit to the head causes an 18-hour delay (Azam et al. 1990; Proctor et al. 1992). When the fish is filleted before or during rigor mortis, rigor mortis has the greatest impact on the business. The body of the fish will be entirely stiff in rigor; the filleting yield will be significantly reduced, and rough handling can cause fillet gaping. If the fillets are separated from the bone

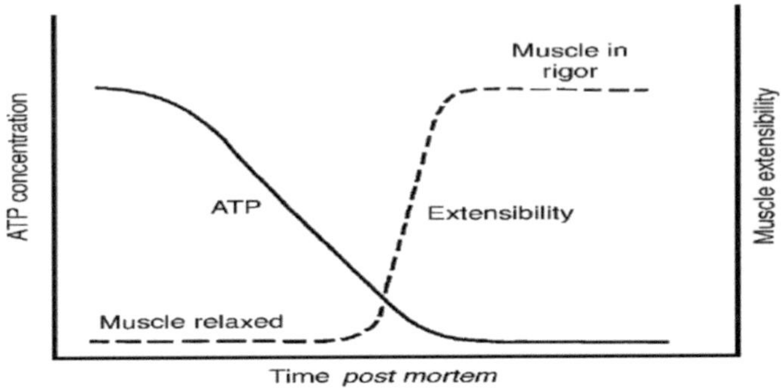

Fig. 11.6 Relation between ATP depletion and the onset of rigor mortis (Adopted from Warriss 2000a)

before the beginning of rigor, the muscle can contract freely, and the fillets will shorten. White muscle can contract up to 52% of the time, while dark muscle can contract up to 52% of the time. White muscle may contract up to 15% of its original length, while dark muscle can contract up to 52% (Buttkus 1963). The texture of the fish will be quite mushy and pasty if it is cooked before rigor. When the fish is cooked rigorously, the texture is rough but not dry. The flesh will become hard, juicy, and stretchy after rigidity.

11.6 Physical Characteristics of Meat

11.6.1 Meat pH

The significance of pH in meat quality is an essential element in deciding the meat quality. The muscles contain a small quantity of muscle-specific carbohydrate known as glycogen (1%), which is mostly broken down to lactic acid in the muscle meat in the first hours (up to 12 h) following slaughtering. This metabolic activity is necessary for the formation of acidity (low pH) in the meat. The accumulation of lactic acid in the muscle causes it to become more acidic, as measured by the pH. At slaughter, the pH of normal muscle is around 7.0, although this will decrease in meat. The pH of a normal animal is between 5.8 and 5.4. The extent to which muscle pH is reduced after slaughter has a substantial influence on the quality of the meat produced. As a result, pH has a significant impact on color, stiffness, and water-holding capacity, as well as minor impacts on taste, tenderness, and postmortem conditioning rate. The pH of muscle has a greater impact on water-holding capacity, which is linked to meat yield and quality. The color of meat is likewise affected by the pH, and while a high pH generates dark meat, both pale and dark meat is unappealing to customers. Glycolysis (the conversion of glycogen into lactic acid) and postmortem metabolism (glycolysis) cause pH changes in meat. A large amount of the variance in meat water-holding capacity and color can be attributed to changes in the rate and/or degree of postmortem glycolysis.

11.6.2 Water Holding Capacity

The ability of meat to retain water throughout production, storage, and cooking is referred to as its "water-holding capacity." High drip loss and poor eating quality are common consequences of low water-holding capacity. Water loss results in a reduction in the amount of product that can be sold. Low water-retention capacity and pale meat color are frequently associated with low meat pH. A quick decline in final meat pH, induced by a combination of factors including genetics, preslaughter stress, and postslaughter handling, generally results in low WHC and pale color. PSE (Pale, Soft, Exudative) or DFD (Dark, Firm, Dry) meats have a high water-holding capacity, which is reflected in the yield. When compared to normal meat, PSE meat yields less and DFD meat yields more (Kirchheim et al. 2001). The metabolism before and after slaughter has a significant impact on the water-holding capacity and pH. A rapid pH drop in the early stages of postmortem has been linked to inadequate water-retention capacity.

The pH of meat is important in predicting the stability, color, stage of rigor mortis as well as it also influences water-holding capacity and several other processing and quality characteristics. Some of the meat's glycogen converts to lactic acid after slaughter. The pH level is decreased as a result. The rate at which the acidity of a mature corpse rises depends on a variety of factors, including the type of animal, breed, upbringing features, and the animal's treatment before slaughter. Accumulation of lactic acid after animal death decreases the pH of muscle from about 7.2 to roughly around 5.5 depending on the species, condition of the animal before slaughter, etc., and if glycogen reserve is low, pH remains high leading to DFD; while, if pH decline is rapid the meat becomes PSE. Meat pH affects water-holding capacity, stiffness, and color, as well as taste, tenderness, and postmortem conditioning rate. The pH of meat is used in predicting the condition of meat and also plays an important role in ultimate meat quality.

11.6.2.1 Factors Affecting Water-Holding Capacity

11.6.2.1.1 Net Charge

It has been reported that during the conversion of muscle to meat there is an accumulation of lactic acid as metabolism undergoes an anaerobic pathway and stored glycogen is converted to lactic acid that lowers the pH of muscle (Offer and Trinick 1983). As the pH reaches the isoelectric pH (pI) of major proteins present in muscle, particularly myosin (pI = 5.4), the net charge becomes zero and, as there is no charge on protein, the interaction between a polar water molecule and protein is not possible; hence, water is not held between muscle proteins. Also as the pH approaches pI there is diminished repulsion between the like charges and the structure becomes more closely packed with no or little space for water molecules. Partial denaturation of protein due to low pH is also found responsible for shrinkage in myofibrillar spacing (Offer and Knight 1988).

11.6.2.1.2 Genetic Factors

According to Fujii et al. (1991), pigs with a mutation in the ryanodine receptor/calcium release channel (halothane gene) in the sarcoplasmic reticulum release calcium under stress as a result of a rapid reduction in pH while the muscle remains warm, causing protein denaturation and resulting in PSE (Pale, Soft, Exudate). This calcium increases the pace of muscle metabolism and pH decline (Bendall and Wismer-Pedersen 1962; Lundström et al. 1989; Huff-Lonergan and Lonergan 2005). The extent of the pH drop in postmortem muscle can also be influenced by other metabolic processes and muscular conditions, such as lactate accumulation.

11.6.2.1.3 Postmortem Proteolysis and Rigor Mortis

As muscle undergoes rigor, there is a reduction in space for water to reside as cross-bridges form between the thick and thin filaments (Offer and Trinick 1983). This decrease in inter-filamental space forces sarcoplasmic fluid to move out. Sarcomeres can also shorten during rigor development, reducing the amount of space accessible for water within the myofibril. Drip loss has been demonstrated to rise linearly with a reduction in the length of sarcomeres in muscle cells (Honikel et al. 1986). According to Bendall and Swatland (1988), there is a loss of volume in the myofibrillar region as well as pH-induced lateral shrinkage of the myofibril, which results in water expulsion. The major enzymes involved in postmortem proteolysis are calpain and calpastatin. The mechanism of water-holding capacity is highly dependent on the muscle proteins, particularly myofibrillar protein. It has a direct effect on pH, ionic strength, and protein oxidation. The main mechanism behind it is that there occur alterations in protein structure that results in muscle cell shrinkage and movement of water in extracellular space (Huff-Lonergan and Lonergan 2005).

11.6.3 Drip Loss in Raw Meat

Offer and Knight (1988) suggested that pH fall at pre-rigor stage induces volume change in myofibrils due to attachment of myosin head to actin filaments at rigor leading, and thus there is shrinkage of myofibrils. Denaturation of protein is also responsible for low WHC in meat mainly when there is a rapid fall in pH at pre-rigor stage (Leygonie et al. 2012). When a muscle is cut, the accumulated fluids between fiber bundles will drain out from the surface under gravity.

11.6.4 Water Loss in Cooked Meat

Heating causes conformational changes in proteins, which cause structural changes such as cell membrane destruction, transverse cell membrane destruction, transverse and longitudinal shrinkage of muscle fibers, sarcoplasmic protein aggregation, and connective tissue shrinkage, all of which result in cooking losses. (Honikel 1998; Kong et al. 2008; Tornberg 2005). The water loss will take different times depending upon the storage and various other conditions to which meat is subjected. Water in extracellular compartments will be lost more easily than that present in deeper compartments. Water retention in muscle is determined by several factors, including the net charge of myofibrillar proteins, the shape of muscle cells and their components, and the amount of extracellular space within the muscle (Huff-Lonergan and Lonergan 2005).

11.6.5 Meat Color

Colour is an important sensory characteristic that can strongly influence the acceptability of food by consumers. The visual appearance and the initial perception will throw more light on the overall characteristics of the product. It has direct control on meat purchasing as represents the noticeable freshness of the product (Mancini and Hunt 2005). In addition to other heme proteins like cytochrome C and hemoglobin, the pigment myoglobin is responsible for flesh color (Miller et al. 2002). The color is affected by factors such as pigment content, pH drop rate postmortem, physical features of the muscle, and additives.

11.6.5.1 Forms of Myoglobin and Chemistry of Meat Color

The color of raw meat is a mixture of myoglobin content and reflection from the protein denaturation as a result of the pH change. Myoglobin (Mb), oxymyoglobin (OMb), and metmyoglobin (MMb) are interrelated from one to the other, depending on the processing of meat and is depicted in Fig. 11.7. Metmyoglobin is the oxidized form of the oxygen-carrying protein myoglobin. It causes the characteristic brown coloration of meat. After slaughter, the meat color is purple-red due to myoglobin pigment which on exposure to air becomes a bright red color due to absorption of oxygen and binds to the iron in myoglobin. The bright red color of oxymyoglobin is extremely desirable for meat consumers.

$$Mb^{2+} + O_2 \rightarrow Mb^{2+}O_2$$

Myoglobin stores oxygen for aerobic metabolism and is a water-soluble protein. It constitutes a protein portion and a non-protein porphyrin ring with a central atom of iron. The oxidation state of iron defines the color status and which compounds are normally attached to the iron portion.

The conversion of myoglobin to oxymyoglobin and vice versa is usually relatively simple. Similarly, brown flesh metmyoglobin is easily formed, whereas reverse metmyoglobin synthesis is more difficult. A dynamic cycle exists in raw meat in such a way Mb, OMb, and MMb pigments are in equilibrium with one another and are interconverted. The intensity of myoglobin within a muscle is influenced by the animal's species, muscle function, and age. The type of iron in myoglobin's porphyrin ring (Fe^{+2} and Fe^{+3}), as well as the compound linked to the myoglobin ligand, are mostly influenced by the meat's storage circumstances. The color intensity of the flesh increases as the myoglobin level increases, from white or

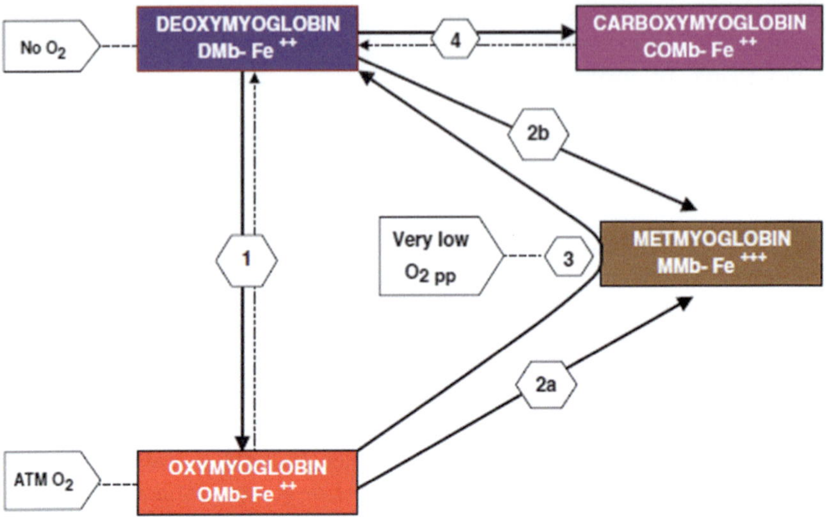

Rx 1 (Oxygenation): DMb + O_2 → OMb
Rx 2a (Oxidation): OMb + [oxygen consumption or low O_2 partial pressure] – e^- → MMb
Rx 2b (Oxidation): [DMb – hydroxyl ion – Hydrogen ion complex]+ O_2 → MMb + O_2^-
Rx 3 (Reduction): MMb + Oxygen consumption + metmyoglobin reducing activty → DMb
Rx 4 (CarboxyMb): DMb + carbon monoxide → COMb

Fig. 11.7 Interconversions of myoglobin on the surface of the meat (Reprinted from Current research in meat color, Meat Science 71(1) (2005) 100–121, Mancini, R. A., and Hunt, M. with permission from Elsevier)

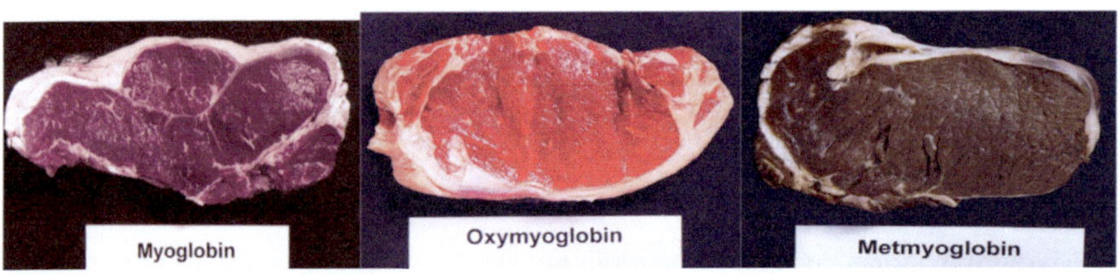

Fig. 11.8 Colour characteristics of meat cuts with conversion of myoglobin forms (*Source:* Mancini and Hunt 2005)

pink to very dark red. (Miller 1994). Meat color is affected by pH, temperature, reaction of myoglobin with NO and CO or metmyoglobin reductase, fat content or preservation and cooking treatments, etc. (King and Whyte 2006). The images of color attributes of meat cuts with myoglobin interconversion forms are represented in Fig.11.8.

The oxidation of ferrous-oxymyoglobin (Fe^{2+}) to ferric-metmyoglobin (Fe^{3+}), which causes discoloration, may be linked to the fluctuation in oxymyoglobin and metmyoglobin proportions in HPP chevon meat (Chaijan 2008). HP treatment separates the ion pairs, causing electrostatic connections to be disrupted, resulting in globin denaturation and/or heme displacement or release (Andrés et al. 2004; Campus et al. 2008).

The major pigments found in fresh, cured, and cooked meat are represented in Table 11.2 and the various studies conducted by different researchers on the impact of different processing conditions and storage are given in Table 11.3.

11.6.5.2 Dark Firm and Dry (DFD)

In meat industry, DFD has been a problem as the cut becomes dark and unattractive on cutting. Muscles that maintain a consistent high pH during postmortem conversion to meat depict a dark, firm, and dry condition that is highly unacceptable among consumers (Sharma 1999). According to Katsaras and Peetz (1990), DFD meat is generally relatively soft, although not in the same way as normal meat, and DFD meat has a lower pH than normal meat. Before slaughter when an animal undergoes chronic stress condition, the glycogen reserves get depleted resulting in less lactic acid formation thus the meat is not normally acidified and the ultimate pH remains high (Viljoen et al. 2002).

Because of the increased intracellular water, which reflects less light, the muscle appears dark. As a result of the increased pH, myoglobin is less denaturated, allowing for more aerobic metabolism at the surface. In addition, the high pH keeps iron in a reduced (ferrous) condition. Because of the great water-holding capacity, the muscle is hard, and the surface feels dry because the water is tightly held within the muscle. Viljoen et al. (2002) studied the consumer acceptability of DFD and normal pH beef steaks and concluded that normal pH beef steaks were more acceptable in comparison to DFD steaks in general appearance, color, and overall acceptability. In a study, this has been reported that DFD muscles on irradiation and then vacuum-packaged are stable and resistant to oxidative changes. DFD pork, highly susceptible to microbial spoilage due to high pH conditions, could benefit the most from irradiation because spoilage microorganisms along with pathogens will be dramatically reduced by irradiation (Ahn et al. 2001).

11.6.5.3 Pale, Soft, and Exudate (PSE)

PSE is caused in pigs by extreme, short-term stress soon before slaughter, such as during offloading, handling, confinement, and stunning. Manhandling, fighting in the enclosures, and poor stunning techniques lead the animals to experience significant anxiety and fear. All of these could cause biochemical changes in the muscle, such as the quick breakdown of muscle glycogen and the meat turning pale, acidic (pH values of 5.4–5.6 shortly after slaughter), and bland. Butchers and meat processors find it difficult, if not impossible, to use this type of meat, and it is frequently discarded. Allowing pigs to rest for 1 hour prior to slaughter and handling them in a quiet manner will greatly lower the danger of PSE. (FAO 1990).

When the pH of meat is below <6 at 45 min after slaughter, PSE is thought to have happened. White muscular fibers have a high glycogen content and are more susceptible to PSE than other muscle fibers. The muscles in the loin area are one example of this. When an animal is exposed to stress that includes the beating of

Table 11.2 Major pigments found in fresh, cured, and cooked meat

Pigment	Mode of formation	State of iron	State of hematin nucleus	State of globin	Color
Myoglobin	Reduction of metmyoglobin, deoxygenation of oxymyoglobin	Fe^{2+}	Intact	Native	Purplish-red
Oxymyoglobin	Oxygenation of myoglobin	Fe^{2+}	Intact	Native	Bright red
Metmyoglobin	Oxidation of myoglobin, oxymyoglobin	Fe^{3+}	Intact	Native	Brown
Nitric oxide myoglobin	Combination of myoglobin with nitric oxide	Fe^{2+}	Intact	Native	Bright red
		Fe^{3+}	Intact	Native	Crimson
Metmyoglobin nitrite	Combination of metmyoglobin with excess nitrite	Fe^{3+}	Intact	Native	Reddish brown
Globin myohemochromogen	Effect of heat, denaturing agents on myoglobin, oxymyoglobin; irradiation of globin hemi chromogen	Fe^{2+}	Intact	Denatured	Dull red
		Fe^{3+}	Intact	Denatured	Brown
Nitric oxide myohemochromogen	Effect of heat, denaturing agents on nitric oxide myoglobin	Fe^{2+}	Intact	Denatured	Bright red
Sulfmyoglobin	Effect of H_2S and oxygen on myoglobin	Fe^{3+}	Intact	Native	Green
Metsulfmyoglobin	Oxidation of sulfmyoglobin	Fe^{3+}	Intact	Native	Red
Choleglobin	Effect of H_2O_2 on myoglobin or oxymyoglobin; effect of ascorbine or other reducing agent on oxymyoglobin	Fe^{2+}	Intact	Native	Green
Verdohaem	Effect of reagents as in 7–9 in excess	Fe^{3+}	Porphyrin ring opened	Absent	Green
Bile pigments		Fe absent	Porphyrin ring destroyed	Absent	Yellow
Nitrihemin	Effect of large excess nitrite	Fe^{3+}	Intact	Absent	Green

Source: Lawrie and Ledward (2006)

Table 11.3 Studies on meat color and its effect

Inference	References
1. Consequence of chill rate of lamb meat during retail display	Jacob and Thomson (2012)
2. Lamb color affected by processing condition and storage temperature	Rosenvold and wiklund (2011)
3. Color measurement by computer image analysis for assessing quality of pork	Chmiel et al. (2011)
4. Broiler breast meat color on pH, emulsification capacity, WHC, and moisture content	Qiao et al. (2001)
5. The impact of pre-dressing medium-voltage electrical stimulation on tenderness and color stability in lamb meat	Toohey et al. (2008)
6. Measuring changes in internal meat color, lightness, and opacity as predictors of cooking time	Pakula and Stamminger (2012)
7. Sodium levulinate was an antimicrobial in fresh pork and fresh Turkey sausage, with no undesirable effects on color. Raw meat redness values decreased with storage time in control and levulinate-treated samples	Vasavada et al. (2003)
8. High-oxygen MAP can gain color stability but reduce odor and flavor stability. Off-odors and flavors may develop before color has deteriorated	Jayasingh et al. (2002)
9. Muscle source, location, and inherent biochemical profile can influence ground beef cooked color	Suman et al. (2010)
10. Adding erythorbate to ground beef increases reducing activity prior to cooking and reduces the frequency of premature browning	Suman et al. (2005)

animals before slaughter, overcrowding of the lairage and fighting among one another before sticking, etc., the rate of acidification is stimulated faster than normal and low pH is reached in the muscle when the temperature of the carcass is still high. This causes denaturation in muscle proteins that are responsible for the reduction in the water-holding capacity as the fluid is expelled into extracellular space (Warriss 2000a). Exudates in significant quantities indicate a lack of water-retaining capacity, as seen in PSE meats. Light scattering from meat surfaces, according to Warriss (2000a), is caused by variations in the refractive indices of the sarcoplasm and myofibrils. The greater the difference, the more scattering occurs, and the flesh seems paler. The amount of light reflected off the meat increases as the myofilament lattice shrinks. The amount of absorbed light is low at high dispersion, and the haem pigments selectively absorb green light, lowering the normal red color. As a result, the meat becomes less red and more yellow as a result. PSE's low pH encourages the oxidation of haem colors such as myoglobin (Mb) and oxymyoglobin (MbO2) to brown metmyoglobin (met Mb) (Adzitey and Nurul 2011).

11.6.6 Texture

During meat consumption, its acceptability is determined by the texture particularly juiciness and tenderness. Meat texture is an important character for both consumers as well as processors. It is highly dependent on three main factors such as the amount of connective tissue, sarcomere length, its degree of cross-linking, and the extent of the proteolytic changes that occur during conditioning postmortem (Warriss 2000a). Several other factors also affect the texture of meat such as water-holding capacity, amount of intramuscular fat, ultimate pH, cooking temperature and time, other zoo-technical characteristics such as handling and feeding characteristics (Aalhus et al. 1998), sex, breed, age, type of muscle (Listrat et al. 2016), and technological characteristics such as electrical stimulation, etc. The texture is a sensory parameter that only a human being can perceive, describe, and quantify (Hyldig and Nielsen 2001).

Multiple factors are considered responsible for the meat flavor and among them, the most important of all is the tenderness that influences the meat palatability (Miller et al. 2001). Huffman et al. (1996) showed that consumers can differentiate between tender and tough meat and will pay for more tender cuts than tough cuts. The three factors that determine meat tenderness are background toughness (exists at the time of slaughter and does not change during the storage period; the resistance to shearing of the unshortened muscle), the toughening phase (caused by sarcomere shortening during rigor development), and the tenderization phase (Koohmaraie and Geesink 2006).

The multicatalytic proteinase complex, lysosomal cathepsins, and the calpain system are three proteolytic systems discovered in the muscle that may play a role in postmortem proteolysis and tenderization (MCP) (Koohmaraie and Geesink 2006). Low-temperature aging has a great impact on meat tenderness due to the activity of proteolytic enzymes.

Tenderness is greatly influenced by the cooking procedure (Sims 1992). Geesink et al. (2011) investigated the effect of pre- and post-rigor meat on tenderness and concluded that pre- and post-rigor meat is extremely tough, and the toughness of early postmortem meat is due to muscle shortening that occurs during the heating process; thus, even if the meat is cooked for an extended period of time, tenderness improves, but not to the extent that meat ages.

11.7 Chemical Characteristics of Meat

11.7.1 Lipid Oxidation

Food lipids are mainly triglycerides, phospholipids, and sterols are a concentrated source of calories and also enhance the organoleptic perception of foods. Lipids are responsible for characteristic color, flavor, texture, and mouthfeel, and hence highly desirable in foods by the consumer. Lipid oxidation occurs in foods with a substantial amount of fat such as milk and meat that changes several quality attributes in food such as flavor changes caused by the production of hydroxyl acids, new volatile odorous compounds are formed that changes aroma. The color change is due to the formation of dark colored compounds due to condensation reaction between oxidation products of lipid and protein and finally due to protein cross-links a new texture is observed in food (Kanner and Rosenthal 1992). The three stages of initiation, propagation, and termination in the autoxidation of lipids as a free radical chain reaction are shown below:

Initiation:	RH+ initiator	$R^{\bullet}$
	RO_2 H	RO'_2
Propagation:	R. + RH	RO'_2
	RO'_2 + RH	RO_2 H + R′
Termination:	R′ + R′	R –R
	RO'_2 + R′	RO_2 R

The radical chain reaction imparts several unique characteristics to lipid oxidation:

(a) Lipid oxidation is autocatalytic.
(b) Small amount of pro or antioxidants causes large rate changes.
(c) The reaction produces multiple intermediates and products that change with reaction conditions and time.

One of the most important non-microbial processes in the degradation of muscle meal is lipid oxidation (Pradhan et al. 2000). Food lipid oxidation control mechanisms are crucial. Lipid oxidation occurs in muscles soon after slaughter, followed by a sequence of metabolic events that result in a loss in antioxidative ability during the conversion of muscle to meat (O'Neill et al. 1998). When unsaturated fatty acids interact with cytosolic pro-oxidants during meat processing, the lipid oxidation process is accelerated. Endogenous pro-oxidants and antioxidants found in meat (Decker and Xu 1998) play a crucial role in the ultimate potential of lipid oxidation in meat. Ferrell myoglobin (activated metmyoglobin or H2O2-activated metmyoglobin) is a key player in the lipid oxidation of preserved meat (Kanner and Harel 1985; Rhee 1988). H_2O_2 is created in meat either through oxymyoglobin oxidation or through other mechanisms, and its amount is affected by various factors that also influence the rate of lipid oxidation in meat. (Pradhan et al. 2000).

Monitoring malonaldehyde (MDA) generation with the thiobarbituric acid (TBA) assay is a standard way to measure the level of lipid oxidation in muscle meals (Fernández et al. 1997). When MDA reacts with TBA, it forms a red-colored complex, the intensity of which is proportional to the amount of MDA present.

11.7.2 Protein Oxidation

Proteins are major components of muscle tissue, having a crucial function in meat and its products in nutritional and sensory aspects (Lawrie 1998). The incidence of protein oxidation in biological systems leads to age-related problems has been studied for years and it has been due to the link (Shacter 2000). Protein oxidation is accountable for decreased amino acid bioavailability and decreased protein digestibility (Gatellier and Santé-Lhoutellier 2009). Liu and Xiong (2000) reported that there is an amide bond formation when carbonyls react with an amino group of non-oxidized amino acids of proteins that leads to the aggregate formation with a decrease in protein digestibility. Due to the formation of disulfide bridges, cysteine oxidation stimulates protein aggregation and decreased the nutritional value of meat (Gatellier and Santé-Lhoutellier 2009).

Proteins undergo irreversible non-enzymatic modification resulting in the production of carbonyl moieties. This process is known as carbonylation which generally happens due to oxidative stresses (Berlett and Stadtman 1997). Carbonyls formation in proteins will take place through following four different pathways, where side chains of amino acids (threonine, lysine, proline, and arginine) undergo direct oxidation (Requena et al. 2001)

(i) Non-enzymatic glycation in the presence of reducing sugars (Akagawa et al. 2005);
(ii) Oxidative cleavage of the peptide backbone via the α-amidation pathway or via oxidation of glutamyl side chains (Berlett and Stadtman 1997)
(iii) Covalent binding to non-protein carbonyl compounds such as 4-hydroxy-2-nonenal or malondialdehyde (Feeney et al. 1974).

Protein oxidation is a complex phenomenon and the type of the products formed is connected to the targets and the initiation of oxidative reactions (Davies 2005). The oxidative damage to proteins can be catalyzed by transition metals and various ROS can participate in the initiation of protein oxidation (Stadtman and Levine 2003). Due to these protein oxidative changes affecting specific amino acid side chains and the peptide backbone, the physical characteristics of proteins can be disturbed. Usually, the protein solubility, aggregation, and fragmentation are affected and can lead to loss of functionality and show declined susceptibility to proteolysis (Xiong 2000). Protein oxidation in meat systems can be enumerated through several manifestations of chemical changes including loss of tryptophan fluorescence (Sun et al. 2011a), gain of carbonyl derivatives (Ganhão et al. 2010), loss of sulfhydryl groups (Frederiksen et al. 2008), and formation of intra- and intermolecular cross-links (Ooizumi and Xiong 2006).

Different thermal and nonthermal processing technologies are known to affect carbonyl formation in meat including irradiation (Rababah et al. 2010), and high-pressure processing (Fuentes et al. 2010), cooking (Ganhão et al. 2010), sausage fermentation (Sun et al. 2011a), and ripening (Ventanas et al. 2006) are represented in Table 11.4.

Table 11.4 Effect of processing techniques on the formation of carbonyls in meat and meat products

Technology	Meat product	Studied effects	Additional analyses	References
Irradiation	Beef meat	Oxidation system	Protein thiol oxidation	Martinaud et al. (1997)
	Chicken breast	Antioxidant effect of plant extracts	TBARS	Rababah et al. (2004)
	Beef sausage	Antioxidant effect of carrot juice	TBARS	Badr and Mahmoud (2011)
High-pressure processing	Sliced dry-cured ham	Pressure and holding time	TBARS	Cava et al. (2009)
		Vacuum-packaging	Volatiles, sensory evaluation	Fuentes et al. (2010)
Cooking	Pork patties	Antioxidant effect of plant phenolics	Hexanal	Vuorela et al. (2005)
		Antioxidant effect of grain meals	Hexanal	Salminen et al. (2006)
		Antioxidant effect of fruit extracts	Tryptophan depletion	Estevez et al. (2008)
			Instrumental color, texture	Ganhão et al. (2010)
	Beef meat	Composition of feeds	Aromatic amino acids,	Gatellier et al. (2010)
Sausage fermentation	Cantonese sausage	Processing	Protein solubility	Sun et al. (2011b)
			Proteolysis, digestibility	Sun et al. (2011a)
Dry-curing	Dry-cured loin	Feeding regime	TBARS, hexanal	Ventanas et al. (2006)
	Dry-cured ham	Feeding regime	TBARS, sensory evaluation	

Source: Estévez (2011)

11.8 Chemistry of Spoilage

After animal death enzyme activity ceases that also prevents glycolysis and ultimately the pH reaches 5.4–5.5. Hence, the indigenous enzymes contribute negligibly to spoilage compared to the microbial action of the microbial flora (Nychas and Tassou 1997; Tsigarida and Nychas 2001). Nychas et al. (2008) said that indigenous enzyme (proteolytic and lipolytic) activity is not sufficient for meat conditioning. Hence, required tenderization in meat is achieved by artificial means such as enzymes, mechanical means, or chemicals are applied (Koohmaraie 1994; Lawrence et al. 2003). Food deterioration is thought to be caused by the accumulation of metabolic by-products rather than the activity of microbial enzymes. (Nychas et al. 2007).

Almost all bacteria in the meat microflora catabolize glucose, lactic acid, and some amino acids, followed by nucleotides, urea, and water-soluble proteins (Gill 1986; McMeekin 1982; Nychas et al. 2007). These chemicals are necessary for microbial growth, and their concentration can influence the sort of microbes present (e.g., saccharolytic, proteolytic, etc.). The rate of spoilage is considered as the principal precursor of those metabolites that are perceived as spoilage agents (Koutsoumanis and Nychas 1999; Nychas et al. 1998; Skandamis and Nychas 2002; Tsigarida and Nychas 2001).

Three sets of substances are involved in the microbial spoilage of products in one or the other way, i.e., (i) compounds contributing to the glycolytic pathway (e.g., glucogen, glucose, glucose-6-phosphate, lactate, etc.), (ii) metabolic products (e.g., gluconate, gluconate-6-phosphate, pyruvate, lactate, etc.), and (iii) nitrogen energy sources (e.g., amino acids, proteins) (Gill 1986; Nychas et al. 1988, 1998).

11.9 Chemistry of Meat Preservation

Due to their varied nutrient content, meat and fish provide an ideal habitat for the growth and proliferation of spoilage microorganisms and common foodborne diseases. Meat is preserved using appropriate preservation processes to ensure its safety and quality (Aymerich et al. 2008). The first and most important step in meat preservation is to prevent microbial spoilage as well as other deteriorating changes such as lipid and protein oxidation. The food processing sector is dominated by the thermal preservation of foods. Food flavor, color, texture, and nutritional qualities such as protein and vitamin degradation are all affected by thermal technologies. Non-thermal methods, on the other hand, have attracted the interest of producers, food scientists, and consumers since they have a minor impact on food sensory and nutritional properties and lengthen the shelf life by suppressing or killing bacteria. Non-thermal food preservation techniques are thought to be more energy-efficient and maintain better quality features than traditional methods (Gould 2000).

Depending on the processes of microbial inactivation, emerging technologies used in food processing and preservation are generally divided into thermal and non-thermal technologies. Non-thermal technologies contain irradiation, high-pressure processing, minimal processing, hurdle technology, pulsed light processing, ultrasound processing, and ozonation while thermal technologies include microwave and ohmic heating, radio-frequency heating, and infrared heating.

11.9.1 Chilling

In addition to other preservation procedures, meat and fish are chilled and then held at 1–5 °C for a few days. Modern packaging techniques, such as carbon dioxide, nitrogen, and vacuum storage, can extend the shelf life of goods by up to 10 weeks. Most pathogens are weakened and the growth of spoilage organisms is slowed when meat is chilled at temperatures close to its freezing point, i.e., −15 °C. Chilling the carcass in the air lowers the surface temperature and aids in drying, reducing microbial growth (Ockerman and Basu 2004).

11.9.2 Freezing

Freezing improves the shelf life of the product by lowering the temperature to achieve lower activity by making no or little available water and solute concentrates. The majority of meat and fish items sold for human consumption are frozen. Commercial preservation at −29 °C and residential preservation at −18 °C are now the norm. The quality of the product is determined by the level of control used during pre-freezing and the stringent temperature control used during post-freezing handling. In most cases, both slow and quick freezing are used. Freezing keeps food safe by reducing the movement of molecules, triggering microbes to enter a dormant stage. Freezing at 0 °C inactivates microbes, such as bacteria, yeasts, and molds present in food. The advantages of using temperatures below freezing point include extending the shelf life of meat while minimizing chemical changes and lowering the microbial load (Lawrie and Ledward 2006). For example, a temperature of −55 °C has been recommended as the optimal storage temperature for frozen meat to avoid any quality alterations (Hansen et al. 2004). At lower temperatures, enzymatic reactions, ice recrystallization, and oxidative rancidity are expected to be negligible.

11.9.3 Smoking

Smoking of meat and fish is a method of preparing red meat and fish which was initiated during prehistoric times. The vital role of smoking is to prolong the shelf life of protein-rich foods that are prone to deterioration during storage. Intensive smoking does lengthen the shelf life by heavier deposition of preservatives and by the drying influence of the hot air. Smoking of meat and fish produces volatile compounds that inhibit bacterial growth and gives a distinct taste to the product (Poligne et al. 2001). Phenolic substances are produced that are important in imparting organoleptic properties to smoked meat (Kjällstrand and Petersson 2001) along with antioxidative (Pöhlmann et al. 2012) and antimicrobial properties (Davidson and Branden 1981). Based on the product's desired characteristics the heat and smoke sources are selected.

11.9.4 Curing

The addition of salt, with or without nitrite or nitrate, to meat and fish is the earliest method of preservation. Nitrite provides consistent red color, works as an antioxidant by sequestering oxygen, inhibits or prevents microbial growth, and imparts a nice flavor to the product. Curing improves the shelf life of the product by preventing it from spoilage microorganisms and it can be applied as a dry process, can be dipped in brine (pickling), injection curing, etc. Oxidative rancidity is prevented with the reduction of nitrite or nitrous acid by microbial action (Honikel 2008).

11.9.5 Drying

The high moisture content of meat and fish makes it highly perishable. Food drying is accomplished using a variety of procedures (Bimbenet et al. 2002) that combine heat or pressure sources to remove water from the product's inner section and mechanical energy to remove water from the product's surface. Drying is a multistep process that involves coordinated mass and heat transfer, as well as physical and structural changes. Heat increases the vapor pressure of water in food products, which aids in moisture evaporation. By diffusion or capillary movement, surface water is withdrawn and replenished with water from within. The concentration gradient created by either liquid or vapor movement acts as the driving force in the diffusion mechanism. A number of drying methods are known such as air drying, vacuum drying, and freeze-drying. The ultimate aim of any process is to maintain quality and should be acceptable by the consumers.

11.9.6 Freeze Drying

Freeze dehydration (FD) is the most advanced and accomplished technology for preserving and storing biological components. FD is a method of preserving a product by removing moisture by the sublimation process under low temperatures and vacuum pressure. Because the entire process is carried out in the absence of liquid water at low temperature and pressure, this technology prevents physicochemical, enzymatic, and microbiological alterations in the result (Ratti 2001). It is a well-known method for attaining superior quality dehydrated foods (Dalgleish 1990). Perishable commodities like juices, meat, and poultry items can be subjected to freeze dehydration process which can result in yielding shelf-stable products which can be easily transported. This whole process includes the primary freezing below eutectic temperature and by reducing the pressure the frozen water is allowed to sublimate without affecting the nutrients of the products.

Freeze-drying (FD) is one of the most successful drying technologies for food preservation, resulting in high-value goods with excellent sensory quality and nutritional retention (Babić et al. 2009; Voda et al. 2012). FD, on the other hand, is a very energy-intensive procedure with a lengthy process cycle (Donsì et al. 2001).

Foods have a longer shelf life and are less likely to deteriorate owing to microbial development or oxidation as a result of this process. Rehydrated FD products will also have the same qualities as fresh FD products. Pre-freezing, primary drying (sublimation), and secondary drying (desorption) are the three separate steps of the process, with drying continuing until the desired moisture content is reached (Baker 1997). These products do not need to be kept cold, and they are just 10–15% of their original weight, making storage, distribution, and commercialization simple.

This method entails freezing a product and then extracting the water content while it is still frozen, resulting in a chemically stable product. Its application has mostly been to biological materials, which are costly to produce and which are highly unstable. However, it is being applied to dehydrate certain foods like meat and meat products, coffee extracts, fruit juices, milk, etc., due to several advantages like high quality, rapid and complete solubility, and retention of flavor and texture of the original foodstuffs.

11.9.6.1 Freeze-Dried Meat Products

When properly packaged, FD meat products can be preserved for indefinite lengths of time while preserving the majority of their physical, biological, chemical, and sensory qualities (Girard and Omolosho 1983). Freeze-drying does not affect the biological value of beef proteins. However, thiamine content can be reduced by up to 30%. In mutton, this drying process induces riboflavin loss, but not in hogs or beef (Lawrie 1985).

Sheridan (1981) conducted studies on a method concerning FD of DFD beef steaks on a weight basis, of glucose or buffers. An evaluation was made between the microflora of beef of normal pH and DFD steaks. An increase in bacterial counts on FD-DFD beef was observed compared to non-freeze-dried steaks. FD had the significant effect of delaying the H2S production in vacuum packs. The procedure allowed the meat's pH to be synced using buffers without the steaks reverting to their previous levels in a short period of time. Several scientific studies have found that freezing conditions have a direct impact on the color of an FD product. Farkas and Singh (1991) discovered that chunks of chicken meat that were frozen quickly kept their white color better than those that were frozen more slowly (Stone Jr and May 1969).

11.9.6.2 Dehydration Effects on Proteins

In freeze-dried products, the final moisture content typically ranges from 2 to 10% by weight. This level of dehydration results in the removal of "bound" water which is intimately hydrogen-bonded at the protein surface. Because protein structure is largely determined by its interaction with water, this dehydration can greatly destabilize the native structure. This in turn can lead to aggregation and loss of gelling functionality. According to the water substitution hypothesis (Carpenter and Crowe 1989), the introduction of a surrogate hydrogen bonding agent may serve to stabilize the proteins as water is removed. They indicated that disaccharides are the agents of choice for this process.

11.9.7 High-Pressure Processing

This is an example of a cutting-edge non-thermal preservation approach that can be used to develop minimally processed and other perishable goods. Product's shelf life is generally extended due to the inactivation of enzymes and microorganisms without affecting product's sensory and nutritional attributes. High-pressure processing can be successfully used in the meat sector by standardizing quality processes for achieving product shelf stability. To achieve the inactivation of germs or to adjust product qualities to match consumer expectations, high-pressure processing can be carried out by selecting and standardizing proper pressures with or without the inclusion of heat. The type of microbe, food content, pH, and water activity all have an impact on the high-pressure processing process.

High-pressure processing results in goods that are nutrient-dense and of high freshness. This method can be used on both acidic and alkaline meals. It is a semi-continuous process that requires expensive equipment, and high pressure can cause textural changes, the formation of free radicals, which cause lipid oxidation, and insensitivity to spores and bacteria resistance. The application of high-pressure processing to food products such as fruit juices and animal products has been expanded (Heinz and Buckow 2010). Because the treatment can be completed near room temperature while assuring safety and stability during refrigerated storage, high-pressure processing techniques can enable significant retention of sensory and nutritional properties of food products.

Consumer demand for fresh-like food products with little sensory and nutritional degradation has sparked research into new non-thermal food treatments. High-pressure processing, also known as cold pasteurization, ultra high-pressure processing, or high hydrostatic processing, has received a lot of attention in recent years as a food preservation method that prevents spoilage and pathogenic microorganisms while preserving the nutritional and sensory qualities of food products (Norton and Sun 2008; Rastogi et al. 2007). In general, food is subjected to high-pressure processing treatment at pressures ranging from 100 to 900 MPa. However, pressures of 400–600 MPa are commonly employed for commercial applications, and it varies by-product as a non-thermal decontamination technique (Jiménez-Colmenero and Borderias 2003). The effects of high-pressure processing differ from the traditional thermal processing techniques such as dehydration, irradiation, and others, which work by inactivating enzymes and lowering microbial load, preventing spoilage, and extending shelflife. Thermal processing processes, on the other hand, have an impact on the product's freshness. Because high-pressure processing is a nonthermal method, it extends shelf life without compromising freshness. High-pressure processing produces homogeneous and practically instantaneous effects throughout the food, making it independent of food geometry and equipment size. Only non-covalent bonds are altered, and thus smaller molecules such as vitamins and flavor compounds may be unaffected (Toepfl et al. 2006).

Commercial applications use pressures up to 800 MPa and temperatures between 5 and 40 °C (Heinz and Buckow 2010). HP processing of meat can cause lipid oxidation depending on the pressure intensity and time (Orlien et al. 2000). It has been recognized that 300 and 600 MPa of pressure levels are critical for encouraging lipid oxidation. Antioxidants can help to reduce lipid oxidation caused by high pressure (Mariutti et al. 2008). Most vegetative bacteria are inactivated by pressures of 300–600 MPa (Smelt 1998). According to López-Caballero et al. (2000), a pressure of 200–400 MPa at 7 °C for 10 min is adequate to kill all targeted bacteria in prawns. Microbes' growth rates are slowed by pressures of 10–50 MPa, while higher pressures can inactivate them (Rademacher 2006). Pressure is more responsive to yeasts and molds than to vegetative bacterial cells (Patterson 2005), while ascospores are exceptionally resistant to pressure treatments (Chapman et al. 2007). The mechanism for microbial destruction in high-pressure processing is by cell membrane alterations (Moussa et al. 2007) dissociation of ribosome (Abe 2007), agglomeration (Farr 2003), and denaturation of proteins (Barbosa-Canovas et al. 1995).

Meat with high moisture content, neutral pH, and high protein content provides an ideal habitat for the growth and proliferation of spoilage microorganisms and foodborne diseases. As a result, appropriate processing and preservation procedures must be used to ensure its quality, safety, and shelf life. High-pressure processing is used as a post-processing step to improve the quality of ready-to-eat beef products and increase their shelf life (Jofré et al. 2009).

Many studies have been conducted to determine the effects of high-pressure processing on meat and meat products, including microbial inactivation (Garriga et al. 2004); texture (Jung et al. 2000); color changes and structural changes of myoglobin in minced beef (Carlez et al. 1995) and pork (Wackerbarth et al. 2009); kinetics of radical formation (Bolumar et al. 2012); and lipid oxidation in poultry (McArdle et al. 2010). Following its initial success in jam and fruit juice (Murchie et al. 2005), HPP has grown in popularity in other food products such as smoothies, rice products, guacamole, salsa, meat, fish, and shellfish. Kaur et al. (2013) found that HPP technology is a suitable processing approach for muscle foods in studies on beef, fish, and shrimp. HPP has been utilized in a variety of meat products from various animals (Souza et al. 2011). Despite its many advantages, this method induces negative changes in texture, color, structure, and lipid oxidation levels in meat products (Ma et al. 2013).

11.9.8 Irradiation

Food radiation processing is a new and potentially useful non-thermal food safety method for enhancing cleanliness and extending storage and distribution life. Ionizing radiation is a technique for assuring the safety of meat products because it can be used to make good changes in foods (Patterson and Stevenson 1995). Radiation affects matter by transferring energy to electrons and ionizing molecules, resulting in positive and negative charges.

Meat processing with ionizing radiation such as electrons, γ-rays, and X-rays is known as radiation processing. γ-rays are emitted by radioisotopes like cobalt-60 and cesium-137, while electrons and X-rays are created by machines. Pathogens and spoilage organisms are killed by exposing packaged meat to effective levels of ionizing radiation. Microbes are inactivated by ionizing radiation that damages nucleic acids both directly through electron and photon contact with DNA and RNA and indirectly through the action of charged ions. During its transition to a stable state of nickel 60, cobalt 60 emits γ-rays (Satin 1996). The benefits of IR include exceptionally effective microbial inactivation, minimal nutritional changes in the food, and the ability to treat after packing.

Irradiation (1–10 kGy) is an effective method for reducing the microbial load in food. *Salmonella, Listeria,* and other hazardous bacteria can infect foods like chicken, pork, eggs, shrimp, and raw milk cheese. Some of these foods can be consumed without the need for further heat treatment to kill the hazardous germs. Irradiation extends the shelf life of meat by reducing spoilage microbes, and, like heat treatment, it may also inactivate enzymes that aid in the rotting of meat.

Numerous studies have been conducted on the effects of radiation on meat products such as bacon, ham, and sausages (Kiss et al. 1990), and beef burgers (Dempster et al. 1985). Irradiation kills harmful bacteria and parasites in meat and meat products, in addition to rotting bacteria. To inactivate 90% of spoilage germs, irradiation doses of around 1–4 kGy are required. Badr (2004) examined the microbiological profile of rabbit meat, as well as the possibility of using irradiation to suppress foodborne pathogenic microorganisms and extend the meat's refrigerated storage life. Irradiated rabbit meat samples (0, 1.5, and 3 kGy) were stored at room temperature. Irradiation greatly raised TBARS levels but had no effect on total volatile nitrogen (TVN) levels, while storage significantly increased both TBARS and TVN levels in both irradiated and nonirradiated samples. Radiation processing had no discernible influence on the sensory qualities of raw beef. Furthermore, burgers made using IR rabbit meat had a high level of sensory acceptance. Irradiation enhanced lipid oxidation when meat and meat products were aerobically packed, according to several studies, resulting in the formation of undesirable color and odor (Ahn et al. 2000, 2001).

Irradiation seems to have the potential to be a preservation method. However, because it changes the nutritional and sensory properties of meat and meat products, it may cause physical-chemical and biochemical changes (Grolichova et al. 2004).

Jayathilakan et al. (2009a, 2009b) studied the development of hurdle processed chicken by the application of gamma irradiation in the presence of lactic acid. The effectiveness of various dosages and concentration of lactic acid was evaluated by establishing various chemical markers like non-heme iron, TBARS, and total carbonyls along with microbiological and sensory attributes. Studies revealed that irradiation at 2 kGy with 2% lactic acid yielded a product with good shelf stability and sensory profile. The product exhibited good physico-chemical and microbiological stability throughout the storage period, as achieved by the application of irradiation and lactic acid.

Plums have been found in irradiated turkey, pre-cooked pig sausage, and roast beef to exhibit antioxidant properties (De Gonzalez et al. 2008). In radiation-processed lamb meat, the usefulness of mint leaves (Kanatt et al. 2007) and chitosan (Kanatt et al. 2006) as natural antioxidants was proven. There have been reports on the beneficial effects of lactic acid (Jayathilakan et al. 2009a, 2009b) and tocopherol in conjunction with sesamol (Nam and Ahn 2003) in preventing quality alterations in irradiated poultry and pork, respectively. Antioxidants can reduce irradiation-induced peroxidation of tallow and lard (Lee et al. 1999). Ahn et al. (1998) conveyed that decrease in the formation of hexanal with irradiated turkey meat was reduced with the addition of α-tocopherol in a dose-dependent fashion.

11.9.9 Hurdle Processing

Hurdle processing employs a mixture of factors, each at an optimal level that is constructively used to convey the essential functional attributes and reduce the spoilage due to microorganisms. Leistner (2000) reported that both microbial safety and stability along with sensory and nutritional properties of the foods are established on the use of several preservative methods (hurdles) such that each imparts a certain level of preservation and does not have any of the unfavorable effects on the quality of food. The concept of hurdle technology for meat products was initially designed by Leistner and Gorris (1995). Hurdle technology helps us to produce safe, stable, and nutritious foods that are both affordable and cost-effective. Besides providing ease in preparation and storage, hurdle processed foods are known to have better sensory attributes along with high nutrient retention. Hurdle technology mainly aims at producing foods with good chemical and microbiological stability by the application of suitable hurdles combined in an intelligent manner and in sequential order.

The effect of heat treatment can be strengthened in modern meat processing by incorporating additional barriers that can limit microbial development. Frequently used hurdles are lowering of water activity (aw), redox potential, control of acidity (pH), irradiation, preservatives (nitrites, sorbate, sulfite, etc.), and competitive microorganisms in a product (Leistner and Gorris 1995). Optimization of preservation techniques along with other hurdles will help in the development of shelf-stable meat products. All of these measures would not stop microbial development on their own, but when combined with heat treatment, they create a variety of barriers that limit the survival of microbes in the product. Naturally, the range of items that may be made shelf-stable by using hurdle technology in the meat industry is restricted, but it may be important in some circumstances, such as when there is no continuous cold chain available.

11.10 Chemistry of Eggs

Humans have eaten eggs for thousands of years. Eggs are produced by female animals of many different kinds, but the chicken egg is by far the most popular food. Eggs are one of the world's healthiest and most nutritious foods. They are one of the most popular animal products on the planet. Many necessary components in a human diet, including proteins, lipids, vitamins, and minerals, are found in eggs.

Because it is easily absorbed and digested by the human body, egg protein is the highest grade dietary protein available. Eggs are also high in lipids, which act as concentrated energy storage and aid in the absorption of fat-soluble vitamins. Eggs have little carbohydrates and no fiber, yet they go well with several dishes. The egg is one of the best and cheapest sources of high-quality protein at a low cost, with a balanced distribution of minerals and vitamins, especially vitamins E, A, B_{12}, B_2, and folate (Surai and Sparks 2001), as well as high amounts of lipids like triacylglycerols, phospholipids, and cholesterol (Watkins 1995).

Eggs are a low-cost source of high-quality protein, vitamins, and minerals such as vitamin A, B_{12}, folic acid, and phosphorous. They are a great source of riboflavin, too. Because it includes all of the required amino acids, the egg is considered a complete protein. The yolk contains the fat, as well as the rest of the protein and the majority of the calories. About 90% of an egg white is water, with the remaining 10% consisting of albumins, mucoproteins, and globulins. Egg yolks are heavy in fat however egg whites are lower in fat and have less than 1% carbohydrate.

The nutrients found in eggs are Vitamin A, Vitamin E, Vitamin B_2, Vitamin B_5, and Vitamin B_{12}, Vitamin D, biotin, choline, iodine, iron, phosphorous, protein, selenium, lutein, and zeaxanthin. There are many health benefits ineggs. The protein in the eggs helps in providing strong muscles. The vitamins and minerals provide enough nerve health and help in the energy production in the body cells. Choline in eggs helps in lower the risk of heart diseases and it also provides enough nutrients during pregnancy. Lutein and zeaxanthin assist in preventing muscle degeneration, which is the major cause of blindness as people grow older. 5.53 g of protein is found in a medium-sized egg weighing 44 g. A large egg has roughly 5 g of fat in it. Unsaturated fat makes up the majority of an egg's fat content, making it the best type to include in a healthy diet. Eggs also include omega-3 fatty acids, which aid in brain function and vision. 164 mg of cholesterol is found in one medium-sized egg weighing 44 gm. Some experts believe that dietary cholesterol raises the ratio of total to HDL cholesterol, negatively affecting the body's lipid profile. However, other studies demonstrate that eating one egg per day does not appear to increase the risk of heart disease in healthy people.

11.10.1 Nutritional Composition of Eggs

Eggs are known to be complete food packed with all the vital nutrients. They are rich sources of high-quality protein, essential vitamins, unsaturated fats, minerals, and antioxidants. Egg protein is the highest quality food protein available as it is easily absorbed and digested by the human body. An average size egg weighing about 50 g provides 6.3 g of protein, 4 g of which are within the egg white and 2.5 g in the yolk (USDA 2010). Nutritional composition of the egg is shown in Table 11.5.

Fats are plentiful in eggs. It contains 4.5 gm of total fat, including 1.5 gm of saturated fat, which raises LDL cholesterol levels in the blood. The monounsaturated fat content of the egg is 2 g. Fats provide concentrated energy storage for the body and aid in the absorption of soluble vitamins. Carbohydrates supply energy to the

Table 11.5 Nutritional composition of egg (per 100 g)

	Whole egg	Albumen	Yolk
Water	75.1	88.3	51.0
Protein	12.5	9.0	16.1
Fat	10.8	Trace	30.5
Carbohydrate	Trace	Trace	Trace
Energy value (kcal)	146	36.56	335

Source: USDA (2010)

body that can be used quickly. Eggs have extremely few carbohydrates (less than 1 g) and no fiber, yet they go well with other foods.

Egg is an important source of nutrients for humans and its composition can be modified to obtain a more functional food through the manipulation of laying hen diet (Van Elswyk et al. 1995; Kubena et al. 1999). Eggs are considered one of nature's perfect foods, and they have been eaten for millennia all across the world. Despite the fact that eggs contain all of the essential elements, egg consumption has decreased in many developed countries due to public perceptions of their high cholesterol level. Current evidence reveals however that there is no link between egg consumption and blood cholesterol levels (Lee and Griffin 2006; Qureshi et al. 2007). The yolk of an egg is high in both nutritive and nonnutritive substances that are beneficial to human health. It is generally known that the nutrition of hens has an impact on the makeup of their yolks. Certain phytochemicals with major health advantages can be enhanced in egg yolk through dietary modification (Surai and Sparks 2001). Eggs are one of the top 25 foods consumed, accounting for 11% of daily calories (Drewnowski 1995). Sadly, eggs are among the top 15 items that contribute to dietary saturated fat and cholesterol, both of which are linked to an elevated risk of cardiovascular disease (CVD), the leading cause of mortality in the United States. One egg also includes about 200 mg of cholesterol (Weggemans et al. 2001), which is close to the American Heart Association's dietary cholesterol intake limit of 300 mg/d. To maintain a healthy heart, daily cholesterol intake should not exceed 200 mg; one big egg per day would surpass this threshold. Dietary cholesterol raises total and LDL cholesterol levels in the blood, both of which are known risk factors for cardiovascular disease (CVD) (Howell et al. 1997). Furthermore, saturated fat makes up about half of an egg's total fat content, which is another contributor to cardiovascular disease (CVD) (Hu et al. 1999).

11.10.1.1 Egg Proteins

There are two types of protein present in an egg that is Egg yolk protein and Egg white protein. The composition of egg proteins is shown in Table 11.6.

11.10.1.1.1 Egg Yolk Proteins

Phosvitin

Phosvitin is the most common phosphoprotein found in egg yolks (it accounts for approximately 16% of egg yolk proteins). It has a phosphorus content of 10%. As a

Table 11.6 Composition of egg proteins

Egg yolk proteins	Composition
Lipovitellins	69%
α-Lipovitellin	58%
β-Lipovitellin	11%
Livetin	12%
β-Livetin (glycoprotein)	5%
α-Livetin (serum albumen)	4%
γ-Livetin (γ-globulin)	2%
Phosvitin	7%
Low-density lipoprotein	12%
Egg white proteins	**Composition**
Ovalbumin	54%
Ovotransferrin	12–13%
Ovomucoid	11%
Ovomucin	1.5–3.5%
Lysozyme	3.4–3.5%
Globulins	8.0%
Ovoinhibitor	1.5%
Ovoglycoprotein	1.0%
Ovoflavoprotein	0.8%
Ovomacroglobulin	0.5%
Avidin	0.05%
Cystatin	0.05%

Source: USDA (2010)

result, phosvitin has a high potential for binding metals (iron and calcium). The pH affects the calcium-binding capabilities of phosvitin. Native phosvitin has calcium-binding capabilities of 20 mol Ca^{++}/mol phosvitin at pH 3.6 and 148 mol Ca^{++}/mol phosvitin at pH 7.0. In phosphatidylcholine liposomes, muscle homogenates, and minced pork, phosphovitin inhibited lipid oxidation. Phosvitin is a great source for making phosphor peptides.

Lipovitellins

These are high-density lipoproteins. They can be separated into two fractions, the α and β-lipovitellins. Each fraction contains 40% neutral lipids and 60% phospholipids. At pH value below 7, lipovitellins occur as a dimer.

Livetin

This consist of three components the α, β and γ-livetins these differ in their molecular weight. α-livetin does not contain hexose and hexosamine, β-livetin contains only hexose, and γ-livetin contains both hexose and hexosamine.

Low-Density Lipoprotein

The density of egg yolk low-density lipoprotein (LDL) is 0.98 g/cc. It is composed of 74% neutral lipids and 26% phospholipids. It can be broken down into two parts: LDL1 and LDL2. The molecular weights of these fractions are high. They are spherical in shape, with a triglyceride core coated with phospholipids and proteins.

11.10.1.1.2 Egg White Proteins

Egg white contains a number of proteins frequently referred to as albumen. All egg white proteins are globular. These globular proteins are important for the foaming properties of egg white. Some of the well-known egg white proteins are the following:

Ovalbumin

This protein is the most important in egg whites, accounting for 55% of the total protein. This is a phosphor glycoprotein that is made up of three parts: A1, A2, and A3. The only difference between them is the amount of phosphorus they contain. A1 has two phosphorous molecules per molecule, A2 has one, and A3 does not have any. Sulfydryl and disulfur groups are present in all forms. The proportions of A1, A2, and A3 components are roughly 35:12:3. Mannose and glucosamine, in a 5:3 ratio, are the carbohydrate components of ovalbumin. Ovalbumin in solution is easily denatured by mechanical agitation, but heat denaturation is difficult. At pH 9 and 62 °C, only 3–5% of the ovalbumin is denatured.

Ovotransferrin

This makes up 13% of the protein in egg albumen. It comes in two varieties, neither of which contains phosphorus or sulfur. Ovotransferrin may be heat coagulated more easily and is less prone to denaturation than ovalbumin.

Ovomucoid

Ovomucoid makes up roughly 10% of the protein in egg whites. It comes in three different forms, each of which is a trypsin inhibitor. Hexoses (galactose and mannose), glucosamine, and sialic acid make up the protein's carbohydrate moiety. There are eight disulfide links in the protein. It is resistant to heat denaturation in acid media, but it denatures quickly in alkaline media.

Ovomucin

Ovomucin accounts for about 1.5–3.5% of total egg white solids. Albumen's thickness is determined by this protein. Its content in thick albumen layers is around four times that of thin layers. It is water-insoluble but soluble in a dilute salt solution. The protein consists of three components. Purified ovomucin in solution is resistant to heat denaturation. Between pH 7.1 and 9.4, heating for 2 h at 90 °C does not bring about any change. Ovomucin and lysozyme can form a water-soluble complex.

Lysozyme

Lysozyme is an enzyme that can lyse (dissolve) bacteria's cell walls. In egg albumen, the enzyme is much more heat-sensitive than when it is present alone. Lysozyme is a well-studied enzyme with four disulfide bonds. Because it is transported to the egg white, the lysozyme content of a laying hen's blood is 10 times higher than in mammals. 129 amino acid residues make up egg white lysozyme, which has a molecular weight of 14.4 kDa.

Avidin

Avidin is composed of three components A, B, and C. It binds biotin and makes the vitamin unavailable. However, avidin is easily denatured by heat and so its presence is not of concern when eggs are cooked. For this and other reasons, it is recommended that eggs be cooked before eating.

Ovoglobulin

Ovoglobulin is a globular protein that constitutes about 1.0% of egg albumen. It consists of two components G1 and G2, which have molecular weights of 36 and 45 kDa, respectively. Both are excellent foaming agents.

Flavoprotein

In a 1:1 ratio, all the riboflavin in egg albumen is bound in the flavoprotein. Ascertain that the riboflavin in the blood serum is transferred to the albumen in the egg white, where it is coupled to an apoprotein termed flavoprotein. With a molecular weight of 32–36 kDa, the apoprotein is acidic and contains a carbohydrate component (14%) made up of mannose, galactose, and glucosamine, as well as 7–8 phosphate groups and 8 disulfide links. One mole of apoprotein binds one mole of riboflavin, but when the protein is exposed to a pH below its isoelectric pH, this binding ability is lost. Ovoinhibitor is a proteolytic enzyme inhibitor that can stop trypsin and chymotrypsin from working.

11.10.1.2 Egg Fat

Most of the fat in the egg is concentrated in the egg yolk. The lipid content in an egg is about 11% of total egg content and about 32% of yolk. The albumen contains a very negligible amount of fat, i.e., only about 0.05%. The yolk fat is divided into three major portions, i.e., lipoproteins, phospholipids, and triglycerides. The triglycerides constitute almost 66% of the egg fat, the phospholipids–28%, and cholesterol about 5%. Lipoproteins are present in conjugation with phospholipids (Kovacs-Nolan et al. 2005). Lecithin is the most common phospholipid found in eggs, and cholesterol is the most common sterol.The fatty acids include monounsaturated (46.5%) > saturated (37.5%) > polyunsaturated (16.5%). Oleic acid accounts for 38.45% of total fatty acids in egg yolk, palmitic acid for 23.5%, linoleic acid for 16.4%, and stearic acid for 14%.

11.10.1.3 Vitamins

The eggs are rich sources of fat-soluble vitamins like A, D, and E and the water-soluble B-vitamins, B_2, B_{12}, and also fair sources of folate, biotin, and choline (Surai and Sparks 2001). The vitamin distribution in egg is represented in Table 11.7.

11.10.1.4 Minerals

Eggs are high in phosphorus, calcium, and potassium, but low in sodium (142 mg per 100 g of whole egg) (Table 11.8). It also contains all essential trace elements, such as copper, iron, magnesium, manganese, selenium, and zinc, with the yolk serving as the primary source of iron and zinc. The calcium in an egg comes primarily from the shell. Egg yolk has a lot of calcium, whereas albumen does not have as much. Magnesium, sodium, chlorine, sulfur, and potassium are other minerals found in yolk and albumen. The mineral composition of chicken feed affects the mineral content of eggs. Table 11.8 shows the mineral distribution in various portions of the egg.

Table 11.7 Composition of vitamins in egg

Vitamin (unit)	Whole egg	Albumen	Yolk
A (IU)	634.4	–	1946
D (IU)	49	–	147.6
E (IU)	1.4	–	4.217
B12 (μg)	1	0.21	3.312
Biotin (μg)	19.96	7.006	45.662
Choline (mg)	430.12	1.257	1301
Folic acid (μg)	46	2.994	144.58
Nicotinic acid (mg)	0.074	0.093	0.012
Pantothenic acid (mg)	1.254	0.12	3.807
Pyridoxine (mg)	0.14	0.003	0.392
Riboflavin(mg)	0.508	0.452	0.639
Thiamin(mg)	0.062	0.006	0.169

Source: USDA (2010)

Table 11.8 Composition of minerals in egg (mg/100 g)

Minerals	Whole egg	Albumin	Yolk
Phosphorous	198	15	390
Sodium	142	166	48
Potassium	138	163	109
Calcium	56	7	129
Magnesium	12	11	5
Iron	1.75	0.08	2.73
Zinc	1.29	0.03	2.30
Copper	0.072	0.023	0.077
Selenium	0.030	0.020	0.056
Manganese	0.028	0.011	0.055
Iodine	0.021	0.002	0.18

11.10.2 Structure of Egg

Natural selection created the egg as a biological structure for reproduction. It shields the growing embryo and offers a full diet, and it is the chick's primary source of nutrition for the first few days of life. An egg weighs about 57 g on average (about 2 ounces). It is made up of 10% shell, 58% white, and 32% yolk. The egg's nutritional content is unaffected by the color of the shell or yolk. The ovum, or yolk (which has germ cells or blastodisk on its surface), albumen, or white (which has thin and thick regions), plus chalazae, the shell membranes, which separate inner and outer layers, and the shell (which is made up of different, distinguishable layers), make up the egg.

The shell is totally filled when the egg is freshly laid. The air cell is generated when the contents contract during cooling and moisture is lost. In a high-quality egg, only a small air cell can be found. The yolk is well-centered in the albumen and encircled by the whitish vitelline membrane. The yolk is linked to the germinal disc, which is where fertilization takes place. Two twisted, whitish cord-like things known as chalazae can be seen on opposite sides of the yolk in Fig. 11.9. Their purpose is to support the yolk in the albumen's core. The size and density of chalazae may vary, although neither affects the cooking performance nor the nutritional value. The albumen is thick for the most part. Two shell membranes and the shell itself surround the albumen. The egg's shell has thousands of pores that allow it to "breathe."

11.10.2.1 Components of Egg and Its Functions

11.10.2.1.1 Eggshell

The eggshell, which is the egg's exterior covering, is the most visible component of the egg. It accounts for 9–11% of an egg's weight. The egg's first line of defense

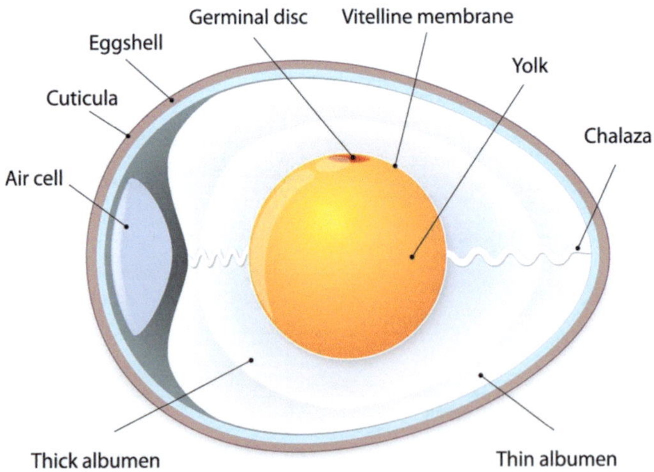

Fig. 11.9 Structure of egg

against bacterial contamination is its shell. Calcium carbonate makes up practically the whole exterior eggshell. The eggshell is a semipermeable membrane with up to 17,000 small pores. The pores in an egg shell allow air and moisture to pass through. The bloom, or cuticle, is a thin outer layer on the shell that helps keep the egg fresh and prevents infection. The inner and outer shell membranes make up the shell. The air cell is normally generated between the shell membranes at the egg's broad end. The outer shell membrane is the air cell membrane.

11.10.2.1.2 Albumen

It is also known as "egg white." The white component of an egg, also known as albumen, accounts for around 58–60% of the total weight of the egg. It is made up of three layers: an outside thin layer; a middle thick layer; and an interior thin layer. The chalaza, which is connected to the chalaziferous layer, is also present. The yolk is surrounded by a chalaziferous layer. More than half of the total protein, potassium, and sodium in an egg is present inside the albumen. The color of the albumen is more opalescent than white. The foggy appearance is caused by CO_2 escaping from the egg as it ages. Therefore, older eggs are clearer than fresh eggs, and albumen tends to thin out as the egg ages.

11.10.2.1.3 Yolk

It is the part of an egg that is yellow. It makes up around a third of the weight of an egg. The germinal disc, latebra, concentric rings of yolk material, and vitelline membrane make up the yolk. The yolk has less water and protein than the white, as well as some fat and the majority of the egg's vitamins and minerals. It also contains lecithin, a powerful emulsifier. Depending on the hen's feed and breed, the yolk color might range from a subtle yellow to a gorgeous rich orange (Fig. 11.10).

11.11 Conclusion

Meat, chicken, fish, and egg form an integral part of the nonvegetarian diet. They are known to deliver important nutrients to our diet like proteins of high biological value, essential fatty acids, vitamins, and minerals which are known for our growth and well-being. The components of muscle have a vital role in determining the final meat quality along with it several factors of both antemortem and postmortem affects the tenderness and processing characteristics. Meat quality indicators like water-holding capacity, color, texture, and toughness strongly influence consumer choice. These qualities can also be controlled to some extent by management practices that have a direct impact on the live animal like minimizing transportation stress, withholding feed before slaughter as well as processing protocols such as controlling the chill rate and use of electrical stimulation. In addition to modification of traditional components of meat, designing meat and eggs is also gaining importance wherein the natural composition of meat and eggs can be altered or designed by altering meat composition for processing and/or during animal production by employing nutritional and genetic approaches. Alteration of the carcass and egg

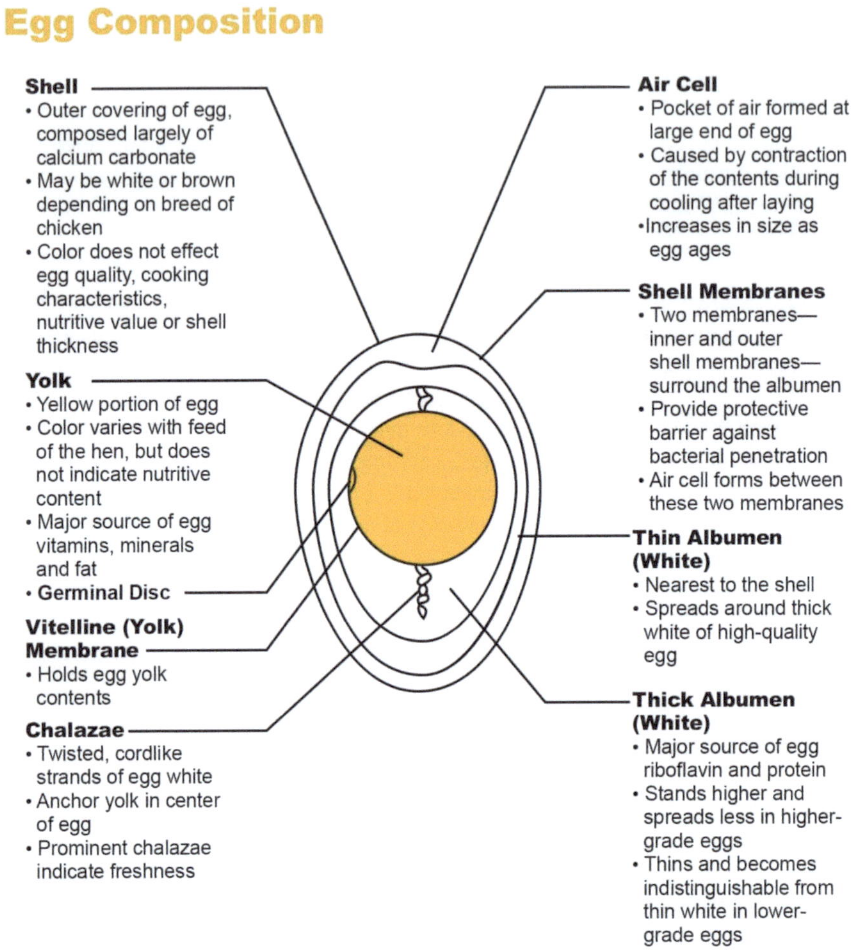

Fig. 11.10 Components of egg and its functions

composition may definitely help in addressing problems like cholesterol content, designer lipid profile, and quality issues like water-holding capacity, PSE meat, DFD meat, tenderness, etc., and reducing the incidence of these quality defects.

References

Aalhus JL, Best DR, Murray AC et al (1998) A comparison of the quality characteristics of pale, soft and exudative beef and pork. J Muscle Foods 9(3):267–280

Abe F (2007) Exploration of the effects of high hydrostatic pressure on microbial growth, physiology and survival: perspectives from piezo physiology. Biosci Biotechnol Biochem 71(10): 2347–2357

Aberle ED, Forrest JC, Gerrard DE et al (2001) Principles of meat science. Hunt Publishing Company, Kendal

Adzitey F, Nurul II (2011) Pale soft exudative (PSE) and dark firm dry (DFD) meats: causes and measures to reduce these incidences-a mini review. Int Food Res J 18(1)

Aggett PJ, Comerford JG (1995) Zinc and human health. Nutr Rev 53(9):S16

Ahn DU, Sell JL, Jo C et al (1998) Effects of dietary vitamin E supplementation on lipid oxidation and volatiles content of irradiated, cooked Turkey meat patties with different packaging. Poultry Sci 77(6):912–920

Ahn DU, Jo C, Du M, Olson DG et al (2000) Quality characteristics of pork patties irradiated and stored in different packaging and storage conditions. Meat Sci 56(2):203–209

Ahn DU, Nam KC, Du M et al (2001) Effect of irradiation and packaging conditions after cooking on the formation of cholesterol and lipid oxidation products in meats during storage. Meat Sci 57(4):413–418

Akagawa M, Sasaki D, Kurota Y et al (2005) Formation of α aminoadipic and γ-glutamic semialdehydes in proteins by the Maillard reaction. Ann N Y Acad Sci 1043(1):129–134

Andrés AI, Møller JK, Adamsen CE et al (2004) High pressure treatment of dry-cured Iberian ham. Effect on radical formation, lipid oxidation and colour. Eur Food Res Technol 219(3):205–210

Aymerich T, Picouet PA, Monfort JM (2008) Decontamination technologies for meat products. Meat Sci 78(1):114–129

Azam K, Mackie IM, Smith J (1990) Effect of stunning method on the time of onset, duration and resolution of rigor in rainbow trout (*Salmo gairdneri*) as measured by visual observation and analysis for lactic acid, nucleotide-degradation products and glycogen. In: Science et Technique du Froid (France). Proceedings of the meeting of commission C2 1.1 .F.-I.I.R, Aberdeen, pp 351–358

Babić J, Cantalejo MJ, Arroqui C (2009) The effects of freeze-drying process parameters on broiler chicken breast meat. LWT-Food Sci Technol 42(8):1325–1334

Badr HM (2004) Use of irradiation to control food borne pathogens and extend the refrigerated market life of rabbit meat. Meat Sci 67(4):541–548

Badr HM, Mahmoud KA (2011) Antioxidant activity of carrot juice in gamma irradiated beef sausage during refrigerated and frozen storage. Food Chem 127(3):1119–1130

Baker CGJ (1997) Dryer selection in industrial drying of foods. Blackie Academic and Professional, London, pp 242–271

Barbosa-Canovas GV, Pothakamury UR, Swanson BG (1995) State of the art technologies for the stabilization of foods by non-thermal processes: physical methods. In: Barbosa-Ca'novas GV, Welt-Chanes J (eds) Food preservation by moisture control. Technomic Publishing Co., Lancaster and Basel, pp 493–532

Belury MA (2002) Dietary conjugated linoleic acid in health: physiological effects and mechanisms of action. Annu Rev Nutr 22(1):505–531

Bendall JR, Swatland HJ (1988) A review of the relationships of pH with physical aspects of pork quality. Meat Sci 24(2):85–126

Bendall JR, Wismer-Pedersen J (1962) Some properties of the fibrillar proteins of normal and watery pork muscle. J Food Sci 27(2):144–159

Berlett BS, Stadtman ER (1997) Protein oxidation in aging, disease, and oxidative stress. J Biol Chem 272(33):20313–20316

Berrazaga I, Micard V, Gueugneau M (2019, 1825) The role of the anabolic properties of plant-versus animal-based protein sources in supporting muscle mass maintenance: a critical review. Nutrients 11(8)

Bimbenet JJ, Bonazzi C, Dumoulin E (2002) Drying of foodstuffs. In: Proceeding of the 13th international drying symposium, Beijing, China, pp 26–22

Bolumar T, Skibsted LH, Orlien V (2012) Kinetics of the formation of radicals in meat during high pressure processing. Food Chem 134(4):2114–2120

Buttkus H (1963) Red and white muscle of fish in relation to rigor mortis. J Fish Res Board Can 20(1):45–58

Campus M, Flores M, Martinez A et al (2008) Effect of high pressure treatment on colour, microbial and chemical characteristics of dry cured loin. Meat Sci 80(4):1174–1181
Carlez A, Veciana-Nogues T, Cheftel JC (1995) Changes in colour and myoglobin of minced beef meat due to high pressure processing. LWT-Food Sci Technol 28(5):528–538
Carpenter JF, Crowe JH (1989) An infrared spectroscopic study of the interactions of carbohydrates with dried proteins. Biochemist 28(9):3916–3922
Cava R, Ladero L, González S, Carrasco A et al (2009) Effect of pressure and holding time on colour, protein and lipid oxidation of sliced dry-cured Iberian ham and loin during refrigerated storage. Innov Food Sci Emerg Technol 10(1):76–81
Chaijan M (2008) Lipid and myoglobin oxidations in muscle foods. Songklanakarin J Sci Technol 30(1):47–53
Chan W, Brown JM, Buss D (eds) (1995) Meat poultry and game: fifth supplement to the fifth edition of McCance and Widdowson's The composition of foods. Royal Society of Chemistry
Chapman B, Winley E, Fong AS et al (2007) Ascospore inactivation and germination by high pressure processing is affected by ascospore age. Innov Food Sci Emerg Technol 8(4):531–534
Chmiel M, Słowiński M, Dasiewicz K (2011) Lightness of the color measured by computer image analysis as a factor for assessing the quality of pork meat. Meat Sci 88(3):566–570
Codex Alimentarius Commission, Code of Hygienic Practice for Meat (2005) https://www.fao.org/fao-who-codexalimentarius/sh-proxy/en/?lnk=1&url=https%253A%252F%252Fworkspace.fao.org%252Fsites%252Fcodex%252FStandards%252FCXC%2B58-2005%252FCXP_058e.pdf
Committee on Medical Aspects of Food and Nutrition Policy (COMA) (2000) Folic acid and the prevention of disease. TSO, London
Dalgleish JM (1990) Freeze-drying for the food industries. Elsevier Science Publishers, pp 57–59
Davidson PM, Branden AL (1981) Antimicrobial activity of non-halogenated phenolic compounds. J Food Prot 44(8):623–632
Davies MJ (2005) The oxidative environment and protein damage. Biochim Biophys Acta -Proteins and Proteomics 1703(2):93–109
De Gonzalez MN, Hafley BS, Boleman RM et al (2008) Antioxidant properties of plum concentrates and powder in precooked roast beef to reduce lipid oxidation. Meat Sci 80(4): 997–1004
Decker EA, Xu Z (1998) Minimizing rancidity in muscle foods. Food Technol (Chicago) 52(10): 54–59
Dempster JF, Hawrysh ZJ, Shand P et al (1985) Effect of low-dose irradiation (radurization) on the shelf life of beef burgers stored at 3°C. Int J Food Sci Technol 20(2):145–154
Donsì G, Ferrari G, Matteo DI (2001) Utilization of combined processes in freeze-drying of shrimps. Food Bioprod Process 79(3):152–159
Dransfield E (1994) Tenderness of meat, poultry and fish. In: Quality attributes and their measurement in meat, poultry and fish products. Springer, Boston, MA, pp 289–315
Drewnowski A (1995) Energy intake and sensory properties of food. Am soc. Clin Nutr 62s:1081Se1085S
Droulez V, Williams PG, Levy G et al (2006) Composition of Australian red meat 2002. 2. Fatty acid profile. Food Aust 58(7):335–341
Dugan LR Jr (1971) Fats. In: Price JE, Schweigert BS (eds) The science of meat and meat products. W.H. Freeman & Company, San Francisco, CA. 2nd edition 133
Enser M, Hallett K, Hewitt B et al (1996) Fatty acid content and composition of English beef, lamb and pork at retail. Meat Sci 42(4):443–456
Estévez M (2011) Protein carbonyls in meat systems: a review. Meat Sci 89(3):259–279
Estevez M, Kylli P, Puolanne E et al (2008) Oxidation of skeletal muscle myofibrillar proteins in oil-in-water emulsions: interaction with lipids and effect of selected phenolic compounds. J Agric Food Chem 56(22):10933–10940
FAO (1990) Food and Agriculture Organisation, FAOSTAT citation database results. FAO. http://faostat.fao.org/faostat

Farkas BE, Singh RP (1991) Physical properties of air-dried and freeze-dried chicken white meat. J Food Sci 56(3):611–615

Farr D (2003) High pressure processing in food industry. Trends Food Sci Technol 1:14–16

Feeney RE, Blankenhorn G, Dixon HB (1974) Carbonyl-amine reactions in protein chemistry. Adv Protein Chem 29:135–203

Fernández J, Pérez-Álvarez JA, Fernández-López JA (1997) Thiobarbituric acid test for monitoring lipid oxidation in meat. Food Chem 59(3):345–353

Frederiksen AM, Lund MN, Andersen ML et al (2008) Oxidation of porcine myosin by hypervalent myoglobin: the role of thiol groups. J Agric Food Chem 56(9):3297–3304

Fuentes V, Ventanas J, Morcuende D et al (2010) Lipid and protein oxidation and sensory properties of vacuum-packaged dry-cured ham subjected to high hydrostatic pressure. Meat Sci 85(3):506–514

Fujii J, Otsu K, Zorzato F et al (1991) Identification of a mutation in porcine ryanodine receptor associated with malignant hyperthermia. Science 253(5018):448–451

Ganhão R, Morcuende D, Estévez M (2010) Protein oxidation in emulsified cooked burger patties with added fruit extracts: influence on colour and texture deterioration during chill storage. Meat Sci 85(3):402–409

Garriga M, Grebol N, Aymerich MT et al (2004) Microbial inactivation after high-pressure processing at 600 MPa in commercial meat products over its shelf life. Innov Food Sci Emerg Technol 5(4):451–457

Gatellier P, Santé-Lhoutellier V (2009) Digestion study of proteins from cooked meat using an enzymatic microreactor. Meat Sci 81(2):405–409

Gatellier P, Kondjoyan A, Portanguen S et al (2010) Effect of cooking on protein oxidation in n-3 polyunsaturated fatty acids enriched beef. Implication on nutritional quality. Meat Sci 85(4): 645–650

Gavino VC, Gavino G, Leblanc MJ et al (2000) An isomeric mixture of conjugated linoleic acids but not pure cis-9, trans-11-octadecadienoic acid affects body weight gain and plasma lipids in hamsters. J Nutr 130(1):27–29

Geesink G, Sujang S, Koohmaraie M (2011) Tenderness of pre-and post rigor lamb longissimus muscle. Meat Sci 88(4):723–726

Gibson SA, Ashwell M (1997) New vitamin D values for meat and their implication for vitamin D intake in adults. Proc Nutr Soc 56:116A

Gill CO (1986) The control of microbial spoilage in fresh meats. In: Pearson AM, Dutson TR (eds) Advances in meat research meat and poultry microbiology. MacMillan, New York, pp 49–88

Girard JP, Omolosho DA (1983) Technological problems related to the drying of raw and cooked meats. In: 29th European congress of meat researcher workers: Salsomaggiore (Parma), 29.8–2.9

Gopalan C, Rama Sastri BV, Balasubramanian SC (1971) Nutritive value of Indian food. National Institute of nutrition, ICMR, Hyderabad

Gould GW (2000) Preservation: past, present and future. Br Med Bull 56(1):84–96

Grolichova M, Dvořák P, Musilova H (2004) Employing ionizing radiation to enhance food safety–a review. Acta Vet Brno 73(1):143–149

Hallberg L, Hulthén L (2000) Prediction of dietary iron absorption: an algorithm for calculating absorption and bioavailability of dietary iron. Am J Clin Nutr 71(5):1147–1160

Hansen E, Juncher D, Henckel P et al (2004) Oxidative stability of chilled pork chops following long term freeze storage. Meat Sci 68(3):479–484

Heinz V, Buckow R (2010) Food preservation by high pressure. J Verbrauch Lebensm 5(1):73–81

Higgs J, Mulvihill B (2002) The nutritional quality of meat. Meat processing: improving quality. In: Kerry J, Kerry J, Ledward D (eds) Woodhead publishing series in food science, technology and nutrition, meat processing. Woodhead Publishing, pp 64–104

Honikel KO (1998) Reference methods for the assessment of physical characteristics of meat. Meat Sci 49(4):447–457

Honikel KO (2008) The use and control of nitrate and nitrite for the processing of meat products. Meat Sci 78(1):68–76
Honikel KO, Kim CJ, Hamm R et al (1986) Sarcomere shortening of prerigor muscles and its influence on drip loss. Meat Sci 16(4):267–282
Howell WH, McNamara DJ, Tosca MA et al (1997) Plasma lipid and lipoprotein responses to dietary fat and cholesterol: a meta analysis. Am J Clin Nutr 65:1747–1764
Hu FB, Stampfer MJ, Manson JE et al (1999) Dietary saturated fats and their food sources in relation to the risk of coronary heart disease in women. Am J Clin Nutr 70(6):1001–1008
Huff-Lonergan E, Lonergan SM (2005) Mechanisms of water-holding capacity of meat: the role of postmortem biochemical and structural changes. Meat Sci 71(1):194–204
Huffman KL, Miller MF, Hoover LC et al (1996) Effect of beef tenderness on consumer satisfaction with steaks consumed in the home and restaurant. J Animal Sci 74(1):91–97
Hui YH, Nip WK, Rogers R et al (2001) Meat composition in meat science and applications. Marcel Dekker, New York
Hyldig G, Nielsen D (2001) A review of sensory and instrumental methods used to evaluate the texture of fish muscle. J Texture Stud:219–242
Igene JO, Yamauchi K, Pearson AM et al (1985) Mechanisms by which nitrite inhibits the development of warmed-over flavour (WOF) in cured meat. Food Chem 18(1):1–8
Jacob RH, Thomson KL (2012) The importance of chill rate when characterising colour change of lamb meat during retail display. Meat Sci 90(2):478–484
Jayasingh P, Cornforth DP, Brennand CP et al (2002) Sensory evaluation of ground beef stored in high-oxygen modified atmosphere packaging. J Food Sci 67(9):3493–3496
Jayathilakan K, Sultana K, Radhakrishna K et al (2009a) Effect of lactic acid and irradiation on the shelf stability characteristics of hurdle processed chicken legs. Int J Poultry Sci 8(7):665–670
Jayathilakan K, Sultana K, Radhakrishna K et al (2009b) Effect of lactic acid and irradiation on the quality characteristics of hurdle processed chicken legs. Int J Poultry Sci 8(7):665–670
Jiménez-Colmenero F, Borderias AJ (2003) High-pressure processing of myosystems. Uncertainties in methodology and their consequences for evaluation of results. Eur Food Res Technol 217(6):461–465
Jofré A, Aymerich T, Grèbol N et al (2009) Efficiency of high hydrostatic pressure at 600 MPa against food-borne microorganisms by challenge tests on convenience meat products. LWT-Food Sci Technol 42(5):924–928
Jung S, Ghoul M, de Lamballerie-Anton M (2000) Changes in lysosomal enzyme activities and shear values of high pressure treated meat during ageing. Meat Sci 56(3):239–246
Kanatt SR, Chander R, Sharma A (2006) Effect of radiation processing of lamb meat on its lipids. Food Chem 97(1):80–86
Kanatt SR, Chander R, Sharma A (2007) Antioxidant potential of mint (*Mentha spicata* L.) in radiation-processed lamb meat. Food Chem 100(2):451–458
Kanner J, Harel S (1985) Initiation of membranal lipid peroxidation by activated metmyoglobin and methemoglobin. Arch Biochem Biophys 237(2):314–321
Kanner J, Rosenthal I (1992) An assessment of lipid oxidation in foods (technical report). Pure Appl Chem 64(12):1959–1964
Katsaras K, Peetz P (1990) Morphological changes in dark cutting beef when heated. Fleischwirtschaft (Germany, FR) 70:68–70
Kaur BP, Kaushik N, Rao PS et al (2013) Effect of high-pressure processing on physical, biochemical, and microbiological characteristics of black tiger shrimp (Penaeus monodon). Food Bioprocess Technol 6(6):1390–1400
King NJ, Whyte R (2006) Does it look cooked? A review of factors that influence cooked meat color. J Food Sci 71(4):R31–R40
Kirchheim U, Kinast C, Schone F (2001) Early post-mortem measurements as indicator of meat quality characteristics. Fleischwirtschaft 81(1):89–90

Kiss IF, Beczner J, Zachariev GY et al (1990) Irradiation of meat products, chicken and use of irradiated spices for sausages. Int J Radiat Appl Instrum Part C Radiat Phys Chem 36(3): 295–299
Kjällstrand J, Petersson G (2001) Phenolic antioxidants in wood smoke. Sci Total Environ 277(1): 69–75
Kong F, Tang J, Lin M et al (2008) Thermal effects on chicken and salmon muscles: tenderness, cook loss, area shrinkage, collagen solubility and microstructure. LWT-Food Sci Technol 41(7): 1210–1222
Koohmaraie M (1994) Muscle proteinases and meat aging. Meat Sci 36(1–2):93–104
Koohmaraie M, Geesink GH (2006) Contribution of postmortem muscle biochemistry to the delivery of consistent meat quality with particular focus on the calpain system. Meat Sci 74(1):34–43
Koutsoumanis K, Nychas GJ (1999) Chemical and sensory changes associated with microbial flora of mediterranean boque (*Boops boops*) stored aerobically at 0, 3, 7, and 10°C. Appl Environ Microbiol 65(2):698–706
Kovacs-Nolan J, Phillips M, Mine Y (2005) Advances in the value of eggs and egg components for human health. J Agric Food Chem 53(22):8421–8431
Kubena LF, Harvey RB, Buckley SA et al (1999) Effects of long-term feeding of diets containing moniliformin, supplied by fusarium fujikuroi culture material, and fumonisin, supplied by fusarium moniliforme culture material, to laying hens. Poult Sci 78(11):1499–1505
Kubow S (1992) Routes of formation and toxic consequences of lipid oxidation products in foods. Free Radic Biol Med 12(1):63–81
Lawrence TE, Dikeman ME, Hunt MC et al (2003) Staged injection marination with calcium lactate, phosphate and salt may improve beef water-binding ability and palatability traits. Meat Sci 65(3):967–972
Lawrie RA (1985) Meat science, 4th edn. Pergamon Press, Oxford
Lawrie RA (1998) The conversion of muscle to meat. In: Lawrie's meat science, 6th edn. Woodhead Publishing Ltd., Cambridge, England, pp 96–118
Lawrie RA, Ledward DA (2006) Chemical and biochemical constitution of muscle. Lawrie's Meat Science CRC Press, New York, In, pp 87–93
Lee A, Griffin B (2006) Dietary cholesterol, eggs and coronary heart disease risk in perspective. Nutr Bull 31:21–27
Lee KH, Yook HS, Lee JW et al (1999) Effects of antioxidants on oxidation of tallow by gamma irradiation. Korean J Food Sci Technol 31(1):7–12
Leistner L (2000) Basic aspects of food preservation by hurdle technology. Int J Food Microbiol 55(1–3):181–186
Leistner L, Gorris LG (1995) Food preservation by hurdle technology. Trends Food Sci Technol 6(2):41–46
Leygonie C, Britz TJ, Hoffman LC (2012) Impact of freezing and thawing on the quality of meat. Meat Sci 91(2):93–98
Listrat A, Lebret B, Louveau I et al (2016) How structure and composition muscle influences meat and flesh quantity. Sci World J 2016:1–4
Liu G, Xiong YL (2000) Oxidatively induced chemical changes and interactions of mixed myosin, β-lactoglobulin and soy 7S globulin. J Sci Food Agric 80(11):1601–1607
López-Caballero ME, Pérez-Mateos M, Borderias JA, Montero P (2000) Extension of the shelf life of prawns (Penaeus japonicus) by vacuum packaging and high-pressure treatment. J Food Prot 63(10):1381–1388
Lundström K, Essen-Gustavsson B, Rundgren M (1989) Effect of halothane genotype on muscle metabolism at slaughter and its relationship with meat quality: a within-litter comparison. Meat Sci 25(4):251–263
Ma F, Chen C, Zheng L et al (2013) Effect of high pressure processing on the gel properties of salt-soluble meat protein containing CaCl2 and κ-carrageenan. Meat Sci 95(1):22–26
Mancini RA, Hunt M (2005) Current research in meat color. Meat Sci 71(1):100–121

Mariutti LR, Orlien V, Bragagnolo N et al (2008) Effect of sage and garlic on lipid oxidation in high-pressure processed chicken meat. Eur Food Res Technol 227(2):337–344

Martinaud A, Mercier Y, Marinova P et al (1997) Comparison of oxidative processes on myofibrillar proteins from beef during maturation and by different model oxidation systems. J Agric Food Chem 45(7):2481–2487

McArdle R, Marcos B, Kerry JP (2010) Monitoring the effects of high pressure processing and temperature on selected beef quality attributes. Meat Sci 86(3):629–634

Mcmeekin TA (1982) Microbial spoilage of meats. In: RDavies. London, UK, Applied Science Publishers p, Developments in Food Microbiology, pp 1–40

Miller RK (1994) Quality characteristics. In Muscle foods Springer, US, pp 296–332

Miller MF, Carr MA, Ramsey CB et al (2001) Consumer thresholds for establishing the value of beef tenderness. J Animal Sci 79(12):3062–3068

Miller RK, Kerry J, Ledward D (2002) Factors affecting the quality of raw meat (meat processing improving quality). Elsevier, pp 27–63

Moussa M, Perrier-Cornet JM, Gervais P (2007) Damage in Escherichia coli cells treated with a combination of high hydrostatic pressure and subzero temperature. Appl Environ Microbiol 73(20):6508–6518

Murchie LW, Cruz-Romero M, Kerry JP et al (2005) High pressure processing of shellfish: a review of microbiological and other quality aspects. Innov Food Sci Emerg Technol 6(3):257–270

Nam KC, Ahn DU (2003) Use of antioxidants to reduce lipid oxidation and off-odor volatiles of irradiated pork homogenates and patties. Meat Sci 63(1):1–8

Neumann C, Harris DM, Rogers LM (2002) Contribution of animal source foods in improving diet quality and function in children in the developing world. Nutr Res 22(1–2):193–220

Norton T, Sun DW (2008) Recent advances in the use of high pressure as an effective processing technique in the food industry. Food Bioprocess Technol 1(1):2–34

Nychas GJ, Tassou CC (1997) Spoilage processes and proteolysis in chicken as detected by HPLC. J Sci Food Agric 74(2):199–208

Nychas GJ, Dillon V, Board RG (1988) Glucose, the key substrate in the microbiological changes occurring in meat and certain meat products. Biotechnol Appl Biochem 10(3):203–231

Nychas GJ, Drosinos EH, Board RG (1998) Chemical changes in stored meat. In: Board RG, Davies AR (eds) The microbiology of meat and poultry London. UK, Blackie Academic and Professional p, pp 288–326

Nychas GJ, Marshall DL, Sofos JN (2007) Meat, poultry and seafood spoilage. In: Doyle MP, Beuchat LR (eds) Food microbiology, fundamentals and Frontiers, 3rd edn. ASM Press, Washington DC, pp 105–140

Nychas GJ, Skandamis PN, Tassou CC et al (2008) Meat spoilage during distribution. Meat Sci 78(1–2):77–89

O'neill LM, Galvin K, Morrissey PA, Buckley DJ (1998) Inhibition of lipid oxidation in chicken by carnosine and dietary α-tocopherol supplementation and its determination by derivative spectrophotometry. Meat Sci 50(4):479–488

Ockerman HW, Basu L (2004) Carcass chilling and boning. In: Encyclopedia of Meat Science, pp 144–149

Offer G, Knight P (1988) The structural basis of water-holding in meat; part 2: drip losses. In: Lawrie R (ed) Developments in meat science, vol 4. Elsevier Applied Science, London, pp 172–243

Offer G, Trinick J (1983) On the mechanism of water holding in meat: the swelling and shrinking of myofibrils. Meat Sci 8(4):245–281

Ooizumi T, Xiong YL (2006) Identification of cross-linking site (s) of myosin heavy chains in oxidatively stressed chicken myofibrils. J Food Sci 71(3):C196–C199

Orlien V, Hansen E, Skibsted LH (2000) Lipid oxidation in high-pressure processed chicken breast muscle during chill storage: critical working pressure in relation to oxidation mechanism. Eur Food Res Technol 211(2):99–104

Ouali A, Herrera-Mendez CH, Coulis G et al (2006) Revisiting the conversion of muscle into meat and the underlying mechanisms. Meat Sci 74(1):44–58

Pakula C, Stamminger R (2012) Measuring changes in internal meat colour, colour lightness and colour opacity as predictors of cooking time. Meat Sci 90(3):721–727

Patterson MF (2005) Microbiology of pressure-treated foods. J Appl Microbiol 98(6):1400–1409

Patterson RL, Stevenson MH (1995) Irradiation-induced off-odour in chicken and its possible control. Br Poult Sci 36(3):425–441

Patterson KY, Bhagwat SA, Williams JR et al (2008) USDA database for the choline content of common foods, release two. Nutrient Data Laboratory, Beltsville Human Nutrition Research Center, ARS, USDA

Pöhlmann M, Hitzel A, Schwägele F et al (2012) Contents of polycyclic aromatic hydrocarbons (PAH) and phenolic substances in Frankfurter-type sausages depending on smoking condition using glow smoke. Meat Sci 90(1):176–184

Poligne I, Collignan A, Trystram G (2001) Characterization of traditional processing of pork meat into boucané. Meat Sci 59(4):377–389

Powell TH, Hunt MC, Dikeman ME (2000) Enzymatic assay to determine collagen thermal denaturation and solubilization. Meat Sci 54(4):307–311

Pradhan AA, Rhee KS, Hernández P (2000) Stability of catalase and its potential role in lipid oxidation in meat. Meat Sci 54(4):385–390

Proctor MR, Ryan JA, McLoughlin JV (1992) The effects of stunning and slaughter methods on changes in skeletal muscle and quality of farmed fish. In: Proceedings from TO. International conference and utilization of fishery products, The Nederland's. 1992 may

Purchas RW, Simcock DC, Knight TW et al (2003) Variation in the form of iron in beef and lamb meat and losses of iron during cooking and storage. Int J Food Sci Technol 38(7):827–837

Qiao M, Fletcher DL, Smith DP et al (2001) The effect of broiler breast meat color on pH, moisture, water-holding capacity, and emulsification capacity. Poultry Sci 80(5):676–680

Qureshi AI, Suri FK, Ahmed S et al (2007) Regular egg consumption does not increase the risk of stroke and cardiovascular diseases. Medical Sci Monit 13:CR1–CR8

Rababah T, Hettiarachchy N, Horax R et al (2004) Effect of electron beam irradiation and storage at 5°C on thiobarbituric acid reactive substances and carbonyl contents in chicken breast meat infused with antioxidants and selected plant extracts. J Agric Food Chem 52(26):8236–8241

Rababah T, Over K, Hettiarachchy NS et al (2010) Infusion of plant extracts during processing to preserve quality attributes of irradiated chicken breasts over 9 months storage at −20°C. J Food Process Pres 34:287–307

Rademacher B (2006) Ultrahochdruckverfahren zur Keiminaktivierung. Chem Ing Tech 78(11): 1674–1681

Rastogi NK, Raghavarao KS, Balasubramaniam VM et al (2007) Opportunities and challenges in high pressure processing of foods. Crit Rev Food Sci Nutr 47(1):69–112

Ratti C (2001) Hot air and freeze-drying of high-value foods: a review. J Food Eng 49(4):311–319

Rayman MP (2000) The importance of selenium to human health. Lancet 356(9225):233–241

Requena JR, Chao CC, Levine RL et al (2001) Glutamic and aminoadipic semialdehydes are the main carbonyl products of metal-catalyzed oxidation of proteins. Proc Natl Acad Sci 98(1): 69–74

Rhee KS (1988) Enzymic and nonenzymic catalysis of lipid oxidation in muscle foods. Food Technol 42(6):127–132

Rosenvold K, Wiklund E (2011) Retail colour display life of chilled lamb as affected by processing conditions and storage temperature. Meat Sci 88(3):354–360

Salminen H, Estévez M, Kivikari R et al (2006) Inhibition of protein and lipid oxidation by rapeseed, camelina and soy meal in cooked pork meat patties. Eur Food Res Technol 223(4): 461–468

Satin M (1996) Food irradiation. In: Food irradiation: a guidebook, 2nd edn. Technomic Publishing Company, Inc., Lancaster, PA, pp 1–25

Schaafsma G (2000) The protein digestibility–corrected amino acid score. J Nutr 130(7):1865S–1867S

Shacter E (2000) Quantification and significance of protein oxidation in biological samples. Drug Metab Rev 32(3–4):307–326

Sharma BD (1999) Meat and meat products technology: (including poultry products technology). Jaypee Brothers Medical Publishers Limited, p 112

Sheridan JJ (1981) The use of freeze-drying in investigations on the microbiology of vacuum-packed DFD beef. In: The problem of dark-cutting in beef. Springer, Dordrecht, pp 476–493

Sims TJ (1992) Structural aspects of cooked meat. Spec Publ R Soc Chem:106–127

Singh P, Sahoo J, Talwar G et al (2015) Development of chicken meat caruncles on the basis of sensory attributes: process optimization using response surface methodology. J Food Sci Technol 52(3):1290–1303

Skandamis PN, Nychas GJ (2002) Preservation of fresh meat with active and modified atmosphere packaging conditions. Int J Food Microbiol 79(1–2):35–45

Smelt JP (1998) Recent advances in the microbiology of high pressure processing. Trends Food Sci Technol 9(4):152–158

Souza CM, Boler DD, Clark DL et al (2011) The effects of high pressure processing on pork quality, palatability, and further processed products. Meat Sci 87(4):419–427

Stadtman ER, Levine RL (2003) Free radical-mediated oxidation of free amino acids and amino acid residues in proteins. Amino Acids 25(3–4):207–218

Stone EW Jr, May KN (1969) The effect of time and temperature of cooking on quality of freeze-dehydrated chicken. Poultry Sci 48(3):813–818

Suman SP, Faustman C, Lee S et al (2005) Effect of erythorbate, storage and high-oxygen packaging on premature browning in ground beef. Meat Sci 69(2):363–369

Suman SP, Mancini RA, Joseph P et al (2010) Color-stabilizing effect of lactate on ground beef is packaging-dependent. Meat Sci 84(3):329–333

Sun W, Cui C, Zhao M et al (2011a) Effects of composition and oxidation of proteins on their solubility, aggregation and proteolytic susceptibility during processing of Cantonese sausage. Food Chem 124(1):336–341

Sun W, Zhao M, Yang B et al (2011b) Oxidation of sarcoplasmic proteins during processing of Cantonese sausage in relation to their aggregation behaviour and in vitro digestibility. Meat Sci 88(3):462–467

Surai PF, Sparks NHC (2001) Designer eggs: from improvement of egg composition to functional food. Trends Food Sci Technol 12:7–16

Swatland HJ (2010) Meat products and consumption culture in the west. Meat Sci 86(1):80–85

Toepfl S, Mathys A, Heinz V et al (2006) Potential of emerging technologies for energy efficient and environmentally friendly food processing. Food Rev Int 22(4):405–423

Toohey ES, Hopkins DL, Stanley DF et al (2008) The impact of new generation pre-dressing medium-voltage electrical stimulation on tenderness and colour stability in lamb meat. Meat Sci 79(4):683–691

Tornberg EV (2005) Effects of heat on meat proteins–implications on structure and quality of meat products. Meat Sci 70(3):493–508

Tsigarida E, Nychas GJ (2001) Ecophysiological attributes of a lactobacillus sp. and a pseudomonas sp. on sterile beef fillets in relation to storage temperature and film permeability. J Appl Microbiol 90(5):696–705

Ulbricht TL, Southgate DA (1991) Coronary heart disease: seven dietary factors. Lancet 338(8773): 985–992

USDA (2010) Food safety inspection service, shell eggs from farm to table. http://www.fsis.usda.gov/PDF/Shell_Eggs_from_Farm_to_Table.pdf

USDA (2011) Food safety and inspection service. Chicken and Turkey nutrition facts, USDA national nutrient database for standard reference, USDA, ARS. https://www.fsis.usda.gov/shared/PDF/Chicken_Turkey_Nutrition_Facts.pdf?utm_source¼RSS&utm_medium¼referral&utm_campaign¼content-partnerships

Van Elswyk ME, Dawson PL, Sams AR (1995) Dietary menhaden oil influences sensory characteristics and headspace volatiles of shell eggs. J Food Sci 60:85–89

Vareltzis K (2000) Fish proteins from unexploited and underdeveloped sources. In: Developments in food science, vol 41. Elsevier, pp 133–159

Vasavada M, Carpenter CE, Cornforth DP, Ghorpade V (2003) Sodium levulinate and sodium lactate effects on microbial growth and stability of fresh pork and Turkey sausages. J Muscle Foods 14(2):119–129

Ventanas S, Estevez M, Tejeda JF et al (2006) Protein and lipid oxidation in longissimus dorsi and dry cured loin from Iberian pigs as affected by crossbreeding and diet. Meat Sci 72(4):647–655

Viljoen HF, De Kock HL, Webb EC (2002) Consumer acceptability of dark, firm and dry (DFD) and normal pH beef steaks. Meat Sci 61(2):181–185

Voda A, Homan N, Witek M et al (2012) The impact of freeze-drying on microstructure and rehydration properties of carrot. Food Res Int 49(2):687–693

Vuorela S, Salminen H, Mäkelä M et al (2005) Effect of plant phenolics on protein and lipid oxidation in cooked pork meat patties. J Agric Food Chem 53(22):8492–8497

Wackerbarth H, Kuhlmann U, Tintchev F et al (2009) Structural changes of myoglobin in pressure-treated pork meat probed by resonance Raman spectroscopy. Food Chem 115(4):1194–1198

Warriss PD (2000a) The chemical composition and structure of meat. In: Warriss PD (ed) Meat science: an introductory text. CABI, Wallingford, pp 37–67

Warriss PD (2000b) Post-mortem changes in muscle and its conversion into meat. In: Warriss PD (ed) Meat science: an introductory text. CABI, Wallingford, pp 93–105

Watkins BA (1995) The nutritive value of the egg. In: Stadelmanand WJ, Cotterill OJ (eds) Egg science and technology, 4th edn. Food Product Press, Binghamton, NY, pp 177–194

Weggemans RM, Zock PL, Katan MB (2001) Dietary cholesterol from eggs increases the ratio of total cholesterol to high-density lipoprotein cholesterol in humans: a meta-analysis. Am J Clin Nutr 73:885–891

Williams P (2007) Nutritional composition of red meat. Nutr Dietetics 64:S113–S119

Xiong YL (2000) Protein oxidation and implications for muscle food quality. Antioxidants Muscle Foods 14:85–111

Zhou GH, Xu XL, Liu Y (2010) Preservation technologies for fresh meat—a review. Meat Sci 86(1):119–128

Chemistry and Physiology of Fruits and Vegetables

12

H. B. Rashmi and P. S. Negi

12.1 Introduction

Composition of fruits and vegetables varies greatly based on the plant type and variety, cultivation practices, maturity at harvest, and the storage conditions. Maturity and ripeness are used for defining the appropriate stage for harvesting and consumption of fruits. Development and maturation of fruit occur while fruit is attached to the plant, however, ripening can happen either before or after the harvest. Fruit ripening is a highly co-ordinated event, which is a genetically programmed phenomenon, and it results in the development of a soft and edible ripe fruit with desirable quality attributes involving various chemical reactions (Giovannoni 2004). In fruits, ripening indicates the completion of development and the commencement of senescence, whereas vegetables usually does not reach ripening stage as they are harvested as and when required, which makes it difficult to differentiate various stages from maturation to senescence in the vegetables.

Processing can alter chemical composition of fruits and vegetables to a great extent depending on the type of the process and fruits and vegetables composition. Significant differences may occur within the same species among various cultivars and even for a single cultivar grown or stored under different conditions. Chemical composition may vary causing further variation in the chemical constituents of processed fruits and vegetables. Processing of fruits and vegetables is known to cause alterations in the colour, texture, flavour, and nutritional quality depending on the processing conditions. Further, a great variability in the stability of nutrients during storage of processed fruits and vegetables has been reported (Solovchenko et al. 2019). The present chapter summarizes the changes in chemical composition of

H. B. Rashmi · P. S. Negi (✉)
Fruit and Vegetable Technology Department, CSIR-Central Food Technological Research Institute, Mysuru, India
e-mail: psnegi@cftri.res.in

O. P. Chauhan (ed.), *Advances in Food Chemistry*,
https://doi.org/10.1007/978-981-19-4796-4_12

various fruits and vegetables during their growth, development, storage, and after processing.

12.2 Chemical Composition of Fruits and Vegetables

12.2.1 Proximate Composition

Most of the fresh fruit and vegetables have high moisture content and are low in protein as well as fat content. Moisture content in most of the fruits and vegetables is greater than 85%. Generally, protein content below 3.5% and fat content of less than 0.5% are reported in fruits and vegetables. However, leguminous vegetables contain higher protein, and significant quantities of fat are found in nuts and avocado fruit. In general, around three fourth of the total solid matter of fruits and vegetables is carbohydrate. The sugars and starches present in fruits and vegetables constitute digestible carbohydrates, whereas indigestible cellulose provides roughage. The total carbohydrate content varies from approximately 2% of the fresh weight in some fruits to around 30% in starchy vegetables. Total carbohydrates primarily consist of simple sugars and polysaccharides, and also include pectic substances and lignin. The major cell wall constituents are cellulose, hemicelluloses, pectins, and lignin. The hexose and pentose sugars constitute a heterogeneous group of polysaccharides, known as hemicelluloses. These polymers are classified as xylans, arabinogalactans, glucomannans, etc. based on the types of predominant sugar residues. Pectin is composed of α-1,4-linked galacturonic acid residues, which are esterified with methanol at different levels. The contents of cell wall constituents of fruits and vegetables vary among plant species, varieties, maturity levels, and storage conditions.

The major carbohydrate of plant tissues is starch, which is a linear (α-1,4) or branched (α-1,4;1,6) polymer of D-glucose. Sucrose, glucose, and fructose are the major sugars present in most of the fruits and vegetables, and reducing sugars are present in higher amount than sucrose. Other sugars present in some fruits and vegetables include xylose, mannose, maltose, and cellobiose. Branched sugar such as D-adiose is also present in many fruits and vegetables. Some fruits also contain sugar alcohols, such as sorbitol and xylitol.

The protein content varies greatly among various fruits and vegetables, ranging from negligible amount in most of the fruits and vegetables to slightly higher amount in leguminous crops, tubers, and bulbs. Protein content in fruits and vegetables may vary significantly based on their geographic origin, as protein content in mangoes is around 0.6% in Colombia (ICBF 2015), 1.5 to 5.5% in Peru; whereas, in Indian cultivars it varies from 0.5 to 1.0% (Saleem-Dar et al. 2016). Fruits and vegetables also contain appreciable quantities of non-protein nitrogen, such as potato tuber or apple fruit contain free amino acids. Although amino acids are the main non-protein nitrogen components of most fruits and vegetables, small amounts of amines, alkaloids, and non-proteinogenic amino acids are also reported. Free amino acids present in fruits and vegetables contribute to their taste. Polyamines, such as spermidine and spermine increase resistance to chilling injury by stabilizing membranes, and delay their ripening by acting as ethylene antagonist (Handa and Mattoo 2010). A small amount of lipid is also present in many fruits and vegetables.

12.2.2 Vitamins and Minerals

Fruit and vegetables are good sources of minerals and vitamins (Table 12.1). The pro-vitamin A (β-carotene and other carotenoids) is present in yellow-orange fruits and vegetables and in the green leafy vegetables. Citrus fruits, green leafy vegetables, tomatoes, cabbage, and green peppers are the excellent sources of vitamin C. Potatoes also provide significant amount of vitamin C for the diets in several countries as despite being low in vitamin C content, consumption of potatoes is higher in those countries.

The total mineral (ash) content in fruits and vegetables ranges from 0.1% to 5% on fresh weight basis. Minerals are present as salts of organic or inorganic acids or as complex organic combinations dissolved in cellular juice. The mineral content in fruit and vegetables generally ranges from 0.60 and 1.80%, and the most abundant minerals in plants are potassium, calcium, magnesium, iron, phosphorus, sulphur, and nitrogen. Vegetables are generally richer in mineral substances as compared to fruits. Spinach, carrots, cabbage, and tomatoes are rich in minerals, whereas among fruits, strawberries, cherries, peaches, and raspberries have higher mineral content. In general, higher quantity of potassium is present in fruits, whereas vegetables are richer in calcium and phosphorus.

12.2.3 Pigments

Pigments are the main constituent responsible for all ranges of colour expression in fruits and vegetables in their developmental stages. Various pigments present in fruits and vegetables (Fig.12.1) include chlorophyll (green), carotenoids (yellow to red), anthocyanin (red to purple), flavonoids (light brown to yellow), tannins (colourless to brown), leucoanthocyanins (colourless), betalains (red), quinones, and xanthones (yellow). The chemical structure of pigments responsible for different colours and biological functions is presented in Fig.12.2.

12.2.3.1 Chlorophyll

Chlorophyll is a cyclic tetrapyrrole ring, in which pyrrole rings (4 Nos.) are joined by methine ($2C_5$) bridges. The nitrogen atoms of pyrrole rings are connected through magnesium. A 20 carbon long hydrophobic phytol (esterified isoprenoid alcohol) group is attached to the tetrapyrrole ring. Chlorophyll can degrade to pheophytins, pheophorbide (via pheophytins or chlorophyllide), and other derivatives such as pyropheophorbides and pyropheophytins (Von Elbe and Schwartz 1996).

Table 12.1 Nutritional composition of selected fruits and vegetables

Sl. No.	Fruit/vegetable	Macro-nutrients (g/100 g)			Vitamins (mg/100 g)						Minerals (mg/100 g)					
		Carbohydrate	Fat	Protein	A	B_1	B_2	B_6	C	E	Calcium	Iron	Phosphorus	Potassium	Sodium	Zinc
1	Apple	13.8	0.26	0.3	0.003	0.017	0.026	0.041	4.6	0.18	6	0.12	107	107	1	0.04
2	Banana	22.84	0.33	1	–	0.031	0.073	0.4	8.7	–	–	0.26	22	358	–	0.15
3	Grapes	18	0.16	1	–	0.069	0.07	0.086	3.2	0.19	10	0.36	20	191	2	0.07
4	Guava	14.32	0.95	2.55	0.031	0.067	0.04	0.11	228.3	–	18	0.26	40	417	2	0.23
5	Kiwi fruit	14.66	0.52	1.14	0.122	0.027	0.025	0.063	92.7	1.46	34	0.31	34	312	3	0.14
6	Mango	16.2	0.34	0.36	0.054	0.028	0.04	0.119	36.4	0.9	11	0.16	14	168	1	0.09
7	Papaya	10.82	0.26	0.47	0.047	0.02	0.03		62	0.3	20	0.25	8.0	182	8	0.08
8	Pineapple	13.12	0.12	0.54	–	0.079	0.032	0.112	47.8	–	13	0.29	8	109	1	0.12
9	Pomegranate	18.7	1.17	1.67	–	0.067	0.053	0.075	10.2	0.6	10	0.3	36	236	3	0.35
10	Sapota	19.96	1.11	0.44	–	–	0.02	0.06	14.7	–	21	0.8	12	193	12	0.1
11	Beetroot	9.56	0.17	1.61	0.002	0.031	0.04	0.07	4.9	–	16	0.8	40	325	78	0.35
12	Bell pepper	4.64	0.17	0.86	0.018	0.057	0.028	0.224	80.4	0.37	10	0.34	20	175	3	0.13
13	Bitter gourd	4.32	0.18	0.84	0.006	0.051	0.053	0.041	33	0.14	9	0.38	36	319	6	0.77
14	Bottle gourd	3.69	0.02	0.6	–	0.029	0.022	0.038	8.5	–	24	0.25	13	170	2	0.7
15	Brinjal	5.88	0.18	0.98	–	0.039	0.037	0.084	2.2	0.3	9	0.23	24	229	–	0.16
16	Carrot	9.6	0.24	0.93	5.011	0.066	0.058	0.138	5.9	0.66	33	0.3	–	–	–	–
17	Cauliflower	5	0.3	1.9	–	0.05	0.06	0.184	48.2	0.08	22	0.42	44	299	30	0.27
18	Chili pepper	8.8	0.4	1.9	0.048	–	–	0.51	144	–	–	1.0	–	322	–	–
19	Cucumber	3.63	0.11	0.65	0.032	0.027	0.033	0.04	2.8	–	16	0.28	24	147	2	0.2
20	French bean	6.97	0.22	1.83	0.035	0.082	0.104	0.141	12.2	–	37	1.03	38	211	–	0.24
21	Lettuce	2.23	0.22	1.35	2.221	0.057	0.062	0.082	3.7	0.18	35	1.24	33	238	5	0.2
22	Okra	7.46	0.19	2.0	0.036	0.2	0.06	–	23	0.27	82	0.62	61	299	–	0.58
23	Onion	9.34	0.1	1.10	–	0.046	0.027	0.12	7.4	–	23	0.21	29	146	–	0.17
24	Pea	14.45	0.4	5.42	0.038	0.266	0.132	0.169	40	0.13	25	1.47	108	244	5	1.24
25	Potato	17	0.09	2.05	0.0006	0.08	0.03	0.30	19.7	0.01	12	0.78	57	421	6	0.29

26	Spinach	3.6	0.4	2.86	2.813	0.078	0.189	0.195	28.1	2	99	2.71	49	558	79	–
27	Sweet potato	20.1	0.1	1.6	0.709	0.078	0.061	0.209	2.4	0.26	30	0.61	47	337	55	0.3
28	Tomato	3.9	0.2	0.9	0.042	0.037	0.594	0.08	14	0.54	10	–	24	237	–	–
29	Turnip	5.1	0.1	0.7	3.476	0.027	0.023	0.067	11.6	–	33	0.18	26	177	16	0.12
30	Yam	27.9	0.17	1.5	0.007	0.112	0.032	0.293	17.1	0.35	17	0.54	55	816	–	0.24

Source: USDA (2018)
'–': Not reported

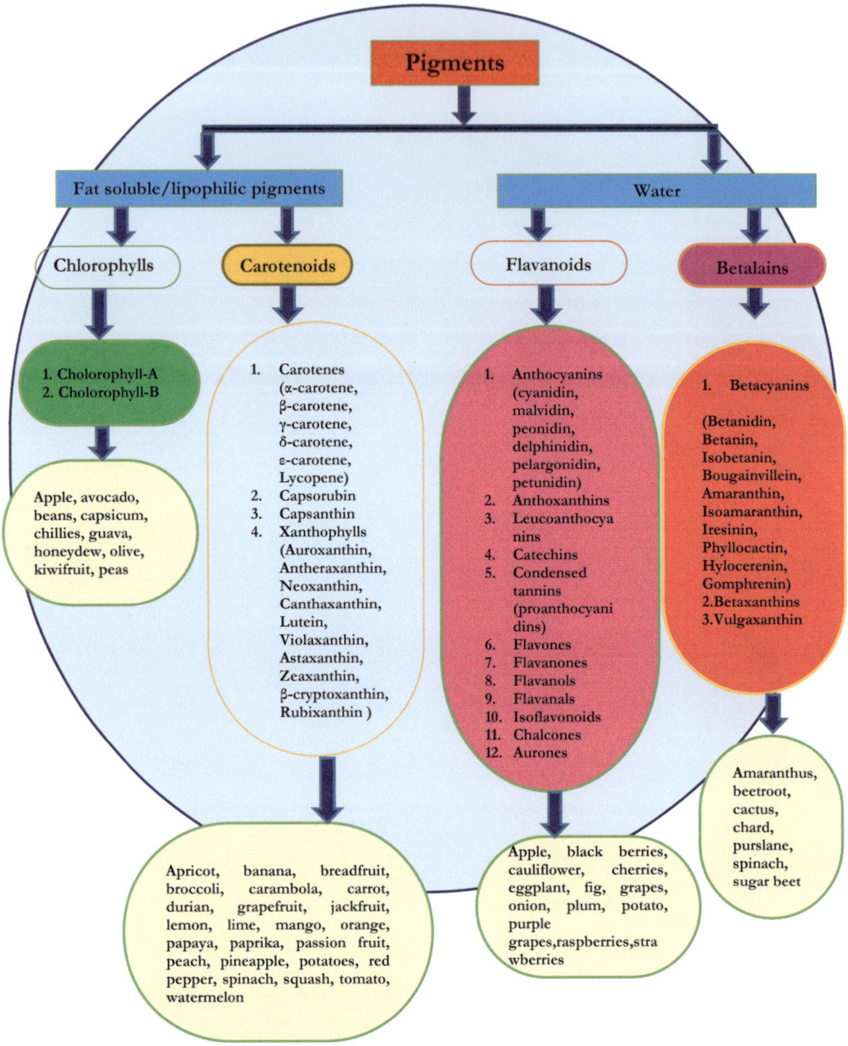

Fig. 12.1 Classification and occurrence of pigments in fruits and vegetables

12.2.3.2 Carotenoids

Carotenoids are ubiquitous in plants and exhibit various shades ranging from yellow to red hues. They often occur along with the chlorophylls in the chloroplasts, and are also present in other chromoplasts. They are lipophilic tetra-terpenoids (C_8) having 40-carbon polyene chain with consecutive double bonds. Hydrocarbon carotenoids have a molecular formula of $C_{40}H_{56}$, while xanthophylls are represented as $C_{40}H_{56}O_2$ or $C_{40}H_{56}O$, which are oxygenated derivatives, and contain oxygen atoms within hydroxy-, epoxy-, or keto- groups. These chemical structures determine the chemical reactivity and colour of carotenoids (Namitha and Negi 2010).

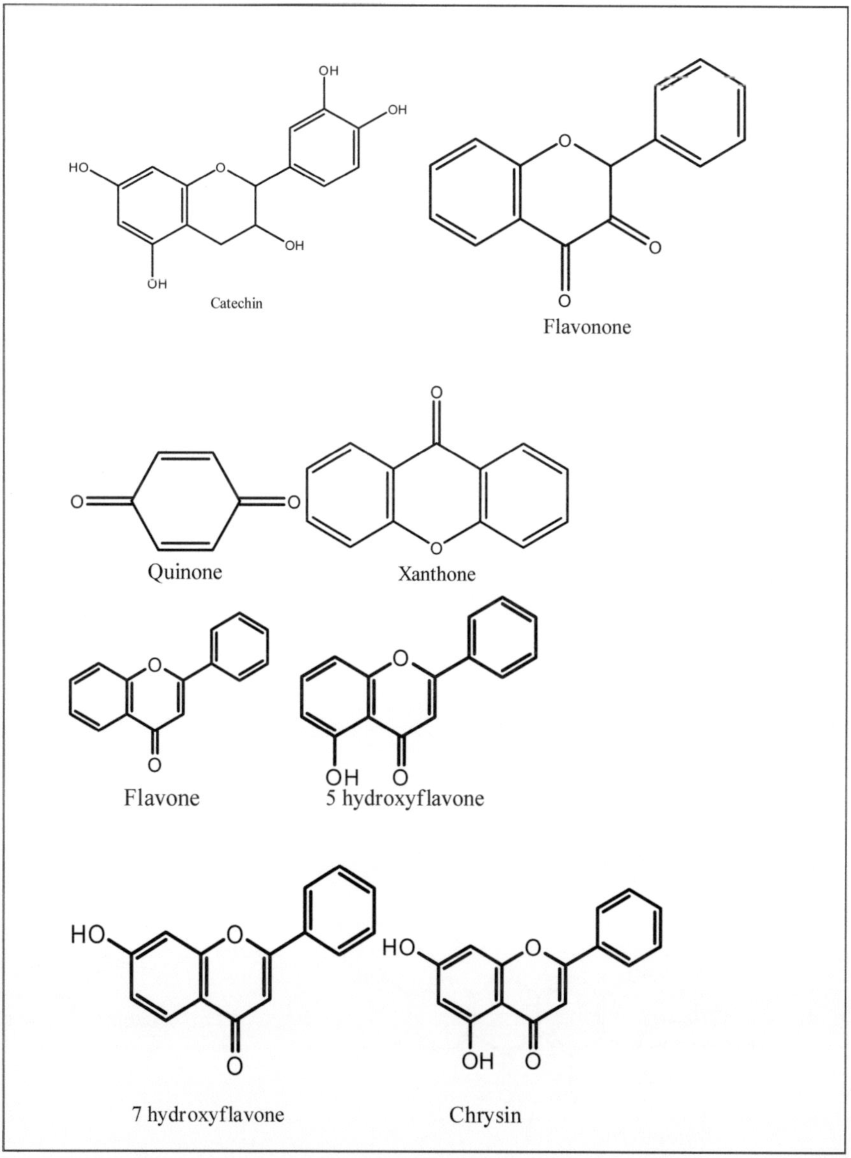

Fig. 12.2 Chemical structures of few pigments found in fruits and vegetables

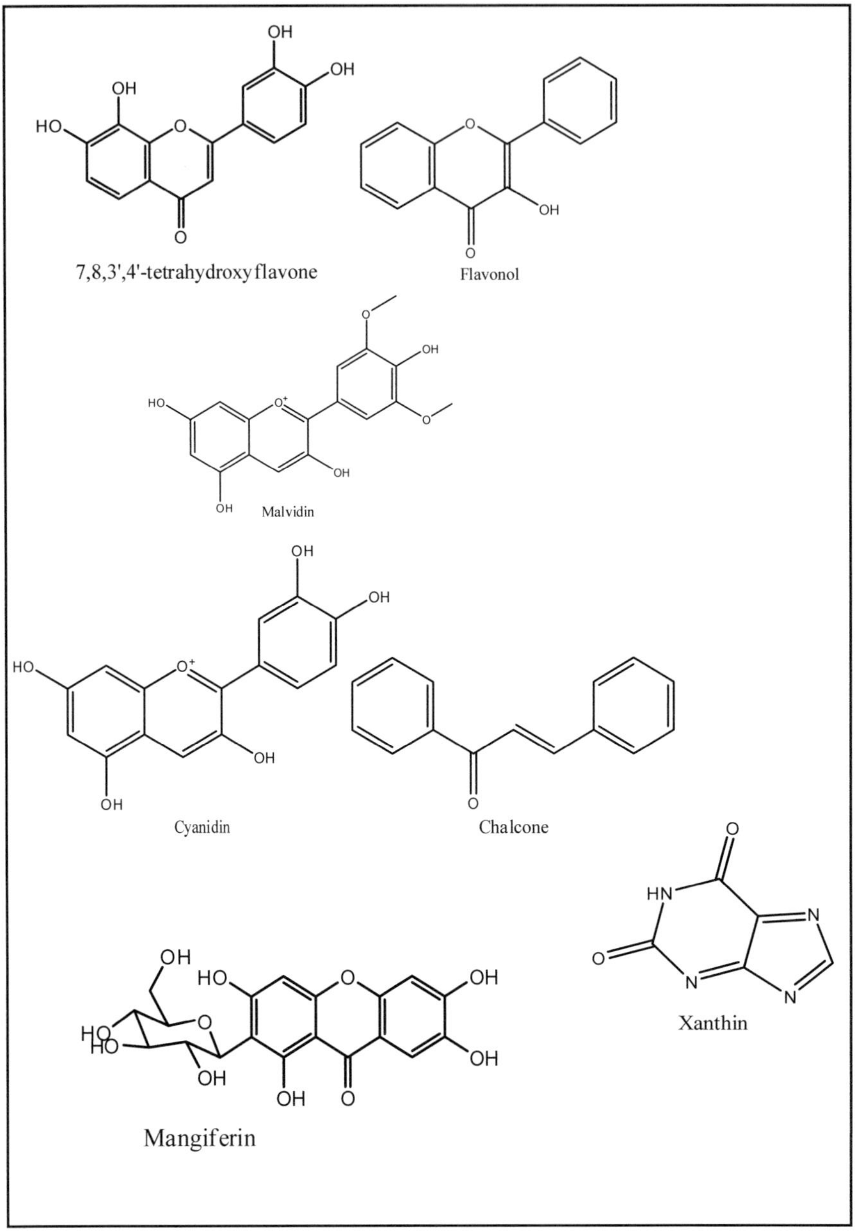

Fig. 12.2 (continued)

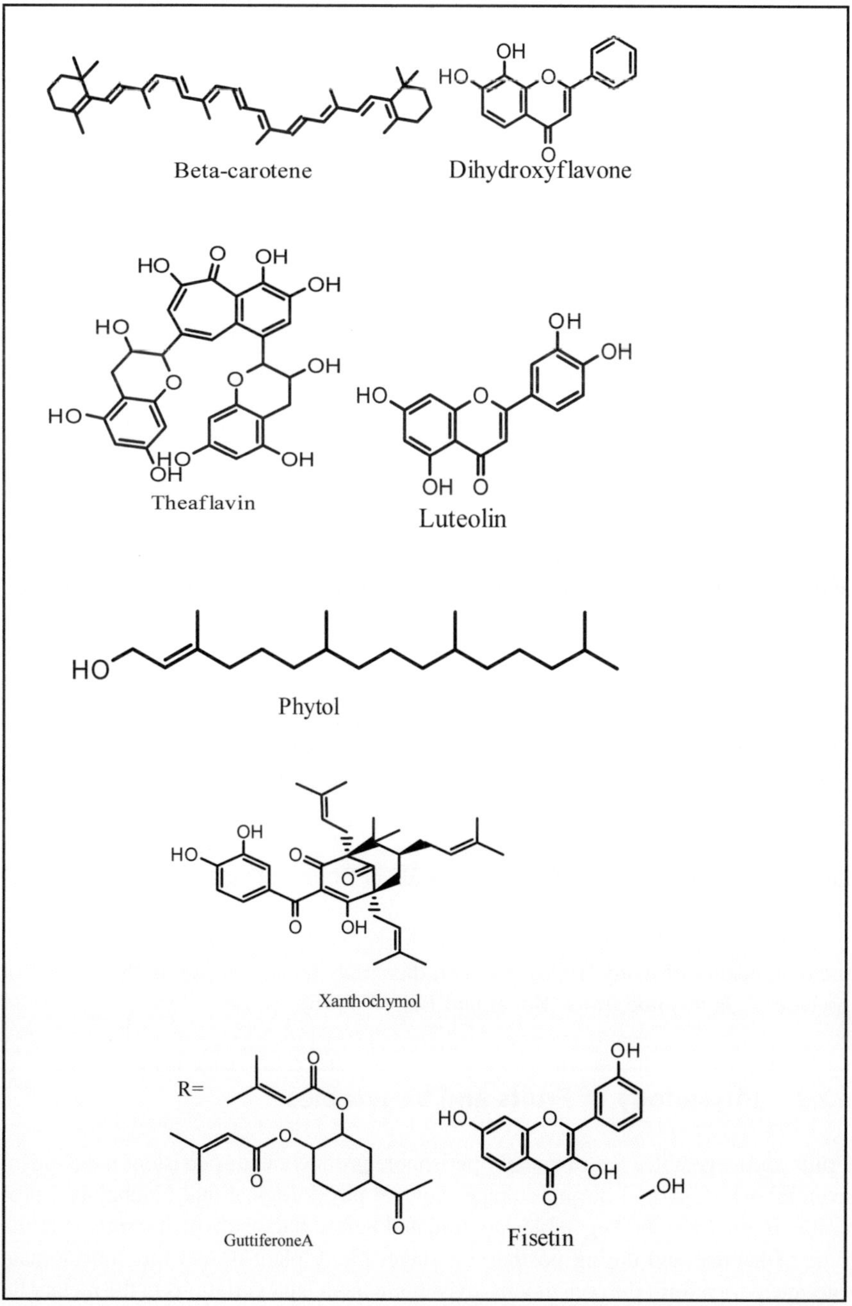

Fig. 12.2 (continued)

12.2.3.3 Flavonoids

Flavonoids are water-soluble pigments with diverse structures and are commonly present in several fruits and vegetables. The flavonoid pigments include the purple, blue, and red anthocyanins; the yellow anthoxanthins, and the colourless catechins and leucoanthocyanins.

The major phenolic compounds in plants include phenols and phenolic acids (C_6–C_1), hydroxycinnamate and other phenylpropanoid derivatives (C_6–C_3), and flavonoids (C_6–C_3–C_6). The substitution in 15-carbon benzo-γ-pyrone (C_6–C_3–C_6) backbone gives rise to anthocyanins (anthocyanidins), condensed tannins (proanthocyanidins), flavonols, flavones, isoflavonoids, chalcones, and aurones (Lattanzio et al. 2008).

12.2.3.4 Betalains

Betalains are indole-derived glycoside pigments containing nitrogen. Betalains are water-soluble, and their colour is stable in a broad pH range (3.5–7.0). The optimum pH for betanin stability is 5.5–5.8, whereas vulgaxanthin-I is stable in a slightly wider pH range (5.0–6.0). Betalains are free-radical scavengers similar to anthocyanins, and are more efficient at alkaline and neutral pH, whereas anthocyanins are active at acidic pH (Von Elbe and Schwartz 1996).

12.2.4 Organic Acids

Organic acids occur in fruits and vegetables in small quantities as metabolic intermediates of tricarboxylic acid cycle, glyoxylate cycle, or shikimic acid pathway and accumulate in vacuoles. The organic acids impart an acidic or sour taste. Organic acids may influence the colour of foods since many plant pigments have different hues based on pH. The most abundant acids in fruits and vegetables are citric and malic acids, however, tartaric acid is the major acid in grapes and avocado, oxalic acid in spinach, and iso-citric acid in blackberry. With ripening, there is a decline in the total acidity of many fruits; however, there may be an increase in the content of specific acids in some fruits (Kader and Barrett 1996).

12.3 Physiology of Fruits and Vegetables

Fruits and vegetables are extremely perishable products with persistent metabolism even after harvest and during storage. Several physiological and biochemical processes in the fruit and vegetables are initiated before harvest, which continue at the time of harvest and during postharvest stage. Many plant tissues are consummate enough to transform the constituents existing in them, and variations in the metabolic shifts, which are explicit to any fruit or vegetables are manifested in ripening, sprouting, browning, toughening, yellowing, and rotting.

The type and intensity of physiological activity in the harvested fruits and vegetables determine their storage longevity. Seeds, fleshy roots, tubers, and bulbs can maintain the tissue in a dormant state, while in fleshy fruits, the maturation is followed by a ripening process followed by senescence (Paltrinieri and Staff 2014).

12.3.1 Transpiration

After harvesting the fruit or vegetable, the loss of water through transpiration may occur based on storage conditions. The water loss to the tune of 5% or more of its original weight can cause shrivelling, and can reduce its marketability and overall quality. They become unappealing because of shrinking, and therefore, the control of transpiration losses is important to maintain marketability and prevent infection, which may arise due to tissue damage (Holcroft 2015).

12.3.2 Respiration

Respiration is a major metabolic process occurring in harvested fruits and vegetables. Respiration is the oxidative breakdown of the more complex materials such as starch, sugars, and organic acids into simpler molecules such as carbon dioxide and water, with the concurrent production of energy. A consistent relationship between storage life and respiration is reported, which can serve as an indicator of the potential storage life of the produce. Several fruits show a variation from the described respiratory pattern as they undergo a pronounced increase in respiration, which coincides with ripening. This increase in respiration is known as a respiratory climacteric and this group of fruits are known as the climacteric fruits. Fruits not exhibiting a respiratory climacteric are known as the non-climacteric fruits, which also shows ripening variations, albeit more slowly than the climacteric fruits (Fig.12.3). All vegetables are considered to have a non-climacteric type of respiratory pattern (Kader and Barrett 1996).

12.3.3 Ethylene Production and Ripening

Ethylene plays an important role in postharvest life of many fruits and vegetables. Ethylene is beneficial as it improves the quality of the products by promoting faster and more uniform ripening before retail distribution, however, it also speeds up senescence and reduces shelf life. During ripening, climacteric fruits manifest upsurge in respiration and ethylene production, whereas low ethylene evolution is reported in non-climacteric fruits (Fig.12.3). The stage of fruit development for ethylene application and response by fruits differs based on their type, climacteric or non-climacteric. Ethylene application at pre-climacteric stage results in enhanced respiration in the climacteric fruits, whereas a non-climacteric fruit responds to ethylene application at all the stages in pre- or post-harvest life.

Ripening is a series of physical, physiological, and biochemical changes happening after termination of growth till the commencement of senescence. A definite

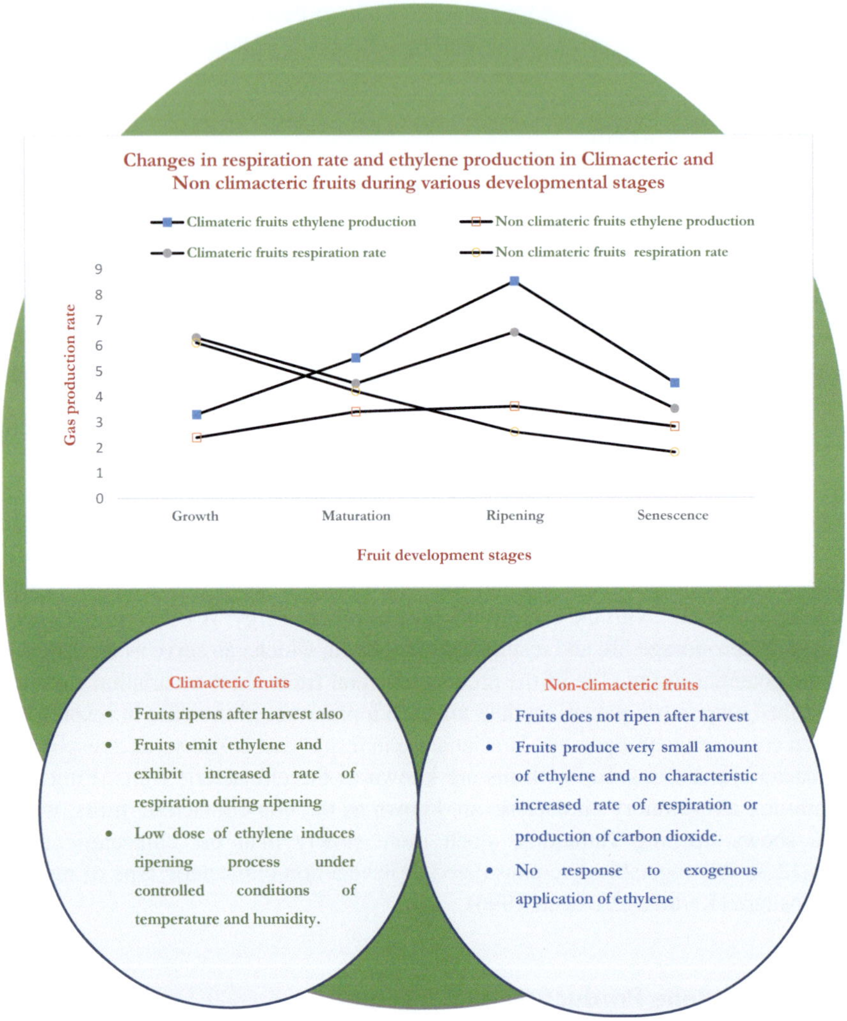

Fig. 12.3 Changes in ethylene and respiration rate during various developmental stages in climacteric and non-climacteric fruits

maturity stage must be attained before ripening can initiate, and ripening process persists while the fruit is intact on the tree, and continues even after the harvest. The ripening changes are associated with the development of optimal quality and synthesis of volatiles in the fruits (Kader and Barrett 1996).

12.3.4 Chemical Changes During Ripening and Senescence

Compositional changes in produce during ripening lead to changes in their colour, texture, firmness, taste, and aroma. The change in colour of fruits and vegetables occurs as the chlorophyll breaks down, and new pigments, such as carotenoids and flavonoids are synthesized. Starch gets converted to sugar, whereas pectin and other polysaccharides will breakdown resulting in the fruit softening during ripening. Ripening process witnesses flavour enhancement due to alteration in organic acids and lipids as well volatile organic compounds synthesis (Kader and Barrett 1996).

12.3.4.1 Change in Carbohydrate

During ripening, significant changes occur in the carbohydrates in fruits and vegetables. Green or raw fruits usually contain starch in abundance, and throughout the ripening transition, the starch is enzymatically converted into sugars. As ripening progresses, majority of the soluble carbohydrates are metabolized. The fully ripened fruit mainly consists of glucose, fructose (invert sugars), and sucrose. A marked decrease in starch in banana occurs, which equals the increase in the contents of glucose, fructose, and saccharose after ripening. Pectic substances and cellulose are also converted to acids and sugars during ripening. There is also a decrease in the degree of methylation of pectins during ripening. Insoluble protopectin is converted to soluble forms, and the soluble pectin binds with polyphenols, causing a decrease in astringent taste and contributes to the mild taste of ripe fruits (Kader and Barrett 1996).

Textural changes in the fruit during ripening are the most prominent change. As the fruit ripens, there is a decrease in total pectic substances, however, soluble pectates and pectinates show increase. There is a decrease in the quantity of protopectin (insoluble) with simultaneous increase in soluble pectin, which turn flesh less firm. These changes in the pectic substances are responsible for the softness of the fruits. The increase in the activities of polygalacturonases and pectin methyl esterases causes decrease in soluble pectin and extensive demethylation of pectin during prolonged storage of apples, pears, peaches, and avocados, which is associated with mealy and soft texture of these crops. Ripe mango comprises 15% of total sugars in which fructose is the main monosaccharide at pre-climacteric stage (Bernardes et al. 2008), but it is replaced by sucrose as the major sugar during ripening stage (USDA 2018).

12.3.4.2 Changes in Organic Acids

The organic acids undergo changes during the progression of ripening. The sourness is associated with organic acids such as citric, malic, succinic, tartaric, and oxalic, and the fruits vary in their content at different stages of development. Further, the pattern of changes of organic acids also varies during fruit development among different fruits. Majority of the fruits show a significant reduction in the acid content during ripening, and proportion of various acids also changes. In few fruits, acids will undergo enzymatic conversion into sugar making them sweet (Mangoes, Oranges), whereas, in some cases, there is no change in acids (Lemons). During ripening in grape, strawberry, mango, and tomato, the amount of malate/citrate decreases, whereas, in lemon, organic acid concentration increases throughout ripening (Kader and Barrett 1996).

12.3.4.3 Changes in Amino Acids and Proteins

Different fruits mature at different rates and, therefore, protein content varies among different species and genotype. Initial fruit development records higher total nitrogen content, which steadily declines owing to the increment of other components (starch, sugar, organic acids). In crops like mango, tomato, and avocado, a slight upsurge in protein was also observed during ripening. Increase in the activity of hydrolases (amylases, cellulases, pectinolytic enzymes, transaminases, peroxidases, and catalases) may cause softening of fruits during ripening in some fruits. Alanine, arginine, glycine, serine, leucine, and isoleucine are reported to be the major amino acids present in the ripe mango, with trace amounts of other amino acids (Tharanathan et al. 2006).

12.3.4.4 Changes in Lipids

The oil composition did not change much during maturation in Avocado, however, considerable increase in total lipid and fatty acid content has been witnessed while mango undergoes ripening. Mango fruits exhibit presence of 17 fatty acids during development and ripening (Deshpande et al. 2016). Palmitic, stearic, oleic, and linoleic acids are present in higher amounts, whereas lower concentration of arachidic, linolenic, and behenic acids is reported in mango kernel (Jahurul et al. 2015).

12.3.4.5 Changes in Pigments

The ripening of fruit is usually accompanied by a change in colour. Fading away of green colour marks the beginning of ripening in most of the fruits. Decline in chlorophyll pigment during ripening is simultaneously accompanied by biosynthesis of supplementary pigments. A remarkable production of carotenoids happens at final stage of ripening. Quantification of carotenoids shows increased content from zero to high levels within few days due to onset of ripening, which is triggered by the ethylene, and carotenoids accumulation depends on temperature and light intensity. The lycopene content in tomatoes and the carotenoids content in citrus fruits and mangoes increases during ripening. The formation of anthocyanins takes place during ripening. Light and temperature influence synthesis of anthocyanin, and the highest increase in intensity of purple colour of red cabbage occurs at and below 10° C storage temperatures. As the fruits ripen, the total polyphenolic and tannin content reduces gradually (Kader and Barrett 1996).

12.3.4.6 Changes in Volatile Compounds

There's a huge demarcation between a ripe and unripe fruit owing to the differences in their flavour intensities. Fruits withhold enormous volatile compounds, which give distinctness among different categories of fruits. Their concentration also varies throughout the stages of fruit development and ripening, which generally increases as fruit ripens. Maturity stage has major effect on aroma production; however, the environmental parameters also influence aroma composition. As fruits overripe, the fermentation sets in leading to the formation of alcohols and esters.

12.3.4.7 Changes in Enzyme Activity

Enzymes are responsible for majority of chemical and physical alterations happening during ripening. Fruit softening, starch conversion to sugar, and colour changes during ripening are caused by enzymes. Activity of oxidative enzymes, such as catalase and peroxidise, increases during mango ripening, as depicted by elevated respiration rate. Activity of glycolytic and hydrolytic enzymes also escalates in ripening mangoes. Increase in the fatty acid synthase activity during ripening and following respiratory climacteric was observed in avocado fruit.

12.4 Chemical Changes During Handling and Storage of Fruits and Vegetables

12.4.1 Pre-Packaging

The pre-packing operations include sorting, sizing, grading, and packing. However, additional operations may also be performed depending on commodity. These activities may have slight influence on chemical composition of produce (Kader and Barrett 1996).

Curing of tuber crops is done prior to storage or sending to the market. Produce is held at slightly higher temperatures and high humidity for few days to facilitate periderm formation. Reduction of moisture content from the outer skin and neck of the onion and garlic during curing helps in reducing the disease infection, minimizing shrinkage and development of skin colour. Potato tubers exposed to direct sun light become green due to the development of solanine, which is bitter in taste, difficult to cook and also poisonous in nature (Haguluha and Natera 2007).

Degreening process removes green colour (chlorophylls) in fruits, and ethylene (C_2H_4) is used to give fruits its characteristic colour. Degreening is carried out under controlled temperatures and humidity conditions at low concentrations of C_2H_4 (~20 ppm) for 24–48 h, and CO_2 is maintained below 1% levels. The best degreening can be achieved around the temperature of 25° C and relative humidity of 85% (Watada 1986).

Quality of fruits and vegetables varies greatly due to genetic, agronomic, pre and post- harvest factors. Grading is done to classify them according to quality in order to enhance their market value. Grades are generally based on parameters such as firmness, size, weight, colour, shape, and maturity.

12.4.2 Pre-Cooling

Pre-cooling is the most essential postharvest operation to get rid of the field heat in order to reduce the respiration and associated changes in the produce. Pre-cooling is carried out to the appropriate storage temperature. Various methods of pre-cooling include room cooling, hydro-cooling, top icing, vacuum cooling, hydrovac cooling, etc. (Kader and Barrett 1996).

12.4.3 Pre-Treatments

12.4.3.1 Waxing

Waxing reduces the respiration and transpiration rates, and increases the shelf life of various fruits and vegetables. The wax is applied to the fruit for protection against decaying organisms. Waxing also enhances the glossiness of fruits and vegetables. Botanicals and fungicides can also be mixed with the wax before coating for controlling microbial spoilage. Carnauba wax, paraffin wax, shellac, bee wax, and wood resins are few of the commercially utilized waxes. Dipping, spray waxing, and foam waxing are few of the methods of wax application (Bourtoom 2008).

12.4.3.2 Sprout Inhibition

Sprouting and root formation hasten deterioration, which affects the saleability of crops like potatoes, yam, garlic, onion. Physical methods such as refrigeration and controlled atmosphere storage reduce sprouting and rooting. Sprouting of potatoes is inhibited by storage at and below the temperature of 5 °C, whereas in yam, the optimum storage temperature was 13°C. Ethylene application suppresses sprout growth by inhibiting leaf blade elongation in onion (Bufler 2009).

CIPC (3-chloro-iso-propyl-N phenylcarbamate/Chlorpropham) inhibits sprout development in potatoes. CIPC application is done after completion of wound healing process, otherwise moisture loss and disease infection can occur because of non-formation of periderm in uncured potatoes (Smith and Bucher 2012). Gamma irradiation of onion bulb, potato, and yam also reduces sprouting.

12.4.3.3 Astringency Removal

Astringency is the dry, puckering mouth feel inflicted by tannins. These astringent compounds impart an unpleasant flavour typically associated with immature fruit (Jiang et al. 2014). CO_2 treatment and ethanol application help to lessen the astringency. There are two methods of astringency removal for persimmon fruits, either by storing in 4% CO_2 at 10 °C for about 2 weeks followed by 6–18 h in 90% CO_2 at 17 °C, or spraying with 35–40% ethanol (1% v/w) followed by 10 days storage at the ambient conditions. Better fruit quality was obtained by ethanol treatment (Bubba et al. 2009). Storage at −20 ° C as well as −80 °C for 60 days also removed the astringency of persimmon fruits (Das and Eun 2020).

12.4.3.4 Disinfestation Treatment

Disinfestation may include removal of insect/ pest and disease causing organisms. The irradiation, vapour heat treatment, and fumigation are the common disinfestation methods. In vapour heat treatment (VHT), the temperature and exposure time are adjusted to kill all stages of the insects. Treatment of citrus, mango, papaya, and pineapples using VHT is done for the control of fruit fly (Hsu et al. 2018).

Hot water treatment for the control of postharvest diseases is done by immersing fruits in hot water before storage or marketing. Hot water treatment at 51–55 °C for 5–30 min is reported to be more effective in combination with fungicides. For control of anthracnose or stem-end rot, the mangoes are exposed at the temperature of 55 °C for 5 min, sweet potato at 40 °C for 2 min, and blue berries at 46–55 °C for 3 min. However, hot water treatment if not done properly, may cause injury to fruit, such as increased weight loss and acceleration of colour development (Konsue et al. 2020).

12.4.4 Transport

Fruits and vegetables are transported to storage facility or market after harvest. Selection of transport method depends on the time taken to reach market; however, proper handling practices must be followed before and during transport to evade damage to the produce. Bruises and injuries during transport reduce the quality and saleability of the fruits and vegetables.

12.4.5 Storage

12.4.5.1 Compositional Factors

12.4.5.1.1 Carbohydrates

There is an initial increase followed by a decrease in sugars during storage, due to the breakdown of polysaccharides. Starch hydrolysis in banana results in equal concentrations of glucose and fructose, and a little sucrose during initial storage period, and the content of these sugars decreases upon prolonged storage. Fruit maturity and storage temperature determine the rate of conversion to reducing sugars. In mangoes, sucrose is converted to reducing sugars during storage. In green mature tomatoes, not much variation in reducing sugars was observed, whereas turning stage fruits had a marked decrease in sweetness with storage (Arthur et al. 2015). In non-climacteric fruits, these changes are slight and slower than climacteric fruits.

Low storage temperatures alter the balance of non-reducing to reducing sugars. Potatoes stored at 4 °C contain more reducing sugars than those stored at 10 °C. A similar trend was observed in cashew apples stored at temperatures ranging from 0 to 25 °C. In carrots, the ratio of non-reducing to reducing sugars exhibited a sharp increase after 14–18 weeks of storage at 1 °C (Phan et al. 1973).

12.4.5.1.2 Acids

Various factors including temperature, light, and plant management practices affect the organic acid content in fruits and vegetables by influencing the source to sink ratio. Changes in acidity during storage may also vary according to the maturity of produce and storage temperature. Green mature and turning tomato fruits show increase in acidity during storage, and the increase coincides with the climacteric pattern. Change in the organic acids and sugars results in increase in TSS and decrease in titratable acidity.

Ascorbic acid loss during storage depends upon both storage temperature and genotype. In most fruits and vegetables, ascorbic acid concentration generally reduces during storage even at optimum storage temperatures. However, storage under MA or CA conditions can retain ascorbic acid for longer duration. MA packaging also helps in retaining vitamin C during storage (Pantastico 1975).

12.4.5.1.3 Lipids

The increase in wax in the cuticle occurs in apple, as the fruit becomes greasy during storage, which depends on the variety. With prolonged storage, the soft wax fraction of the cuticle accumulates increasing amounts of unsaturated compounds. The non-volatile esters are produced rapidly during the early period of storage, while the production of volatile esters increases during prolonged storage. The phospholipid content in tomato fruit was shown to decrease during storage at 20 °C. The lipid composition of avocados changes slightly during storage. The higher oxygen atmosphere increased percentage of polyunsaturated fatty acids, whereas controlled atmosphere storage increased the content of palmitic and palmitoleic acids, but, decreased the content of oleic acid (Prabath Pathirana et al. 2013).

12.4.5.1.4 Pigments

With decrease in chlorophyll content, other pigments increase during storage depending on storage temperature, maturity, and variety. The green coloured chlorophyll changes to olive green and brown (pheophytin) due to ageing. The changes in carotenoids in mature fruits and vegetables are generally small and occur slowly as compared to ripe ones. A decrease in carotene and total carotenoid pigments during storage at low temperatures was observed in sweet potatoes. Immature tomato fruits stored at 10 °C turned yellow later than mature ones. The winter squashes showed increase in β-carotene content during the 60 days storage at 14 °C (Bonina-Noseworthy et al. 2016).

Different pre-harvest and postharvest practices influence accumulation or degradation of phenolic compounds (PCs) in fruits and vegetables. Normally, storage at higher temperatures is better to induce the accumulation of PCs, which depends on the produce and storage conditions as well as the type of PC. The decrease in PC during storage may be due to the oxidation of polyphenols by polyphenol oxidase, and also more complex phenolics can be decomposed to smaller compounds (Rashmi and Negi 2020). The effect of storage temperature on the PC content is higher in the peel, while in the pulp their content remains practically unchanged.

12.4.5.1.5 Pectic Substances

During storage, degradation of insoluble protopectins to the more soluble pectic acid and pectin occurs, which reduces the firmness. Changes in pectic substances are minimal if storage conditions are able to prevent the ripening. In grapefruit, the soluble pectin and proto pectin ratio is unaffected even after 6 weeks storage and no detectable pectic changes occur. Protopectin hydrolysis in CA-stored apples transferred to high temperature storage was very low.

12.4.5.1.6 Volatile Compounds

Storage temperature influences the production of volatile compounds in fruits and vegetables. Almost twice the amount of volatile material was produced in 'Golden Delicious' apples at 2 °C as compared to storage at 0 °C. McIntosh apples produced twice the amount of volatiles at 4 °C than storage at 0 °C. Production of volatile material is delayed in controlled atmosphere storage, however, low O_2 conditions can lead to the development of off-flavours in fruits as a result of accumulation of acetaldehyde or ethyl alcohol.

12.4.5.1.7 Proteins and Amino Acids

Amino acids decrease rapidly during storage, which is temperature dependent. In peas, free amino acid content reduced abruptly after 2 days at 20 °C or 5 days at 6 °C, but were stable at 0 °C up to 28 days (Pantastico 1975).

12.4.5.1.8 Enzymes

Activity of various enzymes depends on the storage temperature and maturity at the start of storage. Higher catalase and pectinesterase activities and a lower oxidase activity in storage were observed in mature fruits as compared to fruits picked at a relatively less mature stage. Starch hydrolytic enzyme activity was higher in less mature fruit of pear than those of later picked fruits. Catalase, pectinesterase, cellulase, and amylase generally show increase in the activity, however, oxidases showed decrease in activity during storage. Increase in amylase activity in pear was observed during storage for 3 months at 0 °C.

12.4.5.2 Physiological Disorders

Physiological disorders are related to exposure of undesirable temperature, humidity, very low oxygen or high carbon dioxide, and nutritional disorders. In general, any breakdown of tissues by means other than invasion of pathogens or mechanical damage is termed as physiological disorder. Many tropical and sub-tropical varieties of fruits and vegetables when stored below 10 °C for a longer time suffer physical and physiological injuries. These injuries include superficial scald, carbon dioxide injury, breakdown of the flesh of the stored commodity (low temperature breakdown or senescence breakdown), water core, bitter pit, freezing injury, and chilling injury. Some of the important physiological responses to these physiological disorders are stimulation of ethylene production and failure of colour development besides loss of nutritional value. A few important physiological disorders that develop during storage in fruits and vegetables and associated chemical changes are listed in Table 12.2.

Table 12.2 Physiological disorders and associated chemical changes during storage

Sl. No.	Commodity	Disorders	Chemical changes	Reference
1	Apple	Superficial scald (storage scald)	Volatile terpene alpha-farnesene oxidized into many conjugated trienes due to ethylene stimulation	Ingle and D'souza (1989)
		Bitter pit	High rate of respiration and ethylene production leads to synthesis of protein and pectin along with dis-balance of potassium and calcium	Ohlendorf (1999)
		Water core	Accumulation of sorbitol-rich solutions in intercellular spaces, ethanol and acetaldehyde, reduced gas diffusion	Itai (2015)
		Internal browning	Accumulation of sorbitol, acetaldehyde, ethanol, and organic acids in intercellular spaces	Matthais 1995
2	Mango	Spongy tissue	Accumulation of low carotene, sugars, potassium, calcium, and sodium	Ram et al. (2020)
		Soft-nose	Lowest mesocarp calcium content	Raymond et al. (1998)
		Stem-end cavity (SEC)	Presence of calcium oxalate crystals in fruit	Raymond et al. (1998)
		Internal necrosis and fruit pitting	Reduction in boron	Shivashankar (2014)
		Chilling injury	Reduction in sucrose, no change in total hexose content with changes in phenolic content	Chhatpar et al. (1971), Chidtragool et al. (2011)
3	Banana	Peel browning	Enzymatic and non-enzymatic oxidation of phenolic compounds	Luyckx et al. (2016)
		Peel bruising	Oxidation of phenolic compounds to quinones	Yang et al. (2001)
		Peel cracking	Peel water content reduces by osmosis (increased sugar content in pulp)	Paull (1996)
		Finger drop	Low soluble and insoluble pectin in peel (increased pectate lyase and pectin methyl esterase activities)	Imsabai et al. (2006)
		Chilling injury	Transition of cell membranes from flexible liquid crystalline phase to solid gel structure	Asghari and Aghdam (2010)
4	Citrus	Fruit cracking	Increase in moisture	Elavarasan and Premalatha (2019)

(continued)

Table 12.2 (continued)

Sl. No.	Commodity	Disorders	Chemical changes	Reference
		Granulation	Decrease in TSS, increase in cellulose, pectic substances, hemicelluloses, starch, and lignin.	Elavarasan and Premalatha (2019)
		Puffiness	Decrease in soluble solids and increase in cell membrane permeability	Wei et al. (2000)
		Chilling injury	Changes in physical properties of cell membrane, structural proteins (tubulin and enzymes), and production of toxic substances such as acetaldehyde	Wills and Golding (2016)
		Oleocellosis	Reduction of α-tocopherol and γ-tocopherol in flavedo	Sawamura et al. (1988)
5	Pomegranate	Chilling injury	External and internal browning due to oxidation of phenolics	Elyatem and Kader (1984)
		Husk scald	Phenolic oxidation	Ben-Arie and Or (1986)
		Aril paleness	Decrease in ascorbic acid and phenolic compounds (especially in monomeric anthocyanins)	Tabar et al. (2009)
6	Pineapple	Black heart	Decrease in ascorbic acid, sinapic acid, phosphatidylinositol but increase in ferulic acid, p-coumaric acid, caffeic acid, and phosphatidic acid	Hassan et al. (2010), Zhou et al. (2014)
		Chilling injury (CI)/internal browning (IB)	Decrease in amino acids and organic acids	Luengwilai et al. (2016, 2018)
7	Papaya	Skin freckles	Increase in fruit calcium levels and soluble solids of the latex with lower K^+ concentration and total soluble solids	Oliveira et al. (2018)
		Pulp softening	Transformation of cellulose, hemicellulose, pectic substances, structural proteins with reduced calcium and firmness	Oliveira et al. (2018)
		Hard lumps in pulp	Inactivation of cell wall degrading enzymes with alteration of skin and pulp colouration	Oliveira et al. (2018)
		Chilling injury	Inferior fruit flavour and reduced carotenoids content, water soaking of flesh, leakage of electrolytes	Oliveira et al. (2018)

12.5 Chemical Changes During Processing of Fruits and Vegetables

The effect of processing on chemical constituents depends on the processing technique and type of fruits and vegetables. Processing of fruits and vegetables induces changes in the colour, texture, flavour, and nutritional quality depending on the type and conditions of processing methods, type of fruit and its initial composition. Ascorbic acid in fruits and vegetables is most sensitive to various processing methods. The carotenoid retention reduces at higher temperature and longer processing time. Degradation of chlorophyll in processed fruits and vegetables occurs due to heat, acid, oxygen, and enzymes. The freezing and canning of fruits and vegetables also affects vitamin E retention. The effect of various processing methods on chemical constituents of selected fruits and vegetables is summarized in Table 12.3.

12.5.1 Preparation

Preparation step involves cutting/slicing, which can expose tissue to oxidation leading to the losses of various nutrients, and sometimes browning. Cutting, shredding, chopping, and pulping of fruits and vegetables induce carotenoid oxidation.

12.5.2 Blanching

Blanching not only inactivates the enzymes, but also removes undesirable flavouring compounds in fruits and vegetables. For many vegetables, leaching of nutrients and pigments has been reported after blanching. Blanching may cause carotenoid losses, but the enzyme inactivation due to heat may reduce losses. The heating of spinach leaves caused degradation of chlorophylls, and the effect was higher during blanching as compared to steaming. Pheophytins were formed after blanching due to acid formation (Von Elbe and Schwartz 1996). Blanching resulted in significant loss of β-carotene and ascorbic acid content of carrots (Negi and Roy 2000a) and leafy vegetables (Negi and Roy 2000b). Losses in vitamin content as a result of blanching, freezing, and canning for several fruits and vegetables have been reported in the literature (Barrett 2007; Rickman et al. 2007a, 2007b).

12.5.3 Drying

The drying of fruits and vegetables may cause significant changes in chemical composition; however, the losses vary with fruit/ vegetable, pre-treatment, and type of dehydration method used. Fats although present in low amounts in vegetables are oxidatively degraded, and oxidation also diminishes odour and flavour. Amino compounds and carbohydrates interaction during drying result in a

Table 12.3 Changes in various chemical constituents during processing of fruits and vegetables

Sl. No.	Commodity	Changes in chemical constituents		Reference
1	Mango	Thermal processing	Alterations in aromatic compounds of pine (α-pinene, β-pinene, terpinolene cis-β-ocimene), herbs (β-phellandrene, phellandrene, eucalyptus), flowers (sabinene, β-myrcene, Para cymene, trans-β-ocimene) and citrus fruit (D-3-carene, α-terpinene, limonene, ϒ-terpinene) in Thai mango	Ounamornas et al. (2019)
			Thermal pasteurization of mango nectar followed by prolonged storage at 27–30 °C showed significant degradation of total carotenoids, ascorbic acid, and colour	Kumar et al. (2013)
		Enzyme treatment	Improved terpenoids by up to a factor of 10 and acetate esters up to a factor of 3 in mango wine, and reduced the formation of medium-chain fatty acids and ethyl esters	Kosseva et al. (2016)
2	Tomato	Baking	Oven baking at 160 °C for 20 min increased lycopene by 75%, β-carotene by 81%, and α-tocopherol by 32%, and the relative proportion of all (E)-lycopene decreased to 52.5% from 75.4%, while that of (5Z)-lycopene increased from 9.4% to 17.9% of total lycopene as compared to fresh tomatoes	Hwang et al. (2012)
			Baking of tomato slurry at 177 °C for 15 min retained 64.1% lycopene, and for 45 min retained 37.3% lycopene, whereas at 218 °C, 51.5% lycopene was retained after 15 min and 25.1% lycopene was retained after 45 min baking	Mayeaux et al. (2006)
		Cooking	Degradation of lycopene (50%) at 100 °C after 60 min, 125 °C after 20 min, and 150 °C after less than 10 min	
		Microwave heating	Microwave heating for 1 min caused degradation of 35.6% of lycopene	

(continued)

Table 12.3 (continued)

Sl. No.	Commodity	Changes in chemical constituents		Reference
		Frying	Frying at 145 °C and 165 °C for 1 min led to the degradation of 63.4 and 64.5% of lycopene, respectively.	
		Paste preparation	Loss of β-carotene (29%) Increase in lycopene (37%)	Abushita et al. (2000)
		Sauce preparation	Loss of lycopene (8%)	Seybold et al. (2004)
		Soup preparation	Loss of β-carotene (56%) Loss of lycopene (48%)	
		Juice canning	Loss of β-carotene (35%) Loss of lycopene (30%)	
3	Pineapple	Osmotically dehydrated	No significant effect on total soluble solid in dehydrated (4 h using sugar of 50 °Brix and sugar/salt solutions (47:3% w/w) followed by oven drying (60 °C, 27 h) was observed, however, osmotic dehydration prior to drying showed better retention of vitamin C	Fasogbon et al. (2013)
4	Cabbage and broccoli	Thermal processing	Destroys myrosinase completely that produces isothiocyanate compounds	Barrett (2007)
5	Broccoli, carrots, green beans, green peas, and spinach	Thermal processing or freezing	No change in fibre content	Barrett (2007)
		Blanching and freezing	Loss of B vitamins (20–60%)	
		Canning	Loss of vitamin B complex (7–70%)	
6	Broccoli	Thermal treatment	Loss of vitamin C (84%)	Murcia et al. (2000)
		Freezing	Limited loss of vitamin C (30%)	Howard et al. (1999)
7	Carrot	Canning	Loss of vitamin C (88%) Loss of β-carotene (50%)	Howard et al. (1999)
		Freezing	Loss of vitamin C (35%) Loss of β-carotene (36%)	
8	Green beans	Canning	Loss of vitamin C (63%) Loss of α- and β-carotene (17%)	Weits et al. (1970)
		Freezing	Loss of vitamin C (28%)	
9	Spinach	Canning	Loss of vitamin C (62%) Loss of α- and β-carotene (17%)	
		Freezing	Loss of vitamin C (61%)	

(continued)

Table 12.3 (continued)

Sl. No.	Commodity	Changes in chemical constituents		Reference
10	Peaches	Canning	Loss of β-carotene (50%) Loss of β-cryptoxanthin (40%)	Lessin et al. (1997)
11	Green peas	Cooking	Loss of sodium (50%) Loss of potassium (44%) Loss of calcium (42%)	Wills et al. (1984)
12	Table beets	Canning	Loss of vitamin C (10%) Increase in total phenolics (5%)	Jiratanan and Liu (2004)

darker colour of the product and development of new aroma substances. Vitamin levels may also reduce drastically during drying and dehydration. β-carotene and the B vitamins remain intact during drying, whereas vitamin C is lost to a great extent. Carotenoids are very sensitive to oxidation which results in colour loss and destruction of vitamin A activity (Namitha and Negi 2010). A drastic reduction in β-carotene, ascorbic acid, and chlorophyll contents of savoy beet, amaranth, and fenugreek leaves by various drying methods was observed, and the losses were least in low temperature drying (Negi and Roy 2000b). The green coloured chlorophyll changes to olive green and brown (pheophytin) due to heat and this conversion is favoured by acidic pH (Von Elbe and Schwartz 1996).

12.5.4 Canning

Carotenoids are fairly resistant to heat, changes in pH and water leaching as they are fat-soluble in nature. The nutritional value in terms of protein and carbohydrate content of fruits and vegetables is not diminished by heating, however, vitamin content is affected. Carotenes are moderately destroyed (5–30%) during retorting. Several studies show that the loss of carotenoids is by isomerization of trans-forms to cis-forms (Namitha and Negi 2010).

Vitamin B_1 in carrots and tomatoes does not decrease significantly, and losses for green beans, peas, and asparagus are to the tune of 10–50%. A high loss (66%) in the content of Vitamin B_1 in spinach are attributed to the large surface area. Vitamin B_2 is lost by leaching during blanching (5–25%), however, further canning process does not aggravate these losses. Nicotinic acid losses during canning are similar to Vitamin B_2 losses. Vitamin C losses are due to its water solubility and enzymatic, heat and chemical degradation, and losses up to 88% during the canning of broccoli, carrot, peas, spinach, and green beans are reported (Rickman et al. 2007a).

Conversion of pheophytin to pyropheophytin occurs during a high heat treatment via decarbomethoxylation. Chlorophyll B is thermally more stable than Chlorophyll A. At the temperature range of 70–100 °C, degradation of green pea Chlorophyll A is reported to be 12–18 times faster than that of Chlorophyll B. Gaur et al. (2006) also summarized that the Chlorophyll B is more stable than Chlorophyll A, and their degradation follows first order kinetics. The thermal degradation of chlorophyll and chlorophyllides results in the formation of the pheophorbides, pyropheophorbides, pheophytins, and pyropheophytins resulting in loss of green pigment (Von Elbe and Schwartz 1996).

12.5.5 Freezing

Freezing preserves nutrients in fruits and vegetables to a great extent. Carotenes are stable in spinach, peas, and beans, however, a moderate reduction in carotenes in asparagus was observed after blanching, freezing, frozen storage, and thawing. The primary processing steps (washing, blanching) affect the losses in the B complex vitamins. However, other freezing steps have no effect on B vitamins. The water or steam blanching of vegetables causes leaching of Vitamin C, however, during freezing and thawing no or insignificant losses occur. Not much change in vitamin E content in spinach and asparagus were noticed during freezing (Rickman et al. 2007b).

12.5.6 Fermentation/Pickling

The fermentation lowers the pH, and the acidic pH of the medium stabilizes vitamin C. Although, the original odour and flavouring substances of the raw material are retained to a large extent in pickled vegetables, development of new aroma is reported during fermentation. The conversion of chlorophyll to corresponding pheophytins is also affected by fermentation due to decrease in pH, which causes changes the colour of chlorophyll. In olives, phytofluene and ζ-carotene are lost, whereas β-carotene and lutein are stable during the fermentation and brine storage, however, the total pigment content remains unchanged (Minguez-Mosquera et al. 1989). Fermentation of fruits and vegetables not only preserves them for longer duration, but also brings about changes in various flavour components such as organic acids and diacetyls (Swain et al. 2014).

12.5.7 Irradiation

Low doses of irradiation increase the content of phenolic compounds in fruits and vegetables, while high doses may cause damage. The accumulation of phenolic compounds is the result of the activation of plant defense against the photo-biological stress. Fruits and vegetables treated with small doses of γ-irradiation (0.25–1.5 kGy) show a dose-dependent increase in the content of total phenolic compounds. Higher doses (>1 kGy) can prevent the onset of fruit ripening. Extensive pulp softening due to activation of pectic enzymes has also been reported by irradiation (Olson 1998). Strawberries treated with up to 900 Gy dose showed very little change in ascorbic acid and anthocyanin content immediately after treatment as well as during 9 days storage at 10 °C (Maraei and Elsawy 2017).

12.5.8 Non-thermal Processing

The non-thermal food processing technologies like pulsed electric fields, ozone, ultrasound, pulse electric field (PEF), and high-pressure processing (HPP) are aimed at curtailing the activity of microorganisms and preserving nutritional quality besides retarding chemical changes due to various enzymes that would otherwise adversely affect the edible quality of foods (Chauhan 2019). The rate of a chemical reaction in a foodstuff processed using non-thermal processing depends on several factors, such as reactant concentration and mobility, pH, oxidation-reduction potential, inhibitors and catalysts as similar to thermal or other processing methods (Erkmen and Faruk Bozoglu 2016). The PEF processing of orange juice inactivated (>80%) pectin methylesterase and peroxidase, however, retained its sugars and ascorbic acid better than conventional treatments (Elez-Martinez et al. 2006). Aguilo-Aguayo et al. (2009) also reported similar effect of PEF on polygalacturonase and pectin methyl esterase activity in strawberry, tomato, and watermelon juices. Degradation of pigments in orange juice (Tiwari et al. 2008) and anthocyanins content of the blackberry juice (Tiwari et al. 2009) by ozone treatment have been reported, however, carotenoids were not changed significantly in sugarcane juice after ozone treatment (Garud et al. 2018). Rajashri et al. (2019) reported a decrease in ascorbic acid, total phenolics, and total flavonoids of tender coconut water after ozone as well as ultrasound treatments.

HPP is used commercially for inactivation of microorganisms and denaturation of enzymes to increase the shelf life of fruit and vegetable products and retain their nutritional and nutraceutical values (Arora and Chauhan 2019). HPP has minimal effect on low molecular weight compounds, and therefore it retains flavour compounds to a greater extent than other processing techniques. HPP is also known to preserve the vitamin and bioactive constituents of various fruits and vegetable products and enhance their bioaccessibility (Oye et al. 2008; Cilla et al. 2020). HPP has been used to modify the proteins (Sim et al. 2019), gelatinize the starches, and enhance the infusion of nutrients in various fruits and vegetables (Balakrishna et al. 2020).

12.6 Conclusion

Fruits and vegetables not only provide nutrition, but also impart numerous health benefits to consumers due to the presence of various phytochemicals, however, a substantial produce is lost due to poor handling and storage operations. Fruit ripening is a genetically programmed biochemical and physiological processes that culminates in desirable changes in the fruit's texture and sensory quality. However, uncontrolled ripening hastens the senescence and makes the produce inedible. Control of ripening process using better storage techniques and genetic modification may help in controlling undesirable changes occurring due to ripening, and may help in keeping produce more aesthetically sound for longer duration and delay the undesirable changes. Processing of fruits and vegetables is done to add value and obtain a diverse range of convenience products for off-season/ off place availability. Optimized processing techniques can help in retaining various nutrients and phytochemicals in fruits and vegetables. Further, use of non-thermal processing techniques looks promising as most of the chemical changes happening during processing are accelerated due to use of heat.

References

Abushita AA, Daood HG, Biacs PA (2000) Change in carotenoids and antioxidant vitamins in tomato as a function of varietal and technological factors. J Agric Food Chem 48:2075–2081

Aguilo-Aguayo I, Soliva-Fortuny R, Martin-Belloso O (2009) Avoiding non enzymatic browning byhigh-intensity pulsed electric fields in strawberry, tomato and watermelon juices. J Food Eng 92:37–43

Arora SK, Chauhan OP (2019) High pressure processing: principles and engineering aspects. In: Chauhan OP (ed) Non-thermal food processing technologies. CRC Press, Boca Raton, FL, pp 1–9

Arthur E, Oduro I, Kumah P (2015) Effect of maturity stage and postharvest calcium chloride treatment on the quality and storage life of tomatoes (*Lycopersicon esculentum* Mill). J Postharvest Technol 3:74–81

Asghari M, Aghdam MS (2010) Impact of salicylic acid on post-harvest physiology of horticultural crops. Trends Food Sci Technol 21(10):502–509

Balakrishna AK, Wazed MA, Farid M (2020) A review on the effect of high pressure processing (HPP) on gelatinization and infusion of nutrients. Molecules 25:2369. https://doi.org/10.3390/molecules25102369

Barrett DM (2007) Maximizing the nutritional value of fruits & vegetables. Food Technol 61(4): 40–44

Ben-Arie R, Or E (1986) The development and control of husk scald on wonderful pomegranate fruit during storage. J Am Soc Hortic Sci 111(3):395–399

Bernardes SAP, Nascimento JRO, Lajolo FM et al (2008) Starch mobilization and sucrose accumulation in the pulp of Keitt mangoes during postharvest ripening. J Food Biochem 32: 384–395. https://doi.org/10.1111/j.1745-4514.2008.00175.x

Bonina-Noseworthy J, Brent Loy J, Curran-Celentano J et al (2016) Carotenoid concentration and composition in winter squash: variability associated with different cultigens, harvest maturities, and storage times. Hortic Sci (51, 5):472–480

Bourtoom T (2008) Edible films and coatings: characteristics and properties. Int Food Res J 15(3): 237–248

Bubba DM, Giordani E, Pippucci L et al (2009) Changes in tannins, ascorbic acid and sugar content in astringent persimmons during on-tree growth and ripening and in response to different postharvest treatments. J Food Comp Anal 22(7–8):668–677

Bufler G (2009) Exogenous ethylene inhibits sprout growth in onion bulbs. Annals Botany 103(1): 23–28

Chauhan OP (ed) (2019) Non-thermal food processing technologies. CRC Press, Boca Raton, FL

Chhatpar HS, Mattoo AK, Modi VV (1971) Biochemical studies on chilling injury in mangoes. Phytochemistry 10(5):1007–1009

Chidtragool S, Ketsa S, Bowen J et al (2011) Chilling injury in mango fruit peel: cultivar differences are related to the activity of phenylalanine ammonia lyase. Postharvest Biol Technol 62(1): 59–63

Cilla A, Rodrigo MJ, de Ancos B et al (2020) Impact of high-pressure processing on the stability and bioaccessibility of bioactive compounds in Clementine mandarin juice and its cytoprotective effect on Caco-2 cells. Food Funct 11:8951–8962

Das PR, Eun JB (2020) Removal of astringency in persimmon fruits (*Diospyros kaki*) subjected to different freezing temperature treatments. J Food Sci Technol. https://doi.org/10.1007/s13197-020-04818-3

Deshpande AB, Chidley HG, Oak PS et al (2016) Data on changes in the fatty acid composition during fruit development and ripening of three mango cultivars (Alphonso, Pairi and Kent) varying in lactone content. J Data Brief 9:480–491

Elavarasan M, Premalatha A (2019) A review: nutrient deficiencies and physiological disorders of citrus. J Pharmacog Phytochem 8(4):1705–1708

Elez-Martinez P, Soliva-Fortuny RC, Martin-Belloso O (2006) Comparative study on shelf life of orange juice processed by high intensity pulsed electric fields or heat treatment. Eur Food Res Technol 222:321–329

Elyatem SM, Kader AA (1984) Post-harvest physiology and storage behaviour of pomegranate fruits. Scientia Hortic 24(3–4):287–298

Erkmen O,Faruk Bozoglu T (eds) (2016)Food Microbiology: Principles into Practice (Volume 1 : Microorganisms Related to Foods, Foodborne Diseases, and Food Spoilage). John Wiley & Sons, Hoboken, NJ

Fasogbon BM, Gbadamosi SO, Taiwo KA (2013) Studies on the chemical and sensory properties of jam from osmotically dehydrated pineapple slices. Curr J Appl Sci Technol 14:1327–1335

Garud SR, Priyanka BS, Rastogi NK et al (2018) Efficacy of ozone and lactic acid as nonthermal hurdles for preservation of sugarcane juice. Ozone Sci Eng 40:198–208

Gaur S, Shivhare US, Ahmed J (2006) Degradation of chlorophyll during processing of green vegetables: a review. Stewart Postharvest Review 5:14

Giovannoni JJ (2004) Genetic regulation of fruit development and ripening. Plant Cell 16(Suppl): S170–S180. https://doi.org/10.1105/tpc

Haguluha V, Natera E (2007) Quality management of fresh produce from the highlands of Papua New Guinea: a postharvest manual. Australian Centre for International Agricultural Research (ACIAR)

Handa AK, Mattoo AK (2010) Differential and functional interactions emphasize the multiple roles of polyamines in plants. Plant Physiol Biochem 48(7):540–546

Hassan A, Wills R, Atan RM et al (2010) Blackheart disorder in fresh pineapple. New trends in postharvest Management of Fresh Produce II. Fresh Produce 4:29–35

Holcroft D (2015) Water relations in harvested fresh produce. PEF White paper (15-01):16

Howard LA, Wong AD, Perry AK et al (1999) β-Carotene and ascorbic acid retention in fresh and processed vegetables. J Food Sci 64:929–936

Hsu YL, Chen SC, Lin KW et al (2018) Quarantine vapor heat treatment of papaya fruit for *Bactrocera dorsalis* and *Bactrocera cucurbitae* (Diptera: Tephritidae). J Eco Entomol 111(5): 2101–2109

Hwang ES, Stacewicz Sapuntzakis M, Bowen PE (2012) Effects of heat treatment on the carotenoid and tocopherol composition of tomato. J Food Sci 77(10):C1109–C1114

Imsabai W, Ketsa S, van Doorn W (2006) Physiological and biochemical changes during banana ripening and finger drop. Postharvest Biol Technol 39(2):211–216

Ingle M, D'souza MC (1989) Physiology and control of superficial scald of apples: a review. HorticSci 24(1):28–31

Instituto Colombiano de Bienestar Familiar (ICBF) (2015). Tabla de composición de alimentos Colombia. Bogotá SAS (ed) Bogotá: Universidad Nacional de Colombia, pp 1–321

Itai A (2015) Watercore in fruits. In: Kanayama Y, Kochetov A (eds) Abiotic stress biology in horticultural plants. Springer, Tokyo, pp 127–145. https://doi.org/10.1007/978-4-431-55251-2_10

Jahurul MHA, Zaidul ISM, Ghafoor K et al (2015) Mango (*Mangifera indica* L.) by products and their valuable components: a review. Food Chem 184:173–180. https://doi.org/10.1016/j.foodchem.2015.03.046

Jiang Y, Gong NN, Matsunami H (2014) Astringency: a more stringent definition. Chem Senses 39(6):467

Jiratanan T, Liu RH (2004) Antioxidant activity of processed table beets (*Beta vulgaris* var. conditiva) and green beans (*Phaseolus vulgaris* L.). J Agric Food Chem 52:2659–2670

Kader AA, Barrett DM (1996) Classification, composition of fruits, and postharvest maintenance of quality. Processing fruits. Sci Technol 1:1–24

Konsue W, Dethoup T, Limtong S (2020) Biological control of fruit rot and anthracnose of postharvest mango by antagonistic yeasts from economic crops leaves. Microorganisms 8(3): 317

Kosseva MR, Joshi VK, Panesar PS (eds) (2016) Science and technology of fruit wine production. Academic Press, Cambridge

Kumar R, Bawa AS, Kathiravan T et al (2013) Thermal processing of mango nectar (*Mangifera indica*) and its effect on chemical, microbiological and sensory quality characteristics. Int J Adv Res 1(8):261–273

Lattanzio V, Kroon PA, Quideau S et al (2008) Plant phenolics—secondary metabolites with diverse functions. Recent Adv Polyphenol Res 1:1–35

Lessin WJ, Catigani GL, Schwartz SJ (1997) Quantification of cis–trans isomers of provitamin A carotenoids in fresh and processed fruits and vegetables. J Agric Food Chem 45:3728–3732

Luengwilai K, Beckles DM, Siriphanich J (2016) Postharvest internal browning of pineapple fruit originates at the phloem. J Plant Physiol 202:121–133

Luengwilai K, Beckles DM, Roessner U et al (2018) Identification of physiological changes and key metabolites coincident with postharvest internal browning of pineapple (*Ananas comosus* L.) fruit. Postharvest Biol Technol 137:56–65

Luyckx A, Lechaudel M, Hubert O et al (2016) Banana physiological post-harvest disorders: a review. MOJ Food Process Technol 3(1):226–231

Maraei RW, Elsawy KM (2017) Chemical quality and nutrient composition of strawberry fruits treated by γ-irradiation. J Radiation Res Appl Sci 10:80–87. https://doi.org/10.1016/j.jrras.2016.12.004

Matthais JP (1995) Factors contributing to internal breakdown of Fuji apples. Washington State University Tree Fruit Postharvest J 6:3–4

Mayeaux M, Xu Z, King JM et al (2006) Effects of cooking conditions on the lycopene content in tomatoes. J Food Sci 71(8):C461–C464

Minguez-Mosquera MI, Garrido-Fernandez J, Gandul-Rojas B (1989) Pigment changes in olives during fermentation and brine storage. J Agric Food Chem 37:8–11

Murcia MA, Lopez-Ayerra B, Martinez-Tome M et al (2000) Evolution of ascorbic acid and peroxidase during industrial processing of broccoli. J Sci FoodAgric 80:1882–1886

Namitha KK, Negi PS (2010) Chemistry and biotechnology of carotenoids. Crit Rev Food Sci Nutr 50(8):728–760

Negi PS, Roy SK (2000a) Effect of low cost storage and packaging on quality and nutritive value of fresh and dehydrated carrots. J Sci Food Agric 80:2169–2175

Negi PS, Roy SK (2000b) Effect of blanching and drying methods on β-carotene, ascorbic acid and chlorophyll retention of leafy vegetables. LWT-Food Sci Technol 33:295–298

Ohlendorf B (1999) Integrated pest management for apples & pears. University of California, Agriculture and Natural Resources

Oliveira JG, Morales LMM, Silva WM (2018) Postharvest physiological disorders in papaya. In: Tonetto de Freitas S, Pareek S (eds) Postharvest physiological disorders in fruits and vegetables (innovations in postharvest technology series), 1st edn. CRC Press, Boca Raton, pp 399–422

Olson DG (1998) Irradiation of food. Food Technol 52(1):56–62

Ounamornas P, Page P, Sommano S (2019) Aromatic profile changes of Thai mangoes during heat processing. Acta Hortic 1244:181–186

Oye I, Van der Plancken I, Van Loey A et al (2008) Does high pressure processing influences nutritional aspects of plant based food systems? Trends Food Sci Technol 19:300–308

Paltrinieri G, Staff FA (2014) Handling of fresh fruits, vegetables and root crops: a training manual for Grenada. http://www.fao.org/3/a-au186e.pdf

Pantastico EB (1975) Post harvest physiology, handling and utilization of tropical and subtropical fruits and vegetables. AVI Pub Co Inc, Westport
Paull R (1996) Ethylene, storage and ripening temperatures affect dwarf Brazilian banana finger drop. Postharvest Biol Technol 8:65–74
Phan C, Hsu H, Sarkar SK (1973) Physical and chemical changes occurring in the carrot root during storage. CanJ Plant Sci 53:635–641
Prabath Pathirana UA, Sekozawa Y, Sugaya S et al (2013) Changes in lipid oxidation stability and antioxidant properties of avocado in response to 1-MCP and low oxygen treatment under low-temperature storage. Int Food Res J 20(3):1065–1075
Rajashri K, Roopa BS, Negi PS et al (2019) Effect of ozone and ultrasound treatments on polyphenol content, browning enzyme activities and shelf life of tender coconut water. J Food Process Preserv. https://doi.org/10.1111/jfpp.14363
Ram RA, Rahim MA, Alam MS (2020) Diagnosis and management of nutrient constraints in mango. In: Srivastava AK, Chengxiao H (eds) Fruit crops. Elsevier, Amsterdam, pp 629–650
Rashmi HB, Negi PS (2020) Phenolic acids from vegetables: a review on processing stability and health benefits. Food Res Int. https://doi.org/10.1016/j.foodres.2020.109298
Raymond L, Schaffer B, Brecht JK et al (1998) Internal breakdown in mango fruit: symptomology and histology of jelly seed, soft nose and stem-end cavity. Postharvest Biol Technol 13(1): 59–70
Rickman JC, Barrett DM, Bruhn CM (2007a) Nutritional comparison of fresh, frozen and canned fruits and vegetables I. Vitamins C and B and phenolic compounds. J Sci Food Agric 87:930–944
Rickman JC, Bruhn CM, Barrett DM (2007b) Nutritional comparison of fresh, frozen and canned fruits and vegetables II. Vitamins A and carotenoids, Vitamin E, minerals and fiber. J Sci Food Agric 87:1185–1196
Saleem-Dar M, Oak P, Chidley H et al (2016) Nutrient and flavor content of mango (*Mangifera indica*L.) cultivars: an appurtenance to the list of staple foods. In: Simmonds MSJ, Preedy VR (eds) Nutritional composition of fruit cultivars, 1st edn. Elsevier, Amsterdam, pp 445–468. https://doi.org/10.1016/B978-0-12-408117-8.00019-2
Sawamura M, Kuriyama T, Li Z (1988) Rind spot, antioxidative activity and tocopherols in the flavedo of citrus fruits. J Hortic Sci 63(4):717–721
Seybold C, Frohlich K, Bitsch R et al (2004) Changes in contents of carotenoids and vitamin E during tomato processing. J Agric Food Chem 52:7005–7010
Shivashankar S (2014) Physiological disorders of mango fruit. In: Janick J (ed) Horticultural reviews, vol 42, pp 313–347
Sim SY, Karwe MV, Moraru CI (2019) High pressure structuring of pea protein concentrates. J Food Process Eng 42(7):e13261
Smith MJ, Bucher G (2012) Tools to study the degradation and loss of the N-phenyl carbamate chlorpropham-a comprehensive review. Environ Int 49:38–50
Solovchenko A, Yahia EM, Chen C (2019) Pigments. In: Yahia EM (ed) Postharvest physiology and biochemistry of fruits and vegetables. Woodhead Publishing, Cambridge, pp 225–252
Swain MR, Anandharaj M, Ray RC et al (2014) Fermented fruits and vegetables of Asia: a potential source of probiotics. Biotechnol Res Int. https://doi.org/10.1155/2014/250424
Tabar SM, Tehranifar A, Davarynejad GH et al (2009) Aril paleness, new physiological disorder in pomegranate fruit (*Punica granatum*): physical and chemical changes during exposure of fruit disorder. Hortic Environ Biotechnol 50(4):300–307
Tharanathan RN, Yashoda HM, Prabha TN (2006) Mango (*Mangifera indica* L.), the king of fruits—an overview. Food Rev Int 22(2):95–123
Tiwari BK, Muthukumarappan K, O'Donnell CP et al (2008) Kinetics of freshly squeezed orange juice quality changes during ozone processing. J Agric Food Chem 56(15):6416–6422
Tiwari BK, O'Donnell CP, Muthukumarappan K et al (2009) Anthocyanin and colour degradation in ozone treated blackberry juice. Innov Food Sci Emerg Technol 10(1):70–75

United States Department of Agriculture, Agricultural Research Service (2018) USDA National nutrient database for standard reference, release 1 April, Nutrient Data Laboratory Home Page, https://ndb.nal.usda.gov/ndb/

Von Elbe JH, Schwartz SJ (1996) Colorants. In: Fennema OR (ed) Food chem, 3rd edn. Marcel Dekker Inc, New York, pp 651–722

Watada AE (1986) Effects of ethylene on the quality of fruits and vegetables. Food Technol 40(5): 82–85

Wei Y, Zhang Q, Zheng Y (2000) Mechanism of puffiness and the methods of control in citrus fruit: I. the rule of early-maturing variety citrus and its physiological characters. J Hunan Agric Univ 26(4):267–270

Weits J, van der Meer MA, Lassche JB et al (1970) Nutritive value and organoleptic properties of three vegetables fresh and preserved in six different ways. IntJ Vitam Nutr Res 40:648–658

Wills RBH, Golding JB (2016) Postharvest: an introduction to the physiology and handling of fruit and vegetables. New South Publishing, Randwick

Wills RB, Evans TJ, Lim JS et al (1984) Composition of Australian foods. 25. Peas and beans. Food Technol Aust 36:512–514

Yang CP, Fujita S, Kohno K et al (2001) Partial purification and characterization of polyphenol oxidase from banana (*Musa sapientum* L.) peel. J Agric Food Chem 49(3):1446–1449

Zhou Y, Pan X, Qu H et al (2014) Low temperature alters plasma membrane lipid composition and ATPase activity of pineapple fruit during blackheart development. J Bioenerg Biomembr 46(1): 59–69

Chemistry of Milk and Milk Products 13

C. T. Manoj Kumar, O. P. Chauhan, M. H. Sathish Kumar, and H. C. Devaraja

13.1 Introduction

Milk is a complex biological fluid secreted by mammals as a sole component of diet for neonates including human infants for a considerable period of time. Milk may be defined as the lacteal secretion obtained by the complete milking of one or more healthy animals, excluding that obtained within 15 before and 5 days after parturition. The principal constituents of milk are fat, protein, lactose (carbohydrate), minerals and vitamins. The composition of milk is variable with days of lactation and species as nutritional requirement specific to age of young mammal and species. During early stage of lactation, milk (colostrum) carries a greater number of immune cells and antibodies (immunoglobulins IgA, IgG and IgM) to protect newborn from disease and infection. As neonate matures, the nutritional requirement leans towards improving physical strength of the body rather than immune. It has been seen that composition of milk varies between species; for example, fat, protein, lactose differ from 1 to 50%, 1 to 20% and 0 to 10%, respectively. Recently, it has also noticed that size and composition of the components of the milk vary with species as nutritional and digestion capacity varies. The milk composition also changes markedly with season (summer, winter and rainy), diet, physical stress and disease (mastitis) condition. Milk is an excellent source of nutrients for all the age groups of human beings, particularly children and adolescents during the period of their intense growth and development. Due to increase in nutritional awareness and disposable income, there is a great demand for milk and milk products worldwide. The global dairy market reached a value of US$ 718.9 billion in 2019. Dairy sector has been

C. T. Manoj Kumar (✉) · M. H. Sathish Kumar · H. C. Devaraja
ICAR-National Dairy Research Institute, Adugodi, Bengaluru, India
e-mail: manoj.ct@icar.gov.in

O. P. Chauhan
Defence Food Research Laboratory, Siddarthanagar, Mysore, India

O. P. Chauhan (ed.), *Advances in Food Chemistry*,
https://doi.org/10.1007/978-981-19-4796-4_13

hailed to play an important role in achieving food security, reducing global poverty, generating employment opportunities for women and providing a regular source of income for rural households. World milk production is expected to be 859 mT in 2020, maximum contribution comes from Asia (367 mT) followed by Europe (225.7 mT), North America (109 mT), South America (61 mT), Africa (47 mT), Oceania (30 mT) and Central America (FAO 2020). India is the largest milk producer in the world records 192 mT in 2019 and this is expected to grow by 5 mT in 2020 (FAO 2020). Out of total milk production, cow milk contribution is 81%, buffalo milk is 15% and 4% for goat, sheep and camel milk combined (FAO 2020). Although milk is highly sensitive and perishable in nature, it has been greatly explored to produce thousands of varieties of dairy products in worldwide such as cheese, yoghurt, milk powder, milk concentrates, ice cream, butter, butter oil (ghee), caseinate, whey protein concentrate, traditional dairy products, etc. During processing, it will undergo several processes such as pasteurization, sterilization, chilling, etc. Understanding the structural, functional, and nutritional properties of each milk component is highly necessary to manifest during processing of milk and milk products.

13.2 Composition of Milk

Milk carries an average of 87.1% water, 4.0% fat, 4.6% lactose, 3.25% protein and 0.7% minerals, 0.17% organic acids and 0.15% miscellaneous compounds. The fat is mostly made up of triglycerides, which constitute nearly 97–98.5% of the total lipid followed by di- and monoglycerides represent in traces. Milk fat contains wide range of fatty acids varying in chain length from C_4 to C_{26}. In addition to that, milk fat comprises phospholipids, sterols, carotenoids, fat soluble vitamins (A, D, E and K) and traces of fatty acids. Lactose is the typical reducing disaccharide present only in milk. It is the major source of energy and involved in brain development for neonates. Milk also contains a negligible amount of sialic acid derivative such as n-acetyl neuraminic acid, which acts as prebiotic. Milk proteins constitute of distinctive casein and whey proteins, casein is the major milk protein representing 80% of total proteins, the rest comprises with whey proteins. The milk proteins also constitute minor fractions of protease-peptone and enzymes. The mineral substances present in milk are Ca, K, Mg, Na, Cl, phosphorous and sulfur. Organic acids are present in ionized or salts form such as citrate. The tree of structural elements of milk is presented in Fig.13.1.

13.2.1 Factors Affecting the Composition of Milk

Milk composition is most dynamic in nature; it is affected by various factors associated with milch animals such as species, breed, genetic, stage of lactation, age, physiological factors, disease and environmental factors. These factors may show individual effects or they may be interdependent. It possesses major challenge

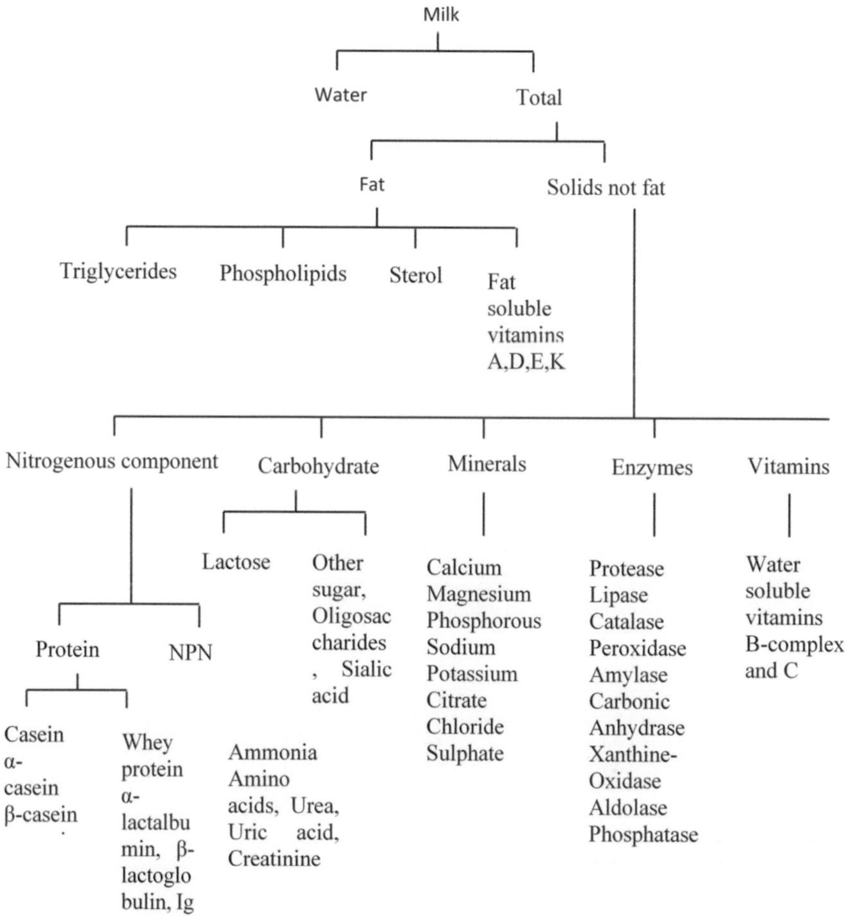

Fig. 13.1 Tree of structural elements of milk

for the dairy industry to maintain homogenous composition of milk every day and may seriously affect economy of milk producer and processor.

13.2.1.1 Species

Milk composition may vary widely with different species/mammals and also may vary periodically with the effect of other factors (Table 13.1). It has been reported that the composition of milk from different species (studied among 150 species) may have dry matter content varying from 8% to 65%, fat 0% to 53%, protein 1% to 19%, carbohydrates 0.1% to 10% and ash 0.1% to 2.3% (Walstra 1999). The change in concentration of milk constituents can be related to requirements of specific species' neonates. It depends on their growth rate and digestion capacity. In addition to that, the organoleptic quality of milk also varies with species, for example, goaty flavour in goat milk due to presence of caproic, caprylic and capric fatty acid.

Table 13.1 Composition of milk from different species

Constituents (%)	Cow	Buffalo	Sheep	Goat	Camel (Bactrian)	Yak	Donkey
Moisture	88.1	83.2	80.1	87.3	84.8	82.6	90.8
Protein	3.2	4.0	6.2	3.4	3.9	5.2	1.6
Fat	3.3	7.4	7.9	3.8	5.0	6.8	0.7
Carbohydrates	5.1	4.4	4.9	4.1	4.2	4.6	6.4
Ash	0.7	0.8	0.9	0.8	0.9	0.8	0.4

Source: Medhammar et al. (2012), Park et al. (2007)

Table 13.2 Composition of milk from different breeds of cow

Constituents (%)	Sahiwal	Kankrej	Tharpakar	Crossbred	Jersey	HF
Total solids	13.99	13.79	14.22	13.69	14.73	11.99
Protein	3.60	3.49	3.92	3.58	3.80	3.15
Fat	4.23	4.88	4.37	3.91	5.14	3.45
Carbohydrates	5.38	4.76	5.35	5.39	5.04	4.65
Ash	0.78	0.76	0.58	0.81	0.75	0.68

Source: Sarkar et al. (2006), Kapadiya et al. (2016), Rai (1980)

Table 13.3 Composition of milk from different breeds of buffalo

Breed	Fat (%)	Total protein (%)	Casein (%)	Solids not fat (%)	Total solids (%)
Bhadawari	7.43	3.92	3.16	8.99	17.70
Mehsana	6.46	3.87	3.07	9.13	15.59
Murrah	7.53	4.03	3.20	9.00	16.53
Surti	6.17	3.93	3.11	8.80	14.96

Source: Misra et al. (2008)

13.2.1.2 Breed

Milk composition may differ between different breeds, between herds of same breed and also greater variation between individual animal of same breed. This variation can be due to both individual genotype and environmental factors. The composition of various breeds of cows and buffaloes is summarized in Tables 13.2 and 13.3. These breeds are from different geographical areas around the world; therefore, any variation in the composition may also be attributed to manage mental and environmental factors.

13.2.1.3 Effect of Different Quarters of Udder

The variations in milk composition can be seen in different quarters of udder, it could be attributed to anatomical structure difference. Kramer et al. (2013) reported that the milk of front udder quarters was found to have significantly higher fat and protein content than milk of rear udder quarter, whereas lactose showed opposite trend (Table 13.4). However, earlier Skrodel (1936) reported that fat content didn't show any significant variation among quarters of udder. The milk yield was higher in hindquarters than forequarters, two forequarters yielded equal amounts of milk,

Table 13.4 Composition of milk from different quarters of udder

Component (%)	Front left	Front right	Rear left	Rear right
Fat	3.73	3.73	3.49	3.51
Protein	4.04	4.05	4.01	3.99
Lactose	4.56	4.48	4.61	4.60

Source: Kramer et al. (2013)

but the right hindquarter yielded more milk than the left. These variations will be disappeared during successive lactation period, machine milking and with different feeding pattern.

13.2.1.4 Stage of Lactation

This is very important factor which affects composition of milk in greater extent. The fat, protein, sodium and chloride content decreased during early stage of lactation (5–60 days) and then gradually increase till end of the lactation. Lactose content remains flat during early stage and then decreases with increase in lactation period. The calcium content decreases to minimum level during early lactation period, remains flat till end mid lactation period (61–210 days) and then increases till end of the lactation period (>210 days). Potassium content increases slightly till mid lactation period, then decreases till end of the lactation.

13.2.1.5 Nutritional and Management Factors

Feeding is one more important factor which has a profound effect on milk composition, but it is indeed that the relationship between feed constituents and milk composition is complex. Energy intake has major effect on SNF and yield of milk. Higher the energy intake, higher will be the SNF content and yield of milk. This is the most common response seen when level of feeding increased. High grain feeding is the most common in US and Europe conditions, which increase the milk yield with composition, remained essentially unchanged. However, higher grain intake depresses the fat content and also alters fatty acid composition. It typically reduces the proportions of milk fatty acids having 6–16 carbons and increases the proportion of 18-carbon unsaturated fatty acids. Grain feeding also showed impact on production of the trans-10 fatty acid isomers by ruminal microorganisms. High concentrate ratio with low roughage/forage increases SNF and protein content, decreases fat content, and milk yield will remain unchanged. Increase in protein content in the diet increases non-protein nitrogen content but not protein content. But high energy ratio with added nitrogen may be more likely to increase protein content in milk. Underfeeding of protein with recommended energy intake will reduce yield of milk without affecting milk protein and SNF content. Milk protein content also varies with the presence of rapidly fermentable dietary carbohydrates. The physical state of ratio, size of roughage, and feeding intervals also affect milk yield and composition. Fat feeding often accompanied by a decline in milk protein content, where the casein fraction declined the most. The fat feeding has limited ability to alter milk fatty acid composition. Because it undergoes bio hydrogenation process, where ruminal microflora converts the excess unsaturated fatty acids to saturated by removing double bonds in a fatty acyl chain. Feeding rumen protected fats or bypass fats, for example, calcium salts of fatty acids, formaldehyde protected oils, protein shell embedded fats, etc., can modify the level of unsaturated fatty acids in milk fat.

13.2.1.6 Disease Conditions

Mastitis is one of the most common diseases in bovines, which leads to severe economic losses to both dairy farming and processing industry. After reviewing literatures, it is clearly reported that there is a rise in the level of whey proteins (particularly in serum albumin and IgG) and sodium content and decrease in casein and lactose content. Rise in serum albumin and IgG could be attributed to inflammatory reactions and presence of pathogens in the udder. However, variation in total protein content, fat and calcium due to mastitis is conflicting in nature. It is also seen that a rise in free fatty acids content attributed to lipolytic activity in udder.

13.3 Physico-Chemical Properties of Milk

Physico-chemical properties of milk depend on composition, structure of components, processing parameters and temperature. The knowledge of these properties is highly important during chemical and biological analysis, determination of milk quality and adulteration, process and engineering design.

13.3.1 The pH, Titratable Acidity and Buffering Properties of Milk

The pH and acidity of milk are influenced by presence of lactose, casein, ions, CO_2, etc. These components vary during the lactation period which eventually influences the pH and acidity of milk. The pH and acidity of normal milk vary from 6.4–6.6 and 0.14–0.16%, respectively. The pH of colostrum (immediately after parturition) and late lactation milk is ~6.0 and up to 7.5 which is different from normal milk. The reduced pH of colostrum milk is due to increased concentration of protein, dihydrogen phosphate, citrate and carbon dioxide. The increase in pH in late lactation milk could be due to increase in Na^+, Cl^- and reduction in the lactose content. The pH and acidity also tend to vary during processing such as heating, concentrating, drying, fermentation, etc. Milk has excellent buffering capacity due to presence of components such as soluble phosphate, colloidal calcium phosphate, citrate, carbonate and proteins. Buffering capacity of colostrum is greater than that of normal milk due to presence of more amounts of milk proteins.

13.3.2 Density and Specific Gravity of Milk

Density (ρ) is defined as mass per unit volume and its unit is kg m^{-3}. Specific gravity is dimensionless quantity; it is defined as ratio of density of product to water. The density of milk is solely depending on the composition and temperature. Specific gravity has the advantage that its numerical value is independent of the units used for density, and its temperature dependence is much lower than that of density. Density is inversely proportional with fat content and temperature of milk and directly proportional with SNF content of milk. The density of whole milk and skim milk is varying from 1.027 to 1.033 and 1.0320 to 1.0365 g/cm^3, respectively. Density and specific gravity are very useful properties while determining adulteration of milk with water and other additives.

13.3.3 Redox Potential

Redox potential (Eh) indicates the potential of oxidation and reduction reactions in milk. Eh of normal milk which is in equilibrium with air varies from +0.25 to +0.35 V at 25 °C and pH 6.6 to 6.7. The Eh of milk is dependent on the concentration of O_2 and Cu^{2+}, temperature, exposure to light and bacterial activity. These factors initiate the oxidation of redox systems present in the milk such as lactate-pyruvate, ascorbate and riboflavin. Normally milk contains 0.3 mmol/L of O_2 after equilibrium with air and complete removal of O_2 lowers the Eh to about −0.12 V. The major components of milk other than water, i.e., fat, lactose and protein, have no effect on Eh of milk.

13.3.4 Surface Tension

The surface tension of milk is a fundamental physical property that reflects the stability of foams and emulsions. It is the cohesive force on the liquid causing reduction in the active surface area. This phenomenon arises when any liquid molecules lack in either hydrophobic or hydrophilic moieties on its surface. Hence, addition of surface-active agents such as surfactants reduces surface tension due to presence of both hydrophobic and hydrophilic groups in it. Milk contains several surface-active components such as casein micelles, phospholipids, whey proteins and fatty acids; salts and lactose do not contribute to surface tension. Generally, surface tension of milk varies from 40 to 60 mNm^{-1} (~52) at 20 °C. Surface tension of milk is influenced with several factors such as fat content, temperature, homogenization and measurement technique.

13.4 Milk Constituents

13.4.1 Carbohydrates

Lactose is the major carbohydrate present typically in milk of most mammals. Milk contains other carbohydrates in trace amount associated with proteins and cerebrosides such as glucide compounds like hexosamines and N-acetylneuraminic acid. These are the oligosaccharides having prebiotic effect present mainly in human

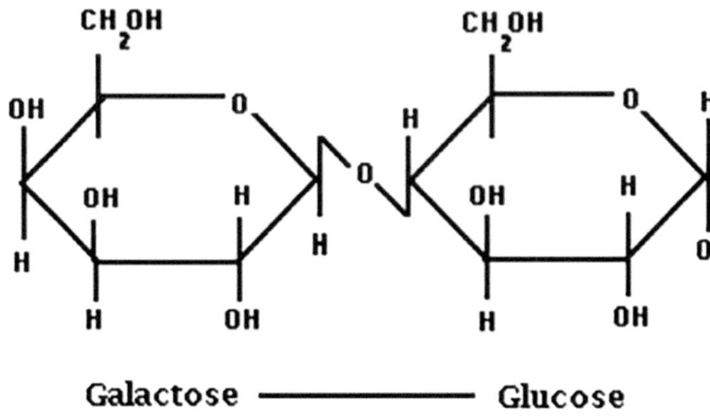

Fig. 13.2 Chemical structure of lactose

milk for neonatal brain development (Urashima et al. 2001). Lactose is a reducing disaccharide comprised of glucose and galactose, linked by a β-1-4-O-glycosidic bond (Fig.13.2). Lactose is a reducing sugar, commonly exists in two anomers form, α and β, which possess significantly different properties and can change from one form to another (mutarotation). It is because of functional aldehyde group at the C-1 position of the glucose moiety exists in the hemiacetal form and, consequently, C-1 is a chiral, asymmetric carbon. Both anomers have different crystallization behaviour, α-lactose crystallizes as a monohydrate, while crystals of β-lactose are anhydrous. The α-lactose is less soluble (7 g/100 mL) than β-lactose (50 g/100 mL) at 20 °C. Hence, α-lactose crystallizes easily due to poor solubility compared to β-lactose. However, α lactose solubility is much more temperature dependent than that of β-lactose and the solubility curves intersect at approximately 93.5 °C (Hui et al. 2008). Above 93.5 °C, α-lactose solubilizes quickly and β-lactose crystallizes from an aqueous solution. Crystals of α-lactose are very hard, slightly hygroscopic, often fairly large, and dissolve slowly, whereas crystals of β-lactose are not very hygroscopic and dissolve quickly. The α-lactose crystals commonly form tomahawk shape kind of crystals. Crystallization of α-lactose happens because of two conditions, that is, when α-lactose dissolved more than 7 g/100 mL and disturbance in α:β ratio of 37:63 giving a final solubility of about 18.2 g/100 mL in aqueous solution. Comparatively, lactose has very less sweetness than other sugars (16% as sweet as sucrose in a 1% solution). This limitation is advantageous to lactose to use it as bulking agent and to modify colour when a high level of sweetness is undesirable. Lactose being a reducing sugar can participate in Maillard reactions; the aldehyde group of lactose reacting with epsilon group of lysine results in release of browning compounds.

Table 13.5 Lipids in milk

Lipid class	Percentage in milk fat (w/w)
Neutral glycerides	98.7
Triglycerides	98.3
Diglycerides	0.3
Monoglycerides	0.03
Free fatty acids	0.1
Phospholipids	0.8
Cholesterol	0.30
Cholesterol esters	0.02
Carotenoids + Vit. A	0.002

Source: Walstra (1999)

Table 13.6 Fatty acid profile of cow and buffalo milk

Component (mole %)	Fatty acid	Cow milk	Buffalo milk
Butyric acid	4:0	8.8	11.4
Caproic	6:0	3.5	3.1
Caprylic	8:0	2.0	1.0
Capric	10:0	3.0	1.6
Lauric	12:0	3.8	2.6
Myristic	14:0	9.9	10.6
Palmitic	16:0	26.1	30.3
Stearic	18:0	9.1	10.5
Higher saturation	20–26.0	1.0	0.7
Unsaturation (mono)	10:1–14:1	1.8	1.0
Lower	16:1	3.6	3.6
Unsaturated (poly)	18:1	26.2	21.6
	Other	3.5	2.0

Source: Mathur et al. (2005)

13.4.2 Milk Fat

Fat or lipid is the second major component in milk, and it is the major factor during pricing of milk. Milk fat normally presents in globule form with variable size of 0.1–20 μm and surrounded by tri-layer fat globule membrane (FGM) of 15 nm thick which maintains fat in emulsion form in milk. Milk fat mainly composed of triglyceride 98.3% and phospholipids 0.8%. Other components such as diglycerides, monoglycerides, sterols, fat soluble vitamins and other fats present in traces (Table 13.5). Most of glycerides present in core of fat globules, FMGs largely composed of phospholipids which act as emulsifiers.

Milk fat mainly composed of 4–18 even numbered carbon chain unbranched fatty acids. The level of short chin fatty acids (4–10 carbon chain) is relatively very high in milk fat compared to any other source of lipid. Particularly, butyric acid (C-4) can be found only in milk. The proportion of saturated fatty acids such as myristic, palmitic and stearic acids is very high in milk fat (~63% w/w). Hence, milk fat solidifies easily at room temperature. The oleic acid occupies maximum level (about 70%) in unsaturated fatty acid category (Table 13.6).

Table 13.7 Physico-chemical properties of cow and buffalo milk fat

Physico-chemical properties	Cow milk fat	Buffalo milk fat
Softening point (°C)	34.3–36.3	33.5–35.9
Melting point (°C)	33.4–46.4	31.5–35.2
Acid value	0.17–0.35	0.26
Refractive index	1.4515–1.4533	1.4498–1.4530
BR reading	41.00–43.50	41.05–42.40
Saponification number	218.23–236.10	221.0–238.0
Iodine value	27.00–33.90	27.70–37.32
Reichert-Meissl value	27.83–35.50	24.60–29.70
Polenske value	0.70–1.60	1.30–1.80
Density (g.mL)	0.905–0.917	0.888–0.911
Unsaponifiable matter (mg/100 mL)	392–398	414–450

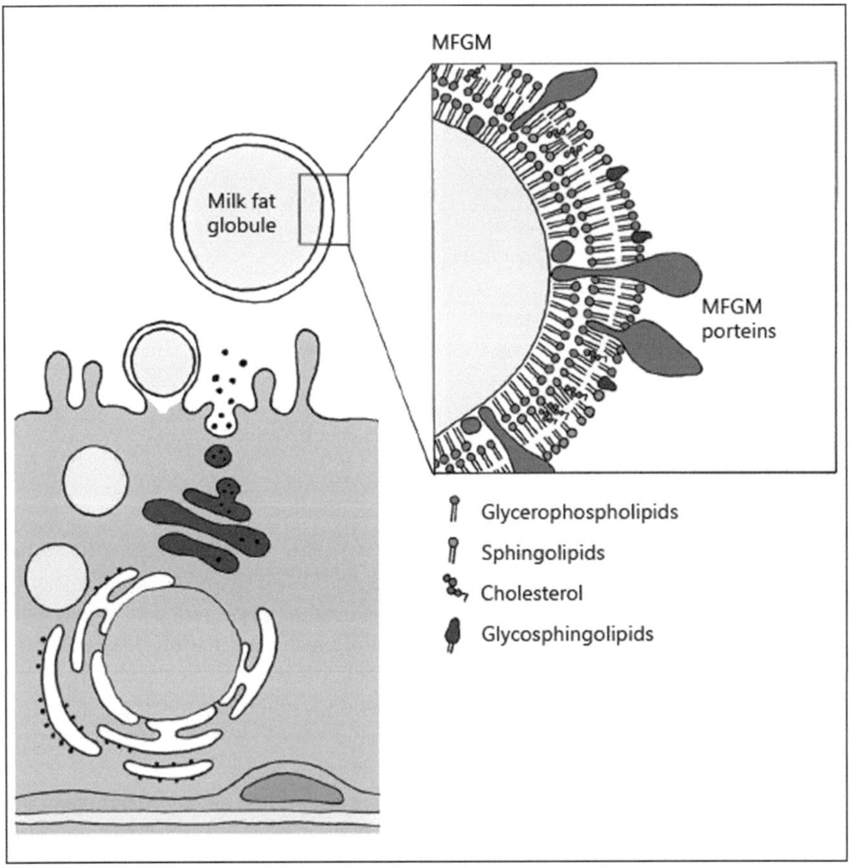

Fig. 13.3 Milk fat globule membrane (*Source:* Hernell et al. 2016)

Milk fat has unique physico-chemical properties which are considered during evaluating the quality and adulteration of milk fat (Table 13.7).

FGM components contain phospholipids, glycolipids (cerebrosides and gangliosides), glycerides, free fatty acids, cholesterol and proteins (Fig.13.3). The lipid and protein content of FGM is approximately around 72% and 22%, respectively. FGM proteins are mucin 1, mucin 15, CD36, butyrophilin, lactadherin, xanthine oxidoreductase, adipophilin and fatty acid binding protein. The phospholipids belong to class of polar lipids containing phosphorous in it. The polarity is due to presence of both lipid (hydrophobic tail) and proteins (hydrophilic tail) which helps in emulsification of fat in milk. Phospholipids majorly contain mono-unsaturated fatty acids such as oleic acid followed by stearic, docosatetraenoic, myristic and arachidonic acids. Milk phospholipids (MPLs) consist of a subclass of polar lipids, namely glycerophospholipids and sphingolipids. Glycerophospholipids comprise a glycerol moiety with two fatty acids esterified at positions sn-1 and sn-2 and a hydroxyl group at sn-3 position, linked to a phosphate group and a polar moiety. The molecular structure of the latter determines the types of glycerophospholipids, namely phosphatidylcholine (PC), phosphatidylserine (PS), phosphatidylethanolamine (PE), phosphatidylinositol (PI), phosphatidylglycerol (PG), and phosphatidic acid (PA). Sphingolipids consist of a sphingosine backbone (2-amino-4-octadecene-1,3-diol) connected to a fatty acid via an amide bond and a polar head. Sphingomyelin (SM), a prominent subclass of sphingolipids, has a phosphocholine residue.

The melting point of milk fat varies from −30 to +40 °C due to wide range of fatty acids having different melting point. The fatty acids having shorter chain length and greater number of double bonds will have lower melting point. The melting point of milk fat also depends on distribution of fatty acid residues over three positions of triglycerides. The asymmetric fatty acid distributed triglyceride has a lower melting point than symmetric one with same fatty acid residues. Crystallization of milk fat is very complicated phenomenon due to its wide range of fatty acids. The triglycerides can crystallize into three different forms, i.e., α, β′ and β. Each form of crystal has characteristic crystal lattice shape and melting point based on type of fatty acid in three positions. The α form of crystal forms at low temperature and has relatively short life than other crystal forms. During heating, milk fat releases aromatic compounds such as methyl ketones, γ- and δ- lactones, which are partly responsible for characteristic flavour of milk fat.

13.4.3 Milk Proteins

Milk proteins, i.e., caseins and whey proteins, have long been known for their nutritional and technological values. The quality of milk proteins in terms of digestibility and bioavailability are superior to proteins from other sources (Table 13.8). They are considered as important constituents of the human diet since they comprise a principal source of nitrogen and essential amino acids

Table 13.8 Quality ranking of proteins from different sources

Protein type	Protein efficiency ratio	Biological value	Net protein utilization	Protein digestibility corrected amino acid score
Beef	2.9	80	73	0.92
Casein	2.5	77	76	1.00
Egg	3.9	100	94	1.00
Milk	2.5	91	82	1.00
Soy protein	2.2	74	61	1.00
Wheat gluten	0.8	64	67	0.25
Whey protein	3.2	104	92	1.00

Source: Hoffman and Falvo (2004)

Table 13.9 Amino acid profile of casein and whey proteins

Particulars	Casein (g/kg)	Whey (g/kg)
Alanine	31	48
Arginine	38	23
Aspartic acid	73	102
Cysteine	4	12
Glutamic acid	223	172
Glycine	19	20
Histidine	32	16
Isoleucine	58	84
Leucine	101	105
Lysine	83	91
Methionine	30	16
Phenylalanine	54	31
Proline	105	61
Serine	63	52
Threonine	46	62
Tryptophan	14	21
Tyrosine	58	24
Valine	74	60

Source: Hall et al. (2003)

(Table 13.9). Milk proteins possess good techno-functional properties such as solubility, water holding, emulsification, foaming, gelling and film formation.

The details pertaining to different milk proteins are summarized in Table 13.10.

13.4.3.1 Casein

As per American dairy science association (ADSA), the caseins in milk can be defined as those phosphoproteins that precipitate from raw skim milk by acidification to pH 4.6 at 20 °C. Caseins form complex with calcium phosphate which eventually represents large colloidal particles, the casein micelles. About 80–95%

Table 13.10 Summary of different proteins in bovine milk

Protein	Composition in skim milk (g/l)	Genetic variant	Molecular weight	Isoionic point	Isoelectric point
αs1-CN	12–15	B C	23,615 23,542	4.92–5.05 5.00–5.35	4.44–4.76 . . .
αs2-CN	3–4	A	25,226	. . .	. . .
β-CN	9–11	A^1 A^2 B	24,023 23,983 24,092	5.41 5.30 5.53	. . . 4.83–5.07 –
κ-CN	2–4	A B	19,037 19,006	5.77 (5.35) 6.07 (5.37)	5.45–5.77 5.3–5.8
β-Lactoglobulin	2–4	A B	18,363 18,277	5.35 5.41	5.13 5.13
α-Lactalbumin	0.6–1.7	B	14,178	–	4.2–4.5
Serum albumin	0.4	A	66,399	5.13	4.7–4.9
Immunoglobulin G1	0.3–0.6	. . .	161,000	–	5.5–6.8
Immunoglobulin G2	0.05	. . .	150,000	–	7.5–8.3
Immunoglobulin A[7]	0.01	. . .	385,000–417,000	. . .	
Immunoglobulin M	0.09	. . .	1,000,000	. . .	. . .
Secretary component	0.02–0.1	. . .	63,750	7.48	. . .
Lactoferrin	0.02–0.1	. . .	76,110	8.95	8.81

Source: Farrell et al. (2004)

of casein is in the form of micelles, which play a major role in stability and various properties of milk and milk products. Colloidal calcium phosphate comprised of calcium, magnesium, phosphate, and citrate, which represents 6% of whole casein micelle. The shape of casein micelles is spherical, ranging in size from 50 to 500 nm in diameter (average about 120 nm) and a molecular mass from 106 to 109 Da. Casein micelles are highly porous in nature and can hold water of 3.7 g/g of casein. Being colloidal particles, it has the ability to scatter light; therefore, the white colour in milk is mainly because of light scattering by casein micelles. Casein consists of several principal components: αs1, αs2, β, and κ-casein, that covers 40, 10, 45 and 5%, respectively, of colloidal micelles stabilized by calcium phosphate bridging (Table 13.11). The limited proteolysis of β-casein by plasmin leads to production of γ_1, γ_2 and γ_3-casein in trace amounts. αs-casein and β-casein are highly phosphorylated caseins, which carry 8–10 and 5 phosphoserine residues, respectively. Severe phosphorylation increases sensitivity to calcium salts and easily precipitates under excess calcium ion atmosphere. Unlike other caseins, κ-caseins

Table 13.11 Approximate composition of casein micelles

Component	Content (g/100 g of micelles)
αs1-CN	35.6
αs1-CN	9.9
β-CN	33.6
κ-CN	11.9
Minor caseins	2.3
Calcium	2.9
Phosphate	2.9
Magnesium	0.1
Sodium	0.1
Potassium	0.3
Citrate	0.4
Sialic acid	0.3
Galactose	0.2
Galactosamine	0.2

Source: McMahon and Brown (1984)

are glycoproteins and they have only one phosphoserine group which makes it more stable in calcium ions rich atmosphere. As κ-casein present in surface of casein micelle, it plays an important role in protecting other caseins from precipitation. Casein is less sensitive to heat; heat treatment at 120 °C for more than 20 min causes casein to coagulate. However, casein is highly sensitive to pH and precipitates at its isoelectric pH.

The structure of casein micelle has been studied extensively and it is highly complex in nature. However, the exact structure of casein micelle is still under debate. Various models for have been proposed to understand the structure of casein micelle. All the proposed models can be classified into three categories: coat-core, submicelle and internal structure models. The model which is most accepted for understanding casein micelle structure is the submicelle model proposed by Walstra in 1999 (Fig.13.4). This model suggests that casein micelles are built of roughly spherical subunits or submicelles having diameter of 12–15 nm. Each submicelle contains 20–25 casein molecules, and their composition is variable. The submicelles are linked through hydrophobic interactions and calcium phosphate linkages. According to this model, two different kinds of submicelles will be there, one mainly consisting hydrophobic caseins such as αs- and β-caseins, which are buried in the centre of the submicelle and these submicelles present inside of the casein micelle. Another type consisting of αs- and κ-caseins, this submicelle is more hydrophilic in nature due to sugar residues on κ-caseins and these submicelles present more towards surface of casein micelles. The κ-caseins are always positioned on surface of the micelle with the hydrophilic part of the C-terminal end protruding from the micelle surface to form a "hairy" layer that will avoid further aggregation of submicelles by steric and electrostatic repulsion. This will keep micelles stable and avoids flocculation (Walstra 1999).

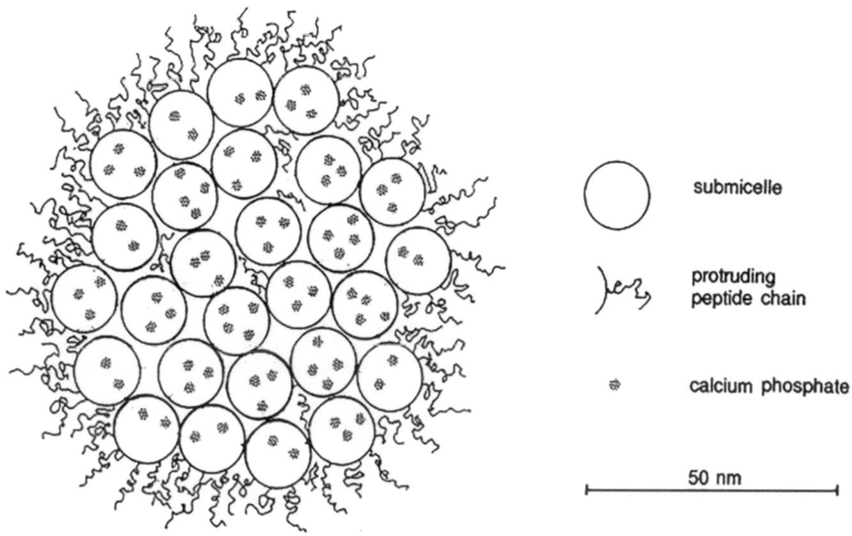

Fig. 13.4 Submicelle model of casein as proposed by Walstra (1999)

13.4.3.2 Whey Proteins

The term whey proteins has been used to describe the group of milk proteins that remain soluble in milk serum or whey after precipitation of CN at pH 4.6 and 20 °C (Farrell Jr et al. 2004). The components of whey proteins are β-LG, α-LA, serum albumin, Ig, lactoferrin, lysozyme and proteose peptone fractions. Most whey proteins have globular confirmation, relatively high hydrophobicity and compactly folded peptide chains. Unlike casein, the whey proteins contain secondary, tertiary and quaternary structures. Hence, whey proteins highly sensitive to heat treatment and get denatured easily up on heating. The denaturation does not result in flocculation, but the proteins precipitate onto the casein micelles and remain dispersed. β-Lactoglobulin is the major whey protein, carries 162 amino acid residues, very hydrophobic similar to casein, but it contains no ester-bound phosphate and fairly little proline. It has two S-S linkages and one free sulfhydryl group, which is very reactive. These reactive sulfhydryl groups involved in formation of β-Lg-κ-CN complex after heating to 90 °C. It is sensitive to pH and ionic strength, but it does not precipitate on acidification of milk. β-Lg is classified as lipocalins due to having similarities with retinol binding proteins. It comprises eight stranded anti-parallel β-sheet (A-H) called as calyx, the outer surface of calyx is flanked by three-turn α-helix and a ninth β-strand forms dimer interface along with the loop connecting strands A and B (Fig.13.5).

The α-lactalbumin is second major whey protein containing 123 amino acid residues. It consists of two domains: a large α-helix and a small β-sheet, which are linked by calcium binding loop (Fig. 13.6). It is structurally identical to lysozyme,

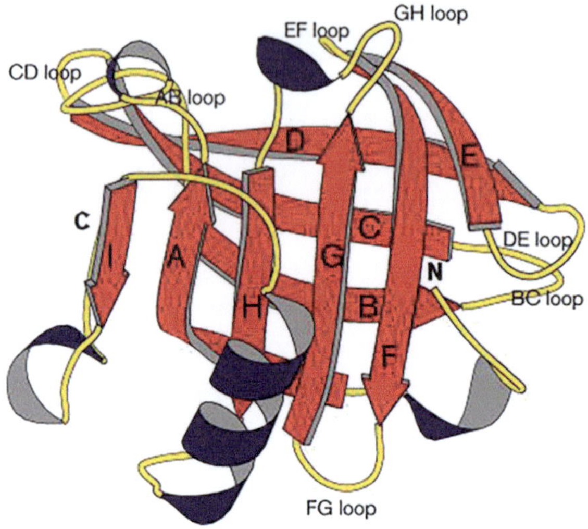

Fig. 13.5 Ribbon diagram of one monomer of bovine β-Lg (*Source:* Brownlow et al. 1997)

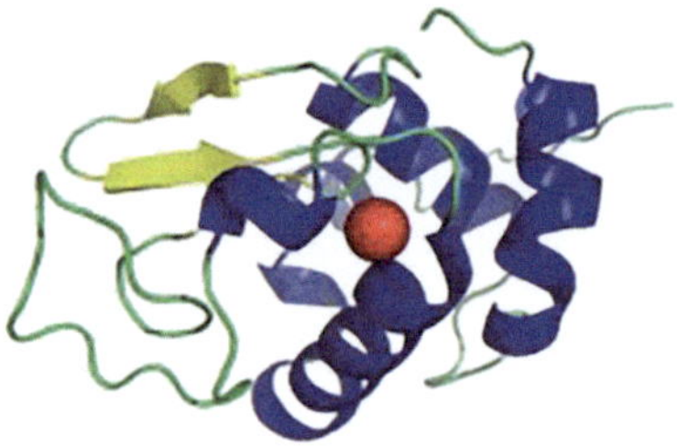

Fig. 13.6 α-LA structure (*Source:* Park et al. 2015)

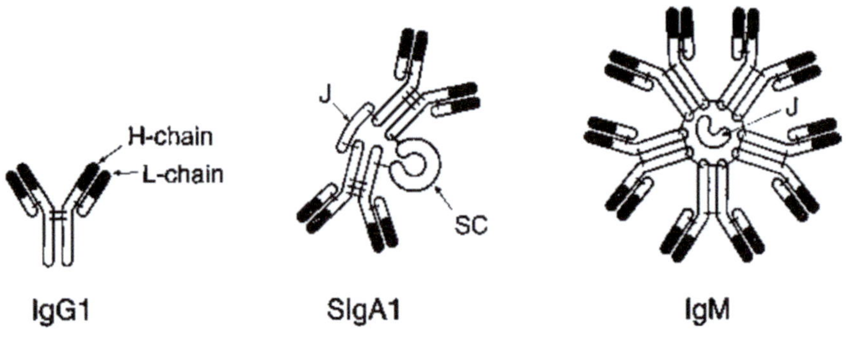

Fig. 13.7 Immunoglobulins G1, A1 and M

but it has no bactericidal effect. It is involved as coenzyme during synthesis of lactose. It is compactly folded, spherical, slightly pH- and salt-dependent.

Bovine serum albumin is present in very low level in milk, i.e., 1–2% of milk protein. It has many double sulfide linkages and α-helix in it. Immunoglobulins are antibodies which act against specific antigens. They are distinguished into various classes: IgG (gammaglobulins), IgA and IgM (macroglobulins). IgG molecules consist of two polypeptide chains: heavy (H) and light (L) (Fig. 13.7). Light chain linked to heavy chain by a disulfide bond, and heavy chains are bonded together by a disulfide bond. The molecule has two identical reactive sites to which antigen binds. IgM is a heavy molecule compared to IgG, it consists of a pentamer of IgG-like molecules joined by a so-called J component (Fig. 13.7).

Protcose peptone is heat and acid stable, phosphoglycoprotein mainly associated with caseins. Hence, proteose peptone will not present in rennet whey, but it is present in acid whey. Lysozyme and lactoferrin are the enzymes having antibacterial activity. Lysozyme attacks on muramic acid of bacterial cell wall, whereas lactoferrin inhibits bacteria by chelating iron ions.

13.4.4 Milk Salts, Vitamins and Trace Elements

Milk contains both organic and inorganic salts (Table 13.12). Milk salts exist in both colloidal and soluble forms. The major portion of salts are present in colloidal form with casein such as calcium, magnesium, phosphate and citrate. These salts play a major role in maintaining milk stability. The stability of milk due to colloidal particles arises with equilibrium between dissolved salts. The major dissolved salts of milk are phosphate, citrate, chloride, sulfate, bicarbonate, sodium, potassium, magnesium and calcium. Several factors affect the salt equilibrium such as temperature, acidification, CO_2 content, total solid content of milk and addition of sequestering agents. Increase in processing temperature will increase the diffusion of salts from soluble to colloidal state. Acidification or decrease in pH increases diffusion of salts from colloidal to soluble form. Increase in CO_2 content in milk leads to increase in soluble salt content. When milk is concentrated, the salts equilibrium will be disturbed by increase in the soluble salt content, it will lead to diffusion of salts to colloidal state to maintain equilibrium. Addition of sequestering

Table 13.12 Salt content in milk

Constituent	Buffalo milk (mg/100 g)	Cow milk (mg/100 g)
Sodium	44.63	58.0
Potassium	103.0	140.0
Calcium	175.5	123.3
Magnesium	19.25	11.1
Phosphorous (Total)	97.6	95.1
Phosphorous (inorganic)	–	–
Chloride	62.84	104.5
Sulfate	–	10
Carbonate (as CO_2)	–	20
Citrate (citric acid)	163.6	176.0

Source: Mathur et al. (2005)

Table 13.13 Vitamins in milk

Vitamins	Concentration (per litre)
Vitamin A	1590 I.U.
Vitamin D	22.1 I.U.
Vitamin E	1.0 mg
Vitamin K	0.04 mg
Thiamine	0.4 mg
Riboflavin	1.5 mg
Pantothenic acid	3.0 mg
Biotin	50 μg
Niacin	0.2–1.2 mg
Pyridoxine	0.7 μg
Folic acid	1.0 μg
Cyanocobalamin	7.0 μg
Ascorbic acid	20 mg
Inositol	180 mg
Choline	150 mg

Source: Mathur et al. (2005)

Table 13.14 Trace elements in milk

Element	Average value
Chlorine	119 mg/100 mL
Sulfur	30 mg/100 mL
Iron	3 ppm
Zinc	3 ppm
Silicon	2 ppm
Copper	0.3 ppm
Fluorine	0.15 ppm
Aluminium	Traces
Manganese	Traces
Iodine	Traces
Boron	Traces

Source: Mathur et al. (2005)

agents such as EDTA and silicates into milk affects salts equilibrium by chelating ions.

Vitamins are essential nutrients required for human beings which cannot be in the body and obtained from different food sources. Milk is a good source of most of the vitamins essentially required for human beings (Table 13.13). Vitamins can be distinguished into two categories based on their solubility: Fat soluble vitamins such as A, D, E and K; water soluble vitamins such as B-complex vitamins include thiamine, riboflavin, niacin, pantothenic acid, pyridoxine, biotin, choline, inositol, folic acid, p-amino benzoic acid, cyanocobalamin and ascorbic acid (vitamin C).

Milk contains several elements in trace level which include metals and metalloids. Major trace elements are boron, copper, iron, zinc and sulfur (Table 13.14). These trace elements may enter in milk through feeds, utensils and equipment and contamination. Some trace elements associated with casein micelle, fat globule membrane and lactoferrin enzyme.

13.5 Milk Products

13.5.1 Liquid Milk Processing

Liquid milk undergoes various steps of processing such as chilling, preheating, cream separation, homogenization, pasteurization, sterilization and ultra-high temperature (UHT) processing. Cream separation process is to remove the excess fat in milk by gravity or centrifugal method. It is based on density difference between fat (0.93 g/cm^3) and skim milk (1.035 g/cm^3). In gravity method, the cream layer is formed on the milk surface when it is allowed to stand for several hours at the velocity as given by Stokes law:

$$V = \frac{2G}{9}\left(\frac{ds - df}{n}\right)r^2$$

where V = velocity at which fat globule rise, G = acceleration due to gravity, ds = density of skim milk, df = density of fat, r = radius of fat globule, n = viscosity of skim milk.

As per Stokes' law, velocity is increased by an increase in the radius of fat globule, increase in density of skim milk and fat and decrease in viscosity of skim milk. In addition to that, temperature also affects fat separation during both gravity and centrifugal method, normally cream is separated at 45 °C.

In centrifugal method, milk is subjected to centrifugal force since fat is being lighter moves towards centre of cream separator and skim milk towards periphery of the cream separator. The velocity at which fat globule separates is as per following formula:

$$V = r^2 \frac{(ds - df)}{n} N^2 RK$$

where V = velocity of fat globule movement, r = radius of fat globule, ds = density of fat globule, df = density of fat, N = speed of bowl (rpm), R = distance of fat globule from the axis of rotation, K = constant, n = velocity of skim milk.

Homogenization is the process to prevent formation of cream layer on milk by reducing size of the fat globule to less than 2 μ. Milk is normally homogenized at two stages with pressure of 2000 psi and 500 psi at first and second stage, respectively. At first stage, the size of fat globules reduced to less than 2 μ and at second stage, the aggregation of fat globules will be prevented. It is essential to maintain the temperature of milk to about 54–65 °C to maintain the fat in liquid phase and to inactivate the lipase enzyme.

Pasteurization is the process of heating milk to 72 °C for 15 s or 65 °C for 30 min to inactivate all the pathogenic organisms and about 90% of spoilage causing organisms. Pasteurization's major aim is to inactivate pathogenic organisms such as *Coxiella burnetii*, *Mycobacterium tuberculosis* and *Mycobacterium paratuberculosis* in milk. Pasteurization has insignificant effect on most of the nutrients present in milk. However, it will cause denaturation of few heat sensitive whey proteins, inactivation of alkaline phosphatase, lipase enzymes and affects few vitamins such as riboflavin (vitamin B_2).

Milk subjected to sterilization and UHT process should have excellent heat stability or very high heat coagulation time (HCT). The heat stability or HCT is determined by subjecting milk to 140 °C or concentrated milk to 120 °C until particles of coagulated proteins are observed in the milk. Heating of milk to extreme temperature leads to the following changes:

(a) During heating at high temperature, the pH of milk reduces due to production of organic acid from lactose and release of H^+ ions during dephosphorylation.
(b) Precipitation of calcium phosphate into hydroxyapatite.
(c) Initiates Maillard reaction between aldehyde group of lactose and epsilon group of lysine.
(d) Heating at extreme temperature for prolong period leads to liberation of non-protein nitrogen.
(e) Dissociation of κ-CN-β-lg complex from casein micelle, which may be reversible or irreversible process.
(f) It may also lead to reduction of zeta potential of casein micelles.

In UHT processing, milk is heated for about 140 °C for few seconds followed by aseptic packaging to avoid contamination in milk. The UHT processed milk comparatively has more brightness because of complex formation between κ-CN-β-lg which eventually increases size of casein micelles. The gelation is the major problem in UHT milk during storage. It may be due to either proteolytic activity or due to chemical changes. The extracellular lipases and proteases produced by psychrotrophic bacteria during raw milk storage induce hydrolysis of proteins and subsequently lead to gelation of milk. On the other hand, the dissociation κ-CN-β-lg complex slowly initiates protein aggregation which subsequently leads to gelation of milk.

Freezing of milk causes some reversible and irreversible changes in structure of milk constituents. Normally, milk starts freezing at −0.54 °C; therefore, the continuous phase of milk forms pure ice and the rest of milk, thus becomes concentrated. As concentration increases, the ionic activity also increases which consequently salt equilibrium of milk affected. After thawing of milk, micellar casein aggregation may happen due to casein salting out. This coagulation is partly irreversible due to diffusion of calcium phosphate to colloidal state and with stirring sometime may redisperse the aggregated casein. However, these changes do not occur during quick freezing of milk below −23 °C. Freezing with subsequent thawing of milk also causes clumping of fat globules and in cream, fat globule membrane will damage due to mechanical action of ice crystals.

13.5.2 Butter

Butter is a fat rich dairy product which contains 80% of fat, 16% moisture, 2.5% of salt (for table butter) and 1.5% of curd particles. Butter is made by churning of cream. Cream is chilled overnight before churning to keep fat in solid state. If fat is present in liquid state, then churning will take long time and huge fat loss can be seen in butter milk. Hence, cream should be aged at 4 °C before churning and temperature of 9–11 °C should be maintained during churning. The mechanism behind formation of butter during churning of cream has been explained by three theories. Phase reversal theory by Fisher and Hookers proposed that churning involves phase reversal of cream. Conversion of fat-in-water type of emulsion, i.e., cream to water-in-fat type of emulsion, i.e., butter is called as phase reversal. During churning, mechanical agitation leads to clumping of fat globules until reduction in surface area to volume ratio to critical level. The fat globules suddenly break and yield butter grains and free butter milk. This theory lacks in explaining the presence of intact fat globules in butter.

Rahn's foam theory proposed that foam with air bubbles dispersed in the cream involved in butter production. The fat globules in clumps encircled the air bubbles, the air bubbles collapse due to increase pressure from fat globules leads to production of butter grains. However, this mechanism doesn't suitable for continuous butter making process.

The King's modern theory falls between the above two theories. It is proposed that fat globules present in clumps and part of the fat is in solid state upon cooling. The clumps of fat globules break during churning and lead to foam formation. The fat globules get concentrated over air bubbles and present in close contact with each other. Due to intense shaking and friction between fat globules leads to breakdown of fat globule membrane. The broken fat globules aggregate together and form large mass appearing like butter grains. The fat in granule is in globular form and encloses air and water in it. During working, the liquid released out of globules which helps in formation of solid block with emulsion of water and air.

During working, salt and moisture will be added into butter to maintain proper moisture content along with uniform distribution of salt. Salt is added only in table butter to give taste and to check growth of spoilage causing organisms.

13.5.3 Yoghurt

Yoghurt is a fermented dairy product obtained by incubating milk with *Lactobacillus bulgaricus* and *Streptococcus thermophilus*. Yoghurt is similar to the curd (*dahi*) prepared in South East Asian countries except the cultures used in preparation, flavour and acidity of the product. The final titratable acidity of yoghurt is 0.8–1.5%, and acetaldehyde is the major flavouring compound. Both cultures work symbiotically, *Streptococcus thermophilus* initiates the production of lactic acid and utilizes oxygen present in the milk. It creates slightly acidic and an oxygenic condition which is favourable for the growth of *Lactobacillus bulgaricus*.

Lactobacillus bulgaricus continues lactic acid fermentation and completes the yoghurt formation. Relatively, yoghurt is easier to digest due to presence of soluble minerals such as calcium and phosphate, free casein, peptides and amino acids.

13.5.4 Ice Cream

Ice cream is a frozen dairy product prepared by mixing milk, skimmed milk powder, cream or butter, sweetening agent, stabilizer and emulsifier with suitable flavours. The average composition of ice cream is given in Table 13.15.

Milk fat is a major component in ice cream which contributes to flavour and texture but impairs whipping ability. MSNF consists of proteins, lactose and minerals which improve the body and texture of ice cream. However, sometimes lactose causes sandiness during storage of ice cream. Stabilizers are used in ice cream to avoid formation of large sized ice crystals especially during storage by adsorbing free moisture. Emulsifiers assist to keep fat in emulsion form, to improve whipping ability and to give dry appearance to frozen ice cream.

Various processes involved in ice cream preparation such as homogenization, pasteurization, ageing, freezing and hardening. During ageing, the protein and stabilizers completely hydrated, it improves viscosity of mix and crystallization fat occurs. The freezing process involves lowering the temperature of mix to freeze portion of water in mix and beating of mix to incorporate air, to distribute the temperature uniformly and to reduce viscosity of mix by beating the gel. The ice cream mix incorporated with 90% of air, and air cell diameter of 60–100 μm gives satisfactory texture to ice cream. The added sugar and other salts depress the freezing point of mix and final temperature of mix after freezing is around −8 °C. After freezing and packing, ice cream is transferred for hardening, freezing process is continued during hardening until the temperature decreases to −18 °C.

13.5.5 Evaporated Milk

Evaporated milk is prepared by evaporation of water from milk till 26% TS using multiple effect evaporator under vacuum. During evaporated milk preparation, forewarming, pilot sterilization test and sterilization are the major processes.

Table 13.15 Composition of ice cream

Components	Average (%)
Weight in grams, per litre (min)	525
Total solids (min)	36
Milk fat (min)	10
Acidity (as lactic acid, max)	0.25
Sucrose (max)	15.0
Protein (min)	3.5
Emulsifier/stabilizer (max)	0.5

Forewarming process carried out at 95 °C/20 min or 120 °C for 2–3 min to enhance the heat stability of the evaporated milk, to reduce viscosity, inactivate enzymes, and kill microorganisms including bacterial spores. The heat stability of the evaporated milk is enhanced by diffusion of casein to colloidal phase and by formation of complex between κ-CN and β-lg. As stability of evaporated milk increases, the casein will not aggregate; hence, viscosity of the product will not increase. However, the role of κ-CN and β-lg complex in increasing heat stability is pH dependent. During pilot sterilization test, the disodium phosphate or trisodium citrate (stabilizers) is used at different concentration to check the heat stability of evaporated milk during sterilization. The right amount of stabilizer will be selected after observing no coagulation in sterilized evaporated milk. The sterilization of evaporated milk is carried out at 115–118 °C for 15 min. The age-thickening is the major problem in sterilized evaporated milk during storage. It happens due to aggregation of casein molecules which forms three-dimensional network looking like casein gel. It is because of storing evaporated milk in cold storage for long time and reduction in pH of evaporated milk.

13.5.6 Dried Milk

Dried milk is commonly prepared by spray drying of concentrated milk. Common processes during production of dried milks (whole milk powder and skim milk powder) are preheating, evaporation, homogenization, spray drying, separation of milk powder and agglomeration. The preheating is very important process because it can cause denaturation of whey proteins. The extent of denaturation is an important quality mark in connection with application of milk powder. The whey protein nitrogen index (WPNI) is generally used to classify milk powder according to intensity of heat treatment given during preparation (Table 13.16).

The low heat powder commonly used for reconstitution of milk and infant powder preparation. Medium heat powder can be used for milk beverage and ice cream preparation. High heat powder can be used for preparation of baked products. High heat treatment leads to generation of -SH groups from amino acids which act as antioxidants and prevent autoxidation in whole milk powder. It is necessary to maintain total solids content of concentrated milk to 45% to get good bulk density and reconstitution properties of prepared milk powder. The homogenization process is essential during preparation of whole milk powder to avoid free fat content. The presence of free fat affects several reconstitution properties such as wetting, flowability and oxidative stability of powder. During homogenization, the casein

Table 13.16 WPNI of milk powder

Milk powder	WPNI (g nitrogen/g of milk powder)	Heat treatment
Low heat	≥6	72 °C/15 s
Medium heat	>1.5–<6	75–80 °C
High heat	≤1.5	90 °C/5 min

particles incorporated into fat globule membrane to cover all the free fat. Atomization is aimed at forming droplets fine enough to dry quickly, but not so fine as to escape with the outlet air after having been dried. Nozzle and centrifugal type of atomizers are commonly used for spraying of milk. The nozzle type atomizer gives irregular size and rough surface milk powder particle, whereas centrifugal type atomizer gives uniform size and smooth surface powder particles. Generally, the milk particle size is maintained above 100 μ to avoid loss of powder through fine powders. The spray drying of milk powder commonly conducted at inlet temperature of 180–220 °C and outlet temperature of 85–90 °C. Generally, dried powder is recovered by using cyclone separators and bag filters. Agglomeration or instantization is done by combining two or more milk particles through fluid bed driers. In case of whole milk powder, lecithination is carried out during fluid bed drying to improve reconstitution properties of whole milk powder.

13.6 Conclusion

Milk is a lacteal secretion obtained from healthy milch animals. It is composed of lactose (carbohydrate), protein (casein and whey proteins), fat, minerals, vitamins, other trace elements and substantial amount of water (~87%). The composition of milk is affected by various factors such as species, breed, genetic, stage of lactation, age, physiological factors, disease and environmental factors. The composition of milk besides other factors plays a major role in governing physico-chemical properties of milk. The lactose being the major component of the milk gives sweetness flavour to milk. The fat along with fat globule membrane maintains emulsion stability in milk and gives characteristic flavour to the milk. The casein and whey proteins are the major proteins in milk. The casein plays an important role in maintaining stability of milk along with calcium, magnesium, phosphate and citrate. The structure of casein is highly complex in nature and it is still under debate. This chapter has elucidated the structure of both casein and major whey proteins. During milk processing and products preparation, the thermal processes such as pasteurization, sterilization, etc., cause denaturation, dephosphorylation and complex formation of proteins. Freezing of milk causes some reversible and irreversible changes in structure of milk constituents. The changes during preparation of milk products are discussed in depth in the present book chapter.

References

Brownlow S, Cabral JHM, Cooper R et al (1997) Bovine β-lactoglobulin at 1.8 Å resolution-still an enigmatic lipocalin. Structure 5(4):481–495

FAO (2020) Dairy market review. March 2020 FAO

Farrell HM Jr, Jimenez-Flores R, Bleck GT et al (2004) Nomenclature of the proteins of cows' milk—sixth revision. J Dairy Sci 87(6):1641–1674

Hall WL, Millward DJ, Long SJ et al (2003) Casein and whey exert different effects on plasma amino acid profiles, gastrointestinal hormone secretion and appetite. Br J Nutr 89(2):239–248

Hernell O, Timby N, Domellof M, Lonnerdal B (2016) Clinical benefits of milk fat globule membranes for infants and children. J Pediatr 173:S60–S65
Hoffman JR, Falvo MJ (2004) Protein–which is best? J Sports Sci Med 3(3):118
Hui YH, Nip WK, Nollet LM et al (eds) (2008) Food biochemistry and food processing. John Wiley & Sons, Hoboken, NJ
Kapadiya DB, Prajapati DB, Jain AK et al (2016) Comparison of Surti goat milk with cow and buffalo milk for gross composition, nitrogen distribution, and selected minerals content. Veterinary World 9(7):710
Kramer M, Erbe M, Bapst B, Bieber A et al (2013) Estimation of genetic parameters for individual udder quarter milk content traits in Brown Swiss cattle. J Dairy Sci 96(9):5965–5976
Mathur MP, Roy DD, Dinakar P (2005) Textbook of dairy chemistry. Directorate of Information and Publication of Agriculture, India
McMahon DJ, Brown RJ (1984) Composition, structure, and integrity of casein micelles: a review. J Dairy Sci 67(3):s499–s512
Medhammar E, WijesinhaBettoni R, Stadlmayr B et al (2012) Composition of milk from minor dairy animals and buffalo breeds: a biodiversity perspective. J Sci Food Agri 92(3):445–474
Misra SS, Sharma A, Bhattacharya TK et al (2008) Association of breed and polymorphism of α-s1and α-s2casein genes with milk quality and daily milk and constituent yield traits of buffaloes (Bubalus bubalis). Buffalo Bull 27:294–301
Park YW, Juarez M, Ramos M et al (2007) Physico-chemical characteristics of goat and sheep milk. Small Ruminant Res 68(1–2):88–113
Park YJ, Kim KH, Lim DW et al (2015) Effects of pH and protein conformation on in-solution complexation between bovine α-lactalbumin and oleic acid: binding trend analysis by using SPR and ITC. Process Biochem 50(9):1379–1387
Rai MM (1980) Dairy chemistry and animal nutrition. Kalyani Publishers, New Delhi, India
Sarkar U, Gupta AK, Sarkar V et al (2006) Factors affecting test day milk yield and milk composition in dairy animals. J Dairying Foods Home Sci 25(2):129–132
Skrodel L (1936) Secretion of milk and milk fat by the separate parts of the udder. MilchwirtschaftlicheForschungen 18:27–45
Urashima T, Saito T, Nakamura T et al (2001) Oligosaccharides of milk and colostrums in nonhuman mammals. Glycoconj J 18:357–371
Walstra P (1999) Dairy technology: principles of milk properties and processes. CRC Press, Boca Raton, FL

Probiotics

14

Anit Kumar, Rakhi Singh, Rachna Sehrawat, Nishant Kumar, and Pratibha

14.1 Introduction

In the last few years, the popularity and usage of probiotics as functional ingredients have increased due to their medicinal and positive impacts on human health. Several scientific studies have shown that probiotics are live microorganisms that help to improve the intestinal microbial balance and confirm the host's health benefit (FAO/WHO 2001; Kechagia et al. 2013). The probiotic word originated from the Greek words *pro* and *biotic* and translates as "for life." The Nobel Prize-winning scientist Elie Metchnikoff first hypothesized the probiotic concept in fermented milk products and also found that the fermented milk products, i.e., yogurt, contain beneficial microorganisms that help to protect the intestine from harmful bacteria (Mackowiak 2013; Tripathi and Giri 2014). In 1907, Metchnikoff reported that the consumption of lactic acid bacteria containing food products, viz., yogurt, sour milk, etc., is excellent for improving health and longevity (Roy 2005; Santacroce et al. 2019). At the beginning of the twentieth century, Tissier stated that the Bifidobacteria act as probiotics in preventing infections in breastfed infants because they are predominant in the intestinal microflora of these infants (Ishibashi and

A. Kumar
Department of Food Technology, School of Chemical Technology, Harcourt Butler Technical University, Kanpur, Uttar Pradesh, India

R. Singh (✉) · N. Kumar
National Institute of Food Technology Entrepreneurship and Management, Kundli, Sonepat, Haryana, India

R. Sehrawat
Department of Food Process Engineering, National Institute of Technology Rourkela, Rourkela, Odisha, India

Pratibha
National Institute of Technology, Delhi, Kurukshetra, Haryana, India

O. P. Chauhan (ed.), *Advances in Food Chemistry*,
https://doi.org/10.1007/978-981-19-4796-4_14

Table 14.1 Microorganisms used in probiotic products

Lactobacilli	Non-LAB	Bifidobacteria	Other LAB
L. acidophilus	*Propionibac. Freudenreichii*	*Bifidobacterium animalis*	*S. diacetylactis*
L. rhamnosus	*Bacillus cereus* var.*toyoi*	*B. adolescentis*	*S. cremoris*
L. plantarum	*Clostridium butyricum*	*B. bifidum*	*P. acidilactici*
L. salvarius	*Saccharomyces boulardii*	*B. lactis*	*E. faecium*
L. johnsonii		*B. infantis*	*S. intermedius*
L. crispatus		*B. longum*	*S. salivarius*
L. reuteri		*B. breve*	*S. thermophilus*
L. casei			*E. faecalis*

Source: Behera and Panda (2020), Hussain (2013), Parvez et al. (2006), Senok et al. (2005), Shortt (1999)

Shimamura 1993; Milani et al. 2017). Heller (2001) also reported the *Lactobacillus acidophilus* (intestinal isolate) bacteria producing "acidophilus-milch" or "reformed yogurt." The viability and applicability of the live microorganisms in food products are significant when used as a prerequisite for an effective immune system (Gill and Rutherfurd 2001). After that, the various definitions and concepts of probiotics have been reported by scholars and scientists. FAO/WHO (2001) confirmed probiotics as live microorganisms for the health benefit of the host through their activities in the human body in adequate amounts. Based on the scientific evidence and clinical trials, the various types of strains of lactic acid bacteria are considered probiotics, which directly impact the gastrointestinal tract of humans (Harzallah and Belhadj 2013).

Various types of probiotic-based products are available in the market. These probiotics are gram-positive bacteria such as *Lactobacillus* and *Bifidobacterium* (FAO/WHO 2001; Balaji et al. 2011; Song et al. 2012). The microorganisms used in probiotic products have been summarized in Table 14.1. Probiotic-rich fermented milk products prepared with lyophilized bacteria for the consumption of adult humans should be recognized as safe for health, produce lactic acid, be resistant to hydrochloric acid, pancreatic bile, and juice with anti-cancer property, reduce permeability of the intestine, and be able to survive in the stomach and duodenum in both acidic and alkaline conditions (Vimala and Dileep 2006; Ghosh et al. 2019). Single and mixed cultures can be used to prepare probiotic live organisms and exhibit beneficial effects on the health of the host (D'Souza et al. 2002; Oyetayo and Oyetayo 2005). Saarela et al. (2000) have reported the process of selection of probiotic microorganisms based on their functional, technological, and safety aspects.

14.2 Major Type of Probiotic

Different types of probiotic live microorganisms have been found in nature, which have a beneficial relationship with human health. The Lactic acid bacteria (LAB) is a widespread microorganism which has been found in any environment rich in carbohydrates, plants, mucosal surfaces of humans; fermented food; marine and terrestrial animals. LAB is recognized as safe for consumption and is a part of normal microflora and microbiota in the human body, which naturally inhabits the gastrointestinal and genitourinary tracts (Florou-Paneri et al. 2013). In addition to their role in preservation, nutritional and therapeutic importance, they are also known for the longevity of human life. LAB can also be used as a "probiotic" in a viable single or mixed culture of microorganisms. They benefit the host (humans or animals) by improving the properties of the indigenous microflora (Havenaar and Huisin't Veld 1992a). Over 100 species of the Lactobacillus genus have been recognized and used commercially as probiotics, including *L. acidophilus*, *L. reuteri*, *L. casei*, *L. rhamnosus*, *L. bulgaricus*, *L. plantarum*, *L. helveticus*, and *L. delbrueckii* (Florou-Paneri et al. 2013).

During the past decade, the investigation into the probiotic field has expanded beyond the live microorganism of the LAB group isolated from fermented dairy food products. This live microorganism studied includes members of the genera Lactobaccilus, S. boulardii, and Bifidobacterium, etc. (McFarland et al. 1994; Fijan 2014).

The various potential advantages of probiotics, such as relatively low cost, antibiotic resistance, and inhibiting pathogens, have been reported, thereby decreasing the chances of development of resistance against the probiotic (Sanders 1999; Rolfe 2000). The Association of American Feed Control Officials (AAFCO) and FDS have recognized probiotics as a safe ingredient for human consumption (AAFCO 1993; Mattia and Merker 2008). Probiotics can be used in whole foods to increase their functional attributes when combined with other healthy ingredients. The consumption of probiotics with nutrient-dense foods, viz., dairy products, can help enhance the nutritional value of the food and consumers can benefit from their consumption (Sanders 1999).

Several types of probiotics have been commercially available and widely used due to their health benefit functions. There are many scientific pieces of evidence available on the efficacy of probiotics in the areas of anti-diarrhoeal effects and the ability to improve digestion of lactose in lactose-intolerant humans (Sanders 1999). At the commercial scale, the major probiotics available contained genera of *Lactobacillus*, *Streptococcus*, *Bifidobacterium*, *Baccilus*, *Pediococcus*, *Leuconostoc*, *Saccharomyces cerevisiae*, *Aspergillus oryzae*, and *Propionibacterium* (Lioyd-Evans 1989; Tobey 1992; Mantere-Alhonem 1995; Verma and Singh 1995). Recent studies with probiotic bacteria have indicated that the selected probiotic strains have the potential for human health, and the studies used on animal models demonstrated the probiotic effects on human disease and their immunological responses. Although scientific investigation and documentation have reported that the *L. casei* probiotic bacteria can reduce pathological processes, i.e., diarrhea, and influence the functions of the immune system in the human body (Erickson and Hubbard 2000).

With respect to the effects of probiotics on the specific immunity of a host, the possible directions include determining. Therefore, probiotics can be used as a vehicle for oral immunization against several types of viruses (Pouwels et al. 1998). The genetically modified *lactobacillus* as probiotics also acts as a carrier for antigens; which helps induce oral tolerance (Maassen 1999). Probiotics are classified based on species, genus, and types of strains.

The same species of probiotic strains also have differences in their traits, i.e., stability, enzyme expression, inhibitors, pattern of carbohydrate fermentation, acid production and resistance ability, colony formation capacity in the gastrointestinal tract, and other clinical properties. The different strains do not prove that they do; therefore, the microbiological occurrence is inflicted upon those attempting to commercialize probiotics (Sanders 2000).

Probiotics are a collection of various strains that provide health benefits. Microorganisms that are biologically active and viable may be required for the target site in the host cell. The natural barrier property of the probiotics must be able to protect against ingested bacteria. Probiotic bacteria genera, i.e., *Lactobacillus* and *Bifidobacterium*, have the potential to combat gastric acid, enzymes (pancreatic), and bile salts. These are contemplated components for the gastrointestinal flora. The LAB microorganism (i.e., lactic acid) has the potential to inhibit the growth of several types of pathogens such as *S. typhimurium*, *S. aureus*, *C. difficule*, *C. perfingens*, and *E. coli* and are also used to treat the gastrointestinal disorders in animal and humans (Silva et al. 1987; Meurman et al. 1995). Salminen and Von Wright (1993) reviewed about the safety concern and pathogenic potentiality of the *Lactobaccilus* and *Bifidobacterium* for the host and reported the lower pathogenic potential ability of *Lactobaccillus* and *Bifidobacterium.* This study has been reported based on the extensive consumption of these microorganisms in fermented foods and food products. They showed negative side effects of the microorganism on the point of feeding in higher levels of immuno and compromised humans (infants, elderly, HIV-infected and chronic disease patients). However, the association of bifidobacteria and lactobacilli potential to compromised health suggest against endocarditis disease in humans (Aguirre and Collins 1993; Gasser 1994; Saxellin et al. 1996). Adams and Mateau (1995) stated the safety concern for the use of *Enterococci* bacteria as probiotics. On the other hand, the association of *Enterococcus* (*faecium* and *faecalis*) with bacterium is responsible to expand the occurrence of antibiotic resistance in these strains to eliminate them from food formulations (Lai 1993; Lai 1996). Enterococci is also used as a probiotic in dietary supplemented formulations and traditionally food fermentation (Giraffa et al. 1997).

In general, the standardization of prebiotic-based products depends on the presumption of culture viability. Thus, the production of a high cell population in the formulations is of primary importance for the strains. In most cases, if viability is not essential, it is correlated with effect due to indicators of the present cell. Previous scientific studies however have demonstrated that some activities have no effect on the viability of the microorganism. Several improvements, such as digestion of lactose (Vesa et al. 1996), anti-hypersensitivity (Maeno et al. 1996), activation of immune system modulation (Perdigon et al. 1986; Tomioka and Saito 1992; Solis Pereyra and Lemonnier 1993; Hosono et al. 1997; Marin et al. 1997), and antihypertensive effects (Maeno et al. 1996) are directly related to the nonviable cells, i.e., fermented products, enzymatic activity, and cell components, etc.

14.3 Mechanism of Probiotics

The probiotic mechanism may contribute to health benefits by stimulating host cell systematic and mucosal immunity; stimulating calcium uptake by enterocytes; degrading mineral complex phytic acid; probiotic anti-arthritic effects; exhibiting mannose-specific adhesion; and releasing cytokines and chemokines (Delcenserie et al. 2005; Katharina et al. 2007). Sanders (2000) reviewed and reported that the prebiotic microorganisms play an important role by providing health benefits, and the mechanism for their beneficial effects is shown in Fig. 14.1.

Several types of probiotic mechanisms are proposed by the researchers, which indicate the disease-reducing ability. Table 14.2 shows the potential effects of the probiotic bacteria on human health (Sanders and Huisin't Veld 1999) by inhibiting colonization. These strains are also known as colonization resistance. In addition, further studies are required to investigate the modes of action and types of mechanism of particular probiotics against specific pathogens.

14.3.1 Colonization Resistance

The diversity of the microbes in the gastrointestinal tract is of importance for the protection of the host. Probiotic bacteria such as *Lactobaccillus* and *Bifidobacetria*

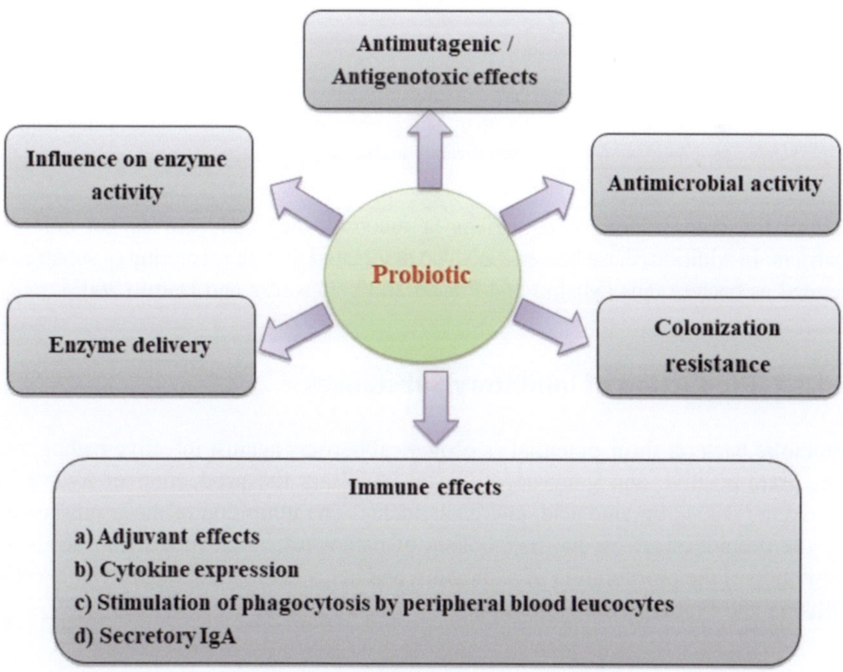

Fig. 14.1 Mechanism action of probiotics on human health

Table 14.2 Health benefits of probiotic bacteria

Target health benefit	Postulated mechanism	Reference
COVID-19 treatment	Probiotic supplementation decreased the viral loads through fighting against COVID-19 by immune modulation and also prolonged the exposure of antibiotics that reduce the risk of secondary infection.	Adnan and Dewi (2020)
Alterations in gut microbiota of patients with COVID-19	Gut microbiome alteration was observed in COVID-19 patients. This dysbiosis persisted even after the symptoms resolved and virus had cleared.	Zuo et al. (2020)
Reducing vitamin D deficiency	Supplementation of probiotics enhanced absorption of vitamin D and increased vitamin D receptor expression.	Costanzo et al. (2018), Hang and Sun (2017)
Influenza infection	Supplementation of probiotics and prebiotics could improve hemagglutination inhibition antibody titers following the influenza vaccination.	Yeh et al. (2018)
Reduce SARS-CoV-2 disease	Probiotics act as a potential blocker to the ACE receptor that acts as a gateway for SARS-CoV-2 to attack GI cells.	Miremadi et al. (2014)
Prevent infection of enteropathogenic coronavirus transmissible gastroenteritis virus	TGEV particles attached to the surface of *E. faecium* might trap virus and prevent infection.	Chai et al. (2013)
Help in lactose digestion	Activity of β-galactosidase significantly increased	He et al. (2008)

inhibit the colonization of pathogens at mucosal sites and help as gut defense barriers. In addition, these bacteria are also associated with the secretion of substrates termed as bacteriocins (Mishra and Prasad 2000; Saavedra and Dattilo 2003).

14.3.2 Production of Inhibitory Substances

Probiotic bacteria show potential as ecological barriers against infective pathogens, i.e., gram-positive and gram-negative, by inhibiting the production of hydrogen peroxide (H_2O_2), organic acid, and bacteriocins. The inhibition of these substances by the probiotics affects the metabolism of pathogens, toxin production, leads to inhibition of the proliferation of pathogens, and helps in the reduction of viable cells (Mishra and Prasad 2000; Rolfe 2000).

14.3.3 Competition for Nutrients

The recommended mechanisms for the competitive prohibition of pathogens are lowering the production of bacteriocins, feuding for nutrition and strength of luminal pH (Collado et al. 2010). Bacteriocins are produced by some species of *Lactobacilli* and *Bifidobacteria* that inhibit the proliferation of target pathogens. For example, the growth of microbial pathogens such as *E. coli*, *Helicobacter*, *rotavirus*, *Shigella* spp., and others is prohibited by the release of *L. acidophilus* and *L. plantarum* in the gastrointestinal tract (Kumar et al. 2016). Nutritional competition is a component of the mechanism by which probiotics and pathogens inhibit and eliminate pathogens from the host. However, in vivo scientific evidence is lacking for this (Rolfe 2000).

14.3.4 Degradation of Toxic Receptor

Toxin receptors are modified by an enzymatic mechanism of probiotics. Toxin receptors of *Clostridium difficile* were degraded by Saccharomyces boulardii in the rabbit ileum and blocked cholera-induced secretion in rat jejunum by producing polyamines and protecting human colonic mucosa (Castagliuolo et al. 1996; Castagliuolo et al. 1999). *C. difficile* colonizes the intestine and releases two potent protein exotoxins, toxin A and toxin B, which mediate diarrhea and colitis caused by this microbe. The host mucosal disaccharidase activity is stimulated by *S. boulardii* and enhances the intestinal mucosal immune response. It may also be involved in the mechanism by which *S. boulardii* reduces the recurrence of *C. difficile* colitis.

14.3.5 Stimulation of Immunity

Probiotics can stimulate the immune system against several pathogenic microflora and antigens. The gut-associated lymphoid tissue (GALT) comes into contact with different bacteria, antigens, and other components exogenous to the body. The oral consumption of probiotics, because of their adherent properties, also has contact with GALT, thereby strengthening the lymphoid tissue (Saloff-Coste 1995). Specific and nonspecific immunity is influenced by probiotics. The oral administration of such an organism activates macrophages, lymphocytes, and immunoglobulins (Mishra and Prasad 2000).

14.4 Probiotic Food Products

Probiotics contain probiotic foods for the improvement of health and affect consumers' choice (Maragkoudakisa et al. 2006). For the production of probiotic-based food products, mostly Lactobacillus groups of live organisms are used. It contributes to the host's physiological needs and easily colonizes the intestines of other animals. Dairy products work as suitable carriers for the survival of live

Table 14.3 Examples of probiotic dairy products available in the world market

Product	Microorganisms	Country of origin
Progurt acidophilus milk	*L. acidophilus*	Several countries
Acidophilus yeast milk	*L. acidophilus*, *S. cerevisiae*, *Saccharomyces fragilis*	Former USSR
Cultra milky	*Bifidobacteria* + *L. acidophilus*	Denmark
Nutrish A/B milk	*Bifidobacteria* + *L. acidophilus*	USA
Biomild	*Bifidobacteria* + *L. acidophilus*	Several countries
Acidophilus yogurt	*L. acidophilus*, *S. thermophilus*, *L. delbrueckii* ssp. *bulgaricus*	Several countries
B-active	*L. acidophilus*, *S. thermophilus*, *B. bifidu*, *L. delbrueckii* ssp. *bulgaricus*	France
Mil-Mil	*L. acidophilus*, *B. breve*, *B. bifidum*	Japan
Yakult	*Lactobacillus casei* strain *Shirota*, *B. bifidum*, *L. acidophilus*	Japan
Actimell	*L. casei*	Germany
Vifit	*L. casei GG* + Oligofructose	Germany
LC-1	*L. acidophilus*	Germany
Mona fysig	*L. acidophilus* + inulin	The Netherlands
Symbalance	*L. acidophilus* + *L. casei* + *L. reuteri* + *Bifidobacteria* + inulin	Switzerland
Probiotic curd	–	India (Heritage Foods Ltd.)
"b-Activ" probiotic curd	–	India (Mother Dairy)
"Nesvita" probiotic yogurt	–	India (Nestle India)
Probiotic ice creams, "Amul prolife," "Prolite," and "AmulSugarfree"	–	India (Amul)

Source: Hussain (2013), Bhadoria and Mahapatra (2011), Lourens-Hattingh and Viljoen (2001)

fermented microorganisms such as *lactobacilli. The Lactobacillus* groups of the microorganism had unique properties, i.e., metabolic growth rate, production of flavor products, and proteolytic activity. *For a long time, the Lactobacillus probiotic group has been associated with dairy food products.* Fermented food products such as dairy products (milk, yogurt, cheese, ice cream, and frozen fermented dairy desserts) are highly acceptable products for delivering viable probiotics to the human GIT (Gardiner et al. 1998). Table 14.3 shows the probiotic dairy food products available in the global market.

In the Indian dairy market, various products, such as yogurt, dahi, lassi, and ice creams, are available, loaded with probiotics. These products are manufactured by reputed industries like Mother Dairy, AMUL, Nestle, and Yakult-Danone. Various studies have been carried out on probiotics and their uses in the development of dairy products. Kumar (2009) developed an ice cream using *L. acidophilus* and *Fructo-oligosaccharide* live microorganisms and investigated the viability of probiotics in ice cream. They revealed that the *L. acidophillus* probiotic viability was more than 10^7cfu/ml during the storage period of 120 days compared to fructo-oligosaccharides.

Rajpal and Kansa (2008) developed probiotic *dahi* using many types of live probiotic cultures such as *L. acidophilus*, *Bifidobacterium bifidum*, and mixed cultures *(L. lactis* sp. *cremoris* and *L. lactis* sp. *lactis* bio var. *Diacetylactis)*. Vibha (2004) also developed probiotic *dahi* using different probiotics in combinations such as *L. acidophilus (NCDC 14), L. lactis* ssp. *lactisbio* var. *diacetylactis* (NCDC 60), *L. casei* (NCDC 19), and mixed culture NCDC 167. During this study, the two combinations of culture NCDC 14 (1%), NCDC 19 (0.4%), NCDC 60 (0.6%), and NCDC 19 (1%) mixed with 0.6% of NCDC 167 culture were found highly acceptable in terms of organoleptic and physicochemical properties with more than 10^7cfu/ml of the probiotic count. The probiotic dahi has an antidiabetic, anticancer, immunomodulatory, and hypocholesterolemic effect.

14.4.1 Functional Dairy Beverages

The dairy beverages that help in enhancing health and imparting physiological benefits are called "functional dairy beverages." It helps prevent and treat various diseases or improves physical and mental health via functional ingredients, biotechnological, and processing modifications (Sloan 2005). Presently, functional beverages cover 60% of the food market. Beverages are forecasted to be the fastest growing segment all over the world. It covered 56% of the total market in 2013. The primary drivers of consumer preferences are the convenience and versatility of beverages with innovative packaging. The New Product Magazine (Stagnito's) mentioned that the prime objectives of the development of beverages are heavily dominated by health (53%), followed by taste and convenience (25 and 23%, respectively). Functional dairy beverages efficiently fulfill all these main drivers (health, taste, and convenience) of functional foods. These are mostly milk-based beverages. The biologically active ingredients are the major source of developing new milk-based functional beverages (Sharma 2005). Functional dairy beverages are targeted by some biological compounds such as plant sterols, phenolic compounds, omega-3 fatty acids, dietary fibers, prebiotics, and probiotics.

14.4.2 Desirable Properties of Probiotics for Incorporation into Food Products

The probiotics as live microorganisms had various desirable properties, i.e., GRAS status, safe, health beneficial, nonpathogenic, antimutagenic, and anti-carcinogenic properties with reducing adhesion of pathogens, tolerance to acid and human gastric juice, phage resistance, bile tolerance, heat and gas tolerance, adherence to the intestinal mucosa, desired metabolic activity, ability to grow in milk, retaining sensory characteristics in products with maximum stability and viability during processing (Kearney et al. 2009; Terpou et al. 2019). Advancement in the formulations of probiotic foods is an important research field for the future of the probiotic food market. The standards for a probiotic strain for food or pharmaceutical applications are constantly emerging and developing. These standards can be roughly classified into four distinct categories: technological, safety, functional, and physiological characteristics.

14.4.3 Probiotic Food Markets

Economic forecasts predict an increase in the global market for probiotic dietary supplements of 3.3–7 billion US dollars from 2015 to 2025, as expected by economic forecasts (Statista 2019). The probiotics market size surpassed USD 4.30 billion in 2020 and is likely to grow at a CAGR of 8.7% between 2021 and 2027 (Insights 2021). Europe accounted for nearly 42%, followed by Asian countries at 30%. European and Asian countries are the most significant contributors, being probiotic culture-based drinks and yogurts. The leading global stakeholders in the probiotic market are India, China, and Japan (Kesavelu et al. 2020). Asia-Pacific shares >40% of the global share market of probiotic food (Elshaghabee et al. 2017). There is no standard for the evaluation of Indian probiotics. However, in 2011, the Indian Council of Medical Research (ICMR) published guidelines on the quality of probiotics in food in 11 different sections such as Scope; Definition of probiotics; Genus, species, and strain identification; In vitro tests to screen potential probiotic strains, in vivo safety studies in animal models; In vivo efficacy studies in animal models; Evaluation of safety of probiotics for human use, Evaluation of efficacy studies in humans, Effective dosage of probiotic strain/strains; Labeling requirements; and manufacturing and handling procedures. The Indian drug formulary (i.e., the Current Index of Medical Specialities) lists over 160 probiotic brands available in India with various single strains and multiple combinations of various strains. A variety of commercial probiotics are available in different forms (e.g., dry powder sachets, capsules, liquid formulations, dry powder syrup, and in combination with antibiotics), and some manufacturers claim their existence in infant formulae and oral rehydration salts (Kesavelu et al. 2020).

14.4.4 Challenges for the Use of Probiotics in Food Products

The probiotic viability range of 106 cfu/g in a product is sufficient to provide health benefits for the consumer's intestinal health (Galdeano and Perdigon 2006). Therefore, developing adequate techniques to maintain the viability and survival of probiotics throughout the product's shelf life is a vital area of research (Shah

2000). Physical and biological parameters such as high temperature, acidity, and oxygen affect the probiotic activity in dairy-based products. So, the primary challenge is to maintain the activity and viability of the probiotic culture in the products. Production of hydrogen peroxide and acid by bacteria, penetration of oxygen, and packaging affect the viability of probiotics in fermented products (Shah 2000). Titratable acidity and low pH values can affect the viability of probiotics during storage (Saarela et al. 2009). Other factors that may affect viability are post-acidification, oxygen level, temperature, food matrix, and interaction with the starter organisms during cold storage of fermented food products (Dave and Shah 1997). The metabolic activities of probiotics depend on the chemical properties of products and must have good compatibility with starters when added to fermented dairy products. The sensory properties of probiotic foods, such as flavor and texture, are the most important nutritional parameters that attract consumers. Hence, it should be standardized before the development of probiotic foods (Kearney et al. 2009).

14.4.5 Suggested Approaches to Maintain the Viability of Probiotics in Food Products

The viability of prebiotic count is more potential to maintain its numbers in food products for a long storage time. Terpou et al. (2019) proposed various approaches to maintain the viability of probiotics in food products, such as probiotic selection, strain adaptation on food matrix and human microenvironment, food packaging system selection, the addition of compounds as probiotic promoters, and probiotic encapsulation by different encapsulating wall materials such as whey protein, soy protein isolate, alginate, and so on.

14.5 Beneficial Properties of Probiotic Dairy Foods

The probiotic-based dairy food products play an important role in the health promotion of human. The health-beneficial effects include probiotics including colonization resistance in GI, respiratory, and urinogenital tracts; reconstruction and formation of microbial ecosystem in respiratory and intestinal tracts; reducing cholesterol level of blood serum and lactose intolerance; anticancer and antimutagenic activity; stimulation of immune systems; improving the synthesis of pre-digested proteins and vitamins; reduced decalcification; improved detoxification activity of toxic substances such as enterotoxins and aflatoxin, respectively (Ram and Bhavadasan 2002).

14.6 Health Benefits of Probiotics

The intestinal microbiota is a complex ecosystem within the human being that exists symbiotically, antagonistically, and competitively. They act as an intermediary to create a strong relationship between microbes and the intake of food. Recently, the scientific community's interest has increased in the field of probiotic-rich nutraceutical food products for disease prevention and health benefits (Hemarajata and Versalovic 2013). Various types of material can be used to encapsulate probiotics in different types of food to improve their viability and storability. The material used for the probiotic encapsulation must be ecofriendly and have the efficiency to create a barrier between the surrounding and internal phase (Chávarri et al. 2012). Previous researchers reported that live microbes regulate the gut microbiota by regular intake of fermented milk and milk products (yogurt, cheese, curds), or fermented vegetables, juice, and others (Pavli et al. 2018; Rezac et al. 2018; Pop et al. 2020; Savaiano and Hutkins 2021). It is proven that probiotics are an important source for human health by influencing nutritional and therapeutic activity. For example, *Bifidobacterium* as a probiotic has the potential to produce water-soluble vitamins such as nicotinic acid, biotin, thiamine, B_{12}, and pyridoxine. Lee and Salminen (1995) reported that the bioavailability of iron compounds is increased due to the encapsulation of *L. acidophilus*. Moreover, probiotics help the alleviation of lactose malabsorption symptoms, resistance against infectious diseases of the intestinal tract, suppression of cancer disease, improved gastrointestinal immunity, control of cholesterol and digestion (Levri et al. 2005; De Vrese and Marteau 2007; Zoumpopoulou et al. 2008; Rowland 2009; Kumar et al. 2010).

14.6.1 Application in Humans

The probiotics regulated and improved the response of the immune system of the host by bidirectional neuronal signaling and activating specific types of genes (Kristensen et al. 2016). The use of probiotics finds several applications during disturbance of indigenous flora, after high doses of irradiation and antibiotics, and inadequate development of flora and neonate intensive care units, respectively (Ram and Bhavadasan 2002).

14.6.1.1 Current Use of Probiotics

At present, probiotic-based functional foods are widely consumed by consumers due to their beneficial health effects. They improve the implantation and survival of microbial dietary proteins, which improves the gastrointestinal tract microbial balance. Numerous types of food products such as dairy-based (fresh milk, yogurt, ice cream, whey drink, whey cheese, cheddar cheese, etc.), soy-based (soy milk, soy cream cheese), juice-based (tomatoes, cabbage, beet, pineapple, carrot, orange, and grapes), edible packaging, and nondairy-based products are available in the markets. These can serve as carriers of probiotics (Nagpal et al. 2012; Song et al. 2012;

Table 14.4 Application of probiotics in dairy products

Organisms	Uses
Yeast	
Candida kefir	In fermented dairy foods; kefir, kumis, for therapeutic purpose like treatment of diarrhea
Kulveromyces fragilis	
Sacchromyces boulardii	
Lactobacillus	
L. acidophilus	As a supplement in dairy food to improve nutritional benefits and used in various processed foods like vegetables, pickles, etc.
L. plantarum	
L. casei subsp. *rhamnosus* GG	Application in dairy products, viz., yogurt, whey drinks, ice cream, and cheese
L. brevis	
L. delbruckii subsp. *bulgaricus*	
Streptococcus thermophiles	
Bifidobacterium	
B. bifidum	As components of new generation of fermented and nonfermented probiotic dairy products as well as in infant formulae preparations
B. infantis	
B. longum	
B. animalis	
B. breve	
Others	
E. faecium	As a constituent of certain health-promoting foods and in the production of buttermilk and certain cheeses
L. lactis subsp. *lactis*	
L. lactis subsp. *cremoris*	

Pech-Canul et al. 2020). Table 14.4 shows the use of some probiotics for the development of functional foods and their health benefits.

14.6.1.2 Prerequisite for Probiotics

The selection of probiotic bacteria must be based on the specific desired properties that can be achieved by genetic engineering technology and biological selection (Ram and Bhavadasan 2002). The nutritional status of the host can be improved by the use of exogenous probiotics. These bacteria are mostly used in the fermentation of food and to improve the bioavailability of macro- and micronutrients. Furthermore, the application of probiotics provides beneficial effects by residing in the gastrointestinal tract. Previous researchers, Havenaar and Huisin't Veld (1992b) reported that the process for selection of microbial strains as probiotics includes several factors, which include viability during processing and storage, in vivo and in vitro resistance, origin of the strain, antimicrobial activity, and colonization. On the other hand, these also provide biological safety during the processing and production, administering mode of probiotic, where the probiotic bacteria must be active. Several types of criteria must be considered during the selection of probiotic strains.

14.6.1.3 Survivability

The probiotic strain can be used as an acid-resistant and rapid acid producer (Lyons 1988). The microbial strains have been introduced as transiently surviving in the small intestine and stomach. The probiotic strains must survive in a lower pH medium and are preferable as buffers in dairy-based fermented and nonfermented food products such as yogurt and milk (Ram and Bhavadasan 2002).

14.6.1.4 Adhesion

The use of probiotic strains of microbes should be efficient to adhere to and can be sustained under prevailing conditions in the intestine. The scientific evidence of previous investigations has reported the range of viable cell populations of between 106 and 109 cfu/mL in the intestinal tract (Tanaka et al. 1982; Kim 1988; Sarkar and Misra 1998). The adhesion of the epithelium is essential for the long-term survival and persistence of probiotics in the intestinal tract. It involves both types of binding mechanisms, such as specific and nonspecific in the intestinal tract, moderated by specific proteins (Conway and Kjellbery 1989).

14.6.1.5 Bile Resistance

Bile resistance is an important property of probiotics, which indicates the survival ability of probiotics in the small intestine tract (Ruiz et al. 2013). The previous study by Ram and Bhavadasan (2002) reported that the higher concentration of bile acid contained in the colon and small intestinal tract has the potential to inhibit the growth of bacteria. In addition, the use of probiotic strains in fermented and nonfermented food products should be positively capable of surviving in gastrointestinal tract bile salt concentrations.

14.6.1.6 Antimicrobial Production

The production of antimicrobial active agents plays an important role in providing microbiological safety due to inhibiting the growth of pathogens (Pyar and Peh 2014; Prabhurajeshwar and Chandrakanth 2019). The different types of antimicrobials, such as bacteriocins, organic acids, acetoin, H_2O_2, and di-acetyl, are produced by probiotics. Table 14.5 shows the important bacteria produced by the probiotics. Therefore, the previous study done by Ram and Bhavadasan (2002) stated that the intestine microbial system is complex and the gastrointestine highly competitive for exogenous microorganisms. They also reported that the probiotic should have the potential to inhibit the growth of pathogens in the intestinal milieu.

14.6.1.7 Stimulation of Immune System

The consumption of probiotics helps in the stimulation of the cellular immune system by the suppression of autoimmune diseases and induces the immune defense system, respectively (Galdeano et al. 2019; Klaenhammer et al. 2012). The activation of cells such as macrophages, release of cytokinins, natural killer, and T

Table 14.5 Bacteriocin produced by lactic acid bacteria

Bacteriocin	Producer strain	Active against
Nisin	*Lactococcus lactis* sp. *lactis*	*Streptococcus*, *Staphylococcus*, *Pediococcus*, *Micrococcus*, *Listeria*, *Mycobacterium*, *Clostridium*, and *Bacillus*
Lactococcin	*Lactococcus lactis* ADR185 L030	*Clostridium tyrobutyricum*, *Lactobacillus helveticus*, and *Streptococcus thermophilus*
Lactin 481	*Lactococcus lactis* 481	Lactic acid bacteria, *Clostridium tyrobutyricum*
Salivaricin	*Streptococcus salvarius*20P3	*Micrococcus luteus*
Acidocin 8912	*Lactobacillus acidophilus*	*Lactobacillus acidophilus*, *Lactobacillus casei*, *Lactococcus lactis*, *Lactobacillus plantarum*
Lactacin F	*Lactobacillus johnsonii* 11088	*Lactobacillus fermentatum*, *Lactobacillus delbrucckii*, *Lactobacillus helveticus*, and *Staphylococcus aureus*
Pediocin ID	*Pediococcusacido lactici*	Listeria monocytogens
Plantarcin S	*Lactobacillus plantarum*	*Leuconostoc*, *Clostridium tyrobutyricum*, *Lactobacillus helveticus*, *Enterococcus*, *Streptococcus*, *Micrococcus*, and *Propionibacterium*
Sakacin A	*Lactobacillus sake* LB 706	*Enterococcus* sp., *Lactobacillus curvatus*, *Lactobacillus brevis*, *Leuconostoc monocytogenes*, *Staphylococcus aureus*, and *Lenconostoc paramesenteroids*
Cascien 80	*Lactobacillus casei* B80	*Lactobacillus casei*
Lacticin A	*Lactobacillus delbrueckii*	*Lactobacillus delbrucekii* spp. *lactic*
Lactocin 27	*Lactobacillus helveticus*	*Lactobacillus acidophilus*, *Lactobacillus helveticus*

Source: Rakhi (2005)

lymphocytes indicated the cellular immune response improved by the probiotics (Havenaar and Huisin't Veld 1992b; Ram and Bhavadasan 2002; Klaenhammer et al. 2012).

14.6.1.8 Biosafety

The safety concern of probiotics in the food and clinical sector is very important. For the biosafety purpose of the probiotics, their cell components and metabolites must be negative for toxicity, allergy, pathogenicity, carcinogenicity, mutagenicity, etc. (Hammes and Tichaczek 1994; Havenaar and Huisin't Veld 1992b).

14.6.1.9 Stability

Probiotic microorganisms must be genetically and gastrointestinal environment stable in order to perform optimally. Plasmid stability is of great concern, especially regarding antibiotic resistance and its transfer (Havenaar and Huisin't Veld 1992b). The stability of probiotics in functional foods is critical to their health-beneficial effects. Numerous researchers have improved the efficiency and stability of probiotics using different technologies and methods. The nature of the matrix, manufacturing process, and temperature directly impact the stability and functionality of the probiotic (Grzeskowiak et al. 2011; Luidalepp et al. 2011).

Shah et al. (2010) improved the stability of three probiotic bacteria (Howaru *L. rhamnosus*, Howaru *B. lactis HNOO1*, and *L. paracasei LPC 37*) using inoculation in a model fruit system. Green tea extract, grape seed extract, vitamins B2, B3, B6, C, and E, among others, are antioxidants and vitamins. On the other hand, the 10 different probiotics (*L. rhamnosus*, *B. longum*, *L. salivarius*, *L. plantarum*, *L. acidophilus*, *L. paracasei*, *B. lactis* (BI-04 and Bi-07), Howaru *L. rhamnosus*, and Howaru *B. bifidum*) stability was improved by Ding and Shah (2009) using different gum-based coating materials such as guar gum, xanthan gum, alginate, loctus bean gum, and carrageenan gum. Various drying techniques such as freeze, spray, and fluidized bed drying were used to improve the stability of *L. paracasei* probiotics in milk powder by Poddar et al. (2014), who reported fluidized bed drying as a potential technique to improve the stability of probiotics under ambient temperature conditions due to reducing the absorption of moisture, low porosity, and large agglomerate size. Similarly, the higher retention and stability of *L. casei* probiotics were also reported in fluidized bed drying compared to freeze drying at 25 °C as reported by Nag and Das (2013). They reported higher viability (2.5 log cfu/g) of probiotics in the fluidized bed drying technique.

14.7 Conclusion

Research evidence has proved the health benefits of probiotics as functional foods to maintain excellent stability and improve intestinal microbiota and resistance against infectious pathogens. Probiotic-based functional foods can help to treat several types of diseases. More investigation should be required to evaluate the effects of probiotics on neurological disorder. At present, there are no legal documents certifying probiotics' use as food supplements and therapeutic agents to prevent various diseases, so required research and documentation can be done by the medical and health authorities. Further investigation is required to establish an adequate level of probiotics and their appropriate dose. At low pH, the growth and stability of the probiotic are restricted, so the new advanced method and procedure are also needed to increase bioavailability at lower pH. The different types of encapsulation technology and biomaterials for improving bioavailability and the stability of probiotics from a commercial point of view should also be investigated to optimize advanced technologies to stabilize their viability during process and storage. Awareness programs among the people for the consumption of probiotics and their health benefits are required. The multidisciplinary approaches (immunology, nutrition, microbiology, pediatric, and medical) to ensure the potential application and cellular and molecular mechanisms of probiotics in adults and infants need in-depth research. Moreover, the incorporation of probiotics in edible coatings and films will be an innovative research arena for the scientific community. Further research is required on the aspects of edible packaging and antifungal activity of probiotics for health safety.

References

AAFCO (1993) Association of American Feed Control Officials. Official Publication

Adams MR, Mateau P (1995) On the safety of lactic acid bacteria from food. Int J Food Microbiol 27:263–264

Adnan ML, Dewi MD (2020) Potential effects immunomodulators on probiotics in COVID-19 preventing infection in the future. A narrative review. Int J Med Stud. https://doi.org/10.5195/ijms.2020.486

Aguirre M, Collins MD (1993) Lactic acid bacteria and human clinical infection. J Appl Bacteriol 73:95–107

Balaji R, Raja KD, Arunachalam (2011) Market potential for probiotic nutritional supplements in India. Afr J Bus Manage 5(14):5418–5423

Behera SS, Panda SK (2020) Ethnic and industrial probiotic foods and beverages: efficacy and acceptance. Curr Opin Food Sci 32:29–36

Bhadoria PBS, Mahapatra SC (2011) Prospects, technological aspects and limitations of probiotics—a worldwide review. Eur J Food Res Rev 1(2):23–42

Castagliuolo I, LaMont JT, Nikulasson ST et al (1996) Saccharomyces boulardii protease inhibits clostridium difficiletoxin a effects in the rat iloum. Infect Immum 64:5225–5232

Castagliuolo I, Riegler MF, Valenick L et al (1999) Saccharomyces boulardii protease inhibits the effects of Clostridium difficile toxin a and B in human colonic mucosa. Infect Immun 67:302–307

Chai W, Burwinkel M, Wang Z (2013) Antiviral effects of a probiotic enterococcus faecium strain against transmissible gastroenteritis coronavirus. Arch Virol 158:799–807

Chávarri M, Marañón I, Villarán MC (2012) Encapsulation technology to protect probiotic bacteria. In probiotics. IntechOpen

Collado MC, Gueimonde M, Salminem S (2010) Probiotics in adhesion of pathogens: mechanisms of action. In: Watson RR, Preedy VR (eds) Bioactive foods in promoting health: probiotics and prebiotics, 1st edn. Academic Press, Elsevier, London, pp 353–370

Conway PL, Kjellbery S (1989) Protein mediated adhesion of L. fermentum strain 737 to mouse stomach squamous epithelium. J Gen Microbiol 135:1175

Costanzo M, Cesi V, Palone F et al (2018) Krill oil, vitamin D and lactobacillus reuteri cooperate to reduce gut inflammation. Benef Microbes 9(3):389–399

D'Souza AL, Rajkumar C, Cooke J et al (2002) Probiotics in prevention of antibiotic associated Diarrhoea: meta-analysis. Brit Med J 324(7350):1361

Dave RI, Shah NP (1997) Viability of yoghurt and probiotic bacteria in yoghurts made from commercial starter cultures. Int Dairy J 7(1):31–41

De Vrese M, Marteau PR (2007) Probiotics and prebiotics: effects on Diarrhoea. JN 137(3):803–811

Delcenserie V, Martel D, Lamoureux M et al (2005) Immunomodulatory effects of probiotics in the intestinal tract. Mol Biol 10:37–54

Ding WK, Shah NP (2009) Effect of various encapsulating materials on the stability of probiotic bacteria. J Food Sci 74(2):100–107

Elshaghabee FMF, Rokana N, Gulhane RD et al (2017) Bacillus as potential probiotics: status, concerns, and future perspectives. Front Microbiol 8:1490

Erickson KL, Hubbard NE (2000) Probitoic immunomodulation in health and disease. J Nutr 130(25):403–405

Food and Agriculture Organization/World Health Organization (FAO/WHO) (2001) Health and nutritional properties of Probiotics in food including powder milk with live lactic acid bacteria.

Report of a Joint FAO/WHO Expert Consultation on Evaluation of Health and Nutritional Properties of Probiotics in Food including Powder Milk with Live lactic acid bacteria, Córdoba, Argentina. http://www.who.int/foodsafety/publications/fs_management/en/probiotics.pdf

Fijan S (2014) Microorganisms with claimed probiotic properties: an overview of recent literature. Int J Environ Res Public Health 11(5):4745–4767

Florou-Paneri P, Christaki E, Bonos E (2013) Lactic acid bacteria as source of functional ingredients.In lactic acid bacteria-R & D for food, health and livestock purposes. IntechOpen

Galdeano CM, Perdigon G (2006) The probiotic bacterium lactobacillus Casei induces activation of the gut mucosal immune system through innate immunity. Clin Vaccine Immunol 13(2): 219–226

Galdeano CM, Cazorla SI, Dumit JML et al (2019) Beneficial effects of probiotic consumption on the immune system. Ann Nutr Metab 74(2):115–124

Gardiner G, Ross RP, Collins JK et al (1998) Development of a probiotic Cheddar cheese containing human-derived lactobacillus Paracasei strains. Appl Environ Microbiol 64(6): 2192–2199

Gasser F (1994) Safety of lactic acid bacteria and their occurrence in human clinical infections. Bull Inst Pasteur 92:45–67

Ghosh T, Beniwal A, Semwal A et al (2019) Mechanistic insights into probiotic properties of lactic acid bacteria associated with ethnic fermented dairy products. Front Microbiol 10:502

Gill HS, Rutherfurd KJ (2001) Probiotic supplementation to enhance natural immunity in the elderly: effects of a newly characterized immune stimulatory strain lactobacillus Rhamnosus HN001 (DR20 (TM)) on leucocyte phagocytosis. Nutr Res 21(1):183–189

Giraffa G, Carminati D, Neviani E (1997) Enterococci isolated from dairy products: a review or risks and potential technological use. J Food Prot 60:732–738

Grzeskowiak L, Isolauri E, Salminen S et al (2011) Manufacturing process influences properties of probiotic bacteria. Br J Nutr 105:887–994

Hammes WP, Tichaczek PS (1994) The potential of lactic acid bacteria for the production of safe and wholesome food. Z Lebensm Unters Forsch 198:193–201

Hang M, Sun J (2017) Vitamin D/VDR, probiotics, and gastrointestinal diseases. Curr Med Chem 24:876–887

Harzallah D, Belhadj H (2013) Lactic acid bacteria as probiotics: characteristics, selection criteria and role in immunomodulation of human GI mucosal barrier. In: Kongo M (ed) Lactic acid bacteria—R & D for food, health and livestock purposes. IntechOpen. https://doi.org/10.5772/50732. Available from: https://www.intechopen.com/books/lactic-acid-bacteria-r-d-for-food-health-and-livestock-purposes/lactic-acid-bacteria-as-probiotics-characteristics-selection-criteria-and-role-in-immunomodulation-o

Havenaar R, Huisin't Veld JHJ (1992a) Probiotics: a general view. In: Wood JB (ed) The lactic acid bacteria, vol 1. Elsevier Applied Science, London, pp 151–170

Havenaar R, Huisin't Veld JHJ (1992b) Probiotics health claims and selection criteria. Lactic 91-Caen, 12–13 Sept. pp. 235–243

He T, Priebe MG, Zhong Y (2008) Effects of yogurt and bifidobacteria supplementation on the colonic microbiota in lactose-intolerant subjects. J Bacteriol Res 104(2):595–604

Heller KJ (2001) Probiotic bacteria in fermented foods: product characteristics and starter organisms. Am J Clin Nutr 73(2):374–379

Hemarajata P, Versalovic J (2013) Effects of probiotics on gut microbiota: mechanisms of intestinal immunomodulation and neuromodulation. Therap Adv Gastroenterol 6(1):39–51

Hosono A, Lee J, Ametanl A et al (1997) Characterization of water soluble polysaccharide fraction with immuno-potentiating activity from Bifidobacteriumadolescentis M 91-4. Biosci Biotechnol Biochem 61:312–316

Hussain SA (2013) Development of a technology for the manufacture of aloe vera supplemented probiotic lassi, PhD Thesis, Dairy Technology Division, National Dairy Research Institute (Deemed University), Karnal, India

Ishibashi N, Shimamura S (1993) Bifidobacteria: research and development in Japan. Fd Tech:126–135
Katharina E, Scholz A, Peter A et al (2007) Prebiotics, probiotics and Synbiotics affect mineral absorption, bone mineral content, and bone Structure. JN 137(3):838–846
Kearney N, Meng XC, Stanton C et al (2009) Development of a spray dried probiotic yoghurt containing lactobacillus Paracasei NFBC 338. Int Dairy J 19(11):684–689
Kechagia M, Basoulis D, Konstantopoulou S et al (2013) Health benefits of probiotics: a review. ISRN Nutr 2:481651. https://doi.org/10.5402/2013/481651
Kesavelu D, SrRohit A, Karunasagar I, Karunasagar I (2020) Composition and laboratory correlation of commercial probiotics in India. Cureus 12(11):e11334. https://doi.org/10.7759/cureus.11334
Kim HS (1988) Characterization of lactobacilli and bifidobacteria as applied to dietary adjuncts. Cult Dairy Prod J 23:6–9
Klaenhammer TR, Kleerebezem M, Kopp MV et al (2012) The impact of probiotics and prebiotics on the immune system. Nat Rev Immunol 12(10):728–734
Kristensen NB, Bryrup T, Allin KH et al (2016) Alterations in fecal microbiota composition by probiotic supplementation in healthy adults: a systematic review of randomized controlled trials. Genome Med 8(1):1–11
Kumar O (2009) Physico-chemical and sensory attributes of symbiotic ice-cream. M. Tech. Thesis Submitted to National Dairy Research Institute, Karnal, India
Kumar R, Grover S, Batish VK (2010) Hypocholesterolaemic effect of dietary inclusion of two putative probiotic bile salt hydrolase-producing lactobacillus Plantarum strains in Sprague-Dawley rats. Br J Nutr 105(4):561–573
Kumar M, Dhaka P, Vijay D et al (2016) Antimicrobial effects of *Lactobacillus plantarum* and *Lactobacillus acidophilus* against multidrug-resistant enteroaggregative *Escherichia coli*. Int J Antimicrobial Agents 48:265–270
Lai HM (1993) Mobility of water in a starch based fat replacer by ^{17}O NMR spectroscopy. J Food Sci 58:1103–1106
Lai KK (1996) Antibiotic-resistant enterococci and the changing face of surgical infections. Arch Surg 156:338–342
Lee YK, Salminen S (1995) Theconing of age of probiotics. Trends Food Sci Technol 6:241–245
Levri KM, Ketvertis K, Deramo M et al (2005) Do probiotics reduce adult lactose intolerance- a systematic review. J Fam Pract 54(7):613–620
Lioyd-Evans LPM (1989) Probiotics. PJB Publications Ltd. Richmond, Survey
Lourens-Hattingh A, Viljoen BC (2001) Yoghurt as probiotic carrier food. Int Dairy J 11(1–2):1–17
Luidalepp H, Joers A, Kaldalu N et al (2011) Age of inoculums strongly influences persister frequency and can mask effects of mutations implicated in altered persistence. J Bacteriol 193:3598–4605
Lyons TP (1988) A biological approach to productivity improvements in the calf and dairy cow. Indian Dairyman 40:115–119
Maassen CBM (1999) A rapid and safe plasmid isolation method for efficient engineering of recombinant lactobacilli expressing immunogenic or tolerogenic epitopes for oral administration. J Immunol Methods 223:131–136
Mackowiak PA (2013) Recycling Metchnikoff: probiotics, the intestinal microbiome and the quest for long life. Front Public Health 1:52
Maeno M, Yamamoto N, Takano T (1996) Identification of antihypertensive peptides from casein hydrolysate produced by a proteinase from Lactobacillus helveticus cp790. J Dairy Sci 73: 1316–1321
Mantere-Alhonem S (1995) Propionibacteria used as probiotics – a review. Lait 75(4/5):447–452
Maragkoudakisa PA, Miarisa C, Rojeza P et al (2006) Production of traditional Greek yogurt using lactobacillus strains with probiotics potential as starter adjuncts. Int Dairy J 16(1):52–60
Marin ML, Lee JH, Murtha J et al (1997) Differential cytokine production in clonal macrophage and T-cell lines cultured with bifidobacteria. J Dairy Sci 80:2713–2720

Mattia A, Merker R (2008) Regulation of probiotic substances as ingredients in foods: premarket approval or "generally recognized as safe" notification. Clin Infect Dis 46(2):115–118

McFarland LV, Surawicz CM, Greenberg RN et al (1994) A randomized placebo-controlled trail of Saccharomyces boulardii in combination with standard antibiotics for Clostridium difficile disease. J Am Med Assoc 27:1913–1918

Meurman HJ, Antila H, Korhonen A et al (1995) Effect of Lactobacillus rhamnosus strain GG (ATCC 53103) on the growth of Streptococcus sobrinus in vitro. Eur J Oral Sci 103:253–258

Milani C, Duranti S, Bottacini F (2017) The first microbial colonizers of the human gut: composition, activities, and health implications of the infant gut microbiota. Microbiol Mol Biol Rev 81(4)

Miremadi F, Ayyash M, Sherkat F et al (2014) Cholesterol reduction mechanisms and fatty acid composition of cellular membranes of probiotic *Lactobacilli* and *Bifidobacteria*. J Funct Foods 9:295–305

Mishra V, Prasad DN (2000) Probiotics and their potential health benefits. Indian Dairyman 52(4): 7–13

Nag A, Das S (2013) Improving ambient temperature stability of probiotics with stress adaptation and fluidized bed drying. J Funct Foods 5(1):170–177

Nagpal R, Kumar A, Kumar M et al (2012) Probiotics, their health benefits and applications for developing healthier foods: a review. FEMS Microbiol Lett 334(1):1–15

Oyetayo VO, Oyetayo FL (2005) Potential of probiotics as biotherapeutic agents targeting the innate immune system. Afr J Biotechnol 4(2):123–127

Parvez S, Malik KA, Ah Kang S et al (2006) Probiotics and their fermented food products are beneficial for health. J Appl Microbiol 100(6):1171–1185

Pavli F, Tassou C, Nychas GJE et al (2018) Probiotic incorporation in edible films and coatings: bioactive solution for functional foods. Int J Mol Sci 19(1):150

Pech-Canul ADLC, Ortega D, García-Triana A et al (2020) A brief review of edible coating materials for the microencapsulation of probiotics. Coatings 10(3):197

Perdigon G, Nader Demacias ME, Alvarez S et al (1986) Effect of perorally administered lactobacilli on macrophage activation in mice. Infect Immun 53:404–410

Poddar D, Das S, Jones G et al (2014) Stability of probiotic lactobacillus paracasei during storage as affected by the drying method. Int Dairy J 39(1):1–7

Pop OL, Pop CR, Dufrechou M et al (2020) Edible films and coatings functionalization by probiotic incorporation: a review. Polymers 12(1):12

Pouwels PH, Leer RJ, Shaw M et al (1998) Lactic acid bacteria as antigen delivery vehicles for oral immunization purposes. Int J Food Microbiol 41:155–167

Prabhurajeshwar C, Chandrakanth K (2019) Evaluation of antimicrobial properties and their substances against pathogenic bacteria in-vitro by probiotic lactobacilli strains isolated from commercial yoghurt. Clin Nutr Exp 23:97–115

Pyar H, Peh KK (2014) Characterization and identification of Lactobacillus acidophilus using biolog rapid identification system. Int J Pharm Pharm Sci 6(1):189–193

Rajpal S, Kansa VK (2008) Buffalo milk probiotic Dahi containing *Lactobacillus acidophilus*, *Bifidobacteriumbifidum* and *Lactococcuslactis* reduces gastrointestinal cancer induced by dimethylhydrazine dihydrochloride in rats. Milchwissenschaft 63:122–125

Rakhi (2005) Process optimization of low calorie probiotic shrikhand and its powder. PhD Thesis. G B Pant University of Agriculture & Technology, Pantnagar

Ram C, Bhavadasan MK (2002) Probiotic dairy foods-present status and future perspectives. Indian Dairyman 54(4):53–57

Rezac S, Kok CR, Heermann M et al (2018) Fermented foods as a dietary source of live organisms. Front Microbiol 9:1785

Rolfe RD (2000) The role of probioitic cultures in the control of gastrointestinal health. J Nutr 130(25):396–400

Rowland IR (2009) The role of the gastrointestinal microbiota in colorectal cancer. Curr Pharm Des 15(13):1524–1527

Roy D (2005) Technological aspects related to the use of Bifidobacteria in dairy products. Lait 85: 39–56

Ruiz L, Margolles A, Sánchez B (2013) Bile resistance mechanisms in lactobacillus and Bifidobacterium. Front Microbiol 4:396

Saarela M, Mogensen G, Fonden R et al (2000) Probiotics in dairy fermented products. J Biotechnol:84197215

Saarela MH, Alakomi HL, Puhakka A et al (2009) Effect of the fermentation pH on the storage stability of lactobacillus Rhamnosus preparations and suitability of in vitro analyses of cell physiological functions to predict it. J Appl Microbiol 106(4):1204–1212

Saavedra JM, Dattilo A (2003) Microbiota of the intestine: prebiotics. Encyclopedia of human nutrition (3rd ed), pp 168–174

Salminen S, Von Wright A (1993) Current probiotics safety assured? Microb Ecol Health Dis 10: 68–77

Saloff-Coste CJ (1995) Nutritional and health benefits of yoghurt and fermented milk. Danone World Newslett 7:2–8

Sanders ME (1999) Probiotics. Food Technol 53(11):67–75

Sanders ME (2000) Consideration for use of probiotic bacteria to modulate human health. J Nutr 130(25):S384–S389

Sanders ME, Huisin't Veld J (1999) Bringing a probiotic-containing functional food to the market: microbiological, product, regulatory and labeling issues. Antonine van Leeuwenhoek 76:293–315

Santacroce L, Charitos IA, Bottalico L (2019) A successful history: probiotics and their potential as antimicrobials. Expert Rev Anti-Infect Ther 17(8):635–645

Sarkar S, Misra AK (1998) Effect of feeding Propiono acid-Bifido (PAB) milk on the nutritional status and excretory pattern in rats and children. Milchwissenschaft 53:666–668

Savaiano DA, Hutkins RW (2021) Yogurt, cultured fermented milk, and health: a systematic review. Nutr Rev 79(5):599–614

Saxellin M, Chuang MH, Chassy B et al (1996) Lactobacilli and bacteremia in southern Finland, 1989-1992. Clin Infect Dis 22:564–566

Senok AC, Ismaeel AY, Botta GA (2005) Probiotics: facts and myths. Clin Microbiol Infect 11(12): 958–966

Shah NP (2000) Probiotic bacteria: selective enumeration and survival in dairy foods. Int J Dairy Sci 83(4):894–907

Shah NP, Ding WK, Fallourd MJ et al (2010) Improving the stability of probiotic bacteria in model fruit juices using vitamins and antioxidants. J Food Sci 75(5):M278–M282

Sharma R (2005) Market trends and opportunities for functional dairy beverages. Aust J Dairy Technol 60(2):195–198

Shortt C (1999) The probiotic century: historical and current perspectives. Trends Food Sci Technol 10(12):411–417

Silva M, Jacobus NV, Deneke C et al (1987) Antimicrobial substance from a human *Lactobacillus* strain. Antimicrob Agents Chemother 31:1231–1233

Sloan E (2005) Top 10 global food trends. Food Technol 59:20–32

Solis Pereyra B, Lemonnier D (1993) Induction of human cytokines by bacteria used in dairy foods. Nutr Res 13:1127–1140

Song D, Ibrahim S, Hayek S (2012) Recent application of probiotics in food and agricultural science. PRO 10:1–34

Tanaka M, Iwanami T, Aai K et al (1982) Cultured milk production method. German Federal Rep Patent De 3048:438

Terpou A, Papadaki A, Lappa IK et al (2019) Probiotics in food systems: significance and emerging strategies towards improved viability and delivery of enhanced beneficial value. Nutrients 11(7): 1591

Tobey JF (1992) A current perspective: The utilization of microbial fermentation products in feed and forage. In Legar DA, Ho SK (eds.). Proceedings of the int round table on animal feed biotechnology, research and scientific regulation agriculture, Canada, Ottawa, Canada. pp. 7–17

Tomioka H, Saito H (1992) Lactic acid bacteria in the support of immuno compromised hosts. In: Wood BJ (ed) The lactic acid bacteria. Vol. 1. The lactic acid bacteria in health and disease. Elsevier Applied Science, London, pp 263–296

Tripathi MK, Giri SK (2014) Probiotic functional foods: survival of probiotics during processing and storage. J Funct Foods 9:225–224

Verma SVS, Singh R (1995) Probiotics and their role in broiler chicken production: an overview. J Anim Sci 10:359–363

Vesa TH, Marteau P, Zidi S et al (1996) Digestion and tolerance of lactose from yoghurt and different semi-solid fermented dairy products- is bacterial lactase important? Eur J Clin Nutr 50: 730–733

Vibha (2004) Effect of standard Lactobacilli on cardiovascular disease risk factors for their potential application as probiotics. Ph.D. Thesis, Thesis submitted to National Dairy Research Institute, Karnal, Deemed University

Vimala Y, Dileep P (2006) Some aspects of probiotics. Indian J Microbiol 46:1–7

Yeh TL, Shih PC, Liu SJ et al (2018) The influence of prebiotic or probiotic supplementation on antibody titers after influenza vaccination: a systematic review and meta-analysis of randomized controlled trials. Drug Des Devel Ther 12:217–230

Zoumpopoulou G, Foligne B, Christodoulou K et al (2008) Lactobacillus Fermentum ACA-DC 179 displays probiotic potential in vitro and protects against Trinitrobenzene sulfonic acid (TNBS)-induced colitis and salmonella infection in murine models. Int J Food Microbiol 121: 18–26

Zuo T, Zhang F, Lui GCY (2020) Alterations in gut microbiota of patients with COVID-19 during time of hospitalization. Gastroenterology 159:944–955

Food Allergies and Toxicity

15

Feby Luckose, Saisree Iyer, Krishna Kumar Ballamoole, and Mamatha Bangera Sheshappa

15.1 Introduction

Food allergy is a major health problem and the prevalence is estimated to be 1.3 and 4–6% in adults and children respectively (WHO 2006). There are above 70 foods listed to cause allergies. Even a little amount of food can trigger allergy and cause serious and fatal reactions among vulnerable persons. According to WHO, food allergy is "A hypersensitivity reaction initiated by proven or strongly suspected immunologic mechanisms". The documents of the Codex Alimentarius Commission, Food and Agriculture Organization (FAO), and World health organization (WHO) specified that if a food constituent has to be a potential allergen it should be a foreign protein (80-amino acid sequence) that has 35% homology with a known allergen. Databases (biologic and molecular) for allergens are crucial for their homology analysis in comparison to a known allergen. While physiological, clinical, and other such allergen information is included under biological databases, molecular databases provide information on the sequences of amino acids and allergen structure.

Allergy is when adverse effects are shown commonly in the skin, eyes, mouth, gut, or rarely in the upper respiratory organs and central nervous system. Besides, an uncontrollable allergic response can lead to the death of the consumer. However, avoiding consumption of foods containing allergens is the only way out. Allergies resulting from food consumption not only affect the life of the consumer (with an allergic response) but also diminish the quality of food and hinder the economic growth of the manufacturer. In foods that involve recombinant DNA technology, primary safety concern is an allergic response. Due to the presence of proteins which

F. Luckose · S. Iyer · K. K. Ballamoole · M. B. Sheshappa (✉)
Department of Biological Sciences, Nitte (Deemed to be) University, Nitte University Centre for Science Education and Research, Mangalore, Karnataka, India
e-mail: mamatha.bs@nitte.edu.in

O. P. Chauhan (ed.), *Advances in Food Chemistry*,
https://doi.org/10.1007/978-981-19-4796-4_15

are genetically modified for its expression, there is a rising concern that allergic consumers could be at risk, on exposure to such human-made alterations. However, procedures to estimate newly expressed proteins for their potential allergenicity, if allergenic, are enlisted by the Codex Alimentarius Commission. Consumers can refer to the International Food Safety Authorities Network (INFOSAN) note on food allergies, to gain basic information regarding the same. In the improvement of lifestyles, the labeling management of the presence of allergens in food is an area that is being increasingly focussed on by all countries. Recently, the effort of researchers has resulted in modern food technological tools which enable allergen identification by consumers (Simons et al. 2005). To recognize the food allergens prior to consumption, mentioning the presence of the allergen in the label on the package is essential. Allergen mention on food labels will help susceptible persons to avoid consumption of allergenic food, which can thereby reduce most food allergies (Soares et al. 2014). Every country has diverse geography and cultures that have led to adaptive eating habits. Hence, the allergic susceptibility of people in different countries depends upon food habits. This distinctness forces every country to establish specific management systems as per the needs of the country.

Industries producing food need to take into account their operating procedures to identify and accordingly establish preventing strategies against unexplored allergens. Manufacturing operations should necessarily be assessed to ascertain the control points. Determining allergen under chemical hazards should be included in HACCP plan. Additionally, personnel from management, researchers, and line production employees must understand the risks associated with allergens. Understanding the potential consequences of food allergy will strengthen the significance of ensuring the appropriate control procedures. It is also necessary to have a continuous tracking system to verify control points.

15.2 Common Food Allergens

Food allergens are substances (biological, chemical, or physical) that can bring about allergic reactions in sensitive individuals. Factors like the type and amount of allergen, entry points, and hypersensitivity level of the individual influence the ability of an allergen to trigger a reaction. Food allergens commonly encountered are proteins that are either water-soluble or salt-soluble (Petruláková and Valik 2015). They can also exist as glycoproteins. Many allergenic food proteins have been characterized and most of them are tolerant to acid or heat treatments. Besides, proteases ranging from 10 to 70 kD in size are ineffective against food allergens (Sicherer and Sampson 2010). During food processing, these allergens do not undergo denaturation or degradation and cause an allergic response on consuming packed food products.

Eight foods/food groups account for the majority (around 90%) of all IgE-mediated food in the world (Taylor and Hefle 2001; Sicherer and Sampson 2010). In 1999, the following foods or food groups—milk, eggs, fish, crustacea (shrimp, crab, lobster, crayfish), peanuts, soybeans, nuts (almonds, walnuts, pecans, cashews, Brazil nuts, pistachios, hazelnuts, etc.), and wheat were listed by the Codex Alimentarius Commission (Taylor and Hefle 2001; Gangal and Malik 2003). Although spices like turmeric, cumin seeds, and some fruits have been discovered to cause mild allergic reactions in the Indian population, these are not included in the Codex food allergen list (Gangal and Malik 2003). The European Union has recommended the following list of foods (as well as food products containing any of these foods as an ingredient) for mandatory marking of allergens (Petruláková and Valik 2015):

- cereals containing gluten (wheat, rye, barley, oats, etc.)
- crustaceans
- eggs
- fish
- peanuts and soybeans
- milk
- nuts (almond, hazelnut, walnut, cashew, pecan nut, Brazil nut, pistachio nut, academia nut, and Queensland nut)
- celery
- mustard and sesame seeds
- sulfites and sulfur dioxide (>10 mg/kg or >10 mg/L)

Food additives like colors, preservatives, flavoring substances, and antioxidants can also trigger allergic reactions and intolerances in some susceptible individuals.

Although seafood is an integral part of human nutrition and health, it can also trigger allergic reactions in some susceptible individuals. Allergic reactions to seafood affect up to 0.2% of the general population, according to available estimates and can range from mild symptoms like itching in the throat, swelling of lips to life-threatening anaphylactic reactions. Some seafood allergies including histamine fish poisoning can trigger clinical symptoms (Prester 2016). Parvalbumins (PV), one of the many components of fish muscle protein, are the most common allergen in fish. It was first discovered in Baltic cod in the 1960s as a fish allergen (Kuehn et al. 2014). PVs are highly stable, calcium-binding proteins with low molecular weight (10–12 kDa). The percentage of PVs in fish differs considerably based on the species and type of muscle. The white muscle has an abundance of parvalbumins up to 5 mg/g of fresh weight (Griesmeier et al. 2010). The study of Kobayashi et al. (2016) showed the relationship between parvalbumin content in different types of fish muscle and their propensity to cause IgE reactivity.

Scombroid syndrome/histamine poisoning is another type of pseudoallergic poisoning caused due to consuming seafood having high amounts of histamine and other biogenic amines (Hungerford 2010). This type of syndrome is triggered by different species of the Scombridae family which have high levels of free amino acid histidine. Histamine is a biogenic amine produced from the amino acid histidine due to the decarboxylation process catalyzed by a histidine decarboxylase synthesized by bacteria (World Health Organization 2013; Mavromatis and Quantick 2002). *Morganella morganii*, *Morganella psychrotolerans*, *Photobacterium damselae*, *Photobacterium phosphoreum*, *Raoultella planticola*, and *Hafnia alvei* are the important bacteria species that participate in the formation of different biogenic amines by producing decarboxylase enzymes (Visciano et al. 2014; World Health Organization 2013). Upon consumption, histamine expresses its effects via the activation of four different types of histamine receptors (H1, H2, H3, and H4) which activate different signaling pathways, resulting in several biological responses such as vasopermeability and vasodilatation, urticaria, hypotension, and headache.

Gluten is a protein which is found in cereals like wheat, barley, rye, etc., that can result in the progression of celiac disease in certain individuals (Petruláková and Valik 2015). Celiac disease is a non IgE cell-mediated reaction. Small intestinal cells remain inflamed and hinder nutrient absorption. The gliadin fraction of gluten is found to be associated with the initiation of celiac disease (Taylor and Hefle 2001; Gangal and Malik 2003). Allergen proteins from cereals like rice, oats, and barley have also been identified and been found to be a part of the α-amylase/trypsin inhibitor family. Food allergy to cereals is mostly seen in Asian countries and not so often in Western countries (Gangal and Malik 2003; Bhattacharya et al. 2018). The trypsin/alpha-amylase inhibitor is a proteinaceous inhibitor from cereals consisting of about 120 amino acids which contain 10 cysteine residues (Gourinath et al. 2000; Ziegler et al. 2019). Trypsin/alpha-amylase inhibitors from wheat have five α-helices that are arranged in an up and down manner and are rich in cysteine residues forming disulfide bonds. These inhibitors are biologically active as they can tolerate food processing conditions and proteolysis in the gastrointestinal (GI) tract. In GI tract, α-amylase/trypsin inhibitors are capable of stimulating immune cells that reside in the lamina propria and mesenteric lymph nodes through TLR4 binding and stimulation thereby initiating Coeliac Disease (CD) and Non-coeliac Wheat Sensitivity (Geisslitz et al. 2021). Legumes are high protein foods and a staple in the Indian diet. Chickpea is a common allergen and an anaphylactic reaction on consumption of chickpea has been reported. It is consumed in boiled, roasted, and flour forms (Devdas et al. 2017). Black gram is a widely consumed legume in India. 1.7% of patients reporting allergic reactions were identified to be sensitized to black gram extracts. Allergens in kidney beans, another common legume consumed mainly in North India, were found to have cross-reactivity with other legumes like black gram, lentil, pea, pigeon pea, and chickpea (Bhattacharya et al. 2018). Allergic reactions to peanuts, although rare in India, are a significant health problem. Unlike hypersensitivity to other foods like egg or milk, hypersensitivity to peanuts is usually life-long and in severe cases can result in anaphylaxis and even death (Maleki and Hurlburt 2002; Devdas et al. 2017). Allergic reactions to peanuts and processed food containing peanuts are a great concern for the peanut industry.

15.3 Prevalence of Food Allergies

Globally about half a billion people suffer from food allergy (Pawankar 2013). Although India shows the least prevalence of food allergy which is as low as 1–2% (Mahesh et al. 2016), urbanization and lifestyle change in recent decades are making Indian people become more prone to allergic reaction to foods like legumes, prawn, eggplant, milk, and egg. It was found that around 22% of the food allergic population in Delhi was sensitive to Kidney bean (*Phaseolus vulgaris*), a commonly consumed plant protein (Bhattacharya et al. 2018; Arakali et al. 2017; Kumar et al. 2013). Children usually outgrow food allergies to most of the foods, with around 50–60% of children allergic to milk and egg demonstrating tolerance by school age (Devdas et al. 2017).

In a survey including 83 World Allergy Organization member countries and 6 non-member countries, only 9 countries had data on food allergy prevalence based on oral food challenges; whereas, 51 countries had no data on food allergy prevalence (Loh and Tang 2018). A few countries had data based on self-reporting, which generally overestimates the prevalence of food allergy as many times a patient or parent mistakes food allergies to other reactions to food like food poisoning or enzyme deficiencies. There is a scarcity of data regarding the trends of food allergy among Indians. In 2016, for the first time, a study was conducted to assess the global prevalence (8 European countries, Hong Kong, Russia, and India) of allergic responses by EuroPrevall-INCO. Clinical symptoms were reported and positive IgE tests were taken as the basis for this study. Participants from South India (11,000 randomly selected in the age group 20–54 years) showed much higher rates of sensitization to the tested foods than participants from European countries (Mahesh et al. 2016). Table 15.1 summarizes the Indian studies on sensitization patterns to usually listed food allergens.

Although the rates of sensitization were high, the allergic prevalence in India was very low at 1.2%. The most common allergens were cow's milk and apple (Arakali et al. 2017, Krishna et al. 2020). Food such as shrimp, buckwheat, and corn that showed sensitization rates above 10%, interestingly did not result in any probable food allergy. Foods such as wheat, sesame, and tomato were associated with a few cases (0.02–0.05% weighted prevalence) of allergic reactions.

This study showed that even in Case-control studies, there is a poor association between patient's history and sensitization patterns. Patients frequently consume foods they are sensitized to without symptoms of food allergy. It is not clear why the Indian population with higher levels of food-specific IgE in comparison to most Western countries have a lesser prevalence of probable food allergy (Mahesh et al. 2016).

It is crucial to note that other than IgE sensitization, additional factors are necessary for allergic response to manifest clinically. Li et al. (2020) suggested that the prevalence of food allergy in developing countries may be relatively low and few food allergens commonly encountered could vary from the ones in developed countries. Furthermore, Arakali et al. (2017) remarked that the trends of food allergy might differ among ethnic groups. Asians follow an exclusive diet that might be composed of different allergens which have still not been investigated or explored. Many factors such as exposure to microbes, dietary preferences, and genetic makeup may induce epigenetic effects in different ethnic groups especially in developing and underdeveloped nations thereby modulating their immune system (Devdas et al. 2017). Interestingly, numerous studies have observed that the timing of the

Table 15.1 Food allergen sensitization patterns in India

Place and samples size	Tested foods	Salient findings	Reference
Tertiary care hospital, New Delhi, North India, $N = 1860$	Rice, black gram, pea, Lentil, citrus fruits, maize, Banana	Sensitization rates: Rice 6.2%; black gram 5.9%; lentil 5.5%; citrus fruits 5.3%; maize 3.8%; pea 3.8%; Banana 3.6%	Kumar et al. (2010)
Clinic in Kolkata, Eastern India, $N = 5161$	46 foods banana, brinjal, lentils, wheat, egg	Sensitization rates: Bananas 32.4%; Brinjal 29.4%; egg 23.1%; wheat 21.7%; lentils 10.4%;	Dey et al. (2014)
Randomly selected General population from Mysore and Bangalore (Part of an international study-EuroPrevall), adults screened $N = 10{,}931$, Case control study, $N = 587$	24 EuroPrevall priority foods—fish, cow's milk, egg, mustard seed, soya bean, peanut, lentil, wheat, buckwheat, walnut, poppy seed, melon, sunflower, corn, banana, sesame, shrimp, tomato, kiwi, carrot, celery, apple, peach, and hazelnut	% Sensitization and probable allergy: Wheat 11.93% and 0.02%; Peanut 8.73% and 0.0%; Egg 2.6% and 0.05%; Milk 2.71% and 0.50%; Fish 0.5% and 0.0%; Shrimp 15.53% and 0.0%; Apple 7.27% and 0.50% (IgE response was used to develop sensitization rate by means of a history of allergic response within 2 h of consumption, specific IgE response or SPT)	Mahesh et al. (2016)
University teaching Hospital in Karnataka, $N = 2219$	31 foods Red gram, Green gram, red kidney beans, wheat, Egg	Sensitization rates: Red gram 12.6%; Green gram 12.5%; Red kidney beans 10.9%; Wheat 9.6%; egg 6.9%	Chogtu et al. (2017)
School children urban and rural schools, $N = 350$	10 foods Prawn, Peanut, Fish, Milk, Banana	Sensitization rates: Prawn 17.7% (urban), 5.7% (rural); Peanut 19.6% (urban), 10.4% (rural); Fish 17.7% (urban), 5.7% (rural); Milk 17.7% (urban), 5.2% (rural); Banana 2.5% (urban), 1.0% (rural)	Gobinaath and AnandArokiaraj (2018)

(continued)

Table 15.1 (continued)

Place and samples size	Tested foods	Salient findings	Reference
Randomly selected general population from Mysore and Bangalore (part of an international study Europrevall, which screened 35,549 in China, India, and Russia), Children screened $N = 5677$, Case control study, $N = 450$	25 EuroPrevall priority foods Hen's egg, cow's milk, peanut, soy, hazelnut, walnut, celery, kiwi, apple, peach, sesame, mustard, wheat, fish, shrimp, buckwheat, corn, carrot, rice, tomato, melon, banana, lentils, sunflower seeds, and poppy seeds.	% Sensitization and probable allergy: Wheat 6.7% and 0.0%; Peanut 6.3% and 0.0% Egg 1.7% and 0.05%; Milk 2.1% and 0.0%; Fish 0.4% and 0.0%; Shrimp 10.3% and 0.0%; Apple 4.2% and 0.0%, (IgE response was used to develop sensitization rate by means of a history of allergic response within 2 h of consumption, specific IgE response or SPT)	Li et al. (2020)

Source: Krishna et al. 2020

introduction of solid foods in infants as a key factor in the prevention of food allergy in children (Rathi and Sharma 2017). Early weaning (prior to 16 weeks) is found to prevent from developing food allergy in children (Greer et al. 2008; Venter et al. 2009). Impeding the introduction of foods considered highly allergenic like peanuts may in fact cause a rise in the incidence of allergy to these foods (Joseph et al. 2011; Robison 2014).

15.4 Pathophysiology of Food Allergic Reactions and Clinical Manifestations

Food allergy is a detrimental health effect as a result of the pathological reaction of the immune system towards food allergens. Ingestion of allergenic foods and/or food containing dietary antigens can set off pathological symptoms such as anaphylaxis, GI disorders, urticaria, and airway inflammation with varying degrees of severity from mild to fatal (Dupont 2011, Vickery et al. 2011). In spite of the immense diversity of the human diet, only a few foods are responsible for the majority of food allergic reactions thereby posing a significant psychosocial burden on affected individuals and their families. The human body has many layers of protection imposed by the immune system with the capability to identify and tolerate whatever belongs to the self and to identify and reject foreign substances. The food allergy is probably the result of the disrupted physiological development of oral tolerance or of disruption of previously formed immune tolerance (Faria and Weiner 2005, Dupont 2011). The columnar epithelial cells of the gastrointestinal tract constitute the major portion of the surface area in the human body and they perform two major functions such as absorption of nutrients and restricting the entry of harmful substances. The intestinal epithelium functions as a barrier to sequester antigens from the mucosa-associated lymphoid tissue (MALT) when exposed to the potential allergen(s), and the leakiness of this barrier has been linked to allergic sensitization. Components of the GI tract participate in the "tolerance" of foreign substances that are not damaging to the body and cooperate with a number of immune cells which play a key role in the development of oral tolerance. Apart from host components, the physical properties of the antigen, the amount and regularity of exposure as well as commensal gut flora can also affect the development of oral tolerance (Chehade and Mayer 2005; Sudo et al. 1997).

15.4.1 IgE-Mediated Hypersensitivity

IgE-mediated hypersensitivity reactions account for the majority of allergy reactions to food allergens, and those who are genetically prone to atopy create particular IgE antibodies to proteins they are exposed to (Dupont 2011; Ho et al. 2014). These allergen-specific IgE antibodies bind to mast cells and basophilic cells that circulate in the bloodstream. When a dietary protein is ingested, the IgE recognizes it on the surface of these cells, triggering the release of mediators (such as histamine), and causes symptoms. The symptoms of IgE-mediated reactions commonly manifest on the skin, respiratory system, and GI tract.

15.4.2 Clinical Manifestations

Unfavorable reactions to food allergen consist of a range of clinical syndromes resulting from immunological to the ingested food allergen. Symptoms can be mild, self-limiting reactions to severe, life-threatening reactions depending on the mechanism. It can affect all organs and differ according to the mechanism involved. The summary of the common clinical manifestation due to food allergy is summarized in Table 15.2.

Table 15.2 Summary of clinical manifestations due to food allergy

Target organ	Clinical manifestation
Skin	Urticaria Angioedema Atopic dermatitis Dermatitis herpetiformis
Respiratory tract	Asthma Allergic rhinitis
Gastrointestinal tract	Oral allergy syndrome Gastrointestinal "anaphylaxis" Allergic eosinophilic gastroenteritis Celiac disease
Systemic	Anaphylactic shock

15.5 Effect of Food Processing on Food Allergens

Understanding the impact of food processing on the allergenic potential of food is essential to manage the risk of allergens in the food chain. Since most food allergens are proteins, modification of their functional properties by altering the structural properties can have an effect on their immunoreactivity (Ekezie et al. 2018).

Various food processing techniques can influence the allergenicity of proteins and thereby favor endurance level in food allergic subjects. There have been few studies on changing food proteins to minimize allergenicity in food processing systems. The ability to tolerate specific allergens at a young age, and thus the capacity of young children to resist food allergies in adulthood, could be achieved by using thermal processing procedures for the specific allergen. It is necessary to develop new pathways in research on processed foods which can possibly have an impact on food allergy prevention and in turn on the quality of life of the patient.

There is a substantial impact on the allergenicity of proteins due to physical, chemical, and biochemical changes during processing. In their denatured state, proteins found in processed foods will aggregate in protein networks during interactions with the food matrix (carbohydrates and lipids), that lowers the allergen content, and hence potentially lowering its sensitizing potential or formation of neoepitopes (Thomas et al. 2007). Thermal processing can reduce allergens by the destruction of predominantly conformational epitopes or by limiting the effect on sequential epitopes or by chemical reactions in the food matrix between proteins, fat, and sugars that allows for lowered protein availability to the immune system. Food processing may affect the food matrix and result in the formation of neoepitopes having lowered digestibility in the stomach. This could lead to a potential increase in the protein allergenicity of the epitopes as they pass through the stomach and react with the immune system in the gut (Mansoor and Sharma 2011). Food is a mixture of allergenic proteins with different physico-chemical qualities, heat and digesting resistance, and propensity to cause IgE sensitization and IgE-mediated hypertensive reactions. Food allergens (e.g., in peanut, wheat, cow's milk, and egg white) that can potentially induce IgE sensitization through the mucosal layer of the GI tract, which are heat-insensitive, acid-stable, water-soluble, and found in the range of 10–70 kDa. Food processing does not affect the class 1 food allergens easily (e.g., Ara h 2 in peanut or Gal d 1 in egg white), although recent discoveries have highlighted the significance of the modification of conformational epitope in egg and cow's milk allergy (Sampson 2004). Many human studies have been conducted in which egg-allergic patients were tested for allergy after consuming cooked and/or uncooked eggs. Approximately 50–85% of children with egg allergy were able to tolerate baked eggs (Bartnikas and Phipatanakul 2013; Cortot et al. 2012; Turner et al. 2014). Studies on the allergenicity of wheat gluten proteins indicated that wheat-derived foodstuffs contain salt-soluble proteins that manifest weaker allergenic potential than those found in raw flour (de Gregorio et al. 2009). There is no effect of exposure to proteolytic digestion and heat processing on the stable structure of fish proteins that makes them capable to sustain their allergenicity (Mondal et al. 2007). Although, when tuna and salmon were exposed to extreme temperature and pressure during canning, it reduced the IgE-binding activity by 100–200-folds (Bernhisel-Broadbent et al. 1992). Further studies concluded that the most effective method to accelerate the digestion of tropomyosin in the GI digestion stage was high-pressure steaming. Appropriate processing of crab can specifically reduce the reactivity of IgG/IgE-binding of TM thereby lowering the number of crab hypersensitivity incidences in humans (Yu et al. 2011).

During food processing, the interaction of proteins with components like fats, sugars, and other proteins of the food matrix result in less available protein for interaction with the immune system. β-lactoglobulin, on high-temperature treatment forms intermolecular disulfide bonds and binding with food matrices, making β-lactoglobulin less allergenic (Thomas et al. 2007). Kato et al. (2000) reported that during bread preparation, egg albumin mixed with gluten (wheat) and wheat flour on heating at 180 °C for 10 min caused a reduction in the ovomucoid solubility. This was found to be due to the polymerization of ovomucoid with gluten-forming complexes of high molecular weight resulting in aggregation and insolubilization of ovomucoid. In a recent survey done on children with egg and milk allergy, a novel way of delivering dietary protein was studied by its interactions with the wheat matrix and heat treatment.

Even though the allergenicity of fish, peanut, and wheat proteins may increase on high-temperature processing, recent studies indicate that food allergies could be combated by applying selective thermal processing regimes on certain allergenic foods (Vissers et al. 2011; Blanc et al. 2011; de Gregorio et al. 2009; Yu et al. 2011). However, thermal processing does not fully eliminate the allergenic potential of allergens. Methods like fermentation and hydrolysis might possibly lower the allergenicity to such an extent that symptoms will not be induced.

Some food allergens are found to be heat-resistant and thermal processing does not effectively reduce their immunoreactivity (Lasekan and Nayak 2016). Non-thermal processing can help in maintaining the sensory properties and improving nutritional properties and as per findings of recent studies, it can also induce a reduction in food immunoreactivity (Meinlschmidt et al. 2016). The most commonly examined non-thermal treatments in food processing are UV irradiation, Ultrasound, High-pressure processing (HPP), cold plasma, and pulsed electric fields. Relatively high intensities of such treatments can alter the structure and functional properties of allergens in food. On exposure to such treatments, hydrophobic groups present in these proteins are affected causing unfolding of the complex structures allowing them to aggregate (Esteghlal et al. 2019). Loss of structural integrity causes reduction in the allergenicity of the proteins. Additionally, inter or intramolecular interactions and disulfide bond breakage has reported to perform similarly (Renzone et al. 2015).

Gamma-irradiation can affect the antibody-binding ability and the allergenicity of proteins by inducing changes in conformational epitopes. Also, these radiations depolymerize the proteins and cause aggregation, which thereby alters the antibody-binding sites in allergens from milk, egg, and shrimp (Esteghlal et al. 2019). Several studies reported reduced allergenicity for 25 kGy and 100 kGy γ-irradiated tropomyosins from shrimp, ovalbumin, and allergens from legumes

(Vaz et al. 2012; Luo et al. 2013). In contrast, no significant reduction in the antigenicity of bread dough and pasta made from wheat, egg proteins (ovalbumin and ovomucoid), as well as β-Lactoglobullin in whey and liquid milk when treated with 10 kGy dose of γ-irradiation (Gomaa and Boye 2015; Ekezie et al. 2018).

HPP (100–800 MPa) is commonly used to inactivate microbes which not only ensures safety of the processed food but also improves its quality (Khan et al. 2019). Similar to other non-thermal processes, the aim is to alter the structure of allergens. Usually, at 200–300 MPa, allergens conformationally change (Kurpiewska et al. 2019). In recent times, due to the denaturation caused by HPP, it has been an effective tool to reduce food allergies (Gharbi and Labbafi 2018). The immunoreactivity of soy protein extract treated to 200–500 MPa for 5–20 min decreased by around 55.5% (Li et al. 2018). Similarly, reduced protein immunoreactivity was noted in walnuts, ginkgo seed, and squid upon subjecting to above 500 MPa (Yang et al. 2017; Zhou et al. 2016; Jin et al. 2015) as well as attenuation of the antigenicity of α-casein in milk (Huang et al. (2014). However, some studies have found that high pressure (up to 600 MPa) is not effective in altering the allergenicity of apple, hazelnut, walnut, and peanuts (Husband et al. 2011; Prieto et al. 2014; Cabanillas et al. 2014; Huang et al. 2014; Ekezie et al. 2018).

Recent studies reported that the application of Pulsed UV light induces protein aggregation of food allergens by conformational modifications. 20–50% reduction in the protein immunoreactivity was seen in almonds, milk, egg, soy, and shrimp (Li et al. 2013; Tammineedi et al. 2013; Manzocco et al. 2013; Shriver and Yang 2011). Pulsed electric field (PEF) can inactivate enzymes and microorganisms and also modify food protein structure and alter their functional characteristics (Han et al. 2018). Unfolding of proteins and conversion into aggregates under high intensive electric field has been reported to lower allergenicity of certain food proteins. (Ekezie et al. 2018). The immunogenic properties of ovalbumin were significantly modified by PEF (high intensities of 25–35 kV/cm) (Yang et al. 2017).

In a recent study, peanut flour and whole peanut (dry and defatted) were treated at 80 kV for 15–60 min and obtained up to 43% antigenicity reduction in samples (Venkataratnam et al. 2019). Similarly, 76% and 37% reduction in immunoreactivity in shrimp and wheat proteins were observed after cold plasma treatment (Nooji 2011).

Use of sonic waves (20 kHz) is an emerging food processing technology. The implosion of sonication bubbles under application of high energy leads to localized high pressure of up to 1000 atm which can cause structural changes of the food allergens (Soria and Villamiel 2010). It can also reduce immunoreactivity in soy, milk, crayfish, and peanuts (Choudhary et al. 2013; Tammineedi et al. 2013; Zhang et al. 2018; Li et al. 2013).

Combining the above-mentioned technologies for food processing helps in effectively reducing the immunoreactivity of allergens (Ekezie et al. 2018). Combined application of high pressure and enzymatic treatment leads to conformational changes of peanut allergens (Vanga et al. 2017, Li et al. 2013). The main allergens in peanut extracts (Ara h 1, Ara h 2) are eliminated as ultrasound-enzymatic (chymotrypsin or trypsin) treatment negatively affects their IgE binding. Boiling and then 25 kGy gamma-irradiation lowered IgE binding in kidney bean, black gram, and peanut proteins. Combination treatment was effective in reducing the potency of allergen proteins in legumes when compared to boiling or irradiation alone. (Kasera et al. 2012).

Both thermal and non-thermal food processing methods have a definite effect on the conformational structure of the protein and thereby on the allergenicity. The novel processing methods have shown promising results for developing hypoallergenic foods. However, further careful evaluation needs to be carried out to determine the influence of these processing methods on allergens (Vanga et al. 2017). Extensive research is needed to enhance the accuracy of methods for the detection of allergens in processed foods and further investigations of the combined outcome of processing methods on antigenicity and allergenicity should be taken up.

15.6 Detection and Quantification of Food Allergens

Reliable detection and quantification methods for the analysis of allergens in foods are important to ensure compliance with food labeling and for better consumer protection. A range of analytical methods with varying degrees of sensitivities is available for the routine screening and quantification of allergens from food samples either by targeting the allergen or a marker that indicates the presence of an allergen in the food being tested. However, the selection of a proper analytical technique to detect food allergens can be difficult due to (i) inherent complexity of food samples, (ii) presence of undeclared allergens, (iii) differences in the labeling requirements, (iv) analytical uncertainty and increased analytical expenses, and (v) amounts of allergenic proteins in study samples (Sharma et al. 2017). The analytical methods employed to estimate allergens in food should ensure the complete absence of allergens in food with a high level of confidence. The most commonly employed method for detecting allergen proteins is based on immunological methods like enzyme-linked immunosorbent assays (ELISA) and lateral flow devices (Planque et al. 2017; Sharma et al. 2017). All the immunological assays developed to detect food allergens employ monoclonal or polyclonal antibodies that specifically target allergenic food proteins as a marker for the allergens. The success of the immunological assay depends on the affinity and specificity of the antibody towards the target analyte. ELISA-based analytical assays, which provide both qualitative and quantitative information on allergens in food, continue to be the most common and perhaps the most reliable methods for routine analysis of food allergens in industry settings. Following allergen extraction from the food sample, ELISA involves multiple steps of incubation and washings followed by allergen level measurement using a standard curve prepared using increasing concentrations of the allergen. Sandwich ELISA and competitive (c-ELISA) ELISA are two widely used ELISA methods for food allergen analysis, and their choice is based on a number of factors, including the food matrix, desired sensitivity, and antibody and target analyte characteristics. Studies have reported that ELISA has high specificity and sensitivity with detection limits of ~0.1–5 mg/kg (Schubert-Ullrich et al. 2009).

Despite the fact that commercially available ELISA kits for the detection of food allergens are thought to have sufficient analytical sensitivity, this assay has a range of drawbacks, including little to no sensitivity for thermally processed food due to epitope degradation, cross-reactivity, and the inability to multiplex the assay (Croote and Quake 2016).

The lateral flow immuno-chromatographic assays are the modified form of ELISA, performed in a paper-based platform to detect and quantify analytes in complex mixtures and the results are displayed within 5–30 min. These assays are low-cost and portable and do not need instrumentation or trained personnel, and hence they can be used for the rapid point care testing of the qualitative or semi-quantitative determination of food allergens (Schubert-Ullrich et al. 2009). Since its introduction, many assays have been developed and commercialized for the detection of many of the priority allergenic foods. Currently, these assays are majorly utilized by food industries and regulatory agencies to monitor the presence of allergens and to avoid food cross-contamination of equipment and food contact surfaces.

Over the last few years, significant progress has been made in standardizing and employing liquid chromatography-mass spectrometry (MS) for the analysis of even trace amounts of a food allergen(s). MS has proven to detect food allergens with a high degree of sensitivity from complex foodstuff and offers the flexibility of multiplexing the quantitation of allergenic proteins (Fig. 15.1). The detection of allergen through MS follows any one of the following proteomic workflows:

(i) Bottom-up approach or peptide-based proteomics: In this approach proteins extracted from the food samples undergo digestion with enzymes and the final peptide fragments are determined by MS (Pandey and Mann 2000)

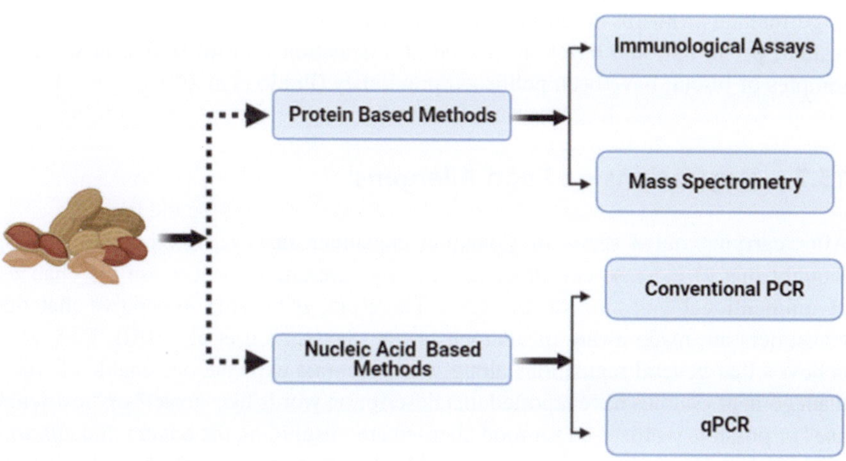

Fig. 15.1 Methods used for detection of allergens

(ii) Top-down approach: Here fragmentation of complete proteins occurs inside the spectrometer, skipping the variable digestion step of protein (McLafferty et al. 2007).

Liquid chromatography is used to separate the peptides from the complex mixture owing to the difference in their relative affinity towards the stationary phase and the mobile phase. Electrospray ionization is used to ionize the eluted peptides and is subsequently evaluated by MS. In both methods, the target analytes are identified and characterized by calculating the theoretical mass values (available in protein databases) against the obtained experimental data.

In recent times, DNA-based methods have been considered promising techniques for the detection of the presence of allergens as it targets the nucleic acid that codes for a protein or other specific DNA marker associated with food allergens such as internal transcribed spacer region (ITS), cytochrome b, and cytochrome oxidase I protein. The detection principle and practical utilities of the DNA-based methods to detect allergenic ingredients in foods have been developed by several authors and have been reviewed by Prado et al. (2016). While analyzing processed foods, DNA-based methods offer higher sensitivity due to the great thermal stability of DNA molecules as compared with proteins. In this method, food allergen is detected using polymerase chain reaction (PCR) to amplify and quantitatively increase the DNA target in the analyzed food thereby providing qualitative results pointing to the presence of allergen. The specificity of PCR depends on the usage of specific oligonucleotide primers that will identify a DNA fragment from the offending ingredient (Prado et al. 2016). Conventional PCR is the simplest endpoint detection method that utilizes agarose gels for the detection of amplicons at the final phase of the PCR reaction and the method is a merely qualitative technique. On the other hand, real-time PCR allows the quantitative analysis (qPCR) of the target DNA while an assay is being performed (García et al. 2017). Over the last decades, several researchers have developed qPCR-based methods for the analysis of food allergens. Experimental protocols describing the application of qPCR with Taq-Man probe for walnut, pecan nut, and hazelnut in food at a detection limit of 0.01% in standard samples of biscuit have been published previously (Prado et al. 2016).

15.7 Regulations on Food Allergens

After carrying out a series of Canadian consumer surveys, Sheth et al. (2000) thought that to increase consumer faith, it was important to decrease the chances of unintentional ingestion of allergens. Therefore, it became imperative that the consumers are made aware of allergen labels (Sakellariou et al. 2010). They also believed that general regulations along with elaborate explanations enable allergen management. Studies have reported that description words like "exact" or "restricted use" or possible words used for food allergen are misguiding the buyers (Sakellariou et al. 2010). Food allergy is a major public health concern globally. Information about the allergens present in the food is indicated by providing a statement of identification on the label as a regulatory risk strategy for allergic people. In order to increase the level of safety of the consumer, prophylactic allergen labeling on majority of food products has been adopted by food manufacturers (Allen et al. 2014). Countries and provinces from different regions show a varying range and level of food sensitization response, probably because of different geographical factors and living habits. Separating food allergens from the other ingredients was thought to be one among the best strategies to avoid food allergy (Wang et al. 2014).

The Food and Drug Administration (FDA) issued the Food Allergen Labelling and Consumer Protection Act (2 August 2004). This act covers the labeling regulations for allergen-containing foods. On October 5, 2005, important information regarding allergens in foods was published. On June 18, 2015, the final documents were released that enabled food manufacturers to register for the allergen exemption mark. In case the requisites of the act are violated, the managers of the violating brand will have to face civil or criminal sanctions or may be detained. However, the FDA entails the food manufacturer to recall the products that contain undeclared allergens. FDA has labeled allergen in the form of legislation, simplifying it so that the masses can easily comprehend the allergens present in food. USFDA had a close watch and deep understanding of the food allergen labels based on the real scenario. If the food materials contain a certain ingredient that causes allergic reactions and is dangerous for allergic people, FDA will issue guidance documents in an appropriate format to put light on this situation. The Codex Alimentarius, GMP, etc., have several levels of regulation on allergen labels thus making for a considerably complete institutional system. Thus, food industries must make products by following the regulations, and the government can play a supervisory role in implementing the system.

The European Food Safety Authority (EFSA) and European Commission (EC) attempt long-standing food allergen management. The EC encourages the food processors and research institutions to investigate and obtain clearance to exempt certain food ingredients from labeling regulations. A specific group of scientists working on dietary, nutritional, and allergic aspects was appointed to investigate possible allergens in food. During 2004–2007, the professional science team reexamined 12 such listed allergens in lupin and mollusks to improve their sensitization (Boyce et al. 2011; Simons et al. 2012). The emergence of allergens in food is categorized by the EU as the one that is added on purpose as an ingredient, or the one that is found by cross-contamination. The EU requires compulsory labeling of allergens in food under categories that include (1) Cereals containing gluten (oats, wheat, rye, and such others); (2) crustaceans; (3) eggs; (4) fish; (5) peanuts; (6) soybeans; (7) milk; (8) nuts; (9) celery; (10) mustard; (11) sesame; (12) sulfur dioxide (>10 mg/kg or >10 mg/mL); (13) lupin; and (14) mollusks. Any products made using the above ingredients also are required to be labeled. Low doses of exposure to allergens do not qualify when sensitization is weak. Hence, EFSA has regarded five classes of sensitizing agents to be spared from the label. The EC has commissioned standard regulations based on the emergence of allergens in food. If the substance is added as one of the ingredients in the food, it has to be duly listed as "xx allergens" or at least the species of the allergen has to be mentioned after the ingredient list.

Even significantly low levels of allergens may sometimes be severely allergic and hence the EU regards controlling cross-contamination as important. The primary production processes sometimes may allow two different foods to share a processing line. Under such circumstances, the presence of an allergen in either of the two foods may contaminate the other because mere washing may not remove the allergen completely. The EU commission on cross-contamination for allergens in food developed regulations to lower residues of such allergens and manage labels to increase awareness among the consumers. In 1994, the EU publicized the "Special Requirements for the Production and Control of Allergen-Containing Food" to regulate the process of food containing allergen in the process line. Afterward, the United Kingdom issued the "Allergen Management and Consumer Information Guide" alerting the consumers to be attentive regarding the allergen information caused by cross-contamination. However, all member countries have different forms of labeling. The United Kingdom in the Allergen Management and Consumer Information Guide are as follows: (1) may contain xx allergens, (2) people who are not suitable for allergies to xx allergens, and (3) the production line of xx food is also used to produce xx allergens. This has warned the consumers and more strongly guaranteed food safety of consumers (Gojkovic et al. 2015).

Food Safety and Standards Authority of India (FSSAI) has set in place detailed rules and regulations to conduct proper analysis and label food products. If foods without any specific allergen labels are found to contain allergens, measures must be taken to prevent their presence. If the contamination cannot be warranted, buyers should be accordingly notified. Under the subtitle "Allergen Management" in the FSSAI guidance document released (June 2019) on "Health Supplement/ Nutraceuticals Processing," details about major food allergens are given. The list stretches into stages so that the producers are enabled to understand and minimize the risk of cross-contamination while making sure that all information regarding the allergen is clearly provided on the labels of the pack. Although, not many companies have made a large attempt to give full disclosure regarding the allergenic compounds in their products and they must be processed in a different facility.

USFDA enforced the Food Allergen Labelling and Consumer Protection Act (2004) ensuring that manufacturers label their products correctly. Nowadays consumer awareness of food allergens is increasing. Also, many companies are working on "allergen-free" products. Identification of allergenic compounds in foods needs testing. FSSAI has begun the process and more work needs to be done. India must actively conduct extensive research to carry out risk evaluation on potential allergens and send in assessment data as the basis to draw a proper regulation. At present Food Safety and Standard for labeling regulation for allergen for infant foods are available. Infant milk substitute container meant for infants with allergy to cow's/buffalo's milk protein or soy protein or label affixed should indicate clearly "HYPOALLERGENIC FORMULA" in capital letters and statement "TO BE TAKEN UNDER MEDICAL ADVICE." Likewise, Indian food safety regulation must fully authorize the Expert Committee on Food Safety Risk Assessment and train a panel of experts on allergen analysis, their nutritive description, and an allergy scientific team for overall assessment of the allergens.

15.8 Allergen Management by Food Companies

15.8.1 Challenges Faced by Food Manufacturers

One of the major challenges faced by food manufacturers is the presence of hidden allergens in food. Hidden allergens unlike true or clear allergens are not ingredients or constituents of the food. Their occurrence in the food is rather unintentional or accidental. Numerous ingredients and production materials that include allergens are used in the preparation and processing of food products in manufacturing plants. In events of cross-contamination, these ingredients may potentially infiltrate the products (Pachołek et al. 2018).

"Cross-contact" refers to the unintentional transfer of remnant or minute amount of an allergic food into other products in spite of following good manufacturing practices (GMP). If the allergen cross-contamination takes place at a significant level, it may pose an unacceptable risk to allergic consumers. For instance, tropomyosin is the main allergen in shellfish. It shares significant homology with arthropods (dust mites and cockroaches). Accidental exposures to the allergenic compound during storage or transportation of food may pose a health risk for sensitive individuals (Patil et al. 2001; Maleki and Hurlburt 2002). Similarly, peanut allergy is a serious health issue posing a major challenge for the peanut industry. Hypersensitive individuals can encounter anaphylaxis that occasionally can be fatal. Complete avoidance of peanut consumption is the safest method to manage peanut allergy. However, the biggest concern for food manufacturers is that peanuts are a common food commodity used in many processed food products. They are commonly used as thickeners agents in foods like stews. Chocolates may contain traces of peanuts even if labeled as free of peanuts due to contamination of machinery used for production (Maleki and Hurlburt 2002).

The occurrence of cross-contamination and its extent can be minimized by modifying the processes if that is practically feasible for the food manufacturer. Nonoccurrence of cross-contamination with allergens during production should be diligently ensured. Statements emphasizing on the lack of allergens present in food components need to be made. For such steps to be implemented, information about the presence of an allergen in the formulation and in potential cross-contamination situations, the quantity of concerned allergens to an appropriate level of detail is required. Except for gluten (20 ppm) and sulfur dioxide (10 ppm), it is difficult to establish a lower limit of permissible ingestion that, if not surpassed, ensures the lack of presence (or a very low chance) of allergy-related adverse effects. Specifying the limits is not easy due to the occurrence of cross contaminations which is highly unpredictable. Moreover, tools to detect and quantify allergens are limited and the techniques to evaluate the allergen concentration are dependent on certain limits. However, quantification of safe levels of major allergens is being attempted in food products using the present medical data as a benchmark. It is important to note that even minute allergen concentrations, if present in the product, can be risky to susceptible individuals. Hence, food manufacturers are trying to completely avoid the usage of allergenic ingredients in their processes (Patil et al. 2001; Sladkevicius et al. 2010).

Malnutrition is quite rampant in developing countries and this becomes an added challenge to food manufacturers who are trying to restrict the use of allergenic compounds in their products. In such financially challenged and developing economies, it is very tough to completely prevent a food allergen (otherwise having high nutritive value) and to replace it with another ingredient that has comparable or average nutritive value to prevent malnutrition. Low levels of literacy and poor labeling legislation pose a great challenge to achieve appropriate food labeling in certain countries. Strategies to address these issues could include implementation of pictorial representative food allergen labels and avoid unwanted contamination of food products (Hossny et al. 2019). Mislabeling, amateur staff, and processes with improper implementation of Good Manufacturing Practices (GMPs) can pose additional challenges and risks to consumers' health. Hence, food allergens management has become an important activity in food industries all over the world. Numerous national regulatory bodies are working towards allergen management by setting laws in place for food producers, mostly concerning labeling requirements.

In this context, the variations in allergen labeling requirements (i.e., list of allergens to be declared) among different nations are a matter of concern for food manufacturers and exporters. Some developed nations like New Zealand, Australia, and Canada have a relatively elaborate list of allergens, while others follow the Codex standards for general labeling. Food exporters need to know of the diverse allergen labeling regulations adopted by different exporting and importing countries. Presently, there is no allergen labeling regulations in place in India. The Food Safety and Standards (Labeling and Display) Regulations, 2018 has enlisted allergens that must be mentioned on the label, in accordance with the Codex guidelines. This lack of a comprehensive food labeling policy for at least the most common allergens in processed and packaged food is an important limiting factor that needs to be addressed (Devdas et al. 2017).

15.8.2 Risk Management Process in Food Allergen Control

Allergen management is a central part of the existing food safety management system. A successful allergen management system considers overall processes from procurement of base materials, production, and packaging to finished products (Hossny et al. 2019).

Risk Analysis helps in understanding allergen management ability and control, but it is not the only sole tool adequate for allergen management. The efficacy of cleaning systems or cleaning validation majorly depends on the analysis results. On-site swabbing tests and dipstick tests are the confirmatory tests to check whether the production line remains free from allergens. Risk management comprises of risk evaluation which needs to take into account the physical form of the allergen (liquid, pieces, powder, etc.), and the quantity of the allergen present, if any. All areas of the supply chain must be accounted for under risk management including procurement, finished goods sales specification, and design of the product. Documented processes for the management and avoidance of any risk or contamination must be instated and easily available to the entire staff in the processing unit (Devdas et al. 2017; Pachołek et al. 2018).

The areas where allergen risks occur and can likely be controlled and managed are as follows:

15.8.2.1 Product Research, Development, and Engineering System

The allergic character of ingredients used should be considered and possibilities of cross-contamination should be explored. Equipment exposure could be minimized by adding allergenic ingredients at the final steps of production thereby lowering equipment exposure. The format and installation procedure of the equipment must ensure easy cleaning, examination, and maintenance.

15.8.2.2 Raw Material/Ingredients Purchasing, Transportation, and Storage

Cross-contact between non-allergenic and allergenic ingredients should be minimized through dedicated raw materials storage. Understand the constituents, processing needs, and rework that constitutes the process of production. Dedicated transportation to bring in ingredients in large quantities or containers for shipping that are reused should be ensured. Specifications of the raw materials should be reviewed. The raw materials should be labeled to indicate the presence of any allergen.

15.8.2.3 Production and Scheduling

Separate lines, equipment, rooms, or facilities for production of allergen-containing products should be used. Products with such allergenic ingredients should be prepared at once or at the final stage of production, after which a thorough clean-up should be performed before running any other products. Cross-contamination by other products next to the carts and conveyors should be avoided. All sampling devices should be sanitized appropriately between the products.

15.8.2.4 Rework

Documents for rework plans that include selected uses for rework, maintaining usage log and other rework controls must be maintained to trace allergens. Rework of products containing certain allergenic ingredients should be avoided. Containers, tags, and plastic liners/bar coding can be used to identify ingredients or products containing allergen. Confirm that refeed systems are correctly controlled.

15.8.2.5 Labeling and Materials Used for Packaging

Incoming ingredients should be rechecked for labels. Stick to compliance of labeling regulations, which necessitates declaration of all the ingredients; except for spices, permitted colors and flavors, and incidental additives at nonsignificant levels or that have no effect. If an incidental additive is derived from an allergenic raw material ingredient, it must be declared on the label. Cross-contamination should be considered for packaging equipment. Management of extra unused packages, removal of out-of-date containers, or expired labels from the processing area must be done regularly. Regular checks for product traceability must be in place, raw materials should be given lot numbers and to the finished products to guarantee recovery of all products in event of a recall, need to conduct a mock recall to verify is very important.

15.8.2.6 Sanitation

Every food industry should need to establish Standardized Procedures for Sanitation Operations (SSOP's) and it should be ensured that they are being implemented. Proper cleaning methods and specifications (for soap and water wash, proper chemicals, proper wiping or scraping, or vacuum tools) should be followed. Focus needs to be given to clean areas, for example, dead spots, valves, pumps, etc. Ensure adequate lighting is placed in a proper site. Movement of the equipment is necessary to access for cleaning without any effort, assess sanitation efficiency; undo the setting if necessary and endorsement of cleaning by sight. Testing like bioluminescence and ELISA should be conducted to ensure the site is devoid of any contamination.

Maintenance tools that are used in raw product areas can potentially cause cross-contamination in the finished product areas. Specify employee practices,washing hands at regular intervals, proper hand washing procedures; clean clothing/aprons, etc., are some of things to be appropriately followed by the personnel involved in the processing of food products.

15.8.2.7 Personnel Training and Education

Ensure all employees understand the allergen prevention program and believe it to be important as an aspect of the food safety facilities. Important pointers: define allergens, outcomes, importance of allergen control, common areas where problems occur and its control measures.

Gojkovic et al. (2015) reviewed the activities undertaken to deal with allergens during production and packaging of various powdered food materials (vegetable food supplement, pudding powder, powdered sugar). They discussed three phases of allergy management strategy: identification, separation, and labeling. Reducing risk to food allergens with which consumers come in contact and identification of food products which contain allergic substances, proper process control of food production and storage (GMP), tools to verify, conduct, and control as explained in regulations of labeling and verification by quick methods for detecting allergen residue in products, can lead to successful allergen management on food industries. A facility risk evaluation should be carried out in order to develop an Allergen Control Plan.

At present, food companies usually adopt HACCP, GMP, SSOP, etc., as standard processing operations norms. However, the cross-contamination of allergens is still a question of concern as there are no specific allergen management standards (Röder et al. 2010). Food manufacturers themselves are required to assess their operating techniques to identify and prepare strategies to manage unknown allergens. To assure specified control points, all aspects of a manufacturing operation should be reviewed. The HACCP plan should include an assessment for allergy risks (chemical hazard). Additionally, training will ensure that all staff involved in management, research, and in-line production understand the risks of allergens. It is critical to have a continuous monitoring system in place to ensure that all of the control points are being fulfilled (Hossny et al. 2019; Managing Food Allergen Risks 2021).

15.9 Conclusion

Food allergy is a growing health concern that affects a significant number of people globally. As a result of its individualistic nature, food allergy is generally not viewed as a major health concern by the public, even though its prevalence is notably increasing in both Western countries as well as developing nations. However, food allergy is an alarming concern to food manufacturers. Food industries need to adopt strategies to reduce the occurrence of allergens in their products wherever possible. In cases where the inclusion of allergen ingredients becomes unavoidable, the industry needs to provide consumers with the necessary details. Ingredient labeling statements are the key to successful allergen management. Manufacturers also need to be aware that certain processing practices like the use of shared equipment and the use of rework can cause undeclared residues of allergenic foods existing in other products. Food manufacturers are required to assess their operating techniques to identify and develop strategies to manage unknown allergens. HACCP, sanitation standard operation procedure (SSOP), good manufacturing practice (GMP), etc., should be adopted as the standards for production operations.

It is also important to acknowledge that there is a lack of rigorously conducted investigations on food allergic diseases. More detailed investigations on objective methods for accurate detection of food allergy and standardization of these methodologies are necessary. The disparity in the incidences of food allergy and the sensitization pattern in India needs to be addressed by further analysis. Also, exploratory studies to understand the disparities in the epidemiology of food allergy between urban and rural populations may help to better understand the causes of rising prevalence of food allergy in developed nations.

References

Allen KJ, Paul JT, Pawankar R et al (2014) Precautionary labelling of foods for allergen content: are we ready for a global framework? World Allergy Organ J 7:1–14

Arakali SR, Green TD, Dinakar C (2017) Prevalence of food allergies in South Asia. Ann Allergy Asthma Immunol 118(1):16–20

Bartnikas LM, Phipatanakul W (2013) Turning up the heat on skin testing for baked egg allergy. Clin Exp Allergy 43:1095–1096

Bernhisel-Broadbent J, Strause D, Sampson HA (1992) Fish hypersensitivity. II: clinical relevance of altered fish allergenicity caused by various preparation methods. J Allergy Clin Immunol 90(4):622–629

Bhattacharya K, Sircar G, Dasgupta A et al (2018) Spectrum of allergens and allergen biology in India. Int Arch Allergy Immunol 177(3):219–237

Blanc F, Vissers YM, Adel-Patient K et al (2011) Boiling peanut Ara h 1 results in the formation of aggregates with reduced allergenicity. Mol Nutr Food Res 55(12):1887–1894

Boyce JA, Assa'a A, Burks AW et al (2011) Guidelines for the diagnosis and management of food allergy in the United States: summary of the NIAID-sponsored expert panel report. Nutrition 27: 253

Cabanillas B, Maleki SJ, Rodríguez J et al (2014) Allergenic properties and differential response of walnut subjected to processing treatments. Food Chem 157:141–147

Chehade M, Mayer L (2005) Oral tolerance and its relation to food hypersensitivities. J Allergy Clin Immunol 115(1):3–12

Chogtu B, Magaji N, Magazine R et al (2017) Pattern of allergen sensitivity among patients with bronchial asthma and/or allergic rhinosinusitis in a tertiary care Centre of southern India. J Clin Diagnostic Res 8:OC01

Choudhary R, Gautam D, Perez-Alvarado G et al (2013) Effect of high intensity ultrasound treatment in reducing the allergenicity of isolated cow's milk and soy proteins. Presented during IFCON proceeding, pp 18–21

Cortot CF, Sheehan WJ, Permaul P et al (2012) Role of specific IgE and skin-prick testing in predicting food challenge results to baked egg. Allergy Asthma Proc 33:275–281

Croote D, Quake SR (2016) Food allergen detection by mass spectrometry: the role of systems biology. Npj Syst Biol Appl 2(1):1–10

Devdas JM, Mckie C, Fox AT et al (2017) Food allergy in children: an overview. Indian J Pediatr 85(5):369–374

Dey D, Ghosh N, Pandey N et al (2014) A hospital-based survey on food allergy in the population of Kolkata, India. Int Arch Allergy Immunol 164(3):218–221

Dupont C (2011) Food allergy: recent advances in pathophysiology and diagnosis. Ann Nutr Metab 59(1):8–18

Ekezie FGC, Cheng JH et al (2018) Effects of nonthermal food processing technologies on food allergens: a review of recent research advances. Trends Food Sci Technol 74:12–25

Esteghlal S, Gahruie HH, Niakousari M et al (2019) Bridging the knowledge gap for the impact of non-thermal processing on proteins and amino acids. Foods 8(7):262

Faria AM, Weiner HL (2005) Oral tolerance. Immunol Rev 206(1):232–259

Gangal SV, Malik BK (2003) Food allergy-how much of a problem really is this in India? J Sci Ind Res 62:755–765

García A, Madrid R, García T et al (2017) Detection of food allergens by Taqman real-time PCR methodology. In: Food Allergens. Humana Press, New York, NY, pp 95–108

Geisslitz S, Shewry P, Brouns F et al (2021) Wheat ATIs: characteristics and role in human disease. Front Nutr 8:265

Gharbi N, Labbafi M (2018) Effect of processing on aggregation mechanism of egg white proteins. Food Chem 252:126–133

Gobinaath TR, AnandArokiaraj L (2018) Food allergy and allergens associated with bronchial asthma among school children in an urban and rural area of Puducherry, India: a cross-sectional study. Int J Contemp Pediatr 5(4):1623–1625

Gojkovic V, Marjanovicbalaban Z, Vukic M et al (2015) Allergens management system in the food production. J Hyg Eng Des 12:76–84

Gomaa A, Boye J (2015) Impact of irradiation and thermal processing on the immunochemical detection of milk and egg allergens in foods. Int Food Res J 74:275–283

Gourinath S, Alam N, Srinivasan A et al (2000) Structure of the bifunctional inhibitor of trypsin and α-amylase from ragi seeds at 2.2 Å resolution. Acta Crystallogr D Biol Crystallogr 56(3): 287–293

Greer FR, Sicherer SH, Burks AW (2008) Effects of early nutritional interventions on the development of atopic disease in infants and children: the role of maternal dietary restriction, breastfeeding, timing of introduction of complementary foods, and hydrolyzed formulas. Pediatrics 121(1):183–191

de Gregorio M, Armentia A, Diaz-Perales A et al (2009) Salt-soluble proteins from wheat-derived foodstuffs show lower allergenic potency than those from raw flour. J Agric Food Chem 57(8): 3325–3330

Griesmeier U, Vázquez-Cortés S, Bublin M et al (2010) Expression levels of parvalbumins determine allergenicity of fish species. Allergy 65(2):191–198

Han Z, Cai MJ, Cheng JH et al (2018) Effects of electric fields and electromagnetic wave on food protein structure and functionality: a review. Trends Food Sci Technol 75:1–9

Ho MH, Wong WH, Chang C (2014) Clinical spectrum of food allergies: a comprehensive review. Clin Rev Allergy Immunol 46(3):225–240

Hossny E, Ebisawa M, El-Gamal Y et al (2019) Challenges of managing food allergy in the developing world. World Allergy Organ J 12(11):100089

Huang HW, Yang BB, Wang CY (2014) Effects of high pressure processing on immunoreactivity and microbiological safety of crushed peanuts. Food Control 42:290–295

Hungerford JM (2010) Scombroid poisoning: a review. Toxicon 56(2):231–243

Husband FA, Aldick T, Van der Plancken I et al (2011) High-pressure treatment reduces the immunoreactivity of the major allergens in apple and celeriac. Mol Nutr Food Res 55(7): 1087–1095

Jin Y, Deng Y, Qian B et al (2015) Allergenic response to squid (Todarodespacificus) tropomyosin Tod p1 structure modifications induced by high hydrostatic pressure. Food Chem Toxicol 76: 86–93

Joseph CL, Ownby DR, Havstad SL et al (2011) Early complementary feeding and risk of food sensitization in a birth cohort. J Allergy Clin Immunol 127(5):1203–1210

Kasera R, Singh AB, Kumar R et al (2012) Effect of thermal processing and γ-irradiation on allergenicity of legume proteins. Food Chem Toxicol 50(10):3456–3461

Kato Y, Watanabe H, Matsuda T (2000) Ovomucoid rendered insoluble by heating with wheat gluten but not with milk casein. Biosci Biotechnol Biochem 64(1):198–201

Khan MU, Ahmed I, Lin H et al (2019) Potential efficacy of processing technologies for mitigating crustacean allergenicity. Crit Rev Food Sci Nutr 59(17):2807–2830

Kobayashi Y, Yang T, Yu CT et al (2016) Quantification of major allergen parvalbumin in 22 species of fish by SDS–PAGE. Food Chem 194:345–353

Krishna MT, Mahesh PA, Vedanthan P et al (2020) An appraisal of allergic disorders in India and an urgent call for action. World Allergy Organ J 13(7):100446

Kuehn A, Swoboda I, Arumugam K et al (2014) Fish allergens at a glance: variable allergenicity of parvalbumins, the major fish allergens. Front Immunol 5:179

Kumar R, Kumari D, Srivastava P et al (2010) Identification of IgE-mediated food allergy and allergens in older children and adults with asthma and allergic rhinitis. Indian J Chest Dis Allied Sci 52(4):217

Kumar S, Verma AK, Das M et al (2013) Clinical complications of kidney bean (Phaseolus vulgaris L.) consumption. Nutrition 29(6):821–827

Kurpiewska K, Biela A, Loch JI et al (2019) Towards understanding the effect of high pressure on food protein allergenicity: β-lactoglobulin structural studies. Food Chem 270:315–321

Lasekan AO, Nayak B (2016) Effects of buffer additives and thermal processing methods on the solubility of shrimp (*Penaeus monodon*) proteins and the immunoreactivity of its major allergen. Food Chem 200:146–153

Li H, Yu J, Ahmedna M et al (2013) Reduction of major peanut allergens Ara h 1 and Ara h 2, in roasted peanuts by ultrasound assisted enzymatic treatment. Food Chem 141(2):762–768

Li H, Jia Y, Peng W et al (2018) High hydrostatic pressure reducing allergenicity of soy protein isolate for infant formula evaluated by ELISA and proteomics via Chinese soy-allergic children's sera. Food Chem 269:311–317

Li J, Ogorodova LM, Mahesh PA et al (2020) Comparative study of food allergies in children from China, India, and Russia: theEuroPrevall-INCO surveys. J Allergy Clin Immunol Pract 8(4): 1349–1358
Loh W, Tang ML (2018) The epidemiology of food allergy in the global context. Int J Environ Res Public Health 15(9):2043
Luo C, Hu C, Gao J et al (2013) A potential practical approach to reduceAra h 6 allergenicity by gamma irradiation. Food Chem 136(3–4):1141–1147
Mahesh PA, Wong GW, Ogorodova L et al (2016) Prevalence of food sensitization and probable food allergy among adults in India: the EuroPrevall INCO study. Allergy 71(7):1010–1019
Maleki SJ, Hurlburt BK (2002) Food allergy: recent advances in food allergy research. ACS Publications, pp 192–204
Managing Food Allergen Risks (2021) A guide for manufacturers, Minnesota Department of Agriculture (MDA) (Managing food allergen risks—a guide for manufacturers, Minnesota Department of Agriculture (state.mn.us))
Mansoor DK, Sharma HP (2011) Clinical presentations of food allergy. Pediatr Clin 58(2):315–326
Manzocco L, Panozzo A, Nicoli MC (2013) Effect of pulsed light on selected properties of egg white. Innov Food Sci Emerg Technol 18:183–189
Mavromatis P, Quantick P (2002) Histamine toxicity and scombroid fish poisoning: a review. In: Alasalvar C, Taylor T (eds) Seafoods—quality, technology and nutraceutical applications. Springer, pp 89–103
McLafferty FW, Breuker K, Jin M et al (2007) Top-down MS, a powerful complement to the high capabilities of proteolysis proteomics. FEBS J 274(24):6256–6268
Meinlschmidt P, Ueberham E, Lehmann J et al (2016) The effects of pulsed ultraviolet light, cold atmospheric pressure plasma and gamma-irradiation on the immunoreactivity of soy protein isolate. Innov Food Sci Emerg Technol 38:374–383
Mondal G, Chatterjee U, Samanta S et al (2007) Role of pepsin in modifying the allergenicity of bhetki (*Latescalcarifer*) and mackerel (*Rastrelligerkanagurta*) fish. Indian J Biochem Biophys 44:94–100
Nooji JK (2011). Reduction of wheat allergen potency by pulsed ultraviolet light, high hydrostatic pressure, and nonthermal plasma. University of Florida, M.Sc. thesis, University of Florida, Florida, United States
Pachołek B, Sady S, Kupińska-Adamczyk E (2018) Management of food allergens in the food industry. J Agric Rural Dev 47(1):73–80
Pandey A, Mann M (2000) Proteomics to study genes and genomes. Nature 405(6788):837–846
Patil SP, Niphadkar PV, Bapat MM (2001) Chickpea: a major food allergen in the Indian subcontinent and its clinical and immunochemical correlation. Ann Allergy Asthma Immunol 87(2):140–145
Pawankar R (2013) Food allergy—a rising global health problem. World Allergy Organisation
Petruláková M, Valik L (2015) Food allergy and intolerance. Acta Chim Slov 8(1):44–51
Planque M, Arnould T, Gillard N (2017) Food allergen analysis: detection, quantification and validation by mass spectrometry. Allergen:7–41
Prado M, Ortea I, Vial S et al (2016) Advanced DNA-and protein-based methods for the detection and investigation of food allergens. Crit Rev Food Sci Nutr 56(15):2511–2542
Prester L (2016) Seafood allergy, toxicity, and intolerance: a review. J Am Coll Nutr 35(3):271–283
Prieto N, Burbano C, Iniesto E et al (2014) A novel proteomic analysis of the modifications induced by high hydrostatic pressure on hazelnut water-soluble proteins. Foods 3(2):279–289
Rathi A, Sharma S (2017) Changing allergy patterns in India. J Allergy Clin Immunol 1(1):1–5, commentary article
Renzone G, Arena S, Scaloni A (2015) Proteomic characterization of intermediate and advanced glycation end-products in commercial milk samples. Proteomics 117:12–23
Robison RG (2014) Food allergy: diagnosis, management & emerging therapies. Indian J Med Res 139(6):805

Röder M, Iris B, Helwig G et al (2010) Allergen sanitation in the food industry: a systematic industrial scale approach to reduce hazelnut cross-contamination of cookies. J Food Prot 73: 1671–1679

Sakellariou A, Sinaniotis A, Damianidou L et al (2010) Food allergen labelling and consumer confusion. Allergy 65:534–553

Sampson HA (2004) Update on food allergy. J Allergy Clin Immunol 113:805–819

Schubert-Ullrich P, Rudolf J, Ansari P et al (2009) Commercialized rapid immunoanalytical tests for determination of allergenic food proteins: an overview. Anal Bioanal Chem 395(1):69–81

Sharma GM, Khuda SE, Parker CH et al (2017) Detection of allergen markers in food: analytical methods. Food Saf Innov Anal Tools Saf Assess, pp 65–121

Sheth B, Moran B, Anderson JM et al (2000) Post-translational control of occludin membrane assembly in mouse trophectoderm: a mechanism to regulate timing of tight junction biogenesis and blastocyst formation. Development 127:831–840

Shriver SK, Yang WW (2011) Thermal and nonthermal methods for food allergen control. Food Eng Rev 3(1):26–43

Sicherer SH, Sampson HA (2010) Food allergy. J Allergy Clin Immunol 125(2):S116–S125

Simons E, Weiss CC, Furlong TJ et al (2005) Impact of ingredient labelling practices on food allergic consumers. J Allergy Clin Immunol 11(5):S243–S243

Simons FER et al (2012) World allergy organization guidelines for the assessment and Management of Anaphylaxis. Curr Opin Allergy Clin Immunol 12:389

Sladkevicius E, Nagy E, Lack G et al (2010) Resource implications and budget impact of managing cow milk allergy in the UK. J Med Econ 13(1):119–128

Soares WK, Takwoingi Y, Panesar SS, Muraro A, Werfel T et al (2014) The diagnosis of food allergy: a systematic review and meta-analysis. Allergy 69:62–75

Soria AC, Villamiel M (2010) Effect of ultrasound on the technological properties and bioactivity of food: a review. Trends Food Sci Technol 21(7):323–331

Sudo N, Sawamura S, Tanaka K et al (1997) The requirement of intestinal bacterial flora for the development of an IgE production system fully susceptible to oral tolerance induction. J Immunol 159:1739–1745

Tammineedi CV, Choudhary R, Perez-Alvarado GC et al (2013) Determining the effect of UV-C, high intensity ultrasound and nonthermal atmospheric plasma treatments on reducing the allergenicity of α-casein and whey proteins. LWT—Food Sci Technol 54(1):35–41

Taylor SL, Hefle SL (2001) Food allergies and other food sensitivities. Food Technol Champaign Then Chicago 55(9):68–84

Thomas K, Herouet-Guicheney C, Ladics G et al (2007) Evaluating the effect of food processing on the potential human allergenicity of novel proteins: international workshop report. Food Chem Toxicol 45(7):1116–1122

Turner PJ, Mehr S, Sayers R et al (2014) Loss of allergenic proteins during boiling explains tolerance to boiled peanut in peanut allergy. J Allergy Clin Immunol 134:751–753

Vanga SK, Singh A, Raghavan V (2017) Review of conventional and novel food processing methods on food allergens. Crit Rev Food Sci Nutr 57(10):2077–2094

Vaz AF, Souza MP, Carneiro-da-Cunha MG et al (2012) Molecular fragmentation of wheat-germ agglutinin induced by food irradiation reduces its allergenicity in sensitised mice. Food Chem 132(2):1033–1039

Venkataratnam H, Sarangapani C, Cahill O et al (2019) Effect of cold plasma treatment on the antigenicity of peanut allergen Ara h Innov. Food Sci Emerg Technol 52:368–375

Venter C, Pereira B, Voigt K et al (2009) Factors associated with maternal dietary intake, feeding and weaning practices, and the development of food hypersensitivity in the infant. Pediatr Allergy Immunol 20(4):320–327

Vickery BP, Chin S, Burks AW (2011) Pathophysiology of food allergy. Pediatr Clin 58(2): 363–376

Visciano P, Schirone M, Tofalo R et al (2014) Histamine poisoning and control measures in fish and fishery products. Front Microbial 5:500

Vissers YM, Blanc F, Skov PS et al (2011) Effect of heating and glycation on the allergenicity of 2S albumins (Ara h 2/6) from peanut. PLoS One 6(8):e23998

Wang MJ, Jiang-Hua LI, Guo LY et al (2014) Food allergen labeling management: a review of the European experience and its inspiration for China. J Food Sci 35:261–265

World Health Organization (2013) Joint FAO/WHO expert meeting on the public health risks of histamine and other biogenic amines from fish and fishery products: meeting report

World Health Organization (2006) The world health report: 2006: working together for health. World Health Organization. https://apps.who.int/iris/handle/10665/43432

Yang X, Sun J, Tao J et al (2017) The allergenic potential of walnuts treated with high pressure and heat in a mouse model of allergy. Innov Food Sci Emerg Technol 39:165–170

Yu HL, Cao MJ, Cai QF et al (2011) Effects of different processing methods on digestibility of *Scylla paramamosain* allergen (tropomyosin). Food Chem Toxicol 49(4):791–798

Zhang Z, Zhang X, Chen W et al (2018) Conformation stability, in vitro digestibility and allergenicity of tropomyosin from shrimp (*Exopalaemonmodestus*) as affected by high intensity ultrasound. Food Chem 245:997–1009

Zhou H, Wang C, Ye J et al (2016) Effects of high hydrostatic pressure treatment on structural, allergenicity, and functional properties of proteins from ginkgo seeds. Innov Food Sci Emerg Technol 34:187–195

Ziegler K, Neumann J, Liu F et al (2019) Nitration of wheat amylase trypsin inhibitors increases their innate and adaptive immunostimulatory potential in vitro. Front Immunol 9:3174

MIX
Papier aus verantwortungsvollen Quellen
Paper from responsible sources
FSC® C105338

If you have any concerns about our products,
you can contact us on
ProductSafety@springernature.com

In case Publisher is established outside the EU,
the EU authorized representative is:
Springer Nature Customer Service Center GmbH
Europaplatz 3, 69115 Heidelberg, Germany

Printed by Libri Plureos GmbH
in Hamburg, Germany